Topics in Chromatic Graph Theory

Chromatic graph theory is a thriving area that uses various ideas of 'colouring' (of vertices, edges, etc.) to explore aspects of graph theory. It has links with other areas of mathematics, including topology, algebra and geometry, and is increasingly used in such areas as computer networks, where colouring algorithms form an important feature.

While other books cover portions of the material, no other title has such a wide scope as this one, in which acknowledged international experts in the field provide a broad survey of the subject. All 15 chapters have been carefully edited, with uniform notation and terminology applied throughout. Bjarne Toft (Odense, Denmark), widely recognized for his substantial contributions to the area, acted as academic consultant.

The book serves as a valuable reference for researchers and graduate students in graph theory and combinatorics and as a useful introduction to the topic for mathematicians in related fields.

LOWELL W. BEINEKE is Schrey Professor of Mathematics at Indiana University–Purdue University Fort Wayne (IPFW), where he has worked since receiving his Ph.D. from the University of Michigan under the guidance of Frank Harary. His graph theory interests include topological graph theory, line graphs, tournaments, decompositions and vulnerability. He has published over 100 papers in graph theory and has served as editor of the *College Mathematics Journal*. With Robin Wilson he has co-edited five books in addition to the three earlier volumes in this series. Recent honours include an award instituted in his name by the College of Arts and Sciences at IPFW and a Certificate of Meritorious Service from the Mathematical Association of America.

ROBIN J. WILSON is Emeritus Professor of Pure Mathematics at the Open University, UK, and Emeritus Professor of Geometry at Gresham College, London. After graduating from Oxford, he received his Ph.D. in number theory from the University of Pennsylvania. He has written and co-edited many books on graph theory and the history of mathematics, including *Introduction to Graph Theory, Four Colors Suffice* and *Combinatorics: Ancient & Modern*. His combinatorial research interests formerly included graph colourings and now focus on the history of combinatorics. An enthusiastic popularizer of mathematics, he has won two awards for his expository writing from the Mathematical Association of America.

ENCYCLOPEDIA OF MATHEMATICS AND ITS APPLICATIONS

All the titles listed below can be obtained from good booksellers or from Cambridge University Press. For a complete series listing visit www.cambridge.org/mathematics.

109 J. M. Borwein and J. D. Vanderwerff *Convex Functions*
110 M.-J. Lai and L. L. Schumaker *Spline Functions on Triangulations*
111 R. T. Curtis *Symmetric Generation of Groups*
112 H. Salzmann *et al. The Classical Fields*
113 S. Peszat and J. Zabczyk *Stochastic Partial Differential Equations with Lévy Noise*
114 J. Beck *Combinatorial Games*
115 L. Barreira and Y. Pesin *Nonuniform Hyperbolicity*
116 D. Z. Arov and H. Dym *J-Contractive Matrix Valued Functions and Related Topics*
117 R. Glowinski, J.-L. Lions and J. He *Exact and Approximate Controllability for Distributed Parameter Systems*
118 A. A. Borovkov and K. A. Borovkov *Asymptotic Analysis of Random Walks*
119 M. Deza and M. Dutour Sikirić *Geometry of Chemical Graphs*
120 T. Nishiura *Absolute Measurable Spaces*
121 M. Prest *Purity, Spectra and Localisation*
122 S. Khrushchev *Orthogonal Polynomials and Continued Fractions*
123 H. Nagamochi and T. Ibaraki *Algorithmic Aspects of Graph Connectivity*
124 F. W. King *Hilbert Transforms I*
125 F. W. King *Hilbert Transforms II*
126 O. Calin and D.-C. *Chang Sub-Riemannian Geometry*
127 M. Grabisch et al. *Aggregation Functions*
128 L. W. Beineke and R. J. Wilson (eds.) with J. L. Gross and T. W. Tucker *Topics in Topological Graph Theory*
129 J. Berstel, D. Perrin and C. Reutenauer *Codes and Automata*
130 T. G. Faticoni *Modules over Endomorphism Rings*
131 H. Morimoto *Stochastic Control and Mathematical Modeling*
132 G. Schmidt *Relational Mathematics*
133 P. Kornerup and D. W. Matula *Finite Precision Number Systems and Arithmetic*
134 Y. Crama and P. L. Hammer (eds.) *Boolean Models and Methods in Mathematics, Computer Science, and Engineering*
135 V. Berthé and M. Rigo (eds.) *Combinatorics, Automata and Number Theory*
136 A. Kristály, V. D. Rădulescu and C. Varga *Variational Principles in Mathematical Physics, Geometry, and Economics*
137 J. Berstel and C. Reutenauer *Noncommutative Rational Series with Applications*
138 B. Courcelle and J. Engelfriet *Graph Structure and Monadic Second-Order Logic*
139 M. Fiedler *Matrices and Graphs in Geometry*
140 N. Vakil *Real Analysis through Modern Infinitesimals*
141 R. B. Paris *Hadamard Expansions and Hyperasymptotic Evaluation*
142 Y. Crama and P. L. Hammer *Boolean Functions*
143 A. Arapostathis, V. S. Borkar and M. K. Ghosh *Ergodic Control of Diffusion Processes*
144 N. Caspard, B. Leclerc and B. Monjardet *Finite Ordered Sets*
145 D. Z. Arov and H. Dym *Bitangential Direct and Inverse Problems for Systems of Integral and Differential Equations*
146 G. Dassios *Ellipsoidal Harmonics*
147 L. W. Beineke and R. J. Wilson (eds.) with O. R. Oellermann *Topics in Structural Graph Theory*
148 L. Berlyand, A. G. Kolpakov and A. Novikov *Introduction to the Network Approximation Method for Materials Modeling*
149 M. Baake and U. Grimm *Aperiodic Order I: A Mathematical Invitation*
150 J. Borwein et al. *Lattice Sums Then and Now*
151 R. Schneider *Convex Bodies: The Brunn–Minkowski Theory (Second Edition)*
152 G. Da Prato and J. Zabczyk *Stochastic Equations in Infinite Dimensions (Second Edition)*
153 D. Hofmann, G. J. Seal and W. Tholen (eds.) *Monoidal Topology*
154 M. Cabrera García and Á. Rodríguez Palacios *Non-Associative Normed Algebras I: The Vidav–Palmer and Gelfand–Naimark Theorems*
155 C. F. Dunkl and Y. Xu *Orthogonal Polynomials of Several Variables (Second Edition)*
156 L. W. Beineke and R. J. Wilson (eds.) with B. Toft *Topics in Chromatic Graph Theory*
157 T. Mora *Solving Polynomial Equation Systems III: Algebraic Solving*
158 T. Mora *Solving Polynomial Equation Systems IV: Buchberger's Theory and Beyond*

Above: Francis Guthrie, who proposed the four-colour problem, and Kenneth Appel and Wolfgang Haken, who solved it. *Below*: Gerhard Ringel (right) and Ted Youngs, who solved the Heawood conjecture. (Courtesy of Robin Wilson.)

Topics in Chromatic Graph Theory

Edited by

LOWELL W. BEINEKE
Indiana University–Purdue University
Fort Wayne

ROBIN J. WILSON
The Open University
and the London School of Economics

Academic Consultant

BJARNE TOFT
University of Southern Denmark, Odense

CAMBRIDGE
UNIVERSITY PRESS

CAMBRIDGE
UNIVERSITY PRESS

University Printing House, Cambridge CB2 8BS, United Kingdom

Cambridge University Press is part of the University of Cambridge.

It furthers the University's mission by disseminating knowledge in the pursuit of education, learning and research at the highest international levels of excellence.

www.cambridge.org
Information on this title: www.cambridge.org/9781107033504

First published 2015

Printed in the United Kingdom by Clays, St Ives plc

A catalogue record for this publication is available from the British Library

Library of Congress Cataloguing in Publication data
Topics in chromatic graph theory / edited by Lowell W. Beineke,
Indiana University-Purdue University, Fort Wayne, Robin J. Wilson,
The Open University and the London School of Economics ;
academic consultant, Bjarne Toft, University of Southern Denmark, Odense.
pages cm. – (Encyclopedia of mathematics and its applications ; 156)
Includes bibliographical references.
ISBN 978-1-107-03350-4 (Hardback)
1. Graph coloring–Data processing. 2. Graph theory–Data processing.
I. Beineke, Lowell W., editor. II. Wilson, Robin J., editor.
QA166.247.T67 2015
511′.56–dc23 2014035297

ISBN 978-1-107-03350-4 Hardback

Contents

Foreword

Bjarne Toft

> The four colour problem is
> the tip of the iceberg,
> the thin end of the wedge
> and the first cuckoo of spring.
> W. T. Tutte, 1978

A fundamental process in mathematics is that of partitioning a set of objects into classes according to certain rules. Chromatic graph theory deals with a situation where the rules are almost as simple as one can imagine: for each pair of objects we are told whether they may be put in the same class or not. However, the simplicity of the rules does not mean that the problems encountered are simple – on the contrary. Starting from the four-colour problem around 1850, the theory has developed into a many-sided body of problems, theories, results and applications, and even though many problems have been solved, sometimes in surprising ways, the number of simply stated but challenging problems remains large and growing. This explains the popularity of the area and why it attracts so many active researchers.

This book presents a picture of this many-sided body as it has evolved so far. Experts from various parts of the area present main ideas, methods and results, and describe what is important. Map-colouring dominated the field for many years, but with authors like K. Wagner, H. Hadwiger, R. L. Brooks, W. T. Tutte, G. A. Dirac, G. Hajós, T. Gallai and P. Erdős, among others, the theory became more general, abstract and applicable.

The chapters cover much ground. The first one outlines the general theory of colouring graphs on surfaces. Other types of graphs, such as perfect graphs, geometric graphs, random graphs and hypergraphs are then treated in chapters of their own, as are special types of colourings, such as edge-colourings, list-colourings and integer flows. Classical topics, such as Brooks's theorem, Hadwiger's conjecture and chromatic polynomials, are described and updated to current knowledge. Applications and relations to other fields, such as scheduling, games and algorithms, are also included. The final chapter presents some 20 unsolved problems: solutions to most of these are probably beyond what can be achieved with current knowledge.

The area continues to surprise, and the achievements of the past few years in particular have witnessed a treasure trove of results, methods, ideas and problems. We now know more of W. T. Tutte's iceberg, even if much still lies hidden below the surface, waiting for discovery!

Preface

The field of graph theory has undergone tremendous growth during the past century. As recently as the 1950s, the graph theory community had few members and most were in Europe and North America; today there are hundreds of graph theorists and they span the globe. By the mid 1970s, the subject had reached the point where we perceived a need for a collection of surveys of various areas of graph theory: the result was our three-volume series *Selected Topics in Graph Theory*, comprising articles written by distinguished experts and then edited into a common style. Since then, the transformation of the subject has continued, with individual branches (such as chromatic graph theory) expanding to the point of having important subdivisions themselves. This inspired us to conceive of a new series of books, each a collection of articles within a particular area of graph theory written by experts within that area. The first three of these books were the companion volumes to the present one, on algebraic graph theory, topological graph theory and structural graph theory. This is thus the fourth volume in the series.

A special feature of these books is the engagement of academic consultants (here, Bjarne Toft) to advise us on topics to be included and authors to be invited. We believe that this has been successful, with the result that the chapters of each book cover the full range of area within the given area. In the present case, the area is chromatic graph theory, with chapters written by authors from around the world. Another important feature is that, where possible, we have imposed uniform terminology and notation throughout, in the belief that this will aid readers in going from one chapter to another. For a similar reason, we have not tried to remove a small amount of material common to some of the chapters.

We hope that these features will facilitate usage of the book in advanced courses and seminars. We sincerely thank the authors for cooperating in these efforts, even though it sometimes required their abandoning some of their favourite conventions – for example, computer scientists commonly use the term *node*, whereas graph theorists use *vertex*; not surprisingly, the graph theorists prevailed on this one. We also asked our contributors to endure the ordeal of having their early versions subjected to detailed critical reading. We believe that as a result the final product is significantly better than it would otherwise have been (as a collection of individual chapters with differing styles and terminology). We want to express our heartfelt appreciation to all of our contributors for their cooperation in these endeavours.

We extend special thanks to Bjarne Toft for his service as Academic Consultant – his advice has been invaluable. We are also grateful to Cambridge University Press for publishing these volumes; in particular, we thank Roger Astley, Charlotte Thomas and Clare Dennison for their advice, support, patience and cooperation. Finally we extend our appreciation to several universities for the ways in which they have assisted with our project: the first editor (LWB) is grateful to his home institution of Indiana University–Purdue University Fort Wayne, while the second editor (RJW) has had the cooperation of the Open University as well as the Mathematical Institute and Pembroke College in Oxford University.

LOWELL W. BEINEKE
ROBIN J. WILSON

Preliminaries

LOWELL W. BEINEKE and ROBIN J. WILSON

1. Graph theory

This section presents the basic definitions, terminology and notation of graph theory, along with some fundamental results. Further information can be found in the many standard books on the subject – for example, Bondy and Murty [1], Chartrand, Lesniak and Zhang [2], Gross and Yellen [3] and West [5], or, for a simpler treatment, Marcus [4] and Wilson [6].

Graphs

A *graph* G is a pair of sets (V, E), where V is a finite non-empty set of elements called *vertices*, and E is a finite set of elements called *edges*, each of which has two associated vertices. The sets V and E are the *vertex-set* and *edge-set* of G, and are sometimes denoted by $V(G)$ and $E(G)$. The number of vertices in G is called the *order* of G and is usually denoted by n (but sometimes by $|G|$); the number of edges is denoted by m. A graph with only one vertex is called *trivial.*

An edge whose vertices coincide is a *loop*, and if two edges have the same pair of associated vertices, they are called *multiple edges*. In this book, unless otherwise specified, graphs are assumed to have neither loops nor multiple edges; that is, they are taken to be *simple*. Hence, an edge e can be considered as its associated pair of vertices, $e = \{v, w\}$, usually shortened to vw. An example of a graph of order 5 is shown in Fig. 1(a).

The *complement* $\overline{G}$ of a graph G has the same vertices as G, but two vertices are adjacent in $\overline{G}$ if and only if they are not adjacent in G. Figure 1(b) shows the complement of the graph in Fig. 1(a).

(a) (b)

Fig. 1.

Adjacency and degrees

The vertices of an edge are its *endpoints* and the edge is said to *join* these vertices. An endpoint v of an edge $e = vw$ and the edge e are *incident* with each other. Two vertices that are joined by an edge are called *neighbours* and are said to be *adjacent*; if v and w are adjacent vertices we sometimes write $v \sim w$, and if they are not adjacent we write $v \nsim w$. Two edges are *adjacent* if they have a vertex in common.

The set $N(v)$ of neighbours of a vertex v is called its *neighbourhood*. If $X \subseteq V$, then $N(X)$ denotes the set of vertices that are adjacent to some vertex of X.

The *degree* $\deg v$, or $d(v)$, of a vertex v is the number of its neighbours; in a non-simple graph, it is the number of occurrences of the vertex as an endpoint of an edge, with loops counted twice. A vertex of degree 0 is an *isolated vertex* and one of degree 1 is an *end-vertex* or *leaf*. A graph is *regular* if all of its vertices have the same degree, and is *k-regular* if that degree is k; a 3-regular graph is sometimes called *cubic*. The maximum degree in a graph G is denoted by $\Delta(G)$, or just Δ, and the minimum degree by $\delta(G)$ or δ.

An *isomorphism* between two graphs G and H is a bijection between their vertex-sets that preserves both adjacency and non-adjacency. The graphs G and H are *isomorphic*, written $G \cong H$, if there exists an isomorphism between them.

Independent sets and cliques

A set of vertices of a graph G is an *independent set* (or *stable set*) if no two vertices are adjacent. The *independence number* (or *stability number*) $\alpha(G)$ is the size of the largest such set.

A set of vertices is a *clique* if all pairs of vertices are adjacent. The *clique number* $\omega(G)$ is the size of a largest clique.

Walks, paths and cycles

A *walk* in a graph is a sequence of vertices and edges $v_0, e_1, v_1, \ldots, e_k, v_k$, in which the edge e_i joins the vertices v_{i-1} and v_i. This walk is said to *go from v_0 to v_k* or to *connect* v_0 *and* v_k, and is called a v_0–v_k walk. It is frequently shortened to $v_0v_1 \cdots v_k$, since the edges can be inferred from this. A walk is *closed* if the first and last vertices are the same. Some important types of walk are the following:

- a *path* is a walk in which no vertex is repeated
- a *cycle* is a non-trivial closed walk in which no vertex is repeated, except the first and last
- a *trail* is a walk in which no edge is repeated
- a *circuit* is a non-trivial closed trail.

Connectedness and distance

A graph is *connected* if it has a path connecting each pair of vertices, and *disconnected* otherwise. A (*connected*) *component* of a graph is a maximal connected subgraph.

The number of occurrences of edges in a walk is called its *length*, and in a connected graph the *distance* $d(v, w)$ from v to w is the length of a shortest v–w path. It is easy to check that distance satisfies the properties of a metric. The *diameter* of a connected graph G is the greatest distance between any pair of vertices in G. If G has a cycle, the *girth* of G is the length of a shortest cycle.

A connected graph is *Eulerian* if it has a closed trail containing all of its edges; such a trail is an *Eulerian trail*. A connected graph G is Eulerian if and only if every vertex of G has even degree. This means that the edge-set of G can be partitioned into cycles.

A graph of order n is *Hamiltonian* if it has a cycle of length n, and is *pancyclic* if it has a cycle of every length from 3 to n. It is *traceable* if it has a path through all vertices. No 'good' characterizations of these properties are known.

Bipartite graphs and trees

If the set of vertices of a graph G can be partitioned into two non-empty subsets so that no edge joins two vertices in the same subset, then G is *bipartite*. The two subsets are called *partite sets*, and if they have orders r and s, G is said to be an $r \times s$ *bipartite graph*. (For convenience, the graph with one vertex and no edges is also called bipartite.) Bipartite graphs are characterized by having no cycles of odd length.

Among the bipartite graphs are *trees*, those connected graphs with no cycles. Any graph without cycles is a *forest*; thus, each component of a forest is a tree. Trees have been characterized in many ways, some of which we give here. For a graph G of order n, the following statements are equivalent:

- G is connected and has no cycles
- G is connected and has $n - 1$ edges
- G has no cycles and has $n - 1$ edges
- G has exactly one path between any two vertices.

The set of trees can also be defined inductively: a single vertex is a tree; and for $n \geq 1$, the trees with $n+1$ vertices are those graphs obtainable from some tree with n vertices by adding a new vertex adjacent to precisely one of its vertices.

This definition has a natural extension to higher dimensions. The *k-dimensional trees*, or *k-trees* for short, are defined as follows: the complete graph on k vertices is a k-tree, and for $n \geq k$, the k-trees with $n+1$ vertices are those graphs obtainable from some k-tree with n vertices by adding a new vertex adjacent to k mutually adjacent vertices in the k-tree. Figure 2 shows a tree and a 2-tree.

Fig. 2.

An important concept in the study of graph minors (introduced later) is the *tree-width* of a graph G, the minimum dimension of any k-tree that contains G as a subgraph.

Special graphs

We now introduce some individual types of graph:

- the *complete graph* K_n has n vertices, each adjacent to all the others
- the *null graph* $\overline{K}_n$ has n vertices and no edges
- the *path graph* P_n consists of the vertices and edges of a path of length $n-1$
- the *cycle graph* C_n consists of the vertices and edges of a cycle of length n
- the *complete bipartite graph* $K_{r,s}$ is the $r \times s$ bipartite graph in which each vertex is adjacent to all of the vertices in the other partite set
- the *complete k-partite graph* $K_{r_1,r_2,\ldots,r_k}$ has its vertices in k sets with orders $r_1, r_2, \ldots, r_k$, and every vertex is adjacent to all of the vertices in the other sets; if the k sets all have order r, the graph is denoted by $K_{k(r)}$.

Examples of these graphs are given in Fig. 3.

Operations on graphs

Let G and H be graphs with disjoint vertex-sets $V(G) = \{v_1, v_2, \ldots, v_r\}$ and $V(H) = \{w_1, w_2, \ldots, w_s\}$.

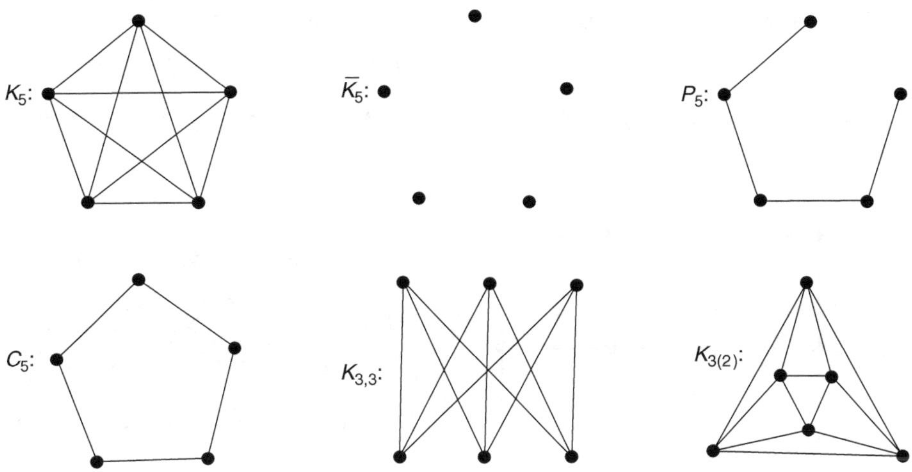

Fig. 3.

The *union* $G \cup H$ has vertex-set $V(G) \cup V(H)$ and edge-set $E(G) \cup E(H)$. The union of k graphs isomorphic to G is denoted by kG.
The *join* $G + H$ is obtained from $G \cup H$ by adding an edge from each vertex in G to each vertex in H.
The *Cartesian product* $G \times H$ (or $G \square H$) has vertex-set $V(G) \times V(H)$, with (v_i, w_j) adjacent to (v_h, w_k) if either v_i is adjacent to v_h in G and $w_j = w_k$, or $v_i = v_h$ and w_j is adjacent to w_k in H; in less formal terms, $G \times H$ can be obtained by taking n copies of H and joining corresponding vertices in different copies whenever there is an edge in G.

Examples of these binary operations are given in Fig. 4.

Subgraphs and minors

If G and H are graphs with $V(H) \subseteq V(G)$ and $E(H) \subseteq E(G)$, then H is a *subgraph* of G, and is a *spanning subgraph* if $V(H) = V(G)$. The subgraph $\langle S \rangle$ (or $G[S]$) *induced* by a non-empty set of S of vertices of G is the subgraph H whose vertex-set is S and whose edge-set consists of those edges of G that join two vertices in S. A subgraph H of G is called an *induced subgraph* if $H = \langle V(H) \rangle$. In Fig. 5, H_1 is a spanning subgraph of G, and H_2 is an induced subgraph.

The *deletion of a vertex* v from a graph G results in the subgraph obtained by removing v and all of its incident edges; it is denoted by $G - v$ and is the subgraph induced by $V - \{v\}$. More generally, if S is any set of vertices in G, then $G - S$ is the graph obtained from G by deleting all of the vertices in S and their incident edges; that is, $G - S = \langle V(G) - S \rangle$. Similarly, the *deletion of an edge e* results in the subgraph $G - e$ and, for any set X of edges, $G - X$ is the graph obtained from G by deleting all the edges in X.

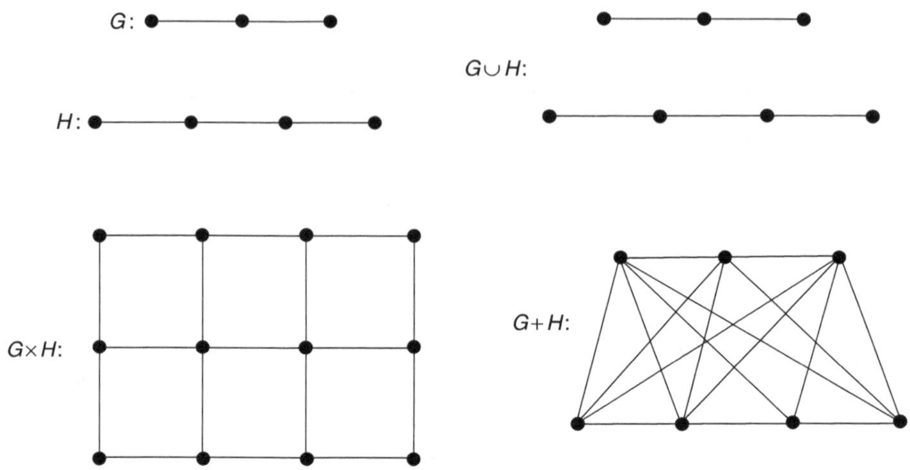

Fig. 4.

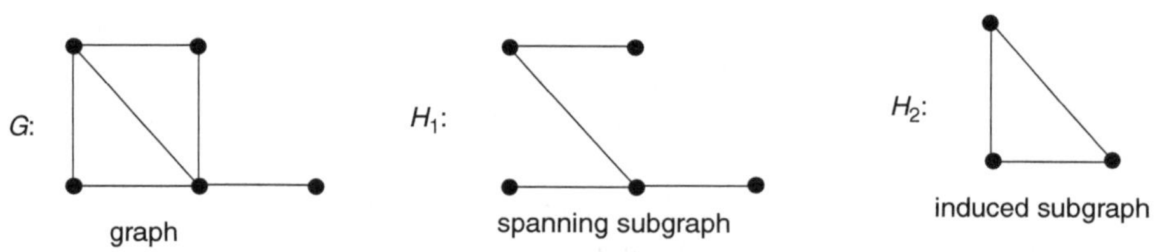

Fig. 5.

If the edge e joins vertices v and w, then the *subdivision* of e replaces e by a new vertex u and two new edges vu and uw. Two graphs are *homeomorphic* if there is some graph from which each can be obtained by a sequence of subdivisions. The *contraction* of e replaces its vertices v and w by a new vertex u, with an edge ux if v or w is adjacent to x in G. The operations of subdivision and contraction are illustrated in Fig. 6.

If H can be obtained from G by a sequence of edge-contractions and the removal of isolated vertices, then G is *contractible* to H. A *minor* of G is any graph that can be obtained from G by a sequence of edge-deletions and edge-contractions, along with deletions of isolated vertices. Note that if G has a subgraph homeomorphic to H, then H is a minor of G.

Digraphs

Digraphs are directed analogues of graphs, and thus have many similarities, as well as some important differences. A *digraph* (or *directed graph*) D is a pair of sets (V, A), where V is a finite non-empty set of elements called *vertices*, and A is a set of ordered pairs of distinct elements of V called *arcs*. Note that the elements of A are ordered,

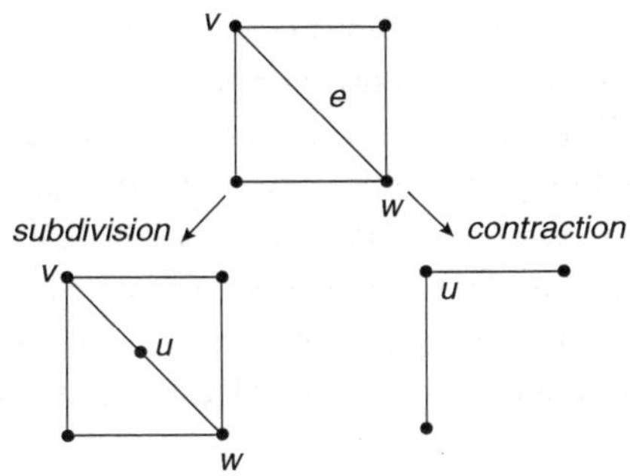

Fig. 6.

which gives each of them a direction. An example of a digraph, with the directions indicated by arrows, is shown in Fig. 7.

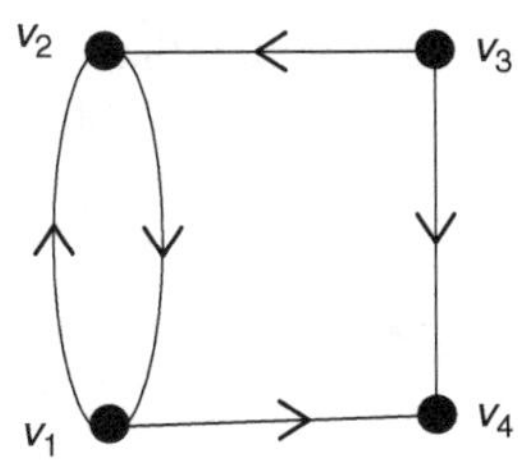

Fig. 7.

Because of the similarities between graphs and digraphs, we mention only the main differences here and do not redefine those concepts that carry over easily. An arc (v, w) in a digraph may be written as vw, and is said to *go from v to w*, or to *go out of v* and *into w*. In the context of digraphs, walks, paths, cycles, trails and circuits are all understood to be directed, unless otherwise indicated. A digraph D is *strongly connected* or *strong* if there is a path from each vertex to each of the others; note that the digraph in Fig. 7 is strong. A *strong component* is a maximal strongly connected subgraph. Every vertex is in at least one strong component, and an edge is in a strong component if and only if it is on a directed cycle.

The *out-degree* $d^+(v)$ of a vertex v in a digraph D is the number of arcs out of v, and the *in-degree* $d^-(v)$ is the number of arcs into v. The minimum out-degree in a digraph is denoted by δ^+, the minimum in-degree δ^-, and the minimum of the two is denoted by δ.

Connectivity

In this section, we give the primary definitions and some of the basic results on connectivity, including two versions of the most important one of all, Menger's theorem.

A vertex v in a graph G is a *cut-vertex* if $G - v$ has more components than G. For a connected graph, this is equivalent to saying that $G - v$ is disconnected, and that there exist vertices u and w, different from v, for which v is on every u–w path. It is easy to see that every non-trivial graph has at least two vertices that are not cut-vertices.

A non-trivial graph is *non-separable* if it is connected and has no cut-vertices. Note that under this definition the graph K_2 is non-separable. There are many characterizations of the other non-separable graphs, as the following statements are all equivalent for a connected graph G with at least three vertices:

- G is non-separable
- every two vertices of G share a cycle
- every two edges of G share a cycle
- for any three vertices u, v and w in G, there is a v–w path that contains u
- for any three vertices u, v and w in G, there is a v–w path that does not contain u.

A *block* in a graph is a maximal non-separable subgraph. Each edge of a graph lies in exactly one block, while each vertex that is not an isolated vertex lies in at least one block, those that are in more than one block being cut-vertices. The graph in Fig. 8 has four blocks.

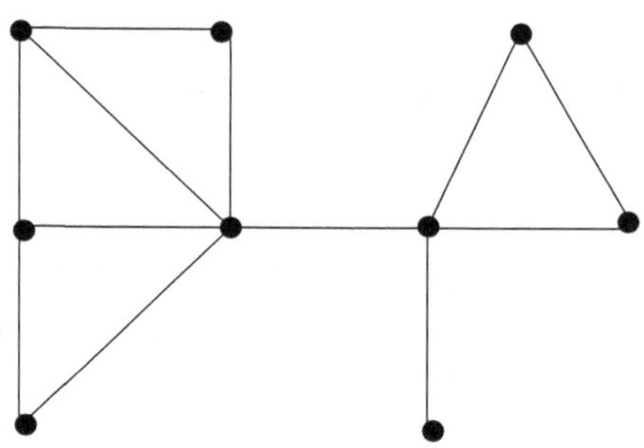

Fig. 8.

The basic idea of non-separability has a natural generalization: a graph G is *k-connected* if the removal of fewer than k vertices always leaves a non-trivial connected graph. The main result on graph connectivity – indeed, it might well be called the Fundamental theorem of connectivity – is Menger's theorem, first published in 1927. It has many equivalent forms, and the first that we give here is the vertex version. Paths joining the same pair of vertices are called *internally disjoint* if they have no other vertices in common.

Menger's theorem (vertex version) *A graph is k-connected if and only if every pair of vertices are joined by k internally disjoint paths.*

The *connectivity* $\kappa(G)$ of a graph G is the maximum non-negative integer k for which G is k-connected; for example, the connectivity of the complete graph K_n is $n - 1$, and a graph has connectivity 0 if and only if it is trivial or disconnected.

There is an analogous body of material that involves edges rather than vertices, and because of the similarities, we treat it in less detail.

An edge e is a *cut-edge* (or *bridge*) of a graph G if $G-e$ has more components than G. (In contrast to the situation with vertices, the removal of an edge cannot increase the number of components by more than 1.) An edge e is a cut-edge if and only if there exist vertices v and w for which e is on every v–w path. The cut-edges in a graph are also characterized by the property of not lying on a cycle; thus, a graph is a forest if and only if every edge is a cut-edge. Graphs with no cut-edges can be characterized in a variety of ways similar to those having no cut-vertices – that is, non-separable graphs. The concepts corresponding to cycles and paths for vertices are circuits and trails for edges.

Moving beyond cut-edges, we have the following definitions. A graph G is *l-edge-connected* if the removal of fewer than l edges always leaves a connected graph. Here is another version of Menger's theorem.

Menger's theorem (edge version) *A graph is l-edge-connected if and only if each pair of its vertices is joined by l edge-disjoint paths.*

The *edge-connectivity* $\lambda(G)$ of a graph G is the greatest non-negative integer l for which G is l-edge-connected. Obviously, $\lambda(G)$ cannot exceed the minimum degree of a vertex of G; furthermore, it is at least as large as the connectivity – that is,

$$\kappa(G) \leq \lambda(G) \leq \delta(G).$$

Along with the undirected versions of Menger's theorem, there are corresponding directed versions (with directed paths and strong connectivity) and weighted versions.

2. Graph colourings

The origins of chromatic graph theory lie in the colouring of maps, a story that is well known. In this section we present some of the definitions and basic results of chromatic graph theory.

Vertex-colourings

A *colouring* of a graph G is an assignment of a colour to each vertex of G so that adjacent vertices always have different colours, and G is said to be *k-colourable* if it has a colouring with k colours. The *chromatic number* $\chi(G)$ is the smallest value of k for which G has a k-colouring.

The fact that computing the chromatic number of a graph is an NP-complete problem has contributed to the attraction of this area of mathematics – in fact, determining whether a graph is 3-colourable is itself NP-complete.

We note that the complete graph K_n has chromatic number n and that a bipartite graph with at least one edge has chromatic number 2, and consequently $\chi(G) \geq 3$ if and only if G contains an odd cycle of odd length. An interesting and useful upper bound on the chromatic number of graphs in general was published by L. Brooks in 1941.

Theorem 2.1 *If G is a connected graph with maximum degree Δ, then*

$$\chi(G) \leq \Delta + 1,$$

with equality if and only if G is a complete graph or a cycle of odd length.

There is also an upper bound on the chromatic number that is in terms of minimum degrees; it is easily proved by induction.

Theorem 2.2 *If G and each of its subgraphs has a vertex of degree δ^* or less, then*

$$\chi(G) \leq \delta^* + 1.$$

As far as lower bounds are concerned, the obvious one is the clique number: $\chi(G) \geq \omega(G)$. However, as was first shown by Blanche Descartes, there are triangle-free graphs with arbitrarily large chromatic numbers. More generally, Paul Erdős proved the following result.

Theorem 2.3 *For all $k \geq 2$ and all g, there exists a k-chromatic graph with girth greater than g.*

We conclude our discussion of bounds with a pair that involve the independence number, the dual concept to the clique number.

Theorem 2.4 *If G is a graph of order n and independence number α, then*

$$\frac{n}{\alpha} \leq \chi(G) \leq n - \alpha + 1.$$

Critical graphs

In the study of the chromatic number, one type of graph that arises quite naturally is that of a *critical graph*, a graph G for which each proper subgraph has chromatic number less than that of G. If $\chi(G) = k$, they are often called *k-critical*; they were first studied by G. A. Dirac. Here are two of his results.

Theorem 2.5 *For $k \geq 2$, every k-critical graph is $(k-1)$-edge-connected.*

Theorem 2.6 *Every k-critical graph is either Hamiltonian or has a cycle of length at least $2k - 2$.*

Colouring graphs on surfaces

Following his publication in 1890 of the flaw in A. B. Kempe's purported proof of the four-colour conjecture, Percy Heawood went on to prove that the analogous result for the torus was a seven-colour theorem. The story of the corresponding result for other surfaces is an interesting one. Given an orientable surface S, a generalization of Euler's polyhedron formula can be used (with the aid of Theorem 2.2) to obtain an upper bound for the chromatic number of every graph embeddable on S. However, it was not until 1968 that this upper bound was also shown to be a lower bound. This breakthrough came in the form of the theorem of G. Ringel and J. W. T. Youngs on the genus $\gamma(K_n)$ of the complete graph – that is, the number of handles that need to be added to the sphere for there to be an embedding of the K_n on the surface:

$$\gamma(K_n) = \lceil \tfrac{1}{12}(n-3)(n-4) \rceil.$$

The maximum chromatic number among all graphs embeddable on a given surface S is called the *chromatic number of* S and is denoted $\chi(S)$.

Theorem 2.7 *The chromatic number of the orientable surface* $S_{\hat{g}}$ *with* g *handles* $(g > 0)$ *is*

$$\chi(S_h) = \lfloor \tfrac{1}{2}(7 + \sqrt{1 + 48g}) \rfloor.$$

We note that the formula also holds when $h = 0$ (the four-colour theorem), but the proof of Ringel and Youngs does not.

There is a corresponding theorem (one that is easier to prove) for non-orientable surfaces; it was found by Ringel in 1954.

Theorem 2.8 *The chromatic number of the non-orientable surface* $N_{\hat{h}}$ *with* h *crosscaps* $(h > 0)$ *is*

$$\chi(N_h) = \lfloor \tfrac{1}{2}(7 + \sqrt{1 + 24h}) \rfloor,$$

except that, for the Klein bottle, $\chi(N_2) = 6$.

Edge-colourings

Interesting problems also arise when the edges of a graph rather than the vertices are being coloured. A graph G is called *k-edge-colourable* if its edges can be coloured with k colours in such a way that all of the edges at each vertex have different colours. The minimum k for which G can be k-edge-coloured is called the *chromatic index* (or the *edge-chromatic number*) of G, denoted by $\chi'(G)$. Clearly the chromatic index of a graph G is at least as large as the maximum degree $\Delta(G)$. The first major result in this area is due to D. König and was discovered in his study of independent sets of edges in graphs.

Theorem 2.9 *If G is a bipartite graph, then* $\chi'(G) = \Delta(G)$.

The main theorem on edge-colourings, due to V. G. Vizing, says that every graph can be edge-coloured with $\Delta(G) + 1$ colours.

Theorem 2.10 *If G is a graph with maximum degree* Δ, *then*

$$\chi'(G) = \Delta \text{ or } \Delta + 1.$$

I. Holyer showed that, even among cubic graphs, deciding which of the two values, 3 or 4, is the correct one is hard. More precisely, he showed that deciding whether a cubic graph is 3-edge-colourable is NP-complete.

List-colourings

Following the resolution of the four-colour conjecture, considerable attention has been given to a variation of graph colouring in which there might be different sets of colours available to different vertices in a graph. Formally, we make the following definitions (the terminology may vary from author to author). If v is a vertex in a graph G, a *colour list* for v is a set $L(v)$ of colours that are permitted at v. If each vertex of G has a colour list, then an *L-colouring* of G is a colouring in which the colour of each vertex comes from its list. If k is an integer for which a graph G has an L-colouring for every list L with $|L(v)| = k$ for all vertices v, then G is said to be *k-list-colourable* or *k-choosable*. The *list-chromatic number* or (*choice number*) $\chi_l(G)$ is the minimum number k for which G is k-list-colourable. Since list-colourings are the subject of Chapter 6, we only mention a few facts about them here.

It is straightforward to show that $\chi_l(G) \leq 1 + \Delta(G)$. In fact, the analogue of Brooks's theorem holds for the list-chromatic number. Clearly, $\chi(G) \leq \chi_l(G)$; however, equality does not always hold. In fact, the list-chromatic number of $K_{3,3}$ is 3. It is also the case that there are planar graphs that are not 4-list-colourable; the following result is due to C. Thomassen.

Theorem 2.11 *Every planar graph is 5-list-colourable.*

References

1. J. A. Bondy and U. S. R. Murty, *Graph Theory*, Springer, 2008.
2. G. Chartrand, L. Lesniak and P. Zhang, *Graphs and Digraphs* (5th edn.), CRC, 2011.
3. J. L. Gross and J. Yellen, *Graph Theory and its Applications* (2nd edn.), CRC, 2005.
4. D. A. Marcus, *Graph Theory*, Math. Assoc. of America, 2008.
5. D. B. West, *Introduction to Graph Theory* (2nd edn.), Pearson, 2001.
6. R. J. Wilson, *Introduction to Graph Theory* (5th edn.), Pearson, 2010.

1
Colouring graphs on surfaces

BOJAN MOHAR

Developments in graph colouring theory were motivated by the four-colour problem and Heawood's theorem. Both of these were originally formulated as map-colouring problems that can be expressed as colouring graphs embedded on surfaces. This chapter gives an overview of the abundance of results concerning the chromatic number of graphs that are embedded on surfaces.

1. Introduction

In 1852 Francis Guthrie asked whether the regions of every planar map can be coloured with four colours in such a way that no two regions of the map with common boundary receive the same colour. In effect, by duality Guthrie was asking whether every planar graph is 4-colourable. This easily stated problem became known as the famous four-colour problem. Attempts to solve it led to many important results in graph theory, but the problem itself remained unsolved for more than a century. It was finally answered in the positive by Appel and Haken [9], [10], [12]. More about its proof comes later in this chapter.

For more than a century, the four-colour problem was one of the driving forces that led to developments in graph theory, and specifically graph colouring theory. Its generalization – the Heawood problem – was the main motivation for developments in topological graph theory (see [51] or [49]). Both of these were originally formulated as map-colouring problems, asking about colouring the faces of an embedded graph so that adjacent faces receive different colours. Every map-colouring problem can be expressed as a graph-colouring problem by considering the dual graph of the map. Its vertices correspond to the faces of the map, and two vertices are adjacent if the corresponding faces are adjacent on the surface.

In this chapter we discuss the colouring of graphs embedded on surfaces. We start by describing some of the ideas in the proof of the four-colour theorem and give a proof of the list-colouring version of the five-colour theorem that is due to Thomassen. Next, we consider a generalization to general surfaces, known as the Heawood problem.

What is the largest chromatic number of a graph that can be embedded in a given surface?

A solution was conjectured by Heawood [36] and eventually proved by Ringel and Youngs [52], [51]. Heawood's conjecture can be reduced to determining the genus and the non-orientable genus of complete graphs. The solution by Ringel and Youngs involves constructions of minimum genus embeddings of these graphs and introduces techniques that are of interest in topological graph theory. Finally, we outline colouring results and problems of a more specialized nature. There is an abundance of new results in this area and we will only give an overview of some of the most important achievements.

2. Planar graphs are 4-colourable and 5-choosable

A direct consequence of Euler's formula is that every planar graph has a vertex of degree 5 or less, from which it follows that every planar graph is 6-colourable. Heawood [36] proved that every planar graph is 5-colourable. In this section we present the extension of this result to list-colourings due to Thomassen [59].

Let G be a graph and suppose that, for each vertex v, some set $L(v)$ of natural numbers is given. We view $L(v)$ as a *list* of colours that can be used when colouring v. An *L-colouring* of G is an assignment of a colour to each vertex of G such that each vertex v receives a colour from its list $L(v)$ and adjacent vertices get different colours. Such a colouring is also called a *list-colouring*. The graph G is *k-choosable* if it admits an L-colouring for every list assignment in which each list $L(v)$ contains at least k colours. The *list-chromatic number* $\chi_l(G)$ is the minimum integer k for which G is k-choosable.

In 1975 Vizing raised the question of whether every planar graph is 5-choosable. Erdős, Rubin and Taylor [31] conjectured that every planar graph is 5-choosable, but not necessarily 4-choosable. In 1993 Voigt [65] proved the last part of this conjecture by exhibiting examples of planar graphs that are not 4-choosable. Subsequently, even 3-colourable non-4-choosable planar graphs have been found (see Voigt and Wirth [67] and Mirzakhani [46]). Thomassen [59] verified the first part of the conjecture, which implies, in particular, that every planar graph is 5-colourable.

Theorem 2.1 *Every planar graph is* 5*-choosable.*

In order to prove Theorem 2.1, it suffices to show that it holds for every planar triangulation (a plane graph with all faces of length 3). Thus, Theorem 2.1 follows immediately from Theorem 2.2 below. We recall that a plane graph G is a *near-triangulation* if it is 2-connected and if all faces, except possibly the outer face, are triangles. Thomassen [59] proved the following result.

Theorem 2.2 *Let G be a near-triangulation with outer cycle $C = v_1v_2 \ldots v_pv_1$. Assume that $L(v_1) = \{1\}$ and $L(v_2) = \{2\}$ and that, for any other vertex v, the list $L(v)$ has at least three colours if $v \in V(C)\backslash\{v_1, v_2\}$ and at least five colours if $v \in V(G - C)$. Then G has an L-colouring.*

Proof (by induction on the number n of vertices in G) If $n = 3$, there is nothing to prove. So we proceed to the induction step.

If C has a chord, say v_iv_j, where $2 \le i \le j - 2 \le p - 1$ (and $v_{p+1} = v_1$), then we apply the induction hypothesis to the cycle $C' = v_1v_2 \ldots v_iv_jv_{j+1} \ldots v_1$ and its interior. We also apply induction to $C'' = v_jv_iv_{i+1} \ldots v_{j-1}v_j$ and its interior, where the list of vertices v_i and v_j consists of the single colour that was obtained by using the induction hypothesis on C' and its interior.

Suppose now that C is a chordless cycle, and let $v_1, u_1, u_2, \ldots, u_m, v_{p-1}$ be the neighbours of v_p in clockwise order around v_p. Since the interior of C is triangulated, G contains the path $P = v_1u_1u_2 \ldots u_mv_{p-1}$. Since C is chordless, $C' = P \cup (C - v_p)$ is a cycle. Let x, y be two colours in $L(v_p)\backslash\{1\}$. Now define $L'(u_i) = L(u_i)\backslash\{x, y\}$, for $1 \le i \le m$, and $L'(v) = L(v)$ if $v \in V(G)\backslash\{u_1, \ldots, u_m\}$. Now apply the induction hypothesis to C' and its interior and the new lists L'. We extend the obtained L'-colouring to an L-colouring of G by assigning x or y to v_p in such a way that v_p and v_{p-1} get distinct colours. ■

The four-colour problem was solved by Appel and Haken [9], [10], [12]. An updated version of their proof may be found in [11], and a simpler and cleaner proof was given by Robertson, Sanders, Seymour and Thomas [54]. An in-depth discussion of the four-colour theorem, comments about its proof and its consequences can be found in [68].

Four-colour theorem *Every planar graph is* 4*-colourable.*

The proof of Appel and Haken is lengthy, and part of it was done by a brute-force computer checking of the millions of possibilities that occurred in their proof.

Initially, there was a lot of controversy about the legitimacy of this proof. It was considered by some mathematicians to be inadequate since it did not qualify as a conventional proof that could be checked by a human in a lifetime. The original proof also had some flaws, but all those that were discovered were subsequently corrected by the authors in their monograph [11].

The proof of Appel and Haken follows the outline proposed by the German mathematician Heinrich Heesch. First, it is shown that every planar graph contains a subgraph from a certain list $\mathcal{C}$ of configurations. The proof of this part is based on a discharging method, a very elaborate use of Euler's polyhedron formula. Secondly, it is shown that no configuration from $\mathcal{C}$ can occur in a minimal counter-example to the four-colour theorem. Having a particular configuration $X \in \mathcal{C}$ contained in a minimal counter-example G, the minimality would ensure that the rest of the graph, when X is removed from G, can be 4-coloured. There may be many possibilities for how the colouring would affect the neighbourhood of X. For some configurations considered by Appel and Haken, several hundred thousands of cases needed to be considered. It is here where the use of computers was necessary. For each such case, it was possible to argue that the colouring can be changed in such a way that it can be extended to X, yielding a contradiction to the assumption that G is a counter-example.

The proof method described above can be illustrated by the following example. A simple corollary of Euler's formula states that every planar graph contains a vertex with at most five neighbours, so the set $\mathcal{C}$ could consist of six configurations, each corresponding to a vertex of degree i, for $0 \leq i \leq 5$. If v is a vertex of degree at most 3, and if $G - v$ has a 4-colouring, then this colouring can be extended to G, since v can be coloured by a colour that does not appear among its neighbours. It can also be shown that a vertex of degree 4 cannot appear in a minimal counter-example, but there is no way to prove the same for vertices of degree 5. This case has to be replaced by hundreds of more complicated configurations.

In 1997 Robertson, Sanders, Seymour and Thomas [54] developed another proof, based on similar principles. The human-checkable part of the proof was made more transparent, and the computer-verified part was supported by a well-structured collection of data that enabled independent checking of the proofs. There may be doubts about the correctness of the compilers and the stability of the hardware used in the computer checking, but this is probably much less susceptible to error than human checking, especially because the latter proof has been independently checked on different platforms.

The complete proof of the four-colour theorem is available on-line with links from [57]. It includes the encoding of the unavoidable set of the 633 near-triangulations, the discharging rules, the computer programs, and other helpful information needed for the proof.

The four-colour theorem is interesting, in that it has many equivalent formulations, some of which do not seem to be related to map colourings at all. Two equivalent versions most frequently used in the literature are dual statements about nowhere-zero flows and edge-colourings, the topics considered in Chapters 5 and 9.

Theorem 2.3 *Every 2-edge-connected planar graph has a nowhere-zero 4-flow.*

Theorem 2.4 *Every 2-connected cubic planar graph is 3-edge-colourable.*

An important side result of the proof of the four-colour theorem is a polynomially bounded algorithm for 4-colouring planar graphs (see [53]).

Theorem 2.5 *A 4-colouring of a given planar graph of order n can be found in $O(n^2)$ steps.*

In contrast to Theorem 2.5, deciding whether a planar graph is 3-colourable is much harder.

Theorem 2.6 *The decision problem as to whether a given planar graph is 3-colourable is NP-complete.*

An interesting (but rather easy) fact about 3-colourings of planar graphs is the following folklore result.

Theorem 2.7 *A planar graph is 3-colourable if and only if it is a spanning subgraph of a 3-colourable triangulation of the plane. A triangulation is 3-colourable if and only if all its vertices have even degree.*

Another folklore result that should be mentioned here concerns 2-colourable planar graphs. By König's theorem, a graph is bipartite if and only if all its cycles have even lengths. For a planar graph, the face-boundaries generate the cycle space, giving the following statement.

Theorem 2.8 *A planar graph is bipartite if and only if all its faces have even length.*

A list-colouring result for planar bipartite graphs was proved by Alon and Tarsi [7] who used the polynomial method that proved to be an indispensable tool for working with the list-chromatic number.

Theorem 2.9 *Every planar bipartite graph is 3-choosable.*

This result is somewhat surprising since, as noted earlier, there are 3-colourable planar graphs whose list-chromatic number is 5 (see Voigt and Wirth [67] and Mirzakhani [46]).

To conclude this section, we present some results about extensions of a partial colouring. Albertson [2] proved that every precolouring of vertices of a planar graph that are far apart from each other can be extended to a 5-colouring of the whole graph.

Theorem 2.10 *If S is a subset of the vertices of a planar graph G, and if the distance between any two vertices in S is at least 4, then any 5-colouring of S can be extended to a 5-colouring of G.*

Proof The proof is based on the four-colour theorem. We start with a 4-colouring of G using colours 1, 2, 3, 4. For each vertex $s \in S$, recolour s according to the given precolouring. If c is the colour given to s, then we recolour all the neighbours of s coloured c with the colour 5. It is easy to see that this gives a 5-colouring of G that extends the given precolouring. ■

Albertson asked whether this theorem can be extended to list-colourings. This problem, which became known as Albertson's conjecture, was solved recently by Dvořák, Lidický, Mohar and Postle [28].

Theorem 2.11 *If G is a planar graph with list-assignment L that gives a list of size* 1 *or* 5 *to each vertex, and if the distance between any pair of vertices with lists of size* 1 *is at least* 1984*, then G is L-colourable.*

Dvořák, Lidický and Mohar [27] obtained another result of a similar flavour.

Theorem 2.12 *Every graph drawn in the plane with some edges crossing each other is* 5*-choosable if all of the crossings are at graph distance* 15 *or more from each other.*

3. Heawood's formula

From now on we consider graphs embedded on arbitrary surfaces. Recall that every surface is homeomorphic to precisely one of the following surfaces:

- the *orientable surface S_g of genus g* homeomorphic to the sphere with g handles ($g \geq 0$)
- the *non-orientable surface N_h of genus h* homeomorphic to the sphere with h crosscaps ($h \geq 1$).

Since adding two crosscaps to a non-orientable surface is equivalent to adding one handle, it makes sense to unify the notion of the genus by introducing the *Euler genus* $\widehat{\mathbf{g}}(S)$ of a surface S by setting $\widehat{\mathbf{g}}(S) = 2g$ if S is homeomorphic to S_g, and $\widehat{\mathbf{g}}(S) = h$ if S is homeomorphic to N_h.

Suppose that G is a graph with n vertices, m edges and f faces, that is embedded in a surface S. Then, by *Euler's formula*,

$$n - m + f \geq 2 - \widehat{\mathbf{g}}(S),$$

where equality holds if and only if the embedding is cellular (all faces are homeomorphic to open discs in the plane).

Heawood [36] proved the following analogue of the four-colour theorem for graphs embedded on arbitrary surfaces.

Theorem 3.1 *Let S be a surface with Euler genus $g > 0$ and let G be a graph embedded in S. Then*

$$\chi(G) \leq \left\lfloor \tfrac{1}{2}\left(7 + \sqrt{1 + 24g}\right) \right\rfloor. \tag{1}$$

Proof Let $r = \chi(G)$. We may assume that each vertex of G has degree at least $r-1$, since if $\deg(v) < r-1$, then $\chi(G-v) = \chi(G)$, and so we may reduce the problem to $G - v$. Euler's formula implies that $m \le 3n - 6 + 3g$. Since $2m \ge (r-1)n$,

$$(r-7)n + 12 - 6g \le 0. \tag{2}$$

Since $g > 0$, the right-hand side of (1) is at least 6, and so we may assume that $r \ge 7$. Since $n \ge r$, (2) implies that $r^2 - 7r + 12 - 6g \le 0$. This proves (1). ■

We note that, by the four-colour theorem, the bound (1) holds also for the sphere.

The proof of Theorem 3.1 shows that every graph embeddable on S has a vertex of degree not exceeding $\frac{1}{2}(5 + \sqrt{1+24g})$. Thus, the Heawood bound also holds for the list-chromatic number:

$$\chi_l(G) \le \left\lfloor \tfrac{1}{2}(7 + \sqrt{1+24g}) \right\rfloor. \tag{3}$$

Heawood conjectured that equality holds in (1). This was eventually verified by Ringel and Youngs [51], [52] who proved that, for each surface except the Klein bottle, there exists a complete graph G for which equality holds in (1). Below we discuss this result in more detail in terms of colour-critical graphs.

A graph G is *k-colour-critical* (or *k-critical*) if $\chi(G) = k$ but $\chi(G') < k$ for every proper subgraph G' of G.

If v is a vertex of a k-critical graph G, then $G - v$ is $(k-1)$-colourable. Since its colouring cannot be extended to a $(k-1)$-colouring of G, all $k-1$ colours must be present on the neighbours of v. In particular, v has degree at least $k-1$. If G has n vertices and m edges, this implies that

$$2m \ge (k-1)n, \tag{4}$$

with equality for K_k. One of Dirac's results [23] on k-critical graphs is the following improvement of (4). If G is a k-critical graph with n vertices and m edges, and if $G \ne K_k$, then

$$2m \ge (k-1)n + k - 3. \tag{5}$$

Let $H(g) = \left\lfloor \frac{1}{2}(7 + \sqrt{1+24g}) \right\rfloor$. Heawood's theorem asserts that every graph G on a surface of Euler genus g has chromatic number at most $H(g)$. Dirac [21], [22] proved the following extension for all values of g except 1 and 3; these cases were proved later by Albertson and Hutchinson [4].

Theorem 3.2 *If G is a graph embeddable in the surface of Euler genus $g \ge 1$, then $\chi(G) < H(g)$ unless G contains the complete graph of order $H(g)$ as a subgraph.*

Proof We present Dirac's proof which works for $g \ne 1$ or 3. Let $h = \chi(G) = H(g)$, and let H be an h-critical subgraph of G. Assume that H has n vertices and m edges, and that $H \neq K_h$. By (5), we have $2m \geq (h-1)n + h - 3$. This inequality, combined with the aforementioned consequence of Euler's formula that $m \leq 3n - 6 + 3g$, gives

$$(h-1)n + h - 3 \leq 6n - 12 + 6g,$$

which implies that

$$(h-7)n \leq 6g - 9 - h. \tag{6}$$

Since $G \neq K_h$, and since there are no h-critical graphs on $h+1$ vertices, $n \geq h+2$. Since $g \geq 2$, $h \geq 7$, and thus (6) implies

$$(h-7)(h+2) \leq 6g - 9 - h. \tag{7}$$

It is easy to see that this inequality leads to a contradiction for $g = 2$ and when $4 \leq g \leq 11$, so we may assume that $g \geq 12$. From (7) we obtain $h^2 - 4h - 5 - 6g \leq 0$, implying that

$$h < 2 + \sqrt{6g+9}.$$

On the other hand, $h = H(g) > \frac{1}{2}(5 + \sqrt{24g+1}\,)$. Hence

$$1 + \sqrt{24g+1} < \sqrt{24g+36}.$$

This implies that $2\sqrt{24g+1} < 34$ and therefore $g < 12$, a contradiction. ■

Since the complete graph K_7 cannot be embedded in the Klein bottle (see, for example, [49]), Theorem 3.2 implies in particular that the Heawood bound can be improved for the Klein bottle.

Theorem 3.3 *If G is a graph embeddable in the Klein bottle, then $\chi(G) \leq 6$.*

Škrekovski [55] considered $(H(g)-1)$-critical graphs on surfaces of given Euler genus g and proved that, for $g \geq 10$, the only such graphs are $K_{H(g)-1}$ and, in some exceptional cases, the join $K_{H(g)-4} + C_5$.

Theorem 3.2 was extended to list-colourings by Böhme, Mohar and Stiebitz [13], except for the case of Euler genus $g = 3$, which was solved later by Král' and Škrekovski [45].

Theorem 3.4 *If G is a graph embeddable in a surface of Euler genus $g \geq 1$, then $\chi_l(G) < H(g)$, unless G contains the complete graph of order $H(g)$ as a subgraph, in which case $\chi_l(G) = H(g)$.*

4. Colouring with few colours

For each surface and each natural number $k \geq 8$, there are only finitely many k-critical graphs embeddable on the surface (see Dirac [24] and Edwards [29]).

Theorem 4.1 *Let S be a surface of Euler genus $g \geq 3$. Every 7-critical graph embeddable on S has at most $69(g-2)$ vertices, and for $k \geq 8$, every k-critical graph on S has at most $6(g-2)/(k-7)$ vertices.*

Since a graph is $(k-1)$-colourable if and only if it contains no k-critical subgraph, Theorem 4.1 implies the following result.

Corollary 4.2 *Let S be a surface and let $k \geq 7$. Then there are only finitely many k-critical graphs on S, and there exists a polynomially bounded algorithm for testing whether an arbitrary S-embeddable graph is $(k-1)$-colourable.*

The algorithm in Corollary 4.2 for testing $(k-1)$-colourability can be extended to a linear-time algorithm which also returns a $(k-1)$-colouring if one exists. Given a graph G on S, remove successively its vertices of degree less than $k-1$ until no such vertex exists. If $k=7$, then also remove a vertex of degree 6 if all neighbours have degree 6 and they induce a cycle. It turns out that the order of the remaining graph H is bounded above by a constant that depends on S only, and every $(k-1)$-colouring of H can be extended (in linear time) to G. Hence G can be $(k-1)$-coloured if and only if H has a $(k-1)$-colouring (which can be discovered in constant time by checking all possibilities). A similar linear-time algorithm was proposed by Edwards [29] who also discussed in detail how to perform the removal of vertices in linear time.

Theorem 4.1 says that a graph on a fixed surface is 6-colourable unless the graph contains a small obstruction. Such an obstruction must contain a non-contractible cycle of length $O(g)$, where g is the Euler genus of the surface. Graphs that do not contain short non-contractible cycles are said to be *locally planar*, since a large neighbourhood of each vertex is planar. To quantify local planarity, we define the notion of a width of the embedded graph as follows (see [49] for more details). The *edge-width* $\mathrm{ew}(G)$ of a cellular embedded graph G is the shortest length of a cycle in G that is not contractible on the surface of this embedding. A related parameter is the *surface non-separating edge-width* $\mathrm{ew}'(G)$ which is the minimum length of a surface non-separating (homologically non-trivial) cycle in G.

We now describe a 6-colour theorem by Fisk and Mohar [33].

Theorem 4.3 *There is a universal constant c for which every graph G embedded in a surface of Euler genus $g > 0$ with $\mathrm{ew}'(G) > c \log g$ is 6-colourable.*

Let us now consider 5-colourings.

If G is k-critical and H is l-critical, then their join $G+H$ is $(k+l)$-critical. Therefore the graphs C_3+C_5, K_2+H_7 (where H_7 is the graph of Fig. 1) are 6-critical. Figure 2 shows that they are toroidal (where the grey vertices of the second graph correspond to K_2). Thomassen [60] characterized those toroidal graphs that are 5-colourable.

Theorem 4.4 *There are precisely four 6-critical toroidal graphs: K_6, C_3+C_5, K_2+H_7 (where H_7 is the graph of Fig. 1), and the graph T_{11} obtained from the 11-cycle $x_0x_1\ldots x_{10}x_0$ by adding all chords x_ix_{i+2} and x_ix_{i+3} for $i=0,1,\ldots,10$ (see Fig. 3).*

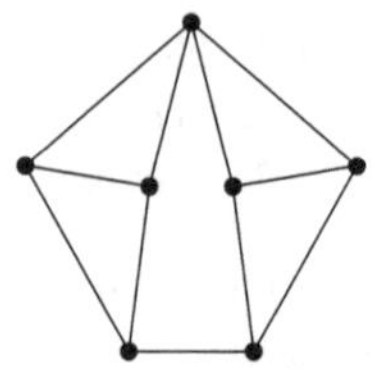

Fig. 1. The 4-critical graph H_7

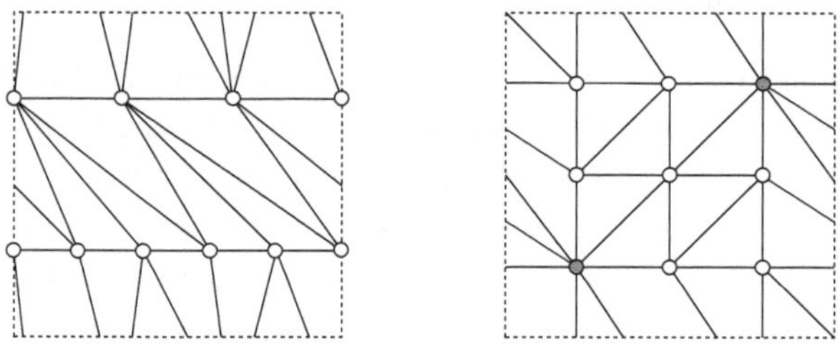

Fig. 2. The graphs $C_3 + C_5$ and $K_2 + H_7$ on the torus

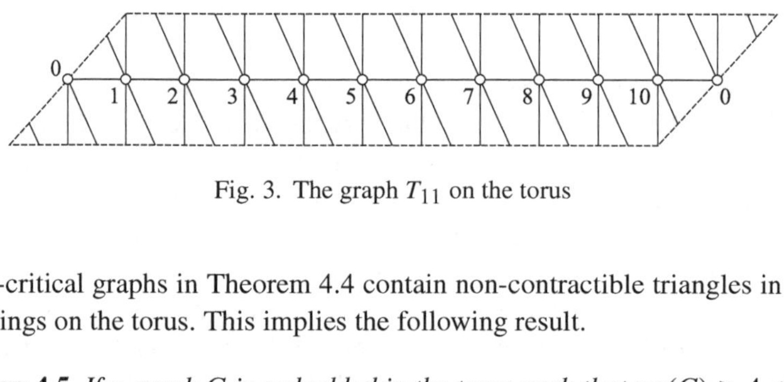

Fig. 3. The graph T_{11} on the torus

All 6-critical graphs in Theorem 4.4 contain non-contractible triangles in all their embeddings on the torus. This implies the following result.

Corollary 4.5 *If a graph G is embedded in the torus such that* $\mathrm{ew}(G) \geq 4$, *then G is 5-colourable.*

Thomassen [58] extended this result to all orientable surfaces.

Theorem 4.6 *If G is embeddable in an orientable surface of genus g so that the edge-width is at least* 2^{14g+6}, *then G is 5-colourable.*

The proof of Theorem 4.6 uses the fact that, for embedded graphs of sufficiently large edge-width, the surface can be cut in such a way that we obtain a planar graph after the cutting, and pairs of vertices that are duplicated by the cutting procedure are far from each other. An extension of the 5-colour theorem for planar graphs is then used, where we take care of colouring the duplicate pairs with the same colour.

Thomassen asked whether his Theorem 4.6 can be extended to list-colourings. His question was answered in the affirmative by DeVos, Kawarabayashi and Mohar [20].

Theorem 4.7 *For each surface S there exists a constant M for which every graph that is embedded in S with edge-width at least M is* 5*-choosable.*

Another extension of Theorem 4.6 was proved by Thomassen [62].

Theorem 4.8 *For each surface S, there are only finitely many* 6*-critical S-embeddable graphs.*

The complete list of 6-critical graphs is known for the sphere (where there are none), the projective plane (where K_6 is the only one, by Theorem 3.2), for the torus (Theorem 4.4), and for the Klein bottle [18], [39]. A generalization of Theorem 4.8 to 5-list-colourings has been obtained by Kawarabayashi and Mohar [40].

Corollary 4.9 *For each fixed surface S, there is a polynomial-time algorithm for deciding whether a graph on S can be* 5*-coloured. The same holds for list-colourings: it can be decided whether a given graph G embedded in S is* 5*-choosable and, given G and a list-assignment L with all lists of size at least* 5*, whether G is L-colourable.*

As mentioned earlier, the problem of 3-colouring graphs on the sphere (and hence any surface) is NP-complete. Corollary 4.9 thus suggests the following problem. Let S be a fixed surface. Does there exist a polynomially bounded algorithm for deciding whether a graph on S can be 4-coloured?

The following result of Fisk [32] implies that there is no 4-colour analogue of Theorem 4.8.

Theorem 4.10 *If G is a triangulation of some surface, and if G has exactly two vertices of odd degree and they are adjacent, then G is not* 4*-colourable.*

The proof of Theorem 4.10 actually works for every graph in which each edge is in precisely two triangles.

Corollary 4.11 *If S is a surface other than the sphere, and if* $k = 3, 4$ *or* 5*, then there are infinitely many k-critical S-embeddable graphs.*

The following problem was proposed by Albertson [1]. Let S be any surface. Does there exist a natural number $q = q(S)$ for which any graph G embedded on S contains a set A of at most q vertices such that $G - A$ is 4-colourable? This problem is open even for the torus where possibly $q = 3$ will do, as conjectured by Albertson [1].

5. Grötzsch's theorem and its generalizations

In this section we discuss the chromatic number of graphs on a fixed surface when conditions on the girth are imposed. We start with a classical result of Grötzsch [35].

Theorem 5.1 *Every planar graph with girth at least* 4 *is* 3*-colourable.*

Concerning a list-colouring version of Grötzsch's theorem, Voigt [66] gave an example of a planar graph of girth 4 that is not 3-choosable. Of course, planar graphs of girth 4 are 4-choosable since they are 3-degenerate. On the other hand, Thomassen [61] proved 3-choosability under a stronger condition.

Theorem 5.2 *Every planar graph of girth at least* 5 *is* 3*-choosable.*

Grötzsch's theorem can be strengthened by allowing triangles, as long as these are far from each other. This was conjectured in 1969 by Havel and proved by Dvořák, Král' and Thomas [26].

Theorem 5.3 *There exists a constant M for which every planar graph whose triangles are at distance at least M from each other is* 3*-colourable.*

The constant M in [26] is extremely large. Dvořák recently succeeded in extending Theorem 5.2 to 3-list-colourings, under the assumption that cycles of length 3 or 4 are far apart.

Inspired by Theorems 5.1 and 4.6, Hutchinson [37] proved the following.

Theorem 5.4 *For each positive integer g there exists a number $f(g)$ for which the following holds. If G is embedded in the orientable surface $\mathbb{S}_g$ so that* $\mathrm{ew}(G) > f(g)$ *and all facial walks have even length, then G is* 3*-colourable.*

Dvořák, Král' and Thomas [25] extended Theorem 5.4 to arbitrary triangle-free graphs on any fixed surface S. As a corollary, they showed that 3-colourability of a triangle-free graph embeddable in S can be tested in polynomial time. In particular, this gives a linear-time algorithm to compute the chromatic number of such graphs.

While the 5-colour result in Theorem 4.6 holds both for orientable and non-orientable surfaces, the 3-colour result of Hutchinson does not extend to non-orientable surfaces because of the following counter-intuitive result of Youngs [69].

Theorem 5.5 *If G is embeddable in the projective plane such that every face is of length* 4*, then G has chromatic number* 2 *or* 4*. Thus, if G is not bipartite, then G is not* 3*-colourable.*

Theorem 5.5 does not extend to toroidal graphs. This follows from Theorem 5.4, or by taking the Cartesian product of two cycles. An easy modification of this example shows also that Theorem 5.5 does not hold for graphs on the Klein bottle. However, Klavžar and Mohar [44] proved that twisted Cartesian products of two cycles of arbitrarily large edge-width on the Klein bottle sometimes need four colours. Mohar and Seymour [47] obtained a counterpart of Theorem 5.4 for non-orientable surfaces, by proving that every non-3-colourable graph of large edge-width with all faces of even length contains a quadrangulation that is not 3-colourable, and then characterizing completely the non-3-colourable quadrangulations. A generalization and full explanation of these phenomena were demystified by DeVos *et al.* [19] and are presented in the section on the duality of colourings and flows.

Gimbel and Thomassen [34] extended Theorem 5.5 to a complete characterization of those graphs in the projective plane which have girth at least 4 and are not 3-colourable.

Theorem 5.6 *Let G be a graph of girth at least* 4 *embedded in the projective plane. Then G is* 3*-colourable if and only if G does not contain a quadrangulation of the projective plane that is not bipartite.*

Gimbel and Thomassen [34] also gave an upper bound for the chromatic numbers of triangle-free graphs on general surfaces, analogous to Theorem 3.1 except that the bound is not sharp. For simplicity we formulate it for orientable surfaces only.

Theorem 5.7 *There exist positive constants c_1 and c_2 for which the following statements hold.*

- *Every triangle-free graph of genus g has chromatic number at most* $c_1\sqrt[3]{g/\log g}$.
- *For each $g \geq 1$, there exists a triangle-free graph of genus g and chromatic number at least* $c_2\sqrt[3]{g}/\log g$.

The following result was proved independently by Fisk and Mohar [33] and by Gimbel and Thomassen [34].

Theorem 5.8 *Let S be a surface.*

- *For each $k \geq 5$, there are only finitely many k-critical triangle-free graphs embeddable on S.*
- *There are only finitely many* 4*-critical graphs with girth at least* 6 *on S.*

Theorem 5.8 implies that, for each positive integer g, there exists an integer $w(g)$ for which any triangle-free graph embedded on a surface of Euler genus g with edge-width exceeding $w(g)$ can be 4-coloured. Fisk and Mohar [33] proved that $w(g) < c \log g$, where c is an absolute constant. An analogous result holds for 3-colourings of graphs with girth at least 6.

Theorem 5.8 also shows that, for each surface S, the chromatic number of a graph of girth 6 embeddable on S can be found in polynomial time. It can also be decided in polynomial time whether a triangle-free graph on S can be 4-coloured. The same applies for 3-colourings if the girth is at least 5 (see Thomassen [63]).

Theorem 5.9 *For each surface S, there are only finitely many* 4*-critical graphs of girth at least* 5 *that can be embedded in S.*

It is known that there are no such graphs on the torus or the projective plane.

Dvořák, Král' and Thomas have announced a structural description of 4-critical triangle-free graphs embedded in a fixed orientable surface, based on which one could obtain a linear-time algorithm for deciding whether they are 3-colourable.

6. Colouring–flow duality

Tutte introduced the notion of a nowhere-zero flow (see Chapter 9) as a notion dual to a colouring. In this section we state the colouring–flow duality for planar graphs, and show that there is an important extension to locally planar graphs embedded on other surfaces.

For a graph G without cut-edges, let $\varphi(G)$ denote its *flow index*, the smallest k for which G admits a nowhere-zero k-flow. Tutte [64] has proved the following result.

Theorem 6.1 *If G is a connected plane graph and if G^* is its dual graph, then $\chi(G) = \varphi(G^*)$.*

This duality extends to the circular version of flows and colourings. Let S_r be a circle in $\mathbb{R}^2$ with circumference $r > 0$. Given a graph G, a mapping $c : V(G) \to S_r$ is called a *circular r-colouring* if, for each edge vw in G, the distance between $c(v)$ and $c(w)$ on the circle S_r is at least 1. The *circular chromatic number* $\chi_c(G)$ is the smallest number r for which G admits a circular r-colouring. It is a fact that the minimum r in the definition is attained, and it turns out that $\chi(G) - 1 < \chi_c(G) \leq \chi(G)$. It is easy to see that $\chi_c(C_{2k+1}) = 2 + (1/k)$, so this coincides with intuition that long odd cycles are close to being 2-colourable.

There is similarly the corresponding notion of a circular flow. Let G be a graph and let us orient the edges of G arbitrarily. A map $\varphi : E(G) \to \mathbb{R}$ is a *flow* if, at each vertex, the sum of the values on the incoming edges is equal to the sum on the outgoing edges. We say that φ is an *α-flow* if $1 \leq |\varphi(e)| \leq \alpha - 1$, for each $e \in E(G)$. Using this notion, we define the *circular flow index* of G as

$$\varphi_c(G) = \inf\{\alpha \in \mathbb{R} : G \text{ admits an } \alpha\text{-flow}\}.$$

As with the circular chromatic number, it turns out that for every graph,

$$\varphi(G) - 1 < \varphi_c(G) \leq \varphi(G).$$

Note that the above notions are independent of the chosen orientation of the edges of G. For example, if we reverse the orientation of an edge e, then by replacing $\varphi(e)$ by $-\varphi(e)$ we preserve the property that φ is an α-flow.

Theorem 6.1 can be extended to circular colourings and flows.

Theorem 6.2 *If G is a connected plane graph and if G^* is its dual graph, then $\chi_c(G) = \varphi_c(G^*)$.*

This duality does not extend beyond planar graphs. It is known that the flow index of every graph is at most 6 (and it is conjectured that it is at most 5), but the dual graph may have arbitrarily large chromatic number. Nevertheless, Theorem 4.6 eliminates such an objection when the graph is locally planar and thus has sufficiently large edge-width. Indeed, DeVos, Goddyn, Mohar, Vertigan and Zhu [19] have generalized Theorem 6.2 to graphs on arbitrary surfaces, by proving the following approximate duality.

Theorem 6.3 *Let S be an orientable surface, and let $\varepsilon > 0$. Then there is an integer M such that, for every graph G embedded in S with edge-width at least M,*

$$\varphi_c(G^*) \leq \chi_c(G) < \varphi_c(G^*) + \varepsilon,$$

where G^* is the dual graph of G in S.

In order to obtain the corresponding duality result for non-orientable surfaces, we have to extend the definition of a flow in the dual graph. This extension gives us an alternative way of defining the circular chromatic number, and we include both definitions for comparison.

Let G be a graph embedded in a surface S, and let us orient the edges of G arbitrarily. A map $\varphi : E(G) \to \mathbb{R}$ is a *tension* if, for each cycle C in G, the sum of the values of φ on the forward edges of C is equal to the corresponding sum on the backward edges of C. This condition is equivalent to a stronger one that, for every closed walk W, the sum of the values of φ on the forward edges of W minus the sum on the backward edges of W is 0. If this condition is satisfied only for every closed walk W in G that defines a contractible curve in the surface, then φ is said to be a *local tension*. If $1 \leq |\varphi(e)| \leq \alpha - 1$ holds for each $e \in E(G)$, we say that φ is a *(local) α-tension*. Using these notions, we have the following:

$$\chi_c(G) = \inf\{\alpha \in \mathbb{R} : G \text{ admits an } \alpha\text{-tension}\}.$$

Analogously to this, we define a new invariant, the *local circular chromatic number*:

$$\chi_{loc}(G) = \inf\{\alpha \in \mathbb{R} : G \text{ admits a local } \alpha\text{-tension}\}.$$

Note that $\chi_{loc}(G) \leq \chi_c(G)$, and that strict inequality is possible. The concepts of tension and local tension involve an auxiliary orientation of the edges, but the values $\chi_c(G)$ and $\chi_{loc}(G)$ are independent of the chosen orientation.

If the surface is orientable, then $\chi_{loc}(G^*) = \varphi_c(G)$ and $\lceil \chi_{loc}(G^*) \rceil = \varphi(G)$, the flow index of G. If the surface is non-orientable, then $\lceil \chi_{loc}(G^*) \rceil$ is the *biflow index* of Bouchet. Thus, χ_{loc} both unifies and refines these two indices for embedded graphs. It can be used to extend Theorem 6.3 to include non-orientable surfaces. The following result is due to DeVos, Goddyn, Mohar, Vertigan and Zhu [19].

Theorem 6.4 *Let S be a surface, and let $\varepsilon > 0$. Then there is an integer M such that, for each graph G embedded in S with edge-width at least M,*

$$\chi_{loc}(G) \leq \chi_c(G) < \chi_{loc}(G) + \varepsilon,$$

where G^* is the dual graph of G in S.

DeVos *et al.* [19] also applied this result to two families of embedded graphs for which χ_{loc} exhibits interesting behaviour. An embedded graph G is *even-faced* if all of its faces have even length. Note that even-faced graphs in the plane are 2-colourable (see Theorem 2.8). It turns out that even-faced embedded graphs come in two types, odd and even, and their chromatic number depends on their type. Another class of graphs whose chromatic number is well understood are planar triangulations (see Theorem 2.7). On surfaces of higher genus, triangulations can be classified as being of even or odd type. In both cases, the *type* of G depends on the 'signature parity' of certain closed walks in the dual G^* (although in strikingly different ways). This distinction is reflected in the value of $\chi_{loc}(G)$ (see [19]).

Theorem 6.5 (Bimodality of χ_{loc}) *Let G be an embedded graph.*
- *If G is even-faced with maximum face-length* 2*r, then*

$$\chi_{loc}(G) = 2 \text{ (even type)} \quad \text{or} \quad \chi_{loc}(G) \geq \frac{2r}{r-1} \text{ (odd type)}.$$

- *If G is a triangulation, then*

$$\chi_{loc}(G) = 3 \text{ (even type)} \quad \text{or} \quad \chi_{loc}(G) \geq 4 \text{ (odd type)}.$$

As a corollary, DeVos *et al.* [19] proved the following.

Corollary 6.6 (Bimodality of χ_c) *For any surface S and any* $\varepsilon > 0$*, there exists an integer M such that, for every S-embedded graph G with* $\text{ew}(G) \geq M$*,*
- *if G is even-faced with maximum face-length* 2*r, then*

$$\chi_c(G) \in [2, 2+\varepsilon] \cup [2r/(r-1), 4]$$

- *if G is a triangulation, then*

$$\chi_c(G) \in [3, 3+\varepsilon] \cup [4, 5].$$

DeVos *et al.* [19] also showed that certain Eulerian triangulations of large edge-width on non-orientable surfaces have circular chromatic number 5, and that there are quadrangulations of non-orientable surfaces with arbitrarily large edge-width, yet with circular chromatic number 4. These results explain Theorem 5.5, that projective plane quadrangulations never have chromatic number 3, and they completely generalize and explain similar phenomena on general surfaces that were observed in the past.

Note that a little more can be said in some special cases.

If S is the plane or projective plane, then the circular chromatic number $\chi_c(G)$ is relatively well behaved. For example, set $\varepsilon = 0$ in Theorem 6.4 and in Corollary 6.6. If G is even-faced, then there is an exact formula for $\chi_c(G)$ (see DeVos *et al.* [19]). In particular, if G has an even-faced embedding in the projective plane, then $\chi_c(G) = 2$ or $\chi_c(G) = 2 + (2/k)$, for some non-negative integer k. Moreover, if a quadrangulation of the projective plane has $\chi_c(G) > 2$, then $\chi_c(G) = 4$; this strengthens Theorem 5.5.

If the girth of G is at least 6, then the upper bound of 4 can be replaced by 3 in the first part of Corollary 6.6 (see [19]). More generally, it may be true that we can replace 4 by $2 + 2/(g-1)$, if the girth is at least $2g$.

The interval $[4, 5]$ in the second part of Corollary 6.6 can be replaced by $\{4\}$ if S is orientable and G is Eulerian. This is implied by a result of Hutchinson *et al.* [38], which states that all triangulations of orientable surfaces of sufficiently large edge-width satisfy $\chi(G) \leq 4$.

7. The acyclic chromatic number

A colouring (or a list-colouring) of a graph G is *acyclic* if every cycle in the graph receives at least three different colours. Equivalently, if the colour classes of the colouring are $V_1, V_2, \ldots, V_k$, then the induced subgraph on $V_i \cup V_j$ is a forest for all $i \neq j$. We define the *acyclic chromatic number* $\chi_a(G)$ of a graph G as the minimum number of colours for which there exists an acyclic colouring of G. This notion was introduced in 1973 by Branko Grünbaum, who proved that every planar graph admits an acyclic 9-colouring. After various improvements of Grünbaum's result, Borodin [14] proved the ultimate version about the acyclic chromatic number of planar graphs.

Theorem 7.1 *Every planar graph is acyclically 5-colourable.*

The bound of 5 is best possible since there are planar graphs whose acyclic chromatic number is 5; an example is depicted in Fig. 4. If we have a 4-colouring of this graph, then the two vertices labelled 1 and 2 must be coloured differently or we obtain a bicoloured 4-cycle, since two of their common neighbours necessarily have the same colour. Thus, any acyclic 4-colouring is as shown in the figure. However, no available colour is left for the vertex z. If we replace each thick edge in this graph by four paths of length 2 joining the same pair of vertices, we obtain an even more striking example, a bipartite 2-degenerate planar graph whose acyclic chromatic number is 5.

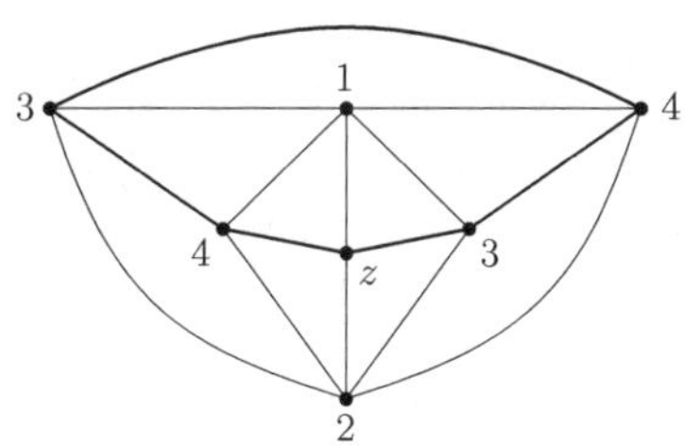

Fig. 4. A planar graph with acyclic chromatic number 5

The notion of acyclic colourings can be extended to list-colourings. We say that G is *acyclically k-choosable* if G admits an acyclic L-colouring for every list-assignment L for which each vertex has k colours in its list. The *acyclic list-chromatic number* $\chi_{a,l}(G)$ is the minimum k such that G is acyclically k-choosable.

There are many other results on acyclic list colourings of planar graphs. Borodin, Fon-der-Flaass, Kostochka, Raspaud and Sopena [16] conjectured that every planar graph is acyclically 5-choosable, and proved the following weaker statement.

Theorem 7.2 *Every planar graph is acyclically 7-choosable.*

If the aforementioned conjecture is true, then it may even be true that the acyclic chromatic number and the acyclic list-chromatic number are equal for every planar graph.

Borodin and Ivanova [17] proved that every planar graph without 4-cycles is acyclically 5-choosable, and Borodin, Chen, Ivanova and Raspaud [15] conjectured that every planar graph with girth at least 5 is acyclically 3-choosable. The following weaker bound was proved by Montassier [50].

Theorem 7.3 *Every planar graph of girth at least* 5 *is acyclically* 4*-choosable.*

Concerning graphs on arbitrary surfaces, Borodin conjectured that, apart from the sphere, the maximum acyclic chromatic number for graphs embeddable on a surface equals the Heawood bound for the usual chromatic number. This conjecture was disproved for all surfaces of large genus by Alon, Mohar and Sanders [6], who proved the following result.

Theorem 7.4 *If G is a graph embeddable on a surface of Euler genus g, then* $\chi_a(G) \leq 100g^{4/7} + 10\,000$.

The corresponding bounds for the acyclic list-chromatic number have not appeared in the literature, but the proof in [6] can be adapted to give the same bounds for the list-chromatic version.

Theorem 7.5 *For each sufficiently large integer g, there is a graph G embeddable on any surface of Euler genus g that satisfies* $\chi_a(G) \geq \frac{1}{10}\, g^{4/7}/(\log g)^{1/7}$.

Examples for Theorem 7.5 were obtained by making small modifications to random graphs (with edge-probability $p = 3(\log n/n)^{1/4}$). Alon, Mohar and Sanders [6] also found Klein bottle graphs whose acyclic chromatic number is 7. For the projective plane, they provided an upper bound given in the next theorem. It is not known whether the bound can be improved to 6.

Theorem 7.6 *Every projective planar graph is acyclically* 7*-colourable.*

For locally planar graphs, the following result of Kawarabayashi and Mohar [41] gives an analogue of Theorem 4.6.

Theorem 7.7 *For each $g \geq 0$, there exists an integer M such that every graph G embeddable on a surface of Euler genus g with edge-width at least M has* $\chi_a(G) \leq 7$.

8. Degenerate colourings

The notion of a degenerate colouring is a strengthening of the notion of an acyclic colouring. A graph G is *k-degenerate* if every subgraph has a vertex of degree at most k. Note that the requirement for the acyclic chromatic number says that the union of any two colour classes induces a forest, which is equivalent to the statement that it induces a 1-degenerate graph. One can impose an even stronger condition on a k-colouring, that the union of t colour classes always induces a $(t-1)$-degenerate subgraph, for every t. Such a colouring is said to be *degenerate*, and one can speak of the *degenerate chromatic number* $\chi_d(G)$ and the *degenerate list-chromatic number* $\chi_{d,l}(G)$. Note that the condition for $t = 1$ is just that of a proper colouring, and the condition for $t = 2$ is the same as that for acyclic colourings. Borodin [14] conjectured that Theorem 7.1 can be strengthened.

Conjecture A *Every planar graph admits a degenerate* 5*-colouring.*

The strongest result in this direction has been obtained by Kierstead, Mohar, Špacapan, Yang and Zhu [43].

Theorem 8.1 *Every planar graph has a degenerate* 9*-colouring. Moreover, the degenerate list-chromatic number of every planar graph is at most* 9.

9. The star chromatic number

A proper colouring of G with no 2-coloured path on four vertices is called a *star colouring*. This is equivalent to saying that the union of any two colour classes induces a star forest – that is, a subgraph in which each component is a star $K_{1,t}$, for some $t \geq 0$. The least number n for which G admits a star colouring with n colours is called the *star chromatic number* of G, denoted by $\chi_s(G)$. Clearly, $\chi_s(G) \geq \chi_a(G)$. If a colouring is both degenerate and star, then we speak of a *degenerate star colouring*. The corresponding chromatic number is denoted by χ_{sd}.

Albertson, Chappell, Kierstead, Kündgen and Ramamurthi [3] proved that every planar graph admits a star colouring with 20 colours.

Theorem 9.1 *If G is a planar graph, then $\chi_s(G) \leq 20$.*

The largest known value of the star chromatic number of a planar graph is 10, so Theorem 9.1 leaves some room for improvement.

Kierstead, Kündgen, and Timmons [42] proved that every bipartite planar graph G satisfies $\chi_s(G) \leq \chi_{s,l}(G) \leq 14$, and they provided examples with $\chi_s(G) \geq 8$. Under a different assumption that the girth is at least 5, Albertson *et al.* [3] proved that $\chi_s(G) \leq 16$. Planar graphs with girth 5 and star chromatic number 6 were found by Timmons [56].

For general surfaces, Mohar and Špacapan [48] improved the linear bound in [3] to a sublinear upper bound.

Theorem 9.2 *If G is a graph embeddable on a surface of Euler genus g, then*

$$\chi_{sd}(G) \leq \lceil 1000 g^{3/5} + 100\,000 \rceil.$$

The bound in Theorem 9.2 is essentially best possible; this is certified by the following result of Mohar and Špacapan [48].

Theorem 9.3 *For each sufficiently large integer g, there is a graph G embeddable in an orientable (or non-orientable) surface of genus g, for which*

$$\chi_{sd}(G) \geq \chi_s(G) \geq \tfrac{1}{32}\, g^{3/5}/(\log g)^{1/5}.$$

The corresponding result for locally planar graphs was proved by Kawarabayashi and Mohar [41].

Theorem 9.4 *For each positive integer g, there is a constant M such that every graph G embeddable on an orientable (or non-orientable) surface of genus g with edge-width at least M satisfies $\chi_s(G) \leq 2s_0^* + 3 \leq 43$, where s_0^* (≤ 20) is the maximum star chromatic number for the class of all planar graphs.*

10. Summary

The known results on the chromatic number of planar and locally planar graphs are collected in the following table. The entries with $\leq$ indicate that improvement is possible. The entries with $\geq$ indicate that there are known examples with the stated chromatic number.

Table 1.

	planar				locally planar
	all	even-faced	triangle-free	girth ≥ 5	ew $\geq M(g)$
χ	4 (4CT)	2 (Th. 2.8)	3 (Th. 5.1)	3 (Th. 5.1)	5 (Th. 4.6)
χ_l	5 (Th. 2.1)	3 (Th. 2.9)	4 (3-degen.)	3 (Th. 5.2)	5 (Th. 4.7)
χ_a	5 (Th. 7.1)	5	5	-	≤ 7 (Th. 7.7)
$\chi_{a,l}$	≤ 7 (Th. 7.2)	-	-	≤ 4 (Th. 7.3)	-
χ_d	≤ 9 (Th. 8.1)	-	-	-	-
χ_s	$10 \leq s_0^* \leq 20$ (Th. 9.1)	$\leq 14,\ \geq 8$ [42]	-	$\leq 16,\ \geq 6$ [3],[56]	$\leq 2s_0^* + 3$ (Th. 9.4)

The dependence of the chromatic number of arbitrary graphs on the genus is summarized in the table below.

Table 2.

	upper bound (all graphs)	lower bound (all graphs)	upper bound (triangle-free)	lower bound (triangle-free)
χ, χ_l	$O(g^{1/2})$ (Th. 3.1)	$\Omega(g^{1/2})$ (Th. 3.2 & 3.4)	$O((g/\log g)^{1/3})$ (Th. 5.7)	$\Omega(g^{1/3}/\log g)$ (Th. 5.7)
χ_a, $\chi_{a,l}$	$O(g^{4/7})$ (Th. 7.4)	$\Omega(g^{4/7}/\log(g)^{1/7})$ (Th. 7.5)	-	-
χ_d	$O(g^{3/5})$ (Th. 9.2)	$\Omega(g^{4/7}/\log(g)^{1/7})$ (bound from χ_a)	-	-
χ_s, χ_{sd}	$O(g^{3/5})$ (Th. 9.2)	$\Omega(g^{3/5}/\log(g)^{1/5})$ (Th. 9.3)	-	-

References

1. M. O. Albertson, Open problem 2, *The Theory and Applications of Graphs* (ed. G. Chartrand *et al.*), Wiley (1981), 609.
2. M. O. Albertson, You can't paint yourself into a corner, *J. Combin. Theory (B)* **73** (1998), 189–194.
3. M. O. Albertson, G. G. Chappell, H. A. Kierstead, A. Kündgen and R. Ramamurthi, Coloring with no 2-colored P_4's, *Electron. J. Combin.* **11** (2004), 13 pp.
4. M. O. Albertson and J. P. Hutchinson, The three excluded cases of Dirac's map-color theorem, *Ann. N. Y. Acad. Sci.* **319** (1979), 7–17.
5. M. O. Albertson and J. P. Hutchinson, On 6-chromatic triangulations, *Proc. London Math. Soc.* **41** (1980), 533–556.
6. N. Alon, B. Mohar and D. P. Sanders, On acyclic colorings of graphs on surfaces, *Israel J. Math.* **94** (1996), 273–283.
7. N. Alon and M. Tarsi, Colorings and orientations of graphs, *Combinatorica* **12** (1992), 125–134.
8. A. Altschuler, Hamiltonian circuits in some maps on the torus, *Discrete Math.* **1** (1972), 299–314.
9. K. Appel and W. Haken, The existence of unavoidable sets of geographically good configurations, *Illinois J. Math.* **20** (1976), 218–297.
10. K. Appel and W. Haken, Every planar map is four colorable. Part I: Discharging, *Illinois J. Math.* **21** (1977), 429–490.
11. K. Appel and W. Haken, *Every Planar Map is Four Colorable*, Contemp. Math. **98**, Amer. Math. Soc., 1989.
12. K. Appel, W. Haken and J. Koch, Every planar map is four colorable. Part II: Reducibility, *Illinois J. Math.* **21** (1977), 491–567.

13. T. Böhme, B. Mohar and M. Stiebitz, Dirac's map-color theorem for choosability, *J. Graph Theory* **32** (1999), 327–339.
14. O. V. Borodin, On acyclic colorings of planar graphs, *Discrete Math.* **25** (1979), 211–236.
15. O. V. Borodin, M. Chen, A. O. Ivanova and A. Raspaud, Acyclic 3-choosability of sparse graphs with girth at least 7, *Discrete Math.* **310** (2010), 2426–2434.
16. O. V. Borodin, D. G. Fon-Der-Flaass, A. V. Kostochka, A. Raspaud and E. Sopena, Acyclic list 7-coloring of planar graphs, *J. Graph Theory* **40** (2002), 83–90.
17. O. V. Borodin and A. O. Ivanova, Acyclic 5-choosability of planar graphs without 4-cycles (in Russian), *Sibirsk. Mat. Zh.* **52** (2011), 522–541.
18. N. Chenette, L. Postle, N. Streib, R. Thomas and C. Yerger, Five-coloring graphs in the Klein bottle, *J. Combin. Theory (B)* **102** (2012), 1067–1098.
19. M. DeVos, L. Goddyn, B. Mohar, D. Vertigan and X. Zhu, Coloring-flow duality of embedded graphs, *Trans. Amer. Math. Soc.* **357** (2005), 3993–4016.
20. M. DeVos, K. Kawarabayashi and B. Mohar, Locally planar graphs are 5-choosable, *J. Combin. Theory (B)* **98** (2008), 1215–1232.
21. G. A. Dirac, Map colour theorems, *Canad. J. Math.* **4** (1952), 480–490.
22. G. A. Dirac, Short proof of the map colour theorem, *Canad. J. Math.* **9** (1957), 225–226.
23. G. A. Dirac, A theorem of R. L. Brooks and a conjecture of H. Hadwiger, *Proc. London Math. Soc.* (3) **7** (1957), 161–195.
24. G. A. Dirac, Map colour theorems related to the Heawood colour formula, *J. London Math. Soc.* **32** (1957), 436–455.
25. Z. Dvořák, D. Král' and R. Thomas, Coloring triangle-free graphs on surfaces, *Proc. 20th Ann. ACM-SIAM Symp. on Discrete Algorithms, SODA 2009*, SIAM (2009), 120–129.
26. Z. Dvořák, D. Král' and R. Thomas, Coloring planar graphs with triangles far apart, *http://arxiv.org/abs/0911.0885*.
27. Z. Dvořák, B. Lidický and B. Mohar, 5-choosability of graphs with crossings far apart, *http://arxiv.org/abs/1201.3014*.
28. Z. Dvořák, B. Lidický, B. Mohar and L. Postle, 5-list-coloring planar graphs with distant precolored vertices, *http://arxiv.org/abs/1209.0366*.
29. K. Edwards, The complexity of some graph colouring problems, *Discrete Appl. Math.* **36** (1992), 131–140.
30. P. Erdős, Graph theory and probability, II, *Canad. J. Math.* **13** (1961), 346–352.
31. P. Erdős, A. L. Rubin and H. Taylor, Choosability in graphs, *Proc. West Coast Conference on Combinatorics, Graph Theory and Computing* (Arcata, California), *Congr. Numer.* **26** (1980), 125–157.
32. S. Fisk, The non-existence of colorings, *J. Combin. Theory (B)* **24** (1978), 247–248.
33. S. Fisk and B. Mohar, Coloring graphs without short non-bounding cycles, *J. Combin. Theory (B)* **60** (1994), 268–276.
34. J. Gimbel and C. Thomassen, Coloring graphs with fixed genus and girth, *Trans. Amer. Math. Soc.* **349** (1997), 4555–4564.
35. H. Grötzsch, Ein Dreifarbensatz für dreikreisfreie Netze auf der Kugel, *Wiss. Z. Martin Luther-Univ. Halle Wittenberg, Math.-Nat. Reihe* **8** (1959), 109–120.
36. P. J. Heawood, Map-colour theorem, *Quart. J. Pure Appl. Math.* **24** (1890), 332–338.
37. J. P. Hutchinson, Three-coloring graphs embedded on surfaces with all faces even-sided, *J. Combin. Theory (B)* **65** (1995), 139–155.
38. J. Hutchinson, R. B. Richter and P. Seymour, Colouring Eulerian triangulations, *J. Combin. Theory (B)* **84** (2002), 225–239.
39. K. Kawarabayashi, D. Král', J. Kynčl and B. Lidický, 6-critical graphs on the Klein bottle, *SIAM J. Discrete Math.* **23** (2009), 372–383.
40. K. Kawarabayashi and B. Mohar, List-color-critical graphs on a fixed surface, *Proc. Twentieth Annual ACM-SIAM Symposium on Discrete Algorithms (SODA'09)*, SIAM (2009), 1156–1165.

41. K.-I. Kawarabayashi and B. Mohar, Star coloring and acyclic coloring of locally planar graphs, *SIAM J. Discrete Math.* **24** (2010), 56–71.
42. H. A. Kierstead, A. Kúndgen and C. Timmons, Star coloring bipartite planar graphs, *J. Graph Theory* **60** (2009), 1–10.
43. H. Kierstead, B. Mohar, S. Špacapan, D. Yang and X. Zhu, The two-coloring number and degenerate colorings of planar graphs, *SIAM J. Discrete Math.* **23** (2009), 1548–1560.
44. S. Klavžar and B. Mohar, The chromatic numbers of graph bundles over cycles, *Discrete Math.* **138** (1995), 301–314.
45. D. Král' and R. Škrekovski, The last excluded case of Dirac's map-color theorem for choosability, *J. Graph Theory* **51** (2006), 319–354.
46. M. Mirzakhani, A small non-4-choosable planar graph, *Bull. Inst. Combin. Appl.* **17** (1996), 15–18.
47. B. Mohar and P. D. Seymour, Coloring locally bipartite graphs on surfaces, *J. Combin. Theory (B)* **84** (2002), 301–310.
48. B. Mohar and S. Špacapan, Degenerate and star colorings of graphs on surfaces, *Europ. J. Combin.* **33** (2012), 340–349.
49. B. Mohar and C. Thomassen, *Graphs on Surfaces*, Johns Hopkins University Press, 2001.
50. M. Montassier, Acyclic 4-choosability of planar graphs with girth at least 5, *Graph Theory in Paris* (ed. A. Bondy *et al.*), Birkhäuser (2007), 299–310.
51. G. Ringel, *Map Color Theorem*, Springer-Verlag, 1974.
52. G. Ringel and J. W. T. Youngs, Solution of the Heawood map-coloring problem, *Proc. Nat. Acad. Sci. U.S.A.* **60** (1968), 438–445.
53. N. Robertson, D. P. Sanders, P. Seymour and R. Thomas, Efficiently four-coloring planar graphs, *Proc. Assoc. Comp. Mach. Symp. Theory Comput.* **28** (1996), 571–575.
54. N. Robertson, D. Sanders, P. Seymour and R. Thomas, The four-colour theorem, *J. Combin. Theory (B)* **70** (1997), 2–44.
55. R. Škrekovski, A theorem on map colorings, *Bull. Inst. Combin. Appl.* **35** (2002), 53–60.
56. C. Timmons, Star coloring high girth planar graphs, *Electron. J. Combin.* **15** (2008), Research Paper 124.
57. R. Thomas, *http://people.math.gatech.edu/~thomas/FC/fourcolor.html*
58. C. Thomassen, Five-coloring maps on surfaces, *J. Combin. Theory (B)* **59** (1993), 89–105.
59. C. Thomassen, Every planar graph is 5-choosable, *J. Combin. Theory (B)* **62** (1994), 180–181.
60. C. Thomassen, Five-coloring graphs on the torus, *J. Combin. Theory (B)* **62** (1994), 11–33.
61. C. Thomassen, 3-list coloring planar graphs of girth 5, *J. Combin. Theory (B)* **64** (1995), 101–107.
62. C. Thomassen, Color-critical graphs on a fixed surface, *J. Combin. Theory (B)* **70** (1997), 67–100.
63. C. Thomassen, The chromatic number of a graph of girth 5 on a fixed surface, *J. Combin. Theory (B)* **87** (2003), 38–71.
64. W. T. Tutte, A contribution to the theory of chromatic polynomials, *Canad. J. Math.* **6** (1954), 80–91.
65. M. Voigt, List colourings of planar graphs, *Discrete Math.* **120** (1993), 215–219.
66. M. Voigt, A not 3-choosable planar graph without 3-cycles, *Discrete Math.* **146** (1995), 325–328.
67. M. Voigt and B. Wirth, On 3-colorable non-4-choosable planar graphs, *J. Graph Theory* **24** (1997), 233–235.
68. R. Wilson, *Four Colors Suffice* (revised color edition), Princeton Science Series, Princeton University Press, 2014.
69. D. A. Youngs, 4-chromatic projective graphs, *J. Graph Theory* **21** (1996), 219–227.

2

Brooks's theorem

MICHAEL STIEBITZ and BJARNE TOFT

R. L. Brooks's seminal paper [4] *of 1941 contains the first result – known as Brooks's theorem – on colouring abstract graphs. That it is worthwhile to study the chromatic number of graphs in general, rather than just planar graphs, was pointed out already by A. B. Kempe in 1879 in his paper* [28] *about map-colouring. However, it was only with the paper by Brooks that vertex-colouring of abstract graphs became a topic of study. Over the years, this topic has developed into a rich theory and, as emphasized by B. Reed in his extensive paper* [51], *Brooks's theorem is just the tip of the iceberg.*

1. Introduction

In this chapter only simple graphs are considered. Brooks's theorem relates the chromatic number to the maximum degree of a graph. In modern terminology Brooks's result is as follows:

Let G be a graph with maximum degree Δ, where $\Delta > 2$, and suppose that no connected component of G is a complete graph $K_{\Delta+1}$. Then it is possible to colour the vertices of G with Δ colours so that no two vertices of the same colour are adjacent, and hence G has chromatic number at most Δ.

Brooks noticed that it suffices to prove the result for connected graphs, since the connected components of a graph can be coloured independently of each other. Furthermore, he observed that every graph with maximum degree Δ admits a

$(\Delta + 1)$-colouring, by giving to each vertex in turn a colour different from all those colours already assigned to vertices to which it is adjacent. The three missing cases, where $\Delta = 0, 1, 2$, can be easily included in Brooks's theorem. So another possible way of formulating Brooks's fundamental result is as follows.

Theorem 1.1 (Brooks's theorem) *Let G be a connected graph of maximum degree* Δ. *Then* $\chi(G) \leq \Delta + 1$, *where equality holds if and only if G is a complete graph or an odd cycle.*

2. Proofs of Brooks's theorem

Over the years many different proofs of Brooks's theorem have been given, but in principle there are just three basic approaches.

A sequential colouring algorithm

Brooks's original proof provides an efficient (quadratic-time) algorithm that produces a Δ-colouring of a connected graph G with maximum degree $\Delta \geq 3$, provided that G is different from $K_{\Delta+1}$. An improved proof, leading to a linear-time colouring algorithm, was given by Lovász [49] in 1975. Lovász's proof, which has become the most popular proof of Brooks's theorem, is based on the following *sequential colouring algorithm*:

Starting from a fixed vertex order $v_1, v_2, \ldots, v_n$ *of G, we consider the vertices in turn and colour each vertex* v_i *with the smallest positive integer not already used to colour any vertex among* $v_1, v_2, \ldots, v_{i-1}$ *adjacent to* v_i.

Since G is connected, we can choose a vertex order $v_1, v_2, \ldots, v_n$ such that each vertex v_i with $1 \leq i \leq n-1$ is adjacent to some vertex v_j with $j > i$. Furthermore, we can choose any vertex v of G as the last vertex v_n. To see this, take a spanning tree of G and delete step by step an endvertex in the remaining tree, where we delete v last. If we apply the sequential colouring algorithm to this ordering, then for each vertex v_i with $1 \leq i \leq n-1$ we use a colour in $\{1, 2, \ldots, \Delta\}$, since v_i is adjacent to at most $\Delta - 1$ of the previous vertices $v_1, v_2, \ldots, v_{i-1}$. This shows, in particular, that $G - v$ admits a Δ-colouring. This Δ-colouring of $G - v$ can be extended to a Δ-colouring of G if the degree of v is at most $\Delta - 1$, or if v has two neighbours coloured with the same colour. So we are done if G contains a vertex of degree less than Δ. We are also done if G is not 2-connected, since

$$\chi(G) = \max\{\chi(H) \ : \ H \text{ is a block of } G\}. \tag{1}$$

Recall that a *block* of a graph is a maximal subgraph with no cut-vertex; alternatively, the edges of a block are the equivalence classes under the relation 'to be equal or to lie in a common cycle'. It remains to consider the case when G is Δ-regular and

2-connected. Since G is not a complete graph and $\Delta \geq 3$, it is easy to show that there are three vertices u, v and w for which $uv, vw \in E(G)$, $uw \notin E(G)$, and $G - u - w$ is connected (if G is 3-connected, this is obvious; if G is not 3-connected, then let v belong to a cutset of two vertices and let u and w belong to different end-blocks of $G - v$, where an *end-block* of a graph is a block containing at most one cut-vertex of the graph). We can thus choose the vertex order $v_1, v_2, \ldots, v_n$ in such a way that $v_1 = u$, $v_2 = w$ and $v_n = v$. The colouring algorithm applied to this ordering then assigns colour 1 to u and w, and hence terminates with a Δ-colouring of G.

If G is 3-connected, Brooks's proof is based on a similar argument, starting from a $(\Delta + 1)$-colouring, where the unwanted colour $\Delta + 1$ is interchanged with a colour on a neighbour, and continuing until it arrives at a vertex v with two neighbours with the same colour, where v may be recoloured, and so $\Delta + 1$ is not used as a colour. To handle the remaining case, where G is not 3-connected, Brooks used induction on the order of G.

A Kempe-change argument

In 1969 Melnikov and Vizing [43] presented a proof of Brooks's theorem using an idea going back to Kempe's attempt to prove the four-colour theorem. The proof is by contradiction.

So we assume that the result is false and consider a smallest counter-example G. Then G is a connected graph with maximum degree $\Delta \geq 3$ for which $G \neq K_{\Delta+1}$ and $\chi(G) = \Delta + 1$. Choose any vertex v of G. Then $G - v$ admits a Δ-colouring, and in each such colouring all Δ colours occur among the neighbours of v. So v has Δ neighbours $v_1, v_2, \ldots, v_\Delta$ and we may choose a Δ-colouring φ of $G - v$ such that $\varphi(v_i) = i$ for $i = 1, 2, \ldots, \Delta$. For $i \neq j$, let $G_{i,j}$ denote the subgraph of $G - v$ induced by the vertices of colour i and j. Let $P_{i,j}$ be the component of $G_{i,j}$ containing the vertex v_i. If we interchange the colours i and j on the vertices of $P_{i,j}$, we obtain a new Δ-colouring φ' of $G - v$. (This recolouring operation is usually called a *Kempe-change*.) We claim that $P_{i,j}$ contains the vertex v_j, implying that $P_{i,j} = P_{j,i}$; for otherwise, we have $\varphi'(v_i) = \varphi'(v_j) = j$, and so we can extend φ' to a Δ-colouring of G, giving a contradiction. Next, it can be shown that each component $P_{i,j}$ is a path. Eventually it can be proved that each component $P_{i,j}$ consists of exactly one edge $v_i v_j$, which implies that $G = K_{\Delta+1}$, giving a contradiction.

Maximum independent sets

Tverberg [60] used an idea developed independently by Catlin [10] and Gerencsér [21] to give a short proof of Brooks's theorem by induction on the maximum degree. Catlin [10] (see also Mitchem [45]) proved that if G is a connected graph with maximum degree $\Delta \geq 1$, which is neither a complete graph nor an odd cycle, then G has a Δ-colouring with a monochromatic maximum independent set.

A degree version of Brooks's theorem

Brooks's theorem tells us that Δ colours suffice for a (proper) colouring of all vertices of a connected graph with maximum degree Δ, unless G is a complete graph or an odd cycle. In 1979 Erdős, Rubin and Taylor [18] extended Brooks's result by proving a degree version, where a list of $d(v)$ colours is assigned to each vertex v of G. Here, a *brick* is a complete graph or an odd cycle.

Theorem 2.1 *Let G be a connected graph and, for each vertex v of G, let $L(v)$ be a set of at least $d(v)$ colours. Unless each block of G is a brick, G admits an L-colouring – that is, there is a colouring of G in which each vertex v of G receives a colour from its list $L(v)$.*

Proof By a *bad pair* we mean a pair (G, L) satisfying the hypothesis of the theorem, for which G has no L-colouring. Our aim is to show that if (G, L) is a bad pair, then each block of G is a brick. The proof is by induction on the order n of G and uses the following reduction. Let v be a non-cut-vertex of G and let $c \in L(v)$ be a colour. Furthermore, let $G' = G - v$ and, for $w \in V(G')$, let $L'(w) = L(w) \setminus \{c\}$ if $vw \in E(G)$, and $L'(w) = L(w)$ otherwise. Then (G', L') is a bad pair and G' has order $n - 1$. In this case we write $(G', L') = (G, L)/(v, c)$.

We now prove by induction on the order n of G that the bad pair (G, L) satisfies the following properties:

(a) $|L(v)| = d_G(v)$ for all $v \in V(G)$;
(b) provided that G has no cut-vertex, $L(v)$ is the same for each vertex v of G and G is regular;
(c) every block of G is a brick.

These are obviously true if $n = 1$, so assume that $n \geq 2$.

For the proof of (a), consider an arbitrary vertex v of G. Since G is connected, there is a non-cut-vertex $w \neq v$ in G. Now consider the bad pair $(G', L') = (G, L)/(w, c)$, where $c \in L(w)$. Then $L(w) \neq \varnothing$ and, by the induction hypothesis, $|L'(v)| = d_{G'}(v)$, which yields $|L(v)| = d_G(v)$. This proves (a).

For the proof of (b), assume that G has no cut-vertex. Suppose, to the contrary, that there are two vertices $v \neq w$ in G for which $L(v) \neq L(w)$. Since G is connected, we may assume that $vw \in E(G)$. Furthermore, we may assume that there is a colour $c \in L(w) \setminus L(v)$. Then $(G', L') = (G, L)/(w, c)$ is a bad pair for which $|L'(v)| > d_{G'}(v)$, contradicting (a). Hence $L(v)$ is the same colour-set for each vertex v of G and so, by (a), G is regular.

For the proof of (c), we consider two cases. First, assume that G has a cut-vertex. Then G has at least two end-blocks B_1 and B_2, and in each B_i there is a vertex v_i such that v_i is not a cut-vertex of G. By the induction hypothesis (using the reduction $(G', L') = (G, L)/(v_i, c_i)$), each block of $G - v_i$ is a brick. Since each block $B \neq B_i$ of G is a block of $G - v_i$, this shows that each block of G is a brick, too. It remains to consider the case that G contains no cut-vertex – that is, G itself is a block. Then it follows from (b) that G is r-regular with $r \geq 1$, and $L(v) = C$ for all $v \in V(G)$, where C is a set of r colours. Let v be a vertex of G. By induction, each block of $G - v$ is a brick. If $G - v$ consists of a single block, then both G and $G - v$ are complete graphs. So suppose that $G - v$ has at least two blocks. Since every end-block of $G - v$ must be $(r-1)$-regular, the degree of v in G is at least $2(r-1)$, and so $r \leq 2$. For $r = 1$, $G = K_2$ (a brick). For $r = 2$, G is a cycle and the same two colours are available at each vertex of G. Since G has no L-colouring, we easily conclude that G is an odd cycle. This proves (c), and completes the proof of the theorem. ■

The above proof combines the sequential colouring argument with a reduction of the lists and can be easily extended to hypergraphs (see [31]). The original proof of the above result by Erdős, Rubin and Taylor [18] was based on a sequential colouring argument and a structural result, that if a graph contains at least two chords in any even cycle, then each of its blocks is a brick. They also gave a characterization of the list assignments that can occur in a bad pair.

Critical graphs

In 1951 Gabriel A. Dirac submitted a Ph.D. thesis entitled *On the Colouring of Graphs* to the University of London. In this thesis Dirac continued the study of colouring abstract graphs started by Brooks, and he introduced the notion of a critical graph. A graph G is *critical* if $\chi(H) < \chi(G)$ whenever H is a proper subgraph of G. A critical graph with chromatic number k is a *k-critical graph*.

That critical graphs form a useful concept relies on the fact that many problems concerning the chromatic number of graphs can be reduced to critical graphs, and critical graphs have stronger structural properties than graphs in general. Let ρ be a *monotone graph invariant* – that is, a mapping that assigns to each graph G a real number $\rho(G)$ such that $\rho(H) \leq \rho(G)$ whenever H is a subgraph of G. If we wish to show that every graph G belonging to a class $\mathcal{G}$ of graphs closed under taking subgraphs satisfies $\chi(G) \leq \rho(G)$, then it suffices to establish this inequality for all critical graphs in $\mathcal{G}$. This follows from the fact that, for any graph $G \in \mathcal{G}$, there is a critical graph $H \in \mathcal{G}$ with $H \subseteq G$ and $\chi(H) = \chi(G)$, and so $\chi(G) = \chi(H) \leq \rho(H) \leq \rho(G)$.

The complete graph K_k is a k-critical graph for every $k \geq 1$; it is the only k-critical graph with k vertices, and for $k = 1$ or 2 there are no others. By König's characterization of bipartite graphs as graphs with $\chi \leq 2$, it follows that the only 3-critical graphs are the odd cycles. However, for any integer $k \geq 4$, a characterization of all k-critical graphs seems unattainable.

Let G be a k-critical graph with $k \geq 3$. By (1), G is 2-connected. Dirac [13] characterized the class of k-critical graphs with a vertex-cut of size 2. Furthermore, he observed that G is $(k-1)$-edge connected and so $\delta(G) \geq k-1$. To see this, consider an arbitrary edge vw of G. Then in any $(k-1)$-colouring of $G - vw$, the vertices v and w receive the same colour c. For any of the remaining $k-2$ colours c', the vertices v and w are joined by a path whose vertices are coloured alternately with c and c'; otherwise, by a Kempe-change, we would get a $(k-1)$-colouring where v and w receive different colours, which would be a $(k-1)$-colouring of G, giving a contradiction. Hence, together with the edge vw, there are $k-1$ edge-disjoint paths between v and w, so G is $(k-1)$-edge connected. Toft [58] characterized the class of k-critical graphs with a non-trivial edge-cut of size $k-1$.

Following Gallai [20], in a k-critical graph G vertices of degree $k-1$ are called *low* and the remaining vertices (of degree at least k) are called *high*. For a critical graph G, we denote by G_L the subgraph induced by the low vertices of G, and by G_H the subgraph induced by the high vertices of G.

Observe that Brooks's theorem is equivalent to the statement that the only critical graphs G with $\chi(G) = \Delta(G) + 1$ (or equivalently with $G_H = \varnothing$) are the complete graphs and the odd cycles. This result was generalized by Gallai. He proved in [20] that if G is a critical graph, then each block of G_L is a brick. This result can be easily deduced from Theorem 2.1. To see this, let $k \geq 1$ and G be a k-critical graph having low vertices, and let G' be a component of G_L. Then there is a colouring φ of $G - V(G')$ with a set C of $k-1$ colours. For $v \in V(G')$, let $L(v) = C \setminus \{\varphi(w) : w \in N_G(v)\}$. Since $\chi(G) = k$, G' has no L-colouring and $|L(v)| \geq d_{G'}(v)$ for all $v \in V(G')$. Hence, Theorem 2.1 implies that each block of G' is a brick.

Gallai [20] also showed that any graph G' with $\Delta(G') \leq k-1$ in which each block is a brick is equal to G_L, for some k-critical graph G. Stiebitz [55] proved that, for any critical graph G having low and high vertices, G_L has at least as many components as G_H, thus answering a question raised by Gallai.

3. Critical graphs with few edges

Dirac was interested in exhibiting structural properties of critical graphs and using these properties to establish colouring properties of general graphs. In particular, he investigated the minimum number $f_k(n)$ of edges in any k-critical graph with n vertices. Let $\mathcal{C}_k(n)$ denote the set of all k-critical graphs with n vertices and $f_k(n)$ edges. Since k-critical graphs with $k \leq 3$ are well characterized, we assume that $k \geq 4$.

First we show that $f_k(n)$ is defined for all $n \geq k$, except for $n = k+1$. To see this, we can use two methods for constructing critical graphs. Let $G = G_1 + G_2$ be the join of two disjoint graphs G_1 and G_2. Then $\chi(G) = \chi(G_1) + \chi(G_2)$, and G is critical if and only if G_1 and G_2 are critical; for instance, $K_p + C_{2h+1}$ is $(p+3)$-critical for all $p \geq 1$.

The next construction was introduced by Hajós [25]. Let G_1 and G_2 be two disjoint k-critical graphs, and let v_1w_1 and v_2w_2 be edges of G_1 and G_2, respectively. Let G be the graph obtained from $G_1 - v_1w_1$ and $G_2 - v_2w_2$ by identifying v_1 and v_2 to a new vertex and by adding the new edge w_1w_2. The resulting graph G is called a *Hajós sum* of G_1 and G_2. Then, as proved by Dirac [13], G is also k-critical.

Based on these two constructions, it is easy to show that there are k-critical graphs with n vertices if $n \geq k$ and $n \neq k+1$. Clearly, there is no k-critical graph with fewer than k vertices and, since no two vertices in a critical graph can have the same neighbourhood, it follows that there is no k-critical graph with $k+1$ vertices.

Since every k-critical graph has minimum degree at least $k-1$, we have $2f_k(n) \geq (k-1)n$, and Brooks's theorem is equivalent to the fact that equality holds if and only if $n = k$. Evidently, $\mathcal{C}_k(k) = \{K_k\}$. In 1957 Dirac [15] extended Brooks's theorem by proving that

$$2f_k(n) \geq (k-1)n + k - 3 \text{ for } n \geq k+2. \tag{2}$$

In 1974 Dirac [16] extended his result further by giving a complete description of the extremal graphs; in particular, he proved that equality holds in (2) if and only if $n = 2k-1$. In 1999 Kostochka and Stiebitz [32] proved that $2f_k(n) \geq (k-1)n + 2(k-3)$, provided that $n \geq k+2$ and $n \neq 2k-1$, with equality when $n = 2k$ – that is, $f_k(2k) = k^2 - 3$.

For $k \geq 3$, let $\mathcal{D}_k$ denote the family of all graphs G whose vertex-set consists of three non-empty pairwise disjoint sets A, B_1 and B_2 with

$$|B_1| + |B_2| = |A| + 1 = k - 1$$

and two additional vertices a and b such that A and $B_1 \cup B_2$ induce cliques in G not joined by any edge, $N(a) = A \cup B_1$ and $N(b) = A \cup B_2$. Each such graph G has $n = 2k-1$ vertices, with $d(v) = k-1$ for all vertices $v \neq a, b$, and $2|E(G)| = (k-1)n + k - 3$. Both Dirac [16] and Gallai [20] proved that $\mathcal{C}_k(2k-1) = \mathcal{D}_k$.

For $k \geq 4$, let $\mathcal{E}_k$ denote the family of all graphs G whose vertex-set consists of four non-empty pairwise disjoint sets A, B_1, B_2 and B_3 with

$$|B_1| + |B_2| + |B_3| = |A| + 1 = k - 1$$

and three additional vertices c_1, c_2 and c_3 for which A and $B_1 \cup B_2 \cup B_3$ are cliques in G not joined by any edge and, for $i \in \{1,2,3\}$, $N(c_i) = A \cup B_i$. Each such graph G has order $n = 2k$ and

$$|E(G)| = \tfrac{1}{2}(k-1)n + (k-3) = k^2 - 3 = f_k(2k).$$

It is easy to check that G is critical, and hence $\mathcal{E}_k \subseteq \mathcal{C}_k(2k)$.

The first improvement (for n large enough) of Dirac's bound (2) was established by Gallai [20] in 1963; he proved that

$$2f_k(n) \geq \left(k - 1 + \frac{k-3}{k^2-3}\right) n \text{ for } n \geq k+2. \tag{3}$$

This result is an easy consequence of Gallai's characterization of the subgraph of a critical graph induced by its low vertices. A further improvement of Gallai's bound was given by Krivelevich [39] in 1997, and by Kostochka and Stiebitz [35] in 2003.

In the second [20] of his two fundamental papers concerning the structure of critical graphs, Gallai proved the following remarkable result. Subsequent proofs were obtained by Molloy [46] and Stehlík [54].

Theorem 3.1 *Let $k \geq 4$, and let G be a k-critical graph of order n. If $n \leq 2k-2$, then G is the join of two non-empty critical graphs. If $n < \frac{5}{3}k$, then one of the two critical graphs may be chosen as a complete graph.*

Based on this result, Gallai [20] found the exact values of the function $f_k(n)$, including a description of the corresponding class $\mathcal{C}_k(n)$, for $k+2 \leq n \leq 2k-1$.

Theorem 3.2 *Let k and n be integers satisfying $k \geq 4$ and $k+2 \leq n \leq 2k-1$. Then $2f_k(n) = (k-1)n + (n-k)(2k-n) - 2$ and $\mathcal{C}_k(n) = \{K_{2k-n-1} + H\colon H \in \mathcal{D}_{n-k+1}\}$.*

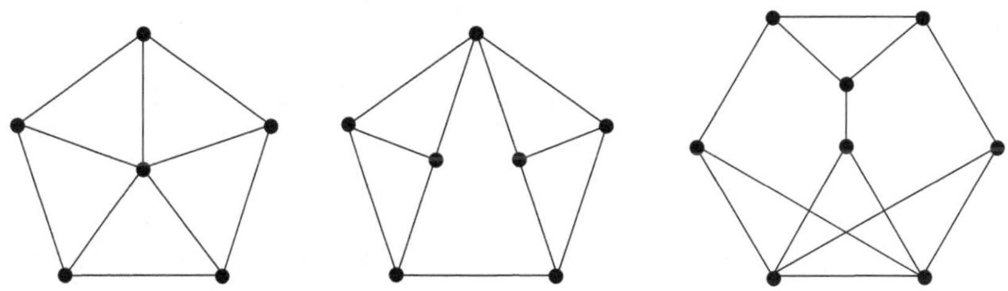

Fig. 1. The graphs in $\mathcal{C}_4(6), \mathcal{C}_4(7)$ and $\mathcal{C}_4(8)$

We next establish an upper bound $g_k(n)$ for $f_k(n)$. For $n = k$, define $g_k(k) = \binom{k}{2}$. If $n \geq k+2$ and $n \equiv p+1 \pmod{(k-1)}$, where $2 \leq p \leq k$, define

$$c_{k,p} = f_k(k+p) - \tfrac{1}{2}\left(k - \frac{2}{k-1}\right)(k+p)$$

and

$$g_k(n) = \tfrac{1}{2}\left(k - \frac{2}{k-1}\right)n + c_{k,p} = \tfrac{1}{2}\left(k - 1 + \frac{k-3}{k-1}\right)n + c_{k,p}.$$

As explained above, the value $f_k(k+p)$ is known, and so $g_k(n)$ is a known function. We claim that, for all feasible values of n, there is a k-critical graph with n vertices and $g_k(n)$ edges, implying that $f_k(n) \leq g_k(n)$. This is evident if $n = k$ or $n = k+p$ and $2 \leq p \leq k$, since then $g_k(n) = f_k(n)$. If G is a k-critical graph with $n \geq k+2$ vertices and $g_k(n)$ edges, then a Hajós sum of G and K_k results in a k-critical graph with $n' = n+k-1$ vertices and

$$m' = |E(G)| + \binom{k}{2} - 1$$

edges. Since $n' \equiv n \pmod{(k-1)}$, an easy calculation yields

$$m' = g_k(n) + \binom{k}{2} - 1 = g_k(n+k-1),$$

which proves our claim. Hajós's construction implies that

$$f_k(n+k-1) \leq f_k(n) + \tfrac{1}{2}k(k-1) - 1.$$

As explained by Jensen and Toft [27], this implies that $\lim_{n\to\infty} f_k(n)/n$ exists for all $k \geq 4$.

Ore [50] conjectured that a Hajós sum of a graph $G \in \mathcal{C}_k(n)$ and that K_k is in $\mathcal{C}_k(n+k-1)$. This implies that $f_k(n) = g_k(n)$ for all integers $n \geq k+2$ and $n = k$.

The function $g_k(n)$ is not linear in n, since the additive term $c_{k,p}$ depends both on k and on the residue class $p+1$ of n (mod $(k-1)$). This may indicate why it is difficult to prove Ore's conjecture that $f_k(n) = g_k(n)$. Recently, Kostochka and Yancey [36] found the best linear approximation for the function $f_k(n)$. Observe that $c_{k,k-1} \leq c_{k,p}$ for $2 \leq p \leq k$.

Theorem 3.3 *Let $k \geq 4$ and let*

$$h_k(n) = \tfrac{1}{2}\left(k - \frac{2}{k-1}\right)n + c_{k,k-1} = \frac{(k+1)(k-2)n - k(k-3)}{2(k-1)}.$$

Then $f_k(n) \geq \lceil h_k(n) \rceil$ whenever $n \geq k$ and $n \neq k+1$.

If $n \equiv 1$ (mod $(k-1)$), then $h_k(n) = g_k(n)$, and therefore $f_k(n) = g_k(n)$. This proves Ore's conjecture for infinitely many values of n for each $k \geq 4$. Since the limit of $f_k(n)/n$ exists, this implies that

$$\lim_{n\to\infty} \frac{f_k(n)}{n} = \tfrac{1}{2}\left(k - \frac{2}{k-1}\right).$$

If $n \equiv p+1$ (mod $(k-1)$) and $n \geq k+2$, then $f_k(n) - h_k(n) \leq c_{k,p} - c_{k,k-1}$. For $k = 4$, we have $c_{4,2} = 0$, $c_{4,3} = -\frac{2}{3}$, and $c_{4,4} = -\frac{1}{3}$. Hence, Theorem 3.3 implies that

$$f_4(n) = \left\lceil \tfrac{1}{3}(5n-2) \right\rceil \tag{4}$$

for all $n \geq 4$ and $n \neq 5$.

The proof of Theorem 3.3 is relatively long and sophisticated. However, as shown by Kostochka and Yancey [37], the special case $k = 4$, that any 4-critical graph G of order n has at least $\frac{1}{3}(5n-2)$ edges, can be handled much more easily, leading to a two-page proof. A very short outline of the proof is as follows.

Assume that G is a counter-example of minimum order. For $X \subseteq V(G)$, define $\rho(X) = 5|X| - 3|E(\langle X \rangle)|$. Since G is a counter-example, $\rho(V(G)) \geq 3$. One can now prove, from the minimality of G, that $\rho(X) \geq 5$ for any proper subset X of $V(G)$; moreover, $\rho(X) = 5$ only if $|X| = 1$, and $\rho(X) = 6$ only if $\langle X \rangle = K_3$. These may then in turn be used to prove that $\Delta(G_L) \leq 1$ – that is, a vertex in G of degree 3 has at most one neighbour of degree 3. A discharging argument may then be used. The initial charge on each vertex v is the degree $d(v)$. Then each vertex of degree greater than 3 sends a charge of $\frac{1}{6}$ to each of its neighbours of degree 3. In this way a vertex of degree 3 gets a new charge of at least $3+\frac{1}{3} = \frac{10}{3}$, and a vertex of degree $d \geq 4$ gets a new charge of at least $d - \frac{1}{6}d = \frac{5}{6}d \geq \frac{10}{3}$. The total amount of charge is $2|E(G)|$, and hence $2|E(G)| \geq \frac{10}{3}|V(G)|$ – that is, $\rho(V(G)) = 5|V(G)| - 3|E(G)| \leq 0$, contradicting the initial assumption that $\rho(V(G)) \geq 3$.

A graph satisfies $\chi \leq k-1$ if each of its induced subgraphs on $n' \geq k$ vertices has less than $f_k(n')$ edges. Lower bounds for $f_k(n)$ can therefore be used to prove colouring theorems.

Dirac [14] used his bound (2) to prove a map-colour theorem related to Heawood's bound. Other applications for bounds of $f_k(n)$ are discussed by Krivelevich [39], Kostochka and Yancey [36], [37], and Borodin *et al.* [3]. As pointed out by Kostochka and Yancey [37], equation (4) can be used to give a short proof of Grötzsch's theorem [22] that *any planar triangle-free graph is* 3*-colourable*. The proof is by contradiction, so we consider a 4-critical graph G without triangles embedded in the plane with n vertices, m edges and f faces. If G contains a face (x, y, z, u) of size 4, then $ab \notin E(G)$ for $(a, b) \in \{(x, z), (y, u)\}$, and the graph G_{ab} obtained from G by identifying a and b has chromatic number at least 4. Since G_{xz} or G_{yu} is triangle-free, this gives a smaller counter-example. If G has no face of size 4, then $5f \leq 2m$ and $n - m + f = 2$ (by Euler's formula), which gives $m \leq \frac{1}{3}(5n - 10)$. However, by (4), $m \geq \frac{1}{3}(5n - 2)$, a contradiction.

For 5-critical graphs, Theorem 3.3 yields $f_5(n) \geq \frac{9}{4}n - \frac{5}{4}$. This implies that any triangle-free graph G embedded on the torus or the projective plane is 4-colourable, since $|E(G)| \leq 2|V(G)|$ (by Euler's formula); the result for the torus was obtained by Kronk and White [40] in 1972. As pointed out by Thomassen [56], there are infinitely many 4-critical triangle-free graphs on the torus and on the projective plane. Furthermore, Thomassen proved a counterpart of Grötzsch's 3-colour theorem for the torus and the Klein bottle: *every graph on the torus or the Klein bottle containing no* C_3 *or* C_4 *is* 3*-colourable*. The proof is by inspection of a smallest counter-example, which is a 4-critical graph of some order $n \geq 6$. Using Euler's formula and (4), we deduce that $|E(G)| = f_4(n)$. However, it is unknown whether there exists a triangle-free graph in $\mathcal{C}_4(n)$, for any $n \geq 6$. Hence the original proof by Thomassen uses an extension of Grötzsch's theorem and is rather involved.

As proved by Kostochka and Stiebitz [33], a triangle-free k-critical graph of order $n \geq k+2$ has at least $(k - o(k))n$ edges. On the other hand, as shown by Abbott, Hare and Zhou [1], there are k-critical graphs without short cycles and with average degree at most $2(k-2)$.

4. Bounding χ by Δ and ω

The *clique number* $\omega(G)$ of a graph G – that is, the order of a largest complete subgraph – is an obvious lower bound for the chromatic number. On the other hand, it is well known that there is no upper bound for the chromatic number of a graph in terms of the clique number. This follows from the fact that, for each integer $k \geq 1$, there are triangle-free k-chromatic graphs. Graphs with these properties were first constructed by Zykov [62] in 1949. Ten years later, Erdős [17] proved by probabilistic arguments that, for all k and ℓ, there exists a k-chromatic graph with no cycles of length less than ℓ.

Brooks's theorem implies that graphs with maximum degree $\Delta \geq 3$ and clique number ω satisfy $\chi \leq \max\{\Delta, \omega\}$, and so provides the first result bounding the chromatic number of a graph from above in terms of the maximum degree and the clique number. Improvements of Brooks's result in this direction were first obtained independently around the same time by Borodin and Kostochka [2], Catlin [9] and Lawrence [41]: if $3 \leq \omega \leq \Delta$, then every graph G with maximum degree Δ and clique number at most ω satisfies

$$\chi(G) \leq \Delta + 1 - \left\lfloor \frac{\Delta + 1}{\omega + 1} \right\rfloor. \tag{5}$$

The inequality (5) is an immediate consequence of Brooks's theorem and a decomposition result of Lovász [42], that any graph G with

$$\Delta(G) \leq d_1 + d_2 + \cdots + d_p + p - 1$$

has a partition $V_1, V_2, \ldots, V_p$ of its vertex-set such that the subgraph $G_i = \langle V_i \rangle$ satisfies $\Delta(G_i) \leq d_i$. Now let $d_i = \omega$ for $1 \leq i \leq p-1$ and $d_p \geq \omega$, so that $d_1 + d_2 + \cdots + d_p = \Delta - p + 1$ and $p = \lfloor (\Delta + 1)/(\omega + 1) \rfloor$. By Lovász's result, there is a partition $G_1, G_2, \ldots, G_p$ of G such that $\Delta(G_i) \leq d_i$, and hence $\chi(G_i) \leq d_i$ by Brooks's theorem when $\omega \geq 3$. Consequently, $\chi(G) \leq \chi(G_1) + \chi(G_2) + \cdots + \chi(G_p) \leq d_1 + d_2 + \cdots + d_p = \Delta - p + 1$. Lovász's decomposition result follows from the basic case $p = 2$ by induction. For $p = 2$, a partition for which $d_1|E(G_2)| + d_2|E(G_1)|$ is minimum has the desired property, since otherwise a vertex of too large a degree could be moved to the other side to obtain a better partition.

For triangle-free graphs, (5) yields the same bound for the chromatic number as for K_4-free graphs, since (5) assumes that $\omega \geq 3$. However, Johanssen (unpublished) has proved that every graph with $\omega(G) = 2$ satisfies

$$\chi(G) \leq \frac{c\Delta(G)}{\log \Delta(G)} \tag{6}$$

for some constant c. A proof of this result by probabilistic means can be found in the book by Molloy and Reed [48].

As an immediate consequence of (5), we deduce that any graph G with $3 \leq \omega(G) \leq \frac{1}{2}(\Delta - 1)$ satisfies $\chi(G) \leq \Delta(G) - 1$. This observation led Borodin and Kostochka [2] to propose the following conjecture.

Conjecture A *Every graph G with $\Delta(G) \geq 9$ and $\omega(G) \leq \Delta(G) - 1$ satisfies $\chi(G) \leq \Delta(G) - 1$.*

As observed by Kostochka [30], it suffices to verify the conjecture for graphs with $\Delta = 9$; this follows from his result that any graph with maximum degree Δ and clique number $\omega \geq \Delta - \sqrt{\Delta} + \frac{3}{2}$ has an independent set meeting all maximum cliques (see also [12]). Reed [52] proved that, for any $\Delta \geq 9$, if G is a smallest counter-example to Conjecture **A** of maximum degree Δ, then the cliques of size $\Delta - 1$ in G are disjoint. Hence, as noticed by Reed, the Borodin–Kostochka conjecture would be a simple consequence of the following statement.

Every graph G with $\Delta(G) = 9$ and all of whose cliques of size 8 are disjoint, has a vertex partition into two graphs G_1 and G_2 such that $\Delta(G_i) \leq 4$ and $\omega(G_i) \leq 4$ for $i = 1, 2$.

This implies, by Brooks's theorem, that $\chi(G) \leq \chi(G_1) + \chi(G_2) \leq 8$.

Reed [52] proved in 1999 that the Borodin–Kostochka conjecture is true for graphs with maximum degree at least $\Delta_0 = 10^{14}$. The large value of Δ_0 is due to the fact that the proof uses the Lovász local lemma in order to find an appropriate partial colouring.

Theorem 4.1 *If G is a graph with $\Delta(G) \geq 10^{14}$ and $\omega(G) \leq \Delta(G) - 1$, then $\chi(G) \leq \Delta(G) - 1$.*

Reed [52] also conjectured the following bounds for the chromatic number, by a convex combination of the clique number and the maximum degree + 1.

Conjecture B $\chi(G) \leq \lceil \frac{1}{2}(\Delta(G) + 1) + \frac{1}{2}\omega(G) \rceil$.

Conjecture C $\chi(G) \leq \frac{2}{3}(\Delta(G) + 1) + \frac{1}{3}\omega(G)$, *provided that* $\Delta(G) \geq 3$.

Gallai [20] constructed an infinite family of 4-critical graphs which are 4-regular and triangle-free. This shows that the rounding-up in Conjecture **B** is necessary. Another family of graphs showing that the rounding-up is necessary for arbitrarily large Δ was given by Kostochka (see [48]): if $G = C_5[K_n]$ is the composition (or lexicographic product) of C_5 and K_n, then $|V(G)| = 5n$, $\Delta(G) = 3n - 1$, $\omega(G) = 2n$ and $\chi(G) = \lceil \frac{5}{2}n \rceil$. In particular, the graph $C_5[K_3]$ shows that the Borodin–Kostochka conjecture does not hold for graphs with maximum degree 8.

Reed's Conjecture **B** is trivial when $\omega(G) = \Delta(G)$, $\omega(G) = \Delta(G) + 1$ and $\omega(G) = 1$, and it follows from Brooks's theorem when $\omega(G) = \Delta(G) - 1$ and $\omega(G) = \Delta(G) - 2$. That this conjecture holds when $\omega(G) = 2$ and $\Delta(G)$ is sufficiently large follows from Johannsen's bound (6). So Conjecture **B** may be considered as a far-reaching extension of Brooks's theorem.

Let (a, b) be a pair of real numbers such that $\chi(G) \leq a(\Delta(G) + 1) + b\omega(G)$ for all graphs with $\Delta(G)$ sufficiently large. By considering complete graphs, we see that $a + b \geq 1$. Furthermore, as proved by Reed [52], $a \geq \frac{1}{2}$.

The following two results, obtained in 1998 by Reed [51], support Conjecture **B**.

Theorem 4.2 *There is a positive constant ε such that*

$$\chi(G) \leq (1 - \varepsilon)(\Delta(G) + 1) + \varepsilon\omega(G)$$

for every graph G with $\Delta(G) \geq 3$.

Theorem 4.3 *For every positive real number b there is a number Δ_b such that $\chi(G) \leq \Delta(G) + 1 - b$ for every graph G with $\omega(G) \leq \Delta(G) + 1 - 2b$ and $\Delta(G) \geq \Delta_b$.*

5. Graphs with χ close to Δ

Brooks's theorem is equivalent to the statement that if $\Delta \geq 3$, then $K_{\Delta+1}$ is the only $(\Delta + 1)$-critical graph with maximum degree Δ. Reed's result (Theorem 4.1) is equivalent to the statement that if $\Delta \geq \Delta_0$, then there is no Δ-critical graph with maximum degree Δ.

For any $\Delta \geq 2$, let h_Δ be the maximum integer h for which $\Delta \geq (h + 1)(h + 2)$, so $h_\Delta \approx \sqrt{\Delta} - 2$. The following result of Molloy and Reed [47] is a strengthening of Theorem 4.1, since no k-critical graph can have order $k + 1$.

Theorem 5.1 *There is a constant Δ' such that, for any $\Delta \geq \Delta'$ and any $h < h_\Delta$, any graph with maximum degree Δ and chromatic number at least $\Delta - h + 1$ contains a $(\Delta - h + 1)$-critical subgraph whose order is at most $\Delta + 1$. In particular, any $(\Delta - h + 1)$-critical graph with maximum degree Δ has order $\Delta + 1$.*

The proof of this remarkable result uses the Lovász local lemma, which requires a large constant Δ'. On the other hand, Molloy and Reed conjectured that the condition $\Delta \geq \Delta'$ is not necessary if we require only that the order of the critical subgraph is bounded by a function $f(\Delta)$.

In contrast to the above theorem, it was also proved by Molloy and Reed that, for any $\Delta \geq 2$ and any $h > h_\Delta$, there are arbitrarily large $(\Delta - h + 1)$-critical graphs with maximum degree Δ. To obtain such graphs, start with a $(\Delta - h + 1)$-critical graph G of arbitrarily large order. If a vertex v in G has degree greater than Δ, split v into an independent set X of size $h + 2$ and join each vertex in X completely to a complete graph $K_{\Delta-h-1}$: this results in another $(\Delta - h + 1)$-critical graph. By repeated application of this operation, we obtain the desired graph. This construction is a special case of a more general construction involving critical hypergraphs, due to Toft [58].

A graph has chromatic number at least k if and only if it contains a k-critical subgraph. So if we study graphs with chromatic number at least $\Delta - h + 1$ for sufficiently large Δ, then Theorem 5.1 implies that the corresponding critical graphs have order close to their chromatic number. For such graph classes, however, Gallai's Theorem 3.1 is applicable.

First, we introduce some notation. Let $\mathcal{G}_k(n)$ be the class of k-critical graphs of order n, and let $\mathcal{G}^*_k(n)$ be the subclass consisting of all graphs in $\mathcal{G}_k(n)$ with no dominating vertices. (Recall that a vertex is *dominating* in G if it is adjacent to all the remaining vertices of G.) For a graph K and a graph class $\mathcal{G}$, define $K + \mathcal{G} = \{K + G : G \in \mathcal{G}\}$ if $\mathcal{G}$ is non-empty, and $K + \mathcal{G} = \varnothing$ otherwise.

Let $k \geq 4$, $2 \leq s < \frac{2}{3}k$ and $G \in \mathcal{G}_k(k + s)$. Clearly, $G = K_{k-\ell} + H$ is the join of a complete graph and a graph H with no dominating vertices (possibly $\ell = k$ or $H = \varnothing$). Since G has order $n \geq k + 2$, $G \neq K_k$, and so H is non-empty. Since G is k-critical, H is ℓ-critical and $|V(H)| = k + s - (k - \ell) = \ell + s$. Since H has no dominating vertices, $\chi(H) \geq 3$ and $|V(H)| \geq 5$. Since $n = k + s < \frac{5}{3}k$, it follows from Theorem 3.1 that $k - \ell \geq 1$ and $\ell + s = |V(H)| \geq \frac{5}{3}\ell$. Consequently, $3 \leq \ell \leq \frac{3}{2}s$

and, since no 3-critical graph has even order,

$$\mathcal{G}_k(k+s) = \bigcup_{\ell=\ell_s}^{\lfloor 3s/2 \rfloor} K_{k-\ell} + \mathcal{G}_\ell^*(\ell+s), \tag{7}$$

where $\ell_s = 3$ if s is even, and $\ell_s = 4$ otherwise. Clearly, $\mathcal{G}_k(k)$ consists only of K_k, and $\mathcal{G}_k(k+1)$ is empty. Lists of small critical graphs were determined by Toft [59], by Jensen and Royle [26], and by Royle [53]. In particular, there are exactly two (non-isomorphic) 4-critical graphs H_1 and H_2 on 7 vertices and neither of which contains a dominating vertex (see Fig. 2). With the help of these lists, one can give an explicit description of $\mathcal{G}_k(k+s)$ for $s = 2, 3, 4$ and 5, provided that $k \geq 4$ and $s < \frac{2}{3}k$. In particular, $K_{k-3} + C_5$ is the only graph in $\mathcal{G}_k(k+2)$ for $k \geq 4$, and $K_{k-4} + H_1$ and $K_{k-4} + H_2$ are the only two graphs in $\mathcal{G}_k(k+3)$ for $k \geq 5$. The class $\mathcal{G}_k(k+4)$ with $k \geq 6$ consists of twenty-two graphs.

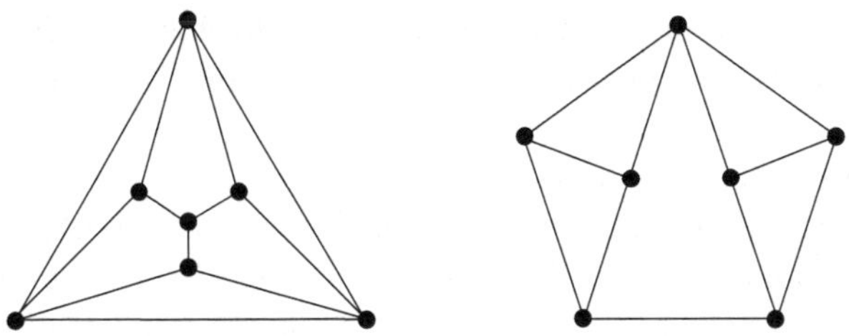

Fig. 2. The only two 4-critical graphs on 7 vertices

Let $\Delta \geq \Delta'$ and $0 \leq h < h_\Delta$, and let $S_{\Delta,h}$ be the set of all non-isomorphic $(\Delta - h + 1)$-critical graphs with order at most $\Delta + 1$. Such a set $S_{\Delta,h}$ consists of the non-isomorphic graphs in the class

$$\mathcal{G} = \mathcal{G}_{\Delta-h+1}(\Delta - h + 1) \cup \bigcup_{s=3}^{h+1} \mathcal{G}_{\Delta-h+1}(\Delta - h + s).$$

From Theorem 5.1 it then follows that every graph G with maximum degree Δ satisfies

(T_h) $\chi(G) \geq \Delta - h + 1$ if and only if G contains a (subgraph isomorphic to a) graph in $S_{\Delta,h}$.

Note that statement (T_0) is Brooks's theorem and statement (T_1) is equivalent to Theorem 4.1.

If we also assume that $\Delta + 1 < \frac{5}{3}(\Delta - h + 1)$ (which is equivalent to $\Delta > \frac{5}{2}h - 1$), then we can apply (7) to determine $S_{\Delta,h}$. In particular, $S_{\Delta,0} = \{K_{\Delta+1}\}$, $S_{\Delta,1} = \{K_\Delta\}$, $S_{\Delta,2} = \{K_{\Delta-1}, K_{\Delta-4}+C_5\}$, and $S_{\Delta,3} = \{K_{\Delta-2}, K_{\Delta-5}+C_5, K_{\Delta-6}+H_1, K_{\Delta-6}+H_2\}$. Furthermore, as shown by Farzad, Molloy and Reed [19], $|S_{\Delta,4}| = 26$ and $|S_{\Delta,5}| = 420$. For $h = 2, 3, 4$ and 5, statement (T_h) was established by Farzad *et al.* Since $\omega(H_1) = \omega(H_2) = 3$, each graph in $S_{\Delta,h}$, for $h = 0, 1, 2$ and 3, contains a $K_{\Delta-h}$. We thus obtain the following result of Farzad *et al.*

(*S*) If $\chi(G) \geq \Delta(G) - 2$ and $\Delta(G)$ is sufficiently large, then $\chi(G) \leq \omega(G) + 1$.

It was also noticed in [19] that the threshold $\Delta - 2$ is sharp. To see this, let $G = K_{\Delta-9} + C_5 + C_5$. Then $\chi(G) = \Delta - 3$ and $\omega(G) = \Delta - 5 = \chi(G) - 2$. This provides a solution to one of the problems proposed in the book by Jensen and Toft [27, Problem 4.7].

6. Notes

Brooks's claim to fame rests on two papers [6] and [4], which to this day continue to stimulate mathematicians. Rowland Leonard Brooks was born on 6 February 1916 in Lincolnshire, England, and died on 18 June 1993. He started his university studies in Cambridge in 1935. On the very first day of lectures he found himself attending an advanced lecture by mistake, together with another freshman, Cedric Smith. Shortly after this he met two other students, A. H. Stone and W. T. Tutte. It was Tutte who in November 1940 communicated Brooks's paper [4] as a research note to the Cambridge Philosophical Society.

After finishing Cambridge, Brooks became an income-tax inspector in London. He continued his interest in mathematics throughout his life (see [5], [7] and [8]). C. A. B. Smith remained a close friend and they met regularly to discuss mathematical problems. Smith told Toft (private communication) that Brooks was a rather shy person who did not want his photograph published, which is why it is difficult to find a photograph of him. He also hesitated to make biographical material available.

The various equivalent formulations of Brooks's theorem, as discussed in this chapter, are the triggers for enhancements and extensions of the fundamental result. Here is another formulation of Brooks's theorem in terms of forbidden subgraphs.

Every connected graph $G \neq K_{k+1}$ ($k \geq 3$) *not containing* $K_{1,k+1}$ *as a subgraph satisfies* $\chi(G) \leq k$.

Taking this formulation into account, it becomes natural to consider arbitrary trees of order $k+2$, and Mihok [44] proved the following result. Here the *colouring number* $\operatorname{col}(G)$ of a graph G is (the maximum minimum degree of the subgraphs of G) + 1.

If $T \neq K_{1,k+1}$ *is a tree of order* $k+2$ *with* $k \geq 3$, *then every connected graph* $G \neq K_{k+1}$ *not containing* T *as a subgraph satisfies* $\delta(G) \leq k-1$, *and hence* $\chi(G) \leq \operatorname{col}(G) \leq k$.

The *list-chromatic number* $\chi_\ell(G)$ of G is the smallest integer $k \geq 0$ such that G is L-colourable for each list assignment L of G satisfying $|L(v)| = k$ for all $v \in V(G)$. Using a sequential colouring algorithm, it is easy to show that $\chi_\ell(G) \leq \operatorname{col}(G)$. Thus the following well-known sequence of inequalities holds for any graph G:

$$\omega(G) \leq \chi(G) \leq \chi_\ell(G) \leq \operatorname{col}(G) \leq \Delta(G) + 1. \tag{8}$$

That Brooks's theorem remains true when the chromatic number χ is replaced by the list-chromatic number χ_ℓ follows from Theorem 2.1 (due to Erdős, Rubin and Taylor [18]). On the other hand, to characterize the class of graphs G for which $\chi(G) = \chi_\ell(G)$ seems very difficult, or even unattainable. Some partial answers are discussed in Chapter 6 on list-colourings.

Another interesting variant of the chromatic number is the *equitable chromatic number* $\chi^{eq}(G)$, defined as the least k for which G has an *equitable k-colouring* – that is, a k-colouring in which the sizes of the colour classes differ by at most 1. As proved by Hajnal and Szemerédi [24], $\chi^{eq}(G) \leq \Delta(G) + 1$. A short proof of this result was recently given by Kierstead *et al.* [29]. It is not known (see [27, Problem 4.10]) whether a Brooks-type result holds for χ^{eq} – that is, whether every connected graph G satisfies $\chi^{eq}(G) \leq \Delta(G)$ if G is neither a complete graph nor an odd cycle. Chen, Lih and Wu [11] conjectured that, for $\Delta \geq 3$, every connected graph with maximum degree Δ has an equitable Δ-colouring, unless $G = K_{\Delta+1}$, or Δ is odd and $G = K_{\Delta,\Delta}$.

Mihok's result about the chromatic number of graphs not containing a specific tree T shows that the most interesting case is when T is a star. However, if we forbid a tree as an induced subgraph, then the corresponding question becomes much more difficult. Gyarfás [23] conjectured that, for every given tree T, there exists a function f_T such that $\chi(G) \leq f_T(\omega(G))$ for every graph G not containing T as an induced subgraph. This conjecture is wide open.

That it is worthwhile to study critical graphs, and especially the function $f_k(n)$, was first emphasized by G. A. Dirac in his thesis, and subsequently by Gallai [20] and Ore [50]. Kostochka and Yancey [36] recently succeeded in determining the best linear approximation to the function $f_k(n)$; their result is a major breakthrough in the study of critical graphs. However, even if we know that $f_4(n) = \lceil \frac{1}{3}(5n-2) \rceil$, our knowledge about the class $\mathcal{C}_4 = \bigcup_n \mathcal{C}_4(n)$ remains incomplete. Is there a decomposition result for graphs in $\mathcal{C}_4$? All known graphs in $\mathcal{C}_4$, except K_4 and $K_1 + C_5$, have a vertex-cut of size 2 or a non-trivial edge-cut of size 3, and can therefore be decomposed into two smaller graphs or hypergraphs belonging to $\mathcal{C}_4$. For instance, the only graph in $\mathcal{C}_4(8)$ can be decomposed into K_4 and a 4-critical hypergraph with five vertices (see Fig. 3). Kostochka and Yancey [38] proved that, if $n \equiv 1 \pmod{(k-1)}$, then any graph in $\mathcal{C}_k(n)$ has a vertex-cut of size 2.

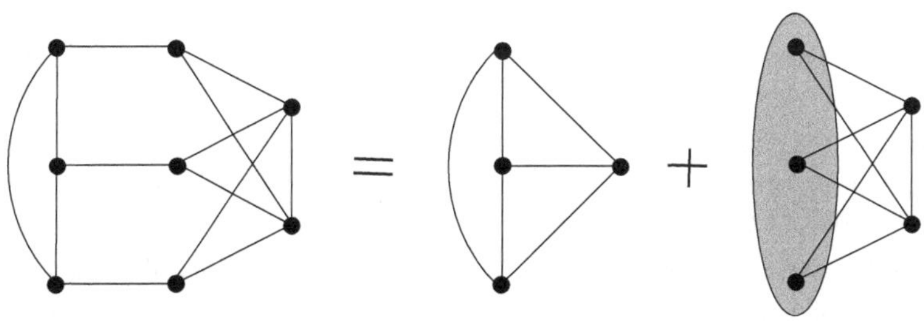

Fig. 3. A decomposition for the graph in $\mathcal{C}_4(8)$

As observed during a discussion between Kostochka and the authors of this chapter, the Kostochka–Yancey bound also holds for critical hypergraphs. Recall that in a (proper) colouring of a hypergraph we require that each edge contains two vertices of different colours. The definition of the chromatic number and criticality can be easily extended to hypergraphs.

Theorem 6.1 *For $k \geq 4$, any k-critical hypergraph with n vertices has at least $h_k(n)$ edges.*

Proof Consider a k-critical hypergraph G on n vertices and m edges. To prove that $m \geq h_k(n)$, we use induction on the number h of hyperedges in G – that is, edges with at least three vertices. For $h = 0$, the result follows from Theorem 3.3. So assume that $h \geq 1$ and let e be a hyperedge of G. We now construct a new hypergraph G' as follows. Let T_k^d denote the graph obtained from d disjoint complete graphs $K^1, K^2, \ldots, K^d$, each with $k-1$ vertices, and an additional vertex u, first by adding $d-1$ edges $x_1y_2, x_2y_3, \ldots, x_{d-1}y_d$, where $x_i \in V(K^i)$, $y_{i+1} \in V(K^{i+1})$ (for $1 \leq i \leq d-1$) and $x_i \neq y_i$ (for $2 \leq i \leq d-1$), and then joining u to the remaining vertices of degree $k-2$. Note that $T_k^1 = K_k$ and that T_k^{d+1} is a Hajós sum of T_k^d and K_k. So T_k^d is a k-critical graph. We now choose d such that $T = T_d^k$ satisfies $d_T(u) \geq |e|$.

Now let G' be the hypergraph obtained from $G - e$ by splitting the vertex u of T into the set e – that is, G is obtained from $G-e$ and $T-u$ by joining each neighbour of u in T to exactly one vertex of e such that each vertex in e has at least one neighbour in $T - u$. It is then not difficult to show that G' is a k-critical hypergraph. This is a special case of a reduction method for critical hypergraphs invented by Toft [58]. For $n' = |V(G')|$ and $m' = |E(G')|$, we have $n' = n + d(k-1)$ and

$$m' = m + \tfrac{1}{2}d(k-2)(k+1).$$

On the other hand, since G' has $h - 1$ hyperedges, the induction hypothesis implies that

$$\begin{aligned} m' \geq h_k(n') &= h_k(n) + \tfrac{1}{2}(k - \frac{2}{k-1})(d(k-1)) \\ &= h_k(n) + \tfrac{1}{2}d(k-2)(k+1), \end{aligned}$$

which yields $m \geq h_k(n)$, as claimed. ■

Note that $n' \equiv n \pmod{(k-1)}$. Thus, if Ore's conjecture that $f_k(n) = g_k(n)$ is true, then we can use the above proof to show that each k-critical hypergraph on n vertices has at least $f_k(n)$ edges.

In addition to Theorem 6.1, Kostochka and Stiebitz [33] proved that if a k-critical hypergraph on n vertices contains no ordinary edge (of size 2), then it has at least $(k-1) - 1/\sqrt[3]{k-1})n$ edges.

It would be also of much interest to know whether the Kostochka–Yancey bound holds for list-colourings. A graph G is called *k-list-critical* if there is a list assignment L with $|L(v)| = k-1$ for all vertices v such that every proper subgraph of G has an L-colouring, but G itself has no L-colouring. Then any k-critical graph is k-list-critical. Furthermore, a graph G is k-list-critical if $\chi_\ell(H) < \chi_\ell(G) = k$ for every proper subgraph H of G. So let $f_k^\ell(n)$ denote the minimum number of edges possible in a k-list-critical graph on n vertices. Not much is known about this function. That Gallai's bound (3) holds for list-critical graphs is a simple consequence of the fact that if G is a k-list-critical graph with $k \geq 4$, then each block of the low-vertex subgraph G_L is a brick; this is a simple consequence of Theorem 2.1 and has been observed by several authors (see, for example, [31], [35] and [57]). That Dirac's bound (2) holds for list-critical graphs – that is, $2f_k^\ell(n) \geq (k-1)n+k-3$ – was proved by Kostochka and Stiebitz [34].

Does there exists a triangle-free graph in $\mathcal{C}_4$? Dirac also considered the function $f_k(n,r)$, equal to the minimum number of edges in any k-critical graph of order n without containing a K_r. Dirac [15] proved that $2f_k(n,r) \geq (k-1)n + 2k - r - 3$. Weinstein [61] improved this bound to $2f_k(n,r) \geq (k-1)n + 3k - 2r - 3$. A Gallai-type bound for the function $f_k(n,r)$ was established by Krivelevich [39]. Kostochka and Stiebitz [33] proved that $f_k(n,3) \geq (k - o(k))n$. This result is a consequence of Johannsen's bound (6) for the chromatic number of triangle-free graphs in terms of the maximum degree; for graphs with small maximum degree, Johannsen's bound is not helpful. Kostochka proved that any triangle-free graph G satisfies $\chi(G) \leq \frac{2}{3}(\Delta(G) + 3)$ (see [27, Problem 4.6]). Note that Reed's conjecture (Conjecture **B**) implies that Kostochka's bound can be improved to $\chi(G) \leq \frac{1}{2}(\Delta(G)+2)$ for triangle-free graphs.

In his 1951 thesis, Dirac considered not only critical graphs with few edges, but also those with many edges. During his doctoral study in London in 1949, Dirac met P. Erdős and told him about the new concept of critical graphs. Erdős immediately asked for the maximum number of edges in such a minimal graph (with respect to χ). Dirac observed that $G = C_{2\ell+1} + C_{2\ell+1}$ is a 6-critical graph with $n = 4\ell + 2$ vertices and $\frac{1}{4}n^2 + n$ edges. Whether there exists a 6-critical graph with n vertices and more than $\frac{1}{4}n^2 + n$ edges is unknown; see also the book of Jensen and Toft [27, Problem 5.1]. Chapter 5 of that book discusses several other problems related to the structure of critical graphs.

References

1. H. L. Abbott, D. R. Hare and B. Zhou, Sparse color-critical graphs and hypergraphs with no short cycles, *J. Graph Theory* **18** (1994), 373–388.
2. O. V. Borodin and A. V. Kostochka, On an upper bound of a graph's chromatic number, depending on the graph's degree and density, *J. Combin. Theory (B)* **23** (1977), 247–250.
3. O. V. Borodin, A. V. Kostochka, B. Lidický and M. Yancey, Short proofs of coloring theorems on planar graphs, manuscript.
4. R. L. Brooks, On colouring the nodes of a network, *Proc. Cambridge Philos. Soc.* **37** (1941), 194–197.
5. R. L. Brooks, A procedure for dissecting a rectangle into squares, and an example for the rectangle whose sides are in the ratio 2:1, *J. Combin. Theory (B)* **8** (1970), 232–242.

6. R. L. Brooks, C. A. B. Smith, A. H. Stone and W. T. Tutte, The dissection of rectangles into squares, *Duke Math. J.* **7** (1940), 312–340.
7. R. L. Brooks, C. A. B. Smith, A. H. Stone and W. T. Tutte, A simple perfect square, *Nederl. Akad. Wetensch. Proc.* **50** (1947), 1300–1301.
8. R. L. Brooks, C. A. B. Smith, A. H. Stone and W. T. Tutte, Determinants and current flows in electric networks, *Discrete Math.* **100** (1992), 291–301.
9. P. A. Catlin, A bound on the chromatic number of a graph, *Discrete Math.* **22** (1978), 81–83.
10. P. A. Catlin, Brooks' graph-coloring theorem and the independence number, *J. Combin. Theory (B)* **27** (1979), 42–48.
11. B. Chen, K. Lih and P. Wu, Equitable coloring and the maximum degree, *Europ. J. Combinatorics* **15** (1994), 443–447.
12. D. W. Cranston and L. Rabern, Conjectures equivalent to the Borodin–Kostochka conjecture that appear weaker, manuscript.
13. G. A. Dirac, The structure of k-chromatic graphs, *Fund. Math.* **40** (1953), 42–55.
14. G. A. Dirac, Map colour theorems related to the Heawood colour formula, *J. London Math. Soc.* **31** (1956), 460–471.
15. G. A. Dirac, A theorem of R. L. Brooks and a conjecture of H. Hadwiger, *Proc. London Math. Soc. (3)* **7** (1957), 161–195.
16. G. A. Dirac, The number of edges in critical graphs, *J. Reine Angew. Math.* **268/269** (1974), 150–164.
17. P. Erdős, Graph theory and probability, *Canad. J. Math.* **11** (1959), 34–38.
18. P. Erdős, A. L. Rubin and H. Taylor, Choosability in graphs, *Proc. West-Coast Conf. on Combinatorics, Graph Theory and Computing, Congr. Numer.* **XXVI** (1979), 125–157.
19. B. Farzad, M. Molloy and B. Reed, $(\Delta - k)$-critical graphs, *Discrete Math.* **93** (2005), 173–185.
20. T. Gallai, Kritische Graphen I, II, *Publ. Math. Inst. Hungar. Acad. Sci.* **8** (1963), 165–192 and 373–395.
21. L. Gerencsér, Szinezési problémákrol, *Mat. Lapok* **16** (1965), 274–277.
22. H. Grötzsch, Ein Dreifarbensatz für dreikreisfreie Netze auf der Kugel, *Wiss. Z. Martin-Luther Univ. Halle-Wittenberg, Math.-Nat. Reihe* **8** (1958/59), 109–120.
23. A. Gyárfás, Problems from the world surrounding perfect graphs, *Zastos. Mat.* **19** (1988), 413–431.
24. A. Hajnal and E. Szemerédi, Proof of a conjecture of P. Erdős, *Combinatorial Theory and its Application,* Vol. II (eds. P. Erdős *et al.*), North-Holland (1970), 601–623.
25. G. Hajós, Über eine Konstruktion nicht n–färbbarer Graphen, *Wiss. Z. Martin-Luther-Univ. Halle-Wittenberg, Math.-Naturw. Reihe* **10** (1961), 116–117.
26. T. R. Jensen and G. F. Royle, Small graphs with chromatic number 5: A computer search, *J. Graph Theory* **19** (1995), 107–116.
27. T. R. Jensen and B. Toft, *Graph Coloring Problems*, Wiley, 1995.
28. A. B. Kempe, On the geographical problem of four colours, *Amer. J. Math.* **2** (1879), 193–200.
29. H. A. Kierstead, A. V. Kostochka, A. V. Mydlarz and E. Szemerédi, A fast algorithm for equitable coloring, *Combinatorica* **30** (2010), 217–224.
30. A. V. Kostochka, Degree, density, and chromatic number (in Russian), *Metody Diskret. Anal.* **35** (1980), 45–70.
31. A. V. Kostochka, M. Stiebitz and B. Wirth, The colour theorems of Brooks and Gallai extended, *Discrete Math.* **191** (1996), 125–137.
32. A. V. Kostochka and M. Stiebitz, Excess in colour-critical graphs, Graph Theory and Combinatorial Biology (Balatonlelle, Hungary, 1996), *Bolyai Society Mathematical Studies* **7** (1999), 87–99.

33. A. V. Kostochka and M. Stiebitz, On the number of edges in colour-critical graphs and hypergraphs, *Combinatorica* **20** (2000), 521–530.
34. A. V. Kostochka and M. Stiebitz, A list version of Dirac's theorem on the number of edges in colour-critical graphs, *J. Graph Theory* **39** (2002), 165–167.
35. A. V. Kostochka and M. Stiebitz, A new lower bound for the number of edges in colour-critical graphs and hypergraphs, *J. Combin. Theory (B)* **20** (2003), 374–402.
36. A. V. Kostochka and M. Yancey, Ore's conjecture on color-critical graphs is almost true, manuscript.
37. A. V. Kostochka and M. Yancey, Ore's conjecture for $k = 4$ and Grötzsch's theorem, *Combinatorica* **34** (2014), 323–329.
38. A. V. Kostochka and M. Yancey, A Brooks-type result for sparse critical graphs, manuscript.
39. M. Krivelevich, On the minimal number of edges in color-critical graphs, *Combinatorica* **17** (1997), 401–426.
40. H. V. Kronk and A. T. White, A 4-color theorem for toroidal graphs, *Proc. Amer. Math. Soc.* **34** (1972), 83–86.
41. J. Lawrence, Covering the vertex set of a graph with subgraphs of smaller degree, *Discrete Math.* **21** (1978), 61–88.
42. L. Lovász, On decomposition of graphs, *Studia Sci. Math. Hungar.* **1** (1966), 237–238.
43. L. S. Melnikov and V. G. Vizing, New proof of Brooks' theorem, *J. Combin. Theory* **7** (1969), 289–290.
44. P. Mihok, An extension of Brooks' theorem, *Ann. Discrete Math.* **51** (1992), 235–236.
45. J. Mitchem, A short proof of Catlin's extension of Brooks' theorem, *Discrete Math.* **21** (1978), 213–214.
46. M. Molloy, Chromatic neighborhood sets, *J. Graph Theory* **31** (1999), 303–311.
47. M. Molloy and B. Reed, Colouring graphs when the number of colours is almost the maximum degree, *Proceedings of the 33rd Annual ACM Symposium on Theory of Computing* (2001), 462–470.
48. M. Molloy and B. Reed, *Graph Coloring and the Probabilistic Method*, Springer, 2002.
49. L. Lovász, Three short proofs in graph theory, *J. Graph Theory* **19** (1975), 269–271.
50. O. Ore, *The Four Colour Problem*, Academic Press, 1967.
51. B. Reed, χ, Δ, and ω, *J. Graph Theory* **27** (1998), 177–213.
52. B. Reed, A strengthening of Brooks's theorem, *J. Combin. Theory (B)* **76** (1999), 136–149.
53. G. F. Royle, Small graphs, *http://school.maths.uwa.edu.au/~gordon/remote/graphs/, 2000.*
54. M. Stehlík, Critical graphs with connected complements, *J. Combin. Theory (B)* **89** (2003), 189–194.
55. M. Stiebitz, Proof of a conjecture of T. Gallai concerning connectivity properties of colour-critical graphs, *Combinatorica* **2** (1982), 315–323.
56. C. Thomassen, Grötzsch's 3-colour theorem and its counterparts for the torus and the projective plane, *J. Combin. Theory (B)* **62** (1994), 268–279.
57. C. Thomassen, Color-critical graphs on a fixed surface, *J. Combin. Theory (B)* **70** (1997), 67–100.
58. B. Toft, Colour-critical graphs and hypergraphs, *J. Combin. Theory (B)* **16** (1974), 145–161.
59. B. Toft, Critical subgraphs of colour critical graphs, *Discrete Math.* **7** (1974), 377–392.
60. H. Tverberg, On Brooks' theorem and some related results, *Math. Scand.* **52** (1983), 37–40.
61. J. Weinstein, Excess in critical graphs, *J. Combin. Theory (B)* **18** (1975), 24–31.
62. A. A. Zykov, On some properties of linear complexes (in Russian), *Mat. Sbornik N. S.* **24** (1949), 163–188; English translation in *Amer. Math. Soc. Transl.* **79** (1952), 418–449.

3

Chromatic polynomials

BILL JACKSON

This chapter is concerned with chromatic polynomials and other related polynomials: flow polynomials, characteristic polynomials, Tutte polynomials and the Potts model partition function. We describe their basic properties and state some recent results and open problems. We also outline some techniques for working with these polynomials.

1. Introduction

The study of chromatic polynomials of graphs was initiated by Birkhoff [3] in 1912 and continued by Whitney [49], [50] in 1932. Inspired by the four-colour conjecture, Birkhoff and Lewis [4] obtained results concerning the distribution of the real zeros of chromatic polynomials of planar graphs and made the stronger conjecture that they have no real zeros greater than or equal to 4. Their hope was that results from analysis and algebra could be used to prove their stronger conjecture, and hence to deduce that the four-colour conjecture was true. This has not yet occurred: indeed, the four-colour conjecture is now a theorem, but the stronger conjecture of Birkhoff and Lewis remains open. Nevertheless, many beautiful results on chromatic polynomials have been obtained, and many other intriguing questions remain unanswered.

Work on chromatic polynomials has received fresh impetus in recent years from an interaction with mathematical physics. The chromatic polynomial is a specialization of the Potts model partition function, used by mathematical physicists to study phase transitions. A combination of ideas and techniques from graph theory and statistical mechanics has led to significant new results on both polynomials. Other recent progress has been made through the use of techniques from algebraic geometry to solve a long-standing open problem concerning the coefficients of chromatic polynomials.

2. Definitions and elementary properties

All graphs considered are finite and (unless stated otherwise) without loops or multiple edges. As usual, we use $G - e$ to denote the graph obtained by deleting an edge e from a graph G and G/e the graph obtained by contracting e. For any positive integer q, the *chromatic polynomial* $P(G, q)$ of G is defined as the number of distinct proper q-colourings of G. It is easy to see that a graph G with n vertices and no edges has chromatic polynomial q^n and that $P(K_n, q) = q(q-1)\dots(q-n+1)$.

Some reduction techniques

The following results give a recursive procedure for calculating chromatic polynomials.

Theorem 2.1 (The deletion/contraction lemma) *Let G be a graph and e be an edge of G. Then*

$$P(G, q) = P(G - e, q) - P(G/e, q).$$

Proof Let $e = vw$. Then $P(G-e, q) = P(G, q) + P(G/e, q)$ for all positive integers q, since $P(G, q)$ counts the number of q-colourings of $G - e$ in which v and w are given different colours and $P(G/e, q)$ counts the number in which v and w are given the same colour. ■

This result can be used recursively to express $P(G, q)$ as a linear combination of chromatic polynomials of graphs with no edges. Alternatively, we can use the 'addition/contraction' form

$$P(G, q) = P(G + e, q) + P(G/e, q),$$

where e is a new edge added between two non-adjacent vertices, to express $P(G, q)$ as a linear combination of chromatic polynomials of complete graphs.

Theorem 2.1 can also be used to deduce many elementary properties of chromatic polynomials. In particular, an easy induction on the number of edges of G gives the following result.

Theorem 2.2 *Let G be a graph with n vertices. Then $P(G, q)$ is a monic polynomial of degree n with zero constant term and with integer coefficients that alternate in sign.*

The recursive calculation of $P(G,q)$ can often be simplified by using the following factorization result.

Theorem 2.3 *Let G be a graph, and let G_1 and G_2 be subgraphs of G for which $G_1 \cup G_2 = G$ and $G_1 \cap G_2 = K_r$. Then*

$$P(G,q) = \frac{P(G_1,q)P(G_2,q)}{q(q-1)\cdots(q-r+1)}.$$

Proof The vertices of K_r must receive distinct colours in any q-colouring of G. There are $P(K_r,q) = q(q-1)\cdots(q-r+1)$ ways to colour K_r and the symmetry between the colours implies that any such colouring of K_r can be extended to exactly $P(G_1,q)/[q(q-1)\cdots(q-r+1)]$ colourings of G_1 and to $P(G_2,q)/[q(q-1)\cdots(q-r+1)]$ colourings of G_2. The result follows. ■

Our third reduction involves a rather special case.

Theorem 2.4 *Let G be a graph, and let v be a vertex adjacent to all other vertices of G. Then*

$$P(G,q) = qP(G-v,q-1).$$

Proof We have q choices to colour v, and this leaves $q-1$ colours to colour $G-v$. ■

Some examples

A tree T with n vertices has $P(T,q) = q(q-1)^{n-1}$.
This can be deduced directly from the definition – we have q choices for colouring any given vertex, then $q-1$ choices for colouring each of its neighbours, and so on. Alternatively, we can express T as $T_1 \cup T_2$, where $T_1 \cap T_2 = K_1$, and apply Theorem 2.3 and induction.
The cycle C_n with n vertices has $P(C_n,q) = (q-1)^n + (-1)^n(q-1)$.
This can be proved by using the deletion/contraction formula and then applying the above result for trees to $C_n - e$ and induction to C_n/e. Alternatively, we can use the addition/contraction formula, and then apply Theorem 2.3 and induction to $C_n + e = C_{n_1} \cup C_{n_2}$ with $C_{n_1} \cap C_{n_1} = K_2$, and $C_n/e = C_{n_1-1} \cup C_{n_2-1}$ with $C_{n_1-1} \cap C_{n_2-1} = K_1$.
The wheel W_n with n vertices has $P(W_n,q) = q[(q-2)^{n-1} + (-1)^{n-1}(q-2)]$.
This follows from the above result for cycles and Theorem 2.4.

Calculating $P(G,q)$ for an arbitrary graph using Theorems 2.1, 2.3 and 2.4 may take an exponential amount of time. This is to be expected, since it is NP-hard to determine the chromatic number of a graph. Since the chromatic number of G is one more than the largest integer zero of $P(G,q)$, it will also be NP-hard to determine $P(G,q)$. More details on algorithms for calculating chromatic polynomials can be found in Chapter 13.

The subgraph expansion

Theorem 2.2 tells us that the coefficients of chromatic polynomials are integers. Whitney [49] gave an interpretation for these values.

Theorem 2.5 *Given a graph $G = (V, E)$,*

$$P(G,q) = \sum_{A\subseteq E}(-1)^{|A|}q^{k(A)}, \tag{1}$$

where $k(A)$ is the number of connected components of the graph (V, A). Equivalently, $P(G,q) = \sum_{k=1}^{n} c_k\, q^k$, where $c_k = c_k^{even}(G) - c_k^{odd}(G)$ and $c_k^{even}(G)$ and $c_k^{odd}(G)$ are respectively the numbers of spanning subgraphs of G with k connected components and with an even and odd number of edges.

Proof Since $P(G,q)$ is the number of colourings of G using colours from $[q] = \{1,2,\ldots,q\}$, we have

$$P(G,q) = \sum_{\sigma:V\to[q]}\prod_{vw\in E}(1-\delta_{\sigma(v),\sigma(w)}),$$

where $\delta_{i,j}$ is the Kronecker delta function. We now expand the product as a sum over subsets of E and then reverse the order of summation to obtain

$$\begin{aligned} P(G,q) &= \sum_{\sigma:V\to[q]}\sum_{A\subseteq E}(-1)^{|A|}\prod_{vw\in A}\delta_{\sigma(v),\sigma(w)} \\ &= \sum_{A\subseteq E}(-1)^{|A|}\sum_{\sigma:V\to[q]}\prod_{vw\in A}\delta_{\sigma(v),\sigma(w)} \\ &= \sum_{A\subseteq E}(-1)^{|A|}\,q^{k(A)}, \end{aligned}$$

where the last equality follows since the product is 1 if σ is constant on each component of (V, A), and is 0 otherwise. ∎

Theorem 2.5 explains why $P(G,q)$ is a monic polynomial of degree n with zero constant term. It also implies that the coefficient of q^{n-1} is $-|E|$. The fact that the coefficients of $P(G,q)$ alternate in sign gives a non-trivial relationship between $c_k^{even}(G)$ and $c_k^{odd}(G)$.

3. Log concavity and other inequalities

One of the most celebrated conjectures on chromatic polynomials, that their coefficients are log concave, was made by Read [36] in 1968. This has recently been proved by Huh [23] using techniques from algebraic geometry.

Theorem 3.1 *If $P(G,q) = \sum_{k=1}^{n} c_k\, q^k$ is the chromatic polynomial of a graph G, then*

$$c_{k-1}c_{k+1} \le c_k^2$$

for all $2 \le k \le n-1$.

A related conjecture of Brenti [8], that $P(G,q)$ is log concave for integer values of $q > \chi(G)$, was verified for several families of graphs in [8] but was shown to be false in general by Seymour [38]. Fadnavis [17] has recently shown that the conjecture is true for integer values of q that are large compared to the maximum degree of G.

Theorem 3.2 *Let G be a graph with maximum degree Δ. Then*

$$P(G,q-1)P(G,q+1) \le P(G,q)^2\,,$$

for all integers $q \ge 9.767\Delta$.

On the other hand, Bartels and Welsh [1] showed that $P(G,q)$ is convex for integers $q > \chi(G)$.

Theorem 3.3 *Let G be a graph. Then for all integers $q > \chi(G)$,*

$$P(G,q-1) + P(G,q+1) \ge 2P(G,q).$$

Given a graph G on n vertices and a positive integer q, the probability that a random q-colouring of the vertices is a proper q-colouring is $P(G,q)/q^n$. It seems reasonable to expect this probability to increase with q, but Colin McDiarmid showed that this is not true in general (see [1]). Dong [10] showed that this probability increases as long as q is large compared to the order of G.

Theorem 3.4 *For all graphs G on n vertices and all real $q \ge n-1$,*

$$\frac{P(G,q)}{q^n} \le \frac{P(G,q+1)}{(q+1)^n}\,.$$

The special case of this result when $q = n-1$ was conjectured by Bartels and Welsh [1]. This conjecture became known as the *shameful conjecture*, because it is equivalent to the 'intuitively obvious' statement that the expected number of colours used in a proper colouring of G is minimized when G has no edges.

Fadnavis [18] has recently used methods from statistical mechanics to show that the probability that a random q-colouring of the vertices of G is a proper q-colouring increases with q if G is claw-free (has no induced $K_{1,3}$), or if q is large compared to the maximum degree of G.

Theorem 3.5 *Let G be a graph with maximum degree Δ and let q be a positive integer. If G is claw-free, or if $q > 400\Delta^{3/2}$, then*

$$\frac{P(G,q)}{q^n} \le \frac{P(G,q+1)}{(q+1)^n}\,.$$

4. Chromatic roots

A zero of a chromatic polynomial of a graph G is called a *chromatic root* of G. Birkhoff and Lewis's original paper on chromatic polynomials stimulated research into finding intervals on the real line that contain no chromatic roots of given families of graphs. Mathematical physicists on the other hand have been concerned with finding regions of the complex plane for which the Potts model partition function is non-zero, because of implications for the location of phase transitions. Regions free from complex chromatic roots are an important special case.

We can view both areas of research as determining real or complex analytic properties of chromatic polynomials. There has been much less work on determining their algebraic properties. In particular, little is known on the fundamental problem of determining whether or not a given algebraic integer is a chromatic root. We describe the main results and open problems in these areas.

Real roots and the sign of the chromatic polynomial

Results which determine intervals on the real line that are free from chromatic roots have invariably determined the sign of the chromatic polynomial in these intervals. The zero-free intervals for the family of all graphs are determined by the following results from [47], [51] and [24].

Theorem 4.1 *Let G be a graph with n vertices, c components and b blocks, with at least two vertices. Then:*

- *$P(G,q)$ is non-zero with sign $(-1)^n$ for $q \in (-\infty, 0)$, and has a zero of multiplicity c at $q = 0$*
- *$P(G,q)$ is non-zero with sign $(-1)^{n+c}$ for $q \in (0,1)$, and has a zero of multiplicity b at $q = 1$*
- *$P(G,q)$ is non-zero with sign $(-1)^{n+c+b}$ for $q \in (1, \frac{32}{27}]$.*

The first part of this theorem follows immediately from Theorem 2.2. The other parts can be deduced from Theorems 2.1 and 2.3 by using increasingly complicated inductive arguments. A conceptually simpler inductive proof for the last part is given in [29], using the added freedom given by the Potts model partition function. We return to this in Section 5.

The fact that Theorem 4.1 gives a complete list of zero-free intervals follows from a result of Thomassen [43], that the chromatic roots of all graphs form a dense subset of $(\frac{32}{27}, \infty)$. A recent related result of Goldberg and Jerrum [16] implies that it is NP-hard to determine the sign of $P(G,q)$ for all $q \in (\frac{32}{27}, \infty)$ other then $q = 2$. (Note that $P(G,2)$ is positive if G is bipartite and is zero otherwise.)

The families of graphs which demonstrate the aforementioned density and complexity results contain many 2-vertex cuts, and it seems likely that Theorem 4.1 can be extended beyond $q = \frac{32}{27}$ when G is 3-connected. In [24] we conjectured that $P(G,q)$ is non-zero with sign $(-1)^n$ for $q \in (1, \alpha]$ when G is 3-connected, where $\alpha \approx 1.781$ is the chromatic root of $K_{3,4}$ in $(1,2)$. Some evidence in favour of this conjecture is given in [11].

We can also consider other special families of graphs. Results of Birkhoff and Lewis [4] and Woodall [51], [52] give an analogue of Theorem 4.1 for plane triangulations which holds throughout the interval $(-\infty, \beta)$, where $\beta \approx 2.546$ is a chromatic root of the octahedron. Thomassen [44] showed that the zero-free interval $(1, \frac{32}{27}]$ can be extended to $(1, \gamma]$ for graphs with a Hamiltonian path, where $\gamma \approx 1.296$ is the real root of $(2 - t)^3 = 4(t - 1)^2$, and he conjectures [42] that it extends all the way to $(1, 2)$ for graphs with a Hamiltonian cycle. Dong and Koh [13] made the same conjecture for the larger family of *independently* 1-*tough graphs* – that is, graphs G with the property that $G - S$ has at most $|S|$ components for all non-empty independent sets of vertices S.

Results have also been obtained on 'upper zero-free intervals'. The idea is that upper bounds on the chromatic number (and hence on the largest integer chromatic root) of a family of graphs often correspond to similar bounds on the largest real chromatic root for the family. A classical example is the result of Birkhoff and Lewis [4] that the five-colour theorem for planar graphs G can be extended to show that $P(G, q) > 0$ for all $q \in [5, \infty)$, and their conjecture that a similar extension holds for the four-colour theorem. An analogous situation occurs for graphs G with maximum degree Δ. The greedy algorithm implies that $\chi(G) \leq \Delta + 1$, and a result of Dong and Koh [12] gives $P(G, q) > 0$ for all $q \in [5.664\Delta, \infty)$. It is an open problem to decide whether $P(G, q) > 0$ for all $q \in (\Delta, \infty)$, although Woodall [53] has shown that this conclusion holds under a stronger hypothesis.

Theorem 4.2 *If every minor of a graph G has maximum degree at most Δ, then $P(G, q) > 0$ for all $q \in (\Delta, \infty)$.*

Note that upper bounds on the chromatic number do not always give rise to upper zero-free intervals: bipartite graphs have chromatic number 2, but can have arbitrarily large real chromatic roots (see Woodall [51]).

Complex roots

Sokal [40] has shown that there are no zero-free regions in the complex plane for the family of all graphs.

Theorem 4.3 *The chromatic roots of all graphs form a dense subset of the complex plane.*

The family of graphs that he used to prove this result is surprisingly simple. The *generalized theta graph* $\Theta_{s,t}$ consists of two vertices joined by t internally disjoint paths of length s. Sokal showed that, if we vary the parameters s and t, then the chromatic roots of these graphs are dense everywhere in the complex plane, with the possible exception of the disc $|q - 1| < 1$. He then covered this disc with chromatic roots by using the families obtained by adding either one vertex, or two adjacent vertices joined to all to other vertices of $\Theta_{s,t}$, and then applying Theorem 2.4.

As for upper zero-free intervals in the real case, we may hope that an upper bound on the chromatic numbers of a family of graphs might correspond to a similar bound on the largest absolute value of chromatic roots for the same family. The generalized theta graphs demonstrate that this does not occur even for the family of planar graphs.

Sokal [39] showed, however, that it does occur for graphs with maximum degree Δ, by proving that all their chromatic roots lie in a disc $|q| < C\Delta$, for some universal constant C – this was a long-standing conjecture of Biggs, Damerell and Sands [2]. His proof used techniques from statistical mechanics, which will be outlined below. Sokal's value $C = 7.963$ was improved to $C = 6.907$ by Fernández and Procacci [21], by strengthening one of the key results from statistical mechanics. The aforementioned result of Dong and Koh that the *real* roots are bounded by 5.664Δ was obtained by making a further improvement in the real case.

The following is a gross oversimplification of Sokal's proof technique from [39]. (In particular, Sokal's proof holds for the more general Potts model partition function.) He constructed a weighted graph, called a *polymer gas*, whose vertices (or *polymers*) are the connected subgraphs of G. Each polymer H has weight $(-1)^{|E(H)|}q^{|V(H)|-1}$, and two polymers are joined by an edge whenever their union is connected. The chromatic polynomial of G can be expressed as the independent-set polynomial of this polymer gas: we have $P(G,q) = \sum_S w(S)$, where the sum extends over all independent sets of vertices S in the polymer gas, and $w(S)$ is the product of the weights of the polymers in S. He used the hypothesis on the maximum degree of G to show that the number of connected n-vertex subgraphs containing a fixed vertex of G is at most $c^n\Delta^n$, for some constant c. This gives rise to a bound on the weighted degree of each polymer in the polymer gas, and this in turn enabled him to bound the absolute values of the zeros of its independent-set polynomial by a linear function of $c\Delta$, using powerful results from statistical mechanics (see [20], [32]).

Algebraic properties

Since chromatic polynomials are monic polynomials with integer coefficients, every chromatic root is an algebraic integer. It is easy to see that if α is a chromatic root of a graph G, then so is every *conjugate* of α (that is, every root of the minimum polynomial of α over $\mathbb{Q}$). Combined with Theorem 4.1, this gives us the following necessary condition for an algebraic integer to be a chromatic root.

Theorem 4.4 *If an algebraic integer α is a chromatic root, then no conjugate of α can belong to* $(-\infty, \frac{32}{27}] \setminus \{0, 1\}$.

Tutte [46] showed that $|P(G, \tau + 1)| \leq \tau^{5-n}$ for all plane triangulations with n vertices, where $\tau = \frac{1}{2}(1 + \sqrt{5})$ is the golden ratio. This indicates that large plane triangulations may have a chromatic root close to $\tau + 1$. On the other hand $\tau + 1$ itself cannot be a chromatic root of any graph; this follows from Theorem 4.4, because $2 - \tau$ is a conjugate of $\tau + 1$ and $2 - \tau \in (0, 1)$.

It seems unlikely that the necessary condition for an algebraic integer to be a chromatic root given in Theorem 4.4 is also sufficient, but no counter-examples are known. The following two problems are particularly attractive (and frustrating) instances of this. The first was suggested by Feng Ming Dong at the Newton Institute Programme on Combinatorics and Statistical Mechanics (CSM) held in Cambridge in 2008; the second came out of conversations between Graham Farr and the author.

Problem *Can a purely imaginary complex number be a chromatic root? In particular, is $i = \sqrt{-1}$ a chromatic root?*

Problem *Are* 1, $e^{2\pi i/3}$, $e^{-2\pi i/3}$ *the only chromatic roots of unity?*

The numbers $e^{2\pi i/3}$, $e^{-2\pi i/3}$ are chromatic roots of C_7. The only root of unity that we know cannot be a chromatic root is -1.

We can also search for conditions that are sufficient for an algebraic integer to be a chromatic root. The following conjectures were formulated by a discussion group on the algebraic properties of chromatic roots that took place during the above mentioned CSM Programme (see [5] and [9]).

Conjecture A *If* α *is an algebraic integer, then* $\alpha + p$ *is a chromatic root for some* $p \in \mathbb{N}$.

Conjecture B *If* α *is a chromatic root, then* αp *is a chromatic root for all* $p \in \mathbb{N}$.

Conjecture **A** was verified for quadratic integers during the CSM programme (see [9]) and for cubic integers by Bohn [6]. Bohn [7] showed that Conjecture **B** is valid for chromatic roots of generalized theta graphs, and hence holds for a set of algebraic integers that is dense in the complex plane, by the remark after Theorem 4.3.

5. Related polynomials

In this section we define flow polynomials, characteristic polynomials of matroids, Potts model partition functions and Tutte polynomials, and outline their connections with the chromatic polynomial. We allow graphs to contain loops and multiple edges, since such edges have a non-trivial effect on these polynomials. This fact also means that we must keep any multiple edges or loops created by the contraction of an edge (in contrast to the situation for the chromatic polynomial). We refer readers who are unfamiliar with flows in graphs to Chapter 9, and readers unfamiliar with matroids to [34].

Flow polynomials

Tutte [45] defined the *flow polynomial* $F(G,q)$ of a graph G to be the number of distinct nowhere-zero Γ-flows in a fixed orientation of G, for a given abelian group Γ of order q, and showed that this number is the same for all orientations of G and all choices of Γ. Note, however, that the number of nowhere-zero q-flows of G (that is, $\mathbb{Z}$-flows which take the values $\{\pm 1, \pm 2, \cdots, \pm(q-1)\}$) is in general not equal to $F(G,t)$ (see Kochol [31]).

The flow polynomial is dual to the chromatic polynomial in the following sense. If G is a connected plane graph and G^* is its planar dual, then

$$F(G,q) = q^{-1}P(G^*,q). \tag{2}$$

This identity tells us that results and conjectures on chromatic polynomials of plane graphs can be restated in terms of flow polynomials of plane graphs. In particular, since generalized theta graphs are planar, their duals show that flow roots of planar graphs are dense everywhere in the complex plane, with the possible exception of the disc $|q-1| < 1$ (by the remark following Theorem 4.3).

For non-planar graphs there are still many similarities between the two polynomials: versions of Theorems 2.1, 2.2 and 2.3 hold for flow polynomials (see, for example, [25]); the log concavity of the coefficients of $F(G,t)$ was also established by Huh in [23]; Wakelin [48] obtained an analogue of Theorem 4.1 which determines the sign of $F(G,q)$ for all $q \in (-\infty, \frac{32}{27}]$.

On the other hand, the behaviour of the flow polynomial for all graphs seems to be closer to the behaviour of the chromatic polynomial for the special case of planar graphs. For example, we showed in [26] and [27] that the results of Woodall on the real chromatic roots of plane triangulations mentioned in Section 4 extend to flow polynomials of *all* cubic graphs. A major open problem for flow polynomials is to decide whether there is an upper zero-free interval for all graphs. Welsh conjectured over ten years ago (see [25]) that $(4,\infty)$ is an upper zero-free interval for flow polynomials, and hence that the Birkhoff–Lewis conjecture for chromatic polynomials of plane graphs holds for flow polynomials of all graphs, as long as $q \neq 4$. Royle [37] showed that this is false by constructing graphs with flow roots in $(4,5)$, but offered the modified conjecture that $[5,\infty)$ is an upper zero-free interval. The value $q = 5$ is of special significance since it corresponds to Tutte's 5-flow conjecture. The best-known general upper bound on flow roots, that $F(G,q) > 0$ whenever G is a bridgeless graph on n vertices and $q > 2\log_2 n$, is given in [25].

Characteristic polynomials of matroids

Similarities between the chromatic and flow polynomials for non-planar graphs can be explained by extending the duality identity (2) using matroids. We denote a *matroid* as $M = (E,r)$, where E is the set of elements and r is the rank function. The *characteristic polynomial* of M is

$$C(M,q) = \sum_{A\subseteq E} (-1)^{|A|} q^{r(E)-r(A)}. \tag{3}$$

The *cycle matroid* of a graph $G = (V,E)$ is the matroid $M_G = (E, r_G)$, where $r_G(A) = |V| - k(A)$ for all $A \subseteq E$. Comparing (1) and (3) we see that

$$P(G,q) = q^{k(G)}\, C(M_G,q), \tag{4}$$

where $k(G)$ is the number of connected components of G.

Given a matroid $M = (E,r)$, its *dual matroid* $M^* = (E,r^*)$ is defined by putting $r^*(A) = |A| - r(E) + r(E \setminus A)$ for all $A \subseteq E$. The *cocycle matroid* M_G^* of a graph G is the dual matroid of its cycle matroid. A similar argument to that used in the derivation of (1) (see, for example, [41]) implies that

$$F(G,q) = \sum_{A\subseteq E}(-1)^{|A|}\, q^{|E\setminus A|-|V|+k(E\setminus A)}, \tag{5}$$

and (3) now gives

$$F(G,q) = C(M_G^*, q). \tag{6}$$

Equations (4) and (6) imply that chromatic and flow polynomials of graphs are both examples of characteristic polynomials of matroids. Furthermore, when G is a connected plane graph and G^* is its planar dual, we have $M_G^* = M_{G^*}$. Hence (4) and (6) also imply (2).

It follows from the above discussion that any result on the characteristic polynomials of a family of matroids which includes both the cycle and cocycle matroids of graphs implies the same result for chromatic and flow polynomials as special cases. For example, Huh showed in [23] that the coefficients of characteristic polynomials of matroids that are representable over a field of characteristic 0 are log concave, and a version of Theorem 4.1 for characteristic polynomials of all matroids was given in [15] and [29].

A challenging open problem is to decide whether some version of the results of Section 4 on complex roots of graphs of bounded degree holds for flow polynomials – or, better still, for characteristic polynomials of a family of matroids which includes the cycle matroids of graphs of maximum degree Δ and their matroid duals. A conjecture and some preliminary results are given in [30] and [28].

The Potts model partition function

Let $G = (V,E)$ be a graph, and let $\mathbf{w} : E \to \mathbb{C}$ be a weight function on E. The *Potts model partition function*, or *multivariate Tutte polynomial*, of the weighted graph $(G,\mathbf{w})$ is

$$Z(G,q,\mathbf{w}) = \sum_{A\subseteq E} q^{k(A)} w_A, \tag{7}$$

where $w_A = \prod_{e\in A} w_e$ and w_e denotes the weight of the edge e. When all edges have the same weight w, we write $Z(G,q,w)$ for the partition function. Comparing (7) with (1), we see that $P(G,q) = Z(G,q,-1)$. A similar argument to that used in the derivation of (1) gives us the following expression for the partition function when q is a positive integer:

$$Z(G,q,\mathbf{w}) = \sum_{\sigma:V\to[q]}\ \prod_{e=vw\in E} \left(1 + w_e \delta_{\sigma(v),\sigma(w)}\right). \tag{8}$$

This expression for $Z(G,q,\mathbf{w})$ is similar to the form given in the original paper of Potts [35]; the equivalence of the two expressions is due to Fortuin and Kasteleyn [22].

Equation (8) expresses $Z(G,q,\mathbf{w})$ as a weighted sum over all q-colourings of the vertices of G, where a particular q-colouring σ has weight $\prod(1+w_e)$ and the product ranges over all edges e that are not properly coloured by σ. This colouring interpretation enables us to extend Theorem 2.1 and a special case of Theorem 2.3 to the partition function, using similar proofs.

Theorem 5.1 *Let $(G,\mathbf{w})$ be a weighted graph and let e be an edge of $G=(V,E)$. Then*

$$Z(G,q,\mathbf{w}) = Z(G-e,q,\mathbf{w}|_{E-e}) + w_e\, Z(G/e,q,\mathbf{w}|_{E-e}).$$

Theorem 5.2 *Let $(G,\mathbf{w})$ be a weighted graph and let $G_1=(V_1,E_1)$ and $G_2=(V_2,E_2)$ be edge-disjoint subgraphs of G with $|V_1\cap V_2|=r$ for some $r\in\{0,1\}$ and with $G_1\cup G_2=G$. Then*

$$Z(G,q,\mathbf{w}) = q^{-r} Z(G_1,q,\mathbf{w}|_{E_1})\, Z(G_2,q,\mathbf{w}|_{E_2})\,.$$

(The hypothesis of Theorem 5.2 that the subgraphs are edge-disjoint is needed in the case when $r=1$ and their common vertex is incident to one or more loops.)

An advantage of introducing edge-weights is that they give rise to new reduction operations. The *parallel reduction operation* replaces two parallel edges by a single edge with the same endpoints. The *series reduction operation* replaces a vertex v of degree 2 and its incident edges by a single edge joining the neighbours of v.

Theorem 5.3 *Let $(G,\mathbf{w})$ be a weighted graph and let e_1,e_2 be edges of G.*

If H is obtained by applying the parallel reduction operation to e_1,e_2, then

$$Z(G,q,\mathbf{w}) = Z(H,q,\mathbf{w}|_{E\setminus\{e_1,e_2\}}, w_{e_1}\|w_{e_2})\,,$$

where $w_{e_1}\|w_{e_2} = w_{e_1}+w_{e_2}+w_{e_1}w_{e_2}$ is the weight of the edge that replaces e_1 and e_2.

If H is obtained by applying the series reduction operation to e_1,e_2, then

$$Z(G,q,\mathbf{w}) = (q+w_{e_1}+w_{e_2}) Z(H,q,\mathbf{w}|_{E\setminus\{e_1,e_2\}}, w_{e_1}\bowtie_q w_{e_2})\,,$$

where $w_{e_1}\bowtie_q w_{e_2} = w_{e_1}w_{e_2}(q+w_{e_1}+w_{e_2})^{-1}$ is the weight of the edge that replaces e_1 and e_2.

We refer the reader to Sokal [41] for a proof of these results. He also gives several impressive applications of how the freedom to change the edge-weights can give simple proofs for $Z(G,q,\mathbf{w})$, and hence give results for $P(G,q)$ that were either new or were obtained by much more difficult proofs. One such example is given below.

The basic idea behind the original proof [24] of the last part of Theorem 4.1 is as follows. We proceed by contradiction. Fix $q\in(1,\frac{32}{27}]$ and suppose that

$(-1)^{n+c+b}P(G,q) \leq 0$ for some graph $G = (V,E)$. Choose G with $|E|$ as small as possible. Theorem 2.3 implies that G is 2-connected and Theorem 2.1 then tells us that either $G - e$ or G/e is not 2-connected for all $e \in E$. This gives structural information about G which can be refined by several more applications of Theorems 2.1 and 2.3 to show that G belongs to a well-defined family of 2-connected graphs called 'generalized triangles'. The proof is completed by an inductive proof that every generalized triangle G satisfies $(-1)^n P(G,q) > 0$.

The proof given in [29] is much simpler. We choose $q \in (1, \frac{32}{27}]$ and use a stronger inductive hypothesis that $(-1)^{n+c+b}Z(G,q,\mathbf{w}) > 0$ for all $\mathbf{w} : E \to I_q$, for some interval $I_q \subseteq (-\infty, -q/2)$ with $-1 \in I_q$. The interval I_q is chosen to be closed under the parallel and series transformations $(w_1, w_2) \mapsto w_1 \| w_2$ and $(w_1, w_2) \mapsto w_1 \bowtie_q w_2$. We use Theorem 5.2 to reduce to the case when G is 2-connected and loopless, and Theorem 5.3 to reduce to the case when G has minimum degree 3 and no multiple edges. We can then use a result of Oxley [33] to deduce that G has an edge e for which $G - e$ and G/e are both 2-connected. The proof is completed by applying Theorem 5.1 using this edge e. This proof also explains the significance of the number $\frac{32}{27}$; it is the largest value of q for which there exists an interval $I_q \subseteq (-\infty, 0)$ such that $-1 \in I_q$ and I_q is closed under the parallel and series transformations.

The proof in [29] is matroidal and gives the same result for the matroid version of the partition function. It therefore provides a simple proof that $(1, \frac{32}{27}]$ is a zero-free interval for characteristic polynomials of matroids, and hence also for flow polynomials of graphs.

Tutte polynomials

The *Tutte polynomial* of a graph $G = (V,E)$ was defined by Tutte [45] as

$$T(G,x,y) = \sum_{A \subseteq E} (x-1)^{k(A)-k(G)} (y-1)^{|A|-|V|+k(A)}. \tag{9}$$

Comparing (9) with (1) and (5) tells us that

$$P(G,q) = (-1)^{|V|-k(G)} \, q^{k(G)} \, T(1-q, 0)$$

and

$$F(G,q) = (-1)^{|E|+|V|+k(G)} \, T(0, 1-q),$$

so the chromatic and flow polynomials are both specializations of the Tutte polynomial. Similarly, a comparison of (9) and (7) implies that

$$T(G,x,y) = (x-1)^{-k(G)} (y-1)^{-|V|} \, Z(G, (x-1)(y-1), y-1)$$

and

$$Z(G,q,w) = q^{k(G)} \, w^{|V|-k(G)} \, T(G, 1 + q/w, 1 + w),$$

so the Tutte polynomial and the partition function with constant edge-weights can be interchanged by a simple change of variables. Note, however, that the variables in these polynomials play different roles. In the partition function, q acts as a global variable representing the number of possible colours or 'states' for each vertex and w represents the weight of an edge. The variables x, y in the Tutte polynomial act as a pair of 'dual' global variables. This duality is best illustrated when we extend the definition of the Tutte polynomial to a matroid $M = (E, r)$: we obtain

$$T(M,x,y) = \sum_{A \subseteq E} (x-1)^{r(E)-r(A)} (y-1)^{r^*(E)-r^*(E \setminus A)}, \qquad (10)$$

where r^* is the rank function for the dual matroid M^*. This immediately gives $T(M,x,y) = T(M^*,y,x)$ and implies that $T(G,x,y) = T(G^*,y,x)$ for any connected planar graph G with planar dual G^*.

The evaluation of the Tutte polynomial of a graph G at the points in the xy-plane (the so-called *Tutte plane*) gives an attractive picture for viewing all of the above polynomials (and many more): the x-axis corresponds to the chromatic polynomial, the y-axis corresponds to the flow polynomial and the hyperbolas $(x-1)(y-1) = q$ give the Potts model partition function for fixed q and all edge-weights equal to $y-1$ (see Fig. 1).

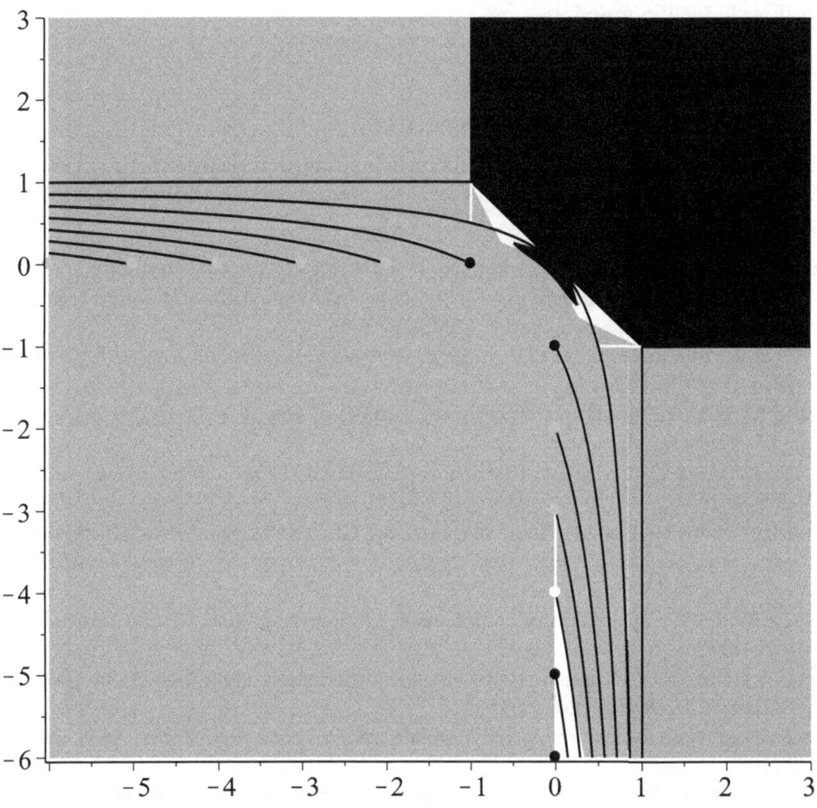

Fig. 1. Determining the sign of $T(x, y)$ in the Tutte plane

The results of [29] can be applied to give a region in the Tutte plane for which the sign of $T(G, x, y)$ is determined by a function of the numbers of vertices, components and blocks of G. Goldberg and Jerrum [16] have recently obtained a complementary result which identifies large regions in the Tutte plane for which it is NP-hard to determine the sign of $T(G, x, y)$. They also point out that in addition to the tractable region implied by [29], we can use (8) to determine the sign of $T(G, x, y)$ at certain points on the hyperbolas $q = (x - 1)(y - 1)$ when q is a positive integer.

Computing the sign of the Tutte polynomial is polynomial at black points and is NP-hard at grey points. The complexity at white points is not known: the white point $(0, -4)$ and black point $(0, -5)$ correspond respectively to Tutte's 5-flow conjecture and Seymour's 6-flow theorem; the lower white region is related to the problem of determining whether there is an upper zero-free interval for flow polynomials; the curved boundary between the black and grey regions is given by the hyperbola $(x - 1)(y - 1) = \frac{32}{27}$; [29, Conjectures 10.1, 10.2] would imply that the black region can be extended to fill most of the two upper white regions.

Acknowledgement I would like to thank Douglas Woodall for stimulating my interest in chromatic polynomials, and Alan Sokal for both his research collaboration and his efforts to increase interaction between the statistical mechanics and graph theory communities. I am also indebted to Leslie Goldberg, Mark Jerrum and Iain Moffat for their helpful comments.

References

We refer the reader to [14], [25] and [37] for more information on chromatic, flow, and characteristic polynomials, to [41] for Potts model partition functions, and to [19] for Tutte polynomials.

1. J. E. Bartels and D. Welsh, The Markov chain of colourings, *Proc. 4th International IPCO Conference on Integer Programming and Combinatorial Optimization*, Lecture Notes in Computer Science **920**, Springer-Verlag (1995), 373–387.
2. N. L. Biggs, R. M. Damerell and D. A. Sands, Recursive families of graphs, *J. Combin. Theory (B)* **12** (1972), 123–131.
3. G. D. Birkhoff, A determinant formula for the number of ways of colouring a map, *Ann. of Math.* **14** (1912), 42–46.
4. G. D. Birkhoff and D. C. Lewis, Chromatic polynomials, *Trans. Amer. Math. Soc.* **60** (1946), 355–451.
5. A. Bohn, Chromatic roots as algebraic integers, *DMTCS Proceedings, North America, July 2012*, *http://www.dmtcs.org/dmtcs-ojs/index.php/proceedings/article/view/dmAR0148/3962*.
6. A. Bohn, Chromatic polynomials of complements of bipartite graphs, *Graphs Combin.* **30** (2014), 287–301.
7. A. Bohn, A dense set of chromatic roots which is closed under multiplication by positive integers, *Discrete Math.* **321** (2014), 45–52.
8. F. Brenti, Expansions of chromatic polynomials and log-concavity, *Trans. Amer. Math. Soc.* **332** (1992), 729–756.
9. P. J. Cameron and K. Morgan, Algebraic properties of chromatic roots, submitted.

10. F. M. Dong, Proof of a chromatic polynomial conjecture, *J. Combin. Theory (B)* **78** (2000), 35–44.
11. F. M. Dong and B. Jackson, A zero-free interval for chromatic polynomials of nearly 3-connected plane graphs, *SIAM J. Discrete Math.* **25** (2011), 1103–1118.
12. F. M. Dong and K. M. Koh, Bounds for the real zeros of chromatic polynomials, *Combin. Probab. Comput.* **17** (2008), 749–759.
13. F. M. Dong and K. M. Koh, On zero-free intervals in $(1,2)$ of chromatic polynomials of some families of graphs, *SIAM J. Discrete Math.* **24** (2010), 370–378.
14. F. M. Dong, K. M. Koh and K. L. Teo, *Chromatic Polynomials and Chromaticity of Graphs*, World Scientific Publishing Co., 2005.
15. H. Edwards, R. Hierons and B. Jackson, The zero-free intervals for characteristic polynomials of matroids, *Combin. Probab. Comput.* **7** (1998), 153–165.
16. L. A. Goldberg and M. Jerrum, The complexity of computing the sign of the Tutte polynomial (and consequent #P-hardness of approximation), *http://arxiv.org/abs/1104.0707v4*.
17. S. Fadnavis, On Brenti's conjecture about the log-concavity of the chromatic polynomial, *Europ. J. Combin.* **33** (2012), 1842–1846.
18. S. Fadnavis, A generalization of the birthday problem and the chromatic polynomial, *http://arxiv.org/abs/1105.0698v2*.
19. G. E. Farr, Tutte–Whitney polynomials: some history and generalizations, *Combinatorics, Complexity, and Chance*, Oxford Lecture Ser. Math. Appl. **34**, Oxford University Press (2007), 28–52.
20. R. Fernández and A. Procacci, Cluster expansion for abstract polymer models: new bounds from an old approach, *Comm. Math. Phys.* **274** (2007), 123–140.
21. R. Fernández and A. Procacci, Regions without complex zeros for chromatic polynomials on graphs with bounded degree, *Combin. Probab. Comput.* **17** (2008), 225–238.
22. C. M. Fortuin and P. W. Kasteleyn, On the random-cluster model. I. Introduction and relation to other models, *Physica* **57** (1972), 536–564.
23. J. Huh, Milnor numbers of projective hypersurfaces and the chromatic polynomial of graphs, *J. Amer. Math. Soc.* **25** (2012), 907–927.
24. B. Jackson, A zero-free interval for chromatic polynomials of graphs, *Combin. Probab. Comput.* **2** (1993), 325–336.
25. B. Jackson, Zeros of chromatic and flow polynomials of graphs, *J. Geometry* **76** (2003), 95–109.
26. B. Jackson, A zero-free interval for flow polynomials of cubic graphs, *J. Combin. Theory (B)* **97** (2007), 127–143.
27. B. Jackson, Zero-free intervals for flow polynomials of near cubic graphs, *Combin. Probab. Comput.* **16** (2007), 85–108.
28. B. Jackson, Counting 2-connected deletion minors of binary matroids, *Discrete Math.* **313** (2013), 1262–1266.
29. B. Jackson and A. D. Sokal, Zero-free regions for multivariate Tutte polynomials (alias Potts-model partition functions) of graphs and matroids, *J. Combin. Theory (B)* **99** (2009), 869–903.
30. B. Jackson and A. D. Sokal, Maxmaxflow and counting subgraphs, *Electron. J. Combin.* **17** (2010), No. 1, Research Paper 99, 46 pp.
31. M. Kochol, Polynomials associated with nowhere zero flows, *J. Combin. Theory (B)* **84** (2002), 260–269.
32. R. Kotecký and D. Preiss, Cluster expansion for abstract polymer models, *Comm. Math. Phys.* **103** (1986), 491–498.
33. J. G. Oxley, On minor-minimally-connected matroids, *Discrete Math.* **51** (1984), 63–72.

34. J. G. Oxley, *Matroid Theory* (2nd edn.), Oxford University Press, 2011.
35. R. B. Potts, Some generalized order–disorder transformations, *Proc. Cambridge Philos. Soc.* **48** (1952), 106–109.
36. R. C. Read, An introduction to chromatic polynomials, *J. Combin. Theory* **4** (1968), 52–71.
37. G. F. Royle, Recent results on chromatic and flow roots of graphs and matroids, *Surveys in Combinatorics 2009* (eds. S. Huczynska, J. D. Mitchell and C. M. Roney-Dougal), Cambridge University Press (2009), 289–327.
38. P. D. Seymour, Two chromatic polynomial conjectures, *J. Combin. Theory (B)* **70** (1997), 184–196.
39. A. D. Sokal, Bounds on the complex zeros of (di)chromatic polynomials and Potts model partition functions, *Combin. Probab. Comput.* **10** (2001), 41–77.
40. A. D. Sokal, Chromatic roots are dense in the whole complex plane, *Combin. Probab. Comput.* **13** (2004), 221–261.
41. A. D. Sokal, The multivariate Tutte polynomial (alias Potts model) for graphs and matroids, *Surveys in Combinatorics, 2005* (ed. B. S. Webb), Cambridge University Press (2005), 173–226.
42. C. Thomassen, On the number of hamiltonian cycles in bipartite graphs, *Combin. Probab. Comput.* **5** (1996), 437–442.
43. C. Thomassen, The zero-free intervals for chromatic polynomials of graphs, *Combin. Probab. Comput.* **6** (1997), 497–506.
44. C. Thomassen, Chromatic roots and hamiltonian paths, *J. Combin. Theory (B)* **80** (2000), 218–224.
45. W. T. Tutte, A contribution to the theory of chromatic polynomials, *Canad. J. Math.* **6** (1954), 80–91.
46. W. T. Tutte, On chromatic polynomials and the golden ratio, *J. Combin. Theory (B)* **9** (1970), 289–296.
47. W. T. Tutte, *Chromials*, Springer Lecture Notes in Math. **411** (1974), 243–266.
48. C. D. Wakelin, *Chromatic Polynomials*, Ph.D. thesis, University of Nottingham, 1994.
49. H. Whitney, A logical expansion in mathematics, *Bull. Amer. Math. Soc.* **38** (1932), 572–579.
50. H. Whitney, The coloring of graphs, *Ann. Math.* **33** (1932), 688–718.
51. D. R. Woodall, Zeros of chromatic polynomials, *Combinatorial Surveys* (ed. P. J. Cameron), Academic Press (1977), 199–223.
52. D. R. Woodall, A zero-free interval for chromatic polynomials, *Discrete Math.* **101** (1992), 333–341.
53. D. R. Woodall, The largest real zero of the chromatic polynomial, *Discrete Math.* **172** (1997), 141–153.

4

Hadwiger's conjecture

KEN-ICHI KAWARABAYASHI

Hadwiger's conjecture states that any graph that does not have the complete graph K_k as a minor is $(k-1)$-colourable. It is well known that the case $k = 5$ is equivalent to the four-colour theorem. In 1993 Robertson, Seymour and Thomas proved that the case $k = 6$ is also equivalent to the four-colour theorem. For $k \geq 7$, the conjecture is still open. Our main focus in this chapter is to present recent results related to minimal counter-examples to Hadwiger's conjecture and some variations. We also consider algorithmic aspects of the conjecture and some of its variants, including the list-colouring version (which is false), the odd case of the conjecture, Hajós's conjecture and totally odd subdivisions, and Hadwiger's conjecture for some special classes of graphs.

1. Introduction

This chapter is motivated by Hadwiger's conjecture from 1943, which suggests a far-reaching generalization of the four-colour theorem. It is among the most challenging open problems in all of graph theory.

Hadwiger's conjecture (strong version) For all k, every k-colourable graph contains the complete graph K_k as a minor.

For $k \leq 3$, Hadwiger's conjecture is easy to prove, and for $k = 4$, it was proved independently by Hadwiger himself [31] and Dirac [21]. For $k = 5$, however, it becomes extremely difficult. In 1937 Wagner [74] proved that this case is equivalent to the four-colour theorem, so, given that result in [2], [3], [56], it follows that the case $k = 5$ also holds. In the deepest theorem in this area so far, Robertson, Seymour and Thomas [62] proved in 1993 that a minimal counter-example to the case $k = 6$ must have a vertex whose removal leaves a planar graph, so this case too follows from the four-colour theorem. For $k \geq 7$, the conjecture is still open. For $k = 7$, Kawarabayashi and Toft [46] proved that every 7-chromatic graph has K_7 or $K_{4,4}$ as a minor, and recently Kawarabayashi [38] proved that every 7-chromatic graph has K_7 or $K_{3,5}$ as a minor.

Because of the current state of affairs, there is considerable interest in a weaker version of Hadwiger's conjecture.

Hadwiger's conjecture (weak version) There exists a constant c such that, for $k \geq 1$, every ck-chromatic graph contains the complete graph K_k as a minor.

The best result of this type currently known is that there exists a constant c for which every $ck\sqrt{\log k}$-chromatic graph has K_k as a minor. This follows from the results in [66], [67], [50] and [49] (see Section 2). It was proved thirty years ago, and no one has been able to improve on the superlinear order $k\sqrt{\log k}$ of the bound on the chromatic number. Hence, it would be of great interest to establish that a linear function of the chromatic number k is sufficient to force K_k as a minor.

In this chapter, we survey recent progress concerning the two conjectures, paying special attention to the following.

Properties of minimal counter-examples, including the question of finiteness (Section 3).
Algorithmic aspects (Sections 4 and 5).
Other variants of Hadwiger's conjecture, including the odd Hadwiger conjecture (Section 6), immersion and Hajós's conjecture (Sections 6 and 8).
Hadwiger's conjecture for special families of graphs, including claw-free graphs and graphs with independence number 2 (Section 7).
The list-colouring version of Hadwiger's conjecture (Section 8).

The first step in attacking Hadwiger's conjecture is to find conditions on the minimum degree in a graph that force the existence of a large complete minor, an approach that has been taken since the 1960s. We survey these classical results in Section 2; for earlier results, see [73].

2. Complete graph minors: early results

Wagner and Mader studied extremal problems concerning the maximum number of edges in K_k-minor-free graphs. Wagner [75] proved that a sufficiently large chromatic number (depending only on k) guarantees K_k as a minor, and Mader [52] showed that a sufficiently large average degree does the same. Later, Kostochka [50], [49] and Thomason [66] independently proved that the correct order of the average degree that forces K_k as a minor is $\Theta(k\sqrt{\log k})$.

Theorem 2.1 *There exist positive constants c_1 and c_2 such that every graph with average degree at least $c_1 k\sqrt{\ln k}$ contains K_k as a minor, while for $k \geq 3$ there exist graphs with average degree at least $c_2 k\sqrt{\ln k}$ that do not.*

More recently, Thomason [67] found the asymptotically sharp value of this 'extremal' function: it is $(\alpha + o(1))k\sqrt{\ln k}$, where $\alpha = 0.319...$ is an explicit constant determined by a particular equation.

These results show that if the minimum degree of a graph G is $o(k\sqrt{\ln k})$, then G does not necessarily contain a K_k-minor. This cannot be improved upon even if we add a connectivity condition; only connectivity of order $\Theta(k\sqrt{\ln k})$ forces the presence of a K_k-minor. However, as Thomason pointed out, the extremal graphs are close to disjoint unions of suitable dense random graphs. Such graphs cannot have too many vertices, a point we return to in Section 4.

Precise extremal numbers of edges for K_k-minors are known only for $k \leq 9$. For $k \leq 7$, these were found by Dirac [23] and Mader [52], while the result for $k = 8$ is due to Jørgensen [34], and the case $k = 9$ was recently settled by Song and Thomas [64].

There is an analogue to Theorem 2.1 for topological minors (recall that a graph H is a topological minor of a graph G if a subgraph of G is homeomorphically reducible to H) [11], [48]. Here the extremal function is of order $\Theta(k^2)$.

Theorem 2.2 *There exist positive constants c_1 and c_2 such that every graph with average degree at least $c_1 k^2$ contains K_k as a topological minor, while for $k \geq 3$ there exist graphs with average degree at least $c_2 k^2$ that do not contain K_k as a topological minor.*

3. Contraction-critical graphs

In attacking Hadwiger's conjecture, it is natural to consider minimal counterexamples, and in this section, we give several properties of such graphs.

Let G be a graph that is k-chromatic, is minimal with respect to the minor-relation in the class of all k-chromatic graphs, and does not contain K_k as a minor. Such a graph is called *k-contraction-critical* and cannot be a complete graph. These graphs were introduced by Dirac [22], [24], and one of his results was that the minimum degree in such a graph is at least k. The following result is shown in [36]; its proof is given below (following Theorem 3.5). Kawarabayashi and Yu [47] have recently shown that the fraction $\frac{2}{27}$ can be replaced by $\frac{1}{9}$.

Theorem 3.1 *Every k-contraction-critical graph is $\frac{2}{27}k$-connected.*

In order to prove Theorem 3.1, we need to define 'highly linked' subgraphs, which will be discussed in the next subsection.

The next result is a variation of an old theorem of Mader [53]; its non-algorithmic counterpart appeared in [9]. For completeness, we include its full proof, needed to establish the algorithmic part. The following concept will be useful in the proof. Given a graph G, a pair (A, B) of induced subgraphs is called a *separation* of G if $G = A \cup B$. Its *order* is $|V(A \cap B)|$. It is also useful to have a name for a $2k$-connected graph H that has at least $5k|V(H)|$ edges; we call it *extra-2k-connected.*

Theorem 3.2 *Let G be a graph with n vertices and m edges, and let k be such that $n \geq \frac{5}{2}k$ and $m \geq \frac{25}{4}kn - \frac{25}{2}k^2$. Then $n \geq 10k + 2$ and G contains an extra-2k-connected subgraph H. Furthermore, H can be found in time $O(n^3)$.*

Proof Clearly, if G is a graph of order n with at least $\frac{25}{4}kn - \frac{25}{2}k^2$ edges, then $\frac{25}{4}kn - \frac{25}{2}k^2 \leq \binom{n}{2}$. Hence, either

$$n \leq \tfrac{25}{4}k + \tfrac{1}{2} - \tfrac{1}{4}\sqrt{(25k+2)^2 - 400k^2} < \tfrac{5}{2}k$$

or

$$n \geq \tfrac{25}{4}k + \tfrac{1}{2} + \tfrac{1}{4}\sqrt{(25k+2)^2 - 400k^2} > 10k + 1.$$

Since $n \geq \frac{5}{2}k$, it follows that $n \geq 10k + 2$.

Suppose now that the theorem is false. Let G be a graph with n vertices and m edges, and let k satisfy the conditions of the theorem. Assume, moreover, that n is the least order of such a graph with the property that no subgraph is extra-$2k$-connected.

We now show that the minimum degree of G satisfies $\delta(G) > \frac{25}{4}k$. Suppose that G has a vertex v with degree at most $\frac{25}{4}k$, let $G' = G - v$, and let n' and m' be the numbers of vertices and edges in G'. By our choice of G, G' has no extra-$2k$-connected subgraphs. It follows from the fact that $n \geq 10k+2$ that $n' = n - 1 \geq \frac{5}{2}k$. Also, $m' \geq m - \frac{25}{4}k \geq \frac{25}{4}kn' - \frac{25}{2}k^2$. Since $n' < n$, this contradicts the minimality of n and thus establishes that $\delta(G) > \frac{25}{4}k$.

Since $m \geq \frac{25}{4}kn - \frac{25}{2}k^2$ and $n \geq 10k+2$, we have $m \geq 5kn$. But no subgraph of G is extra-$2k$-connected, so G cannot be $2k$-connected. Since $n > 2k$, this implies that G has a separation (A_1, A_2) for which both $A_1 - A_2$ and $A_2 - A_1$ are non-empty and $|A_1 \cap A_2| \leq 2k - 1$. By the condition on δ, $|A_i| \geq \frac{25}{4}k + 1$. For $i = 1, 2$, let G_i be a subgraph of G with vertex-set A_i for which $G = G_1 \cup G_2$ and $E(G_1 \cap G_2) = \varnothing$. Further, let n_i be the number of vertices and m_i the number of edges in G_i, and suppose that $m_i < \frac{25}{4}kn_i - \frac{25}{2}k^2$ for $i = 1, 2$. Then

$$\frac{25}{4}kn - \frac{25}{2}k^2 \leq m = m_1 + m_2 < \frac{25}{4}k(n + |A_1 \cap A_2|) - 25k^2 \leq \frac{25}{4}kn - \frac{25}{2}k^2,$$

a contradiction. Hence, we may assume that $m_1 \geq \frac{25}{4}kn_1 - \frac{25}{2}k^2$. Since $n > n_1 \geq \frac{25}{4}k+1$ and G_1 has no extra-$2k$-connected subgraph H, this contradicts the minimality of n, and the existence of H is established.

The above proof yields an $O(n^3)$ algorithm for finding H. First, we remove vertices of degree at most $\frac{25}{4}k$ as long as the minimum degree remains less than $\frac{25}{4}k$; this can be done in linear time – first we list those vertices whose degree is at most $\frac{25}{4}k$, and then we remove them one by one. At each step, degrees change only at the neighbours of the removed vertex, and it takes constant time to lower their degrees by 1 and to move those whose degree drops to $\lfloor\frac{25}{4}k\rfloor$ into the list of vertices to be removed.

Next, we check whether G is $2k$-connected. By an algorithm of Henzinger, Rao and Gabow [32], this can be done in $O(n^2)$ time. At the same time we find a separation (A_1, A_2) of order less than $2k$ if one exists. As shown above, one of the corresponding subgraphs G_1 or G_2 can be used to continue the process. The recursion brings another factor of n to the time complexity, so the subgraph H can be found in $O(n^3)$ time. ■

A graph G is said to be *k-linked* if it has at least $2k$ vertices, and for each pair of k-tuples $(v_1, v_2, \ldots, v_k)$ and $(w_1, w_2, \ldots, w_k)$ of $2k$ distinct vertices, there exist k disjoint paths $P_1, P_2, \ldots, P_k$ with P_i a v_i–w_i path.

An important tool in what follows is the following theorem of Thomas and Wollan [65].

Theorem 3.3 *Every extra-2k-connected graph G is k-linked.*

This theorem implies that every $10k$-connected graph is k-linked. We note that Bollobás and Thomason [11] had proved earlier that every $22k$-connected graph is k-linked, and that Kawarabayashi, Kostochka and Yu [39] improved this to $12k$-connected.

Combining Theorems 3.2 and 3.3, we have the following result.

Corollary 3.4 *Let G be a graph with n vertices and m edges, and let k be such that* $n \geq \frac{5}{2}k$ *and* $m \geq \frac{25}{4}kn - \frac{25}{2}k^2$. *Then G contains a k-linked subgraph, and such a subgraph can be found in* $O(n^3)$ *time.*

Our next goal is to prove Theorem 3.1. To this end, it is useful to have a further definition and a lemma. Let (A, B) be a separation in a graph G and let $\mathbf{P} = \{S_1, S_2, \ldots, S_r\}$ be a partition of $S = V(A) \cap V(B)$. Then we say that A is *contractible* over $\{S_1, S_2, \ldots, S_r\}$ if there is a family $\mathbf{H} = \{H_1, H_2, \ldots, H_r\}$ of disjoint connected subgraphs of A such that $S_i \subseteq V(H_i)$ for each i. The following result is the lemma that we need.

Theorem 3.5 *Let G be a graph with n vertices and minimum degree d. Let* (A, B) *be a separation in G of minimum order, and let* $S = V(A) \cap V(B)$ *and* $s = |S|$. *If* $s \leq \lfloor\frac{2}{27}d\rfloor$ *and* $\mathbf{P}$ *is any partition of S, then A is contractible over* $\mathbf{P}$, *and the corresponding family* $\mathbf{H}$ *of subgraphs can be found in* $O(n^3)$ *time.*

Proof We may assume that the subgraph $A_1 = A - S$ is connected. Note that each vertex in A_1 has degree at least $d - s \geq \frac{25}{27}d$. Therefore, A_1 has at least $\frac{25}{27}d$ vertices and at least $\frac{25}{54}d|V(A_1)|$ edges. Let $k = \lfloor\frac{2}{27}d\rfloor$. By Theorem 3.2, A_1 contains an extra-$2k$-connected subgraph H_1, which (by Corollary 3.4) is k-linked. By the minimality of S and Menger's theorem, there are s disjoint paths $P_1, P_2, \ldots, P_s$ from S to $V(H_1)$, with

P_i joining $v_i \in S$ and $w_i \in H_i$. Since H_1 is $2k$-connected, it contains s independent edges $e_1, e_2, \ldots, e_s$ with $e_i = w_i w'_i$ for $i = 1, 2, \ldots, s$.

We now assume that the vertices $v_1, v_2, \ldots, v_s$ of S are listed so that the first $|S_1|$ are in S_1, the next $|S_2|$ are in S_2, and so on. Since H_1 is k-linked and $s \leq k$, there are $s - 1$ disjoint paths $P'_1, P'_2, \ldots, P'_{s-1}$ in H_1 such that P'_i joins w_i and w'_{i+1}. The subgraph of A consisting of the s paths P_i, the s edges e_i, and the $s-1$ paths P'_i is a tree whose contraction gives rise to identification of all vertices of S_1 into a single vertex. Similarly, there are such trees for $S_2, S_3, \ldots, S_r$, and they are all disjoint. Hence A is contractible for $\{S_1, S_2, \ldots, S_r\}$.

As far as the algorithm is concerned, by Theorem 3.2, we can find H_1 in $O(n^3)$ time. To find paths $P_1, P_1, \ldots, P_s$, we can apply an augmenting path algorithm $s(< 2k)$ times, which can be done in $O(n^2)$ time. Finally, the disjoint connecting paths $P'_1, P'_2, \ldots, P'_{s-1}$ can be found by the graph minors algorithm of Robertson and Seymour [57] in $O(n^3)$ time. This completes the proof of the lemma. ■

We now show that this result implies Theorem 3.1.

Proof of Theorem 3.1 Take a minimal cutset S of G. Then $|S| < \lceil \frac{2}{27}k \rceil$. Let A_1 be a component of $G - S$ and let $A_2 = G - A_1 - S$. Then both $A_1 \cup S$ and $A_2 \cup S$ have chromatic number at most $k - 1$. Let S_1 be a maximum independent set in $\langle S \rangle$, let S_2 be a maximum independent set in $\langle S - S_1 \rangle$, and so on, thereby obtaining a partition $\mathbf{P} = \{S_1, S_2, \cdots, S_r\}$ of S.

By Theorem 3.5, we can contract both A_1 and A_2 over $\mathbf{P}$ so that each of the resulting graphs on S is complete on S. Now let G_1 be the resulting graph together with A_2 and let G_2 be the resulting graph together with A_1. Then by the minimality of G, $\chi(G_1)$ and $\chi(G_2) \leq k - 1$. But clearly we can combine the colourings of G_1 and G_2 to give a colouring of the whole graph G with at most $k - 1$ colours. This contradiction completes the proof. ■

Returning to the algorithmic portion of the result, we note that we can use Robertson and Seymour's algorithm from [57] to find the disjoint connecting paths $P'_1, P'_2, \ldots, P'_{s-1}$. However, it turns out that when we apply Theorem 3.5 in the algorithm for Theorem 4.1 in the next section, we may assume that the k-linked subgraph H_1 is of bounded size; that is, if $N(k)$ is the constant from Theorem 3.6 below and $|V(H_1)| \geq N(k)$, then we know that H_1 contains K_k as a minor. Since this is one of the possible outcomes of the algorithm, we may assume that $|V(H_1)| < N(k)$, and hence the paths $P'_1, P'_2, \ldots, P'_{s-1}$ can be found in constant time.

The following result is proved in [9].

Theorem 3.6 *For each integer l, there exists a constant $N(l)$ such that every $2l$-connected graph with at least $N(l)$ vertices and minimum degree at least $\frac{31}{2}l$ contains K_l as a minor.*

Actually, the main result proved in [9] is stronger: For any integers s, k and l, there is a constant $N(s, k, l)$ such that every $(3l + 2)$-connected graph of order at least

$N(s, k, l)$ and minimum degree at least d contains either $K_{sk,l}$ as a topological minor or a minor isomorphic to s copies of $K_{k,l}$. Since G could be the complete bipartite graph $K_{d,m}$ with m arbitrarily large, it is necessary to include the possibility of having a subdivision of $K_{sk,l}$.

The next result follows from Theorems 3.1 and 3.6.

Theorem 3.7 *If G is a $\frac{31}{2}k$-contraction-critical graph of order n, then there is a function $f(k)$ for which $n \le f(k)$.*

This theorem tells us that there are only finitely many minimal counter-examples to the weak conjecture.

4. Algorithmic aspects of the weak conjecture

Although the weaker version of Hadwiger's conjecture is still open, we can show that from an algorithmic point of view, this problem can be 'decided' in polynomial time (with complexity $O(n^3)$).

Theorem 4.1 *For each $k \in \mathbb{N}$, there is a polynomial-time algorithm for deciding that*

(1) *a given graph G is $27k$-colorable*

(2) *G contains a K_k-minor*

or

(3) *there is a function $f(k)$ for which G contains a minor H of order at most $f(k)$ that does not contain a K_k-minor and is not $27k$-colourable.*

Note that if (3) holds, then H is a counter-example to Hadwiger's conjecture – in fact, to the weaker conjecture that every $27k$-chromatic graph has K_k as a minor. If (1) holds, then the graph can be coloured with $27k$ colours, while if (c) holds, then we can exhibit the minor H.

To prove this theorem, we need Theorem 3.6, but we do not need any result in the graph minor series. However, the proof of Theorem 3.6 given in [9] does depend on Robertson and Seymour's deep results (see [9], [59], [58]).

Before we give an algorithm for Theorem 4.1, we need some additional definitions. A *tree decomposition* of a graph G is a pair (T, Y), where T is a tree and Y is a family $\{Y_t : t \in V(T)\}$ of sets of vertices of G for which the following two properties hold.

(W1) $\bigcup_{t \in V(T)} Y_t = V(G)$, and each edge of G has both ends in some Y_t.

(W2) If $t, t', t'' \in V(T)$ and if t' lies on the t–t'' path in T, then $Y_t \cap Y_{t''} \subseteq Y_{t'}$.

The *width* of a tree decomposition (T, Y) is $\max_{t \in V(T)}(|Y_t| - 1)$.

We are now ready to describe our algorithm.

Algorithm for Theorem 4.1
Input: A graph G
Output: As described in Theorem 4.1.

Description:

Step 1 If G has a vertex of degree less than $27k$, we delete it. We continue this procedure until there are no vertices of degree less than $27k$. (This can be done in linear time.) Let G' be the resulting graph. Proceed to Step 2.

Step 2 Test whether the tree-width of G' is small or not, say smaller than some value $g(k)$. For simplicity in later steps, we assume that $g(k) \geq N(k)$, where $N(k)$ is as in Theorem 3.6: this can be done in linear time by an algorithm of Bodlaender [8]. If the tree-width is at least $g(k)$, go to Step 3. Otherwise, use the linear-time algorithm of Arnborg and Proskurowski [4] to colour G'. If G' can be coloured with $27k$ colours, then colour $G-G'$ greedily, and output the colouring of G. If G' is not $27k$-colourable, check whether it contains a K_k-minor; again, this can be done using the algorithm of Arnborg and Proskurowski (or that of Robertson and Seymour [57]). If G' contains K_k as a minor, output that G contains K_k as a minor. If G' does not contain K_k as a minor, then proceed as argued below. (Up to this point, the whole process can be done in linear time.)

Let (T, Y) be the tree-decomposition found above. The dynamic programming approach of Arnborg and Proskurowski assumes that T is a rooted tree whose edges are directed away from the root. For $tt' \in E(T)$ (where the root is closer to t than to t'), define $S(t,t') = Y_t \cap Y_{t'}$ and let $G'(t,t')$ be the induced subgraph $G'[\bigcup Y_s]$, where the union runs over those vertices s in the component of $T - tt'$ that does not contain the root. The algorithm of Arnborg and Proskurowski starts at all leaves of T and for each edge tt' in T, computes the set $C(t,t')$ of all $27k$-colourings of $S(t,t')$ that can be extended to the whole of $G'(t,t')$. If T has a vertex t of very large degree, then for two neighbours t' and t'', $S(t,t') = S(t,t'')$ and $C(t,t') = C(t,t'')$. Then $G'(t,t')$ can be deleted, and we still have a graph of bounded tree-width without a K_k minor and with no $27k$-colourings.

If all vertices of T have bounded degree, then T has a long path and there are two edges t_1t_1' and t_2t_2' on this path (with the second further from the root) for which $|S(t,t')| = |S(t,t'')|$ and $C(t,t') = C(t,t'')$. In the same way as argued in [9], we may assume that there are $|S(t,t')|$ disjoint paths joining $S(t,t')$ and $S(t,t'')$. By contracting these paths and replacing $G'(t,t')$ by $G'(t,t'')$, we obtain a minor of G' that is still of bounded tree-width, without a K_k minor and without $27k$-colourings. Repeating this, we eventually end up with the desired minor of G' of bounded size. This bound is actually a doubly exponential value expressed in terms of k.

Step 3 Test whether G' is $2k$-connected. By the assumption in Step 2, if G' is $2k$-connected, then $|G'| \geq g(k) \geq N(k)$; it then follows from Theorem 3.6 that G' contains K_k as a minor. So output that G has K_k as a minor. If G' is not $2k$-connected, go to Step 4.

Step 4 If G' is not $2k$-connected, find a minimal separation (A, B); this can be done in polynomial time by standard methods. The best known algorithm for this is due to Henzinger, Rao and Gabow [32] and needs time $O(n^2)$.

Let $S = V(A) \cap V(B)$. Then $|S| < 2k$. As in the proof of Theorem 3.5, let A'_1 be a component of $G' - S$ and $A'_2 = G' - A'_1 - S$. Let S_1 be a maximal independent set in $G'[S]$. Let S_i be a maximal independent set in $G'(S) - \bigcup_{j=1}^{i-1} S_l$, for $i = 2, 3, \ldots$. These maximal independent sets can be found greedily (or by other means) in constant time (since $|S| < 2k$). Next, we collapse each non-empty set S_i to a single vertex s_i; then the resulting graph on $S' = \{s_1, s_2, \ldots, s_r\}$ is complete. Let A_1, A_2 be the corresponding graphs obtained from A'_1, A'_2 by adding the clique S' and the corresponding edges between A'_i and S'.

Finally, we test A_1 and A_2 (recursively), starting from Step 1. If both A_1 and A_2 are $27k$-colourable, they give rise to a $27k$-colouring of their union since S' is a clique. Since the sets S_i that were collapsed into single vertices in S' are independent in G', this colouring gives rise to a colouring of G' which can then be extended to all of G.

If one of the graphs, say A_1, contains a K_k-minor, then we obtain a K_k-minor in G' (after contracting A_2 onto S') by using Theorem 3.5 with $d = 27k$. Similarly, if outcome (c) occurs for A_1, we get the same outcome for G' by using a contraction of A_2 onto S'.

This algorithm stops when the order of the current graph is small or when the current graph is $2k$-connected with minimum degree at least $27k$. The correctness follows from Theorem 3.5.

Now we estimate the time-complexity of the algorithm. All steps except the application of Theorem 3.5 can be done in time proportional to n^2. Another factor of n arises because of the recursion in Step 4. Finally, Theorem 3.5 is applied only when we backtrack from the recursion. If we apply it on the graph A_1 of order n_1, then we spend $O(n_1^3)$ time, but we never use it again on the same vertices. Therefore applications of Theorem 3.5 use only $O(n^3)$ time in total. This completes the proof of the correctness and of the stated time-complexity of the algorithm.

5. Algorithmic aspects of the strong conjecture

Robertson and Seymour found the result stated in the next theorem; it gives rise to a polynomial-time algorithm for k-colouring K_k-minor-free graphs if Hadwiger's conjecture is true. They did not publish a proof, but one can be found in Kawarabayashi and Reed [44].

Theorem 5.1 *For each $k \in \mathbb{N}$, there is a polynomial-time algorithm for deciding that*

(1) *a given graph G is k-colourable.*
(2) *G contains a K_{k+1}-minor.*

or

(3) *G contains a minor H of order at most $N(k)$ that is K_{k+1}-minor-free and is not k-colourable.*

Robertson has pointed out that in order to prove this theorem, he and Seymour used the following result. It is not only of independent interest, but is arguably the strongest known result in this direction.

Theorem 5.2 *For $k \geq 4$ and any graph G with no K_{k+1}-minor, one of the following statements holds:*

(1) *There exists an integer $f(k)$ such that G has tree-width at most $f(k)$.*
(2) *G has a vertex of degree at most k.*
(3) *G has a vertex v of degree $k + 1$ with three mutually non-adjacent neighbours.*
(4) *G has a separation (A, B) of order at most k and with $A \neq G$ such that A can be contracted to a clique K on $A \cap B$ with each vertex of $A \cap B$ in a different vertex of K.*
(5) *G has a set S of at most $k - 4$ vertices for which $G - S$ is planar.*

Note that (2), (3) and (4) cannot happen in a minimal counter-example to Hadwiger's conjecture, and that a graph satisfying (5) is also not a counter-example, by the four-colour theorem. The proof is complicated and uses the graph minor structure theory heavily, but does not use the well-quasi-ordering result (see [59] or [58]). The proof in [44] is much simpler, but still depends on the graph minor structure theory. Seymour has also pointed out that property (3) can be eliminated at the expense of a considerably longer proof.

6. The odd conjecture

Recently, the concept of an 'odd minor' has received considerable attention from researchers. We say that *G* has an *odd complete minor* of size *k* if it contains *k* disjoint trees with any two having adjacent vertices and which can be properly 2-coloured in such a way that the edges between trees have both vertices the same colour (hence the vertices of all trivial trees must receive the same colour). We say that *G* has an *odd K_k minor* if it has an odd complete minor of size *k*. It is easy to see that any graph with an odd K_k minor contains K_k as a minor.

Gerards and Seymour (see [33]) conjectured an analogue of Hadwiger's conjecture for odd complete graphs.

The odd conjecture For each $k \in \mathbb{N}$, every graph with no odd K_{k+1} minor is *k*-colorable.

This conjecture is trivially true for $k = 1$ and 2. In fact, for $k = 2$, it says simply that every graph with no odd cycles is 2-colourable. The case $k = 3$ was proved by Catlin [13]. Recently, Bertrand Guenin announced a solution of the $k = 4$ case, which would imply the four-colour theorem since a graph that has an odd K_5-minor certainly has a K_5-minor. The conjecture remains open for $k \geq 5$.

The following theorem, first proved by Geelen *et al.* [29], is clearly a generalization of Theorem 2.1.

Theorem 6.1 *Any graph with no odd K_k-minor is $O(k\sqrt{\log k})$-colourable.*

Our proof [37] uses induction on the number of vertices. To this end, we prove the stronger statement below. We let c_1 be as in Theorem 2.1; thus, any graph with minimum degree at least $c_1 k\sqrt{\log k}$ contains a K_k-minor.

Theorem 6.2 *If G is a graph of order n and if $k \leq \frac{1}{4}n$, then either G has an odd K_k-minor or every precolouring of an induced subgraph of order $4k$ can be extended to a $\lfloor 64c_1 k\sqrt{\log k} + 4k\rfloor$-colouring of G.*

Proof Suppose that the result does not hold. We assume that $4k < n$ since otherwise there is nothing to prove. Let G be a counter-example of minimum order, and let Z be a set of $4k$ vertices of G. If $G - Z$ has a vertex v of degree at most $64c_1 k\sqrt{\log k} + 4k - 1$, then, by the minimality of G, $G - v$ has a $\lfloor 64c_1 k\sqrt{\log k} + 4k\rfloor$-colouring. This colouring can clearly be extended to a colouring of G, which is a contradiction. Hence, we assume that $\delta(G - Z) \geq 64c_1 k\sqrt{\log k} + 4k$.

We prove the theorem by establishing five claims.

(1) *G does not have a separation (A, B) of order $2k$ or less for which both $A - B - Z$ and $B - A - Z$ are non-empty.*

Suppose that there is such a separation (A, B) of order at most $2k$, and let $S = A \cap B$. Since $|S| \leq 2k$ and $|Z| \leq 4k$, it follows that either $|S \cup (A \cap Z)| \leq 4k$ or $|S \cup (B \cap Z)| \leq 4k$, say the former. By the minimality of G, there is a colouring of $G[B \cup Z]$ with Z precoloured. Note that the subgraph of G induced on $B \cup Z$ is smaller than G.

Again, by the minimality of G, we can colour $G[A]$ with $Z' = S \cup (A \cap Z)$ precoloured, where the precolouring of vertices in S comes from the colouring of $G[B \cup Z]$. Recall that $|Z'| \leq 4k$ and that the subgraph induced on A is smaller than G. Hence, there is a colouring of $G[A]$ with Z' precoloured. A combination of the obtained colourings of $G[B]$ and $G[A]$ yields a desired colouring of G, giving the required contradiction and proving the claim.

We choose a spanning bipartite graph H of $G-Z$ whose minimum degree is as large as possible. An old theorem of Erdős says that every graph of minimum degree at least $2l$ has a spanning bipartite graph with minimum degree at least l. Since $G-Z$ has minimum degree at least $64c_1 k\sqrt{\log k}$, H has minimum degree at least $32c_1 k\sqrt{\log k}$. The following result now follows by Mader's theorem [53].

(2) *H has an $8c_1 k\sqrt{\log k}$-connected subgraph L.*

We say that P is a *parity-breaking path* for L if P is disjoint from L except for its endpoints, and L together with P has an odd cycle. We need a generalization of

Mader's well-known S-path theorem for paths of odd length, and for this we use the following result of Geelen *et al.* [29].

Theorem 6.3 *For any set S of vertices in a graph G and for any integer k, either there are k disjoint paths of odd length and both ends in S, or there is a set X of at most $2k-2$ vertices for which $G-X$ has no such set of paths.*

Since L is $8c_1k\sqrt{\log k}$-connected, each of the partite sets of L has at least $8c_1k\sqrt{\log k}$ vertices.

(3) *There are at least k disjoint parity-breaking paths for L.*

Take one partite set S of L and apply Theorem 6.3 to G and S. If there are at least k disjoint odd S paths in G, then we can clearly find k disjoint parity-breaking paths for L, since otherwise, by Theorem 6.3, there is a separation (A,B) of order at most $2k-2$ such that A contains S and $A-B$ has no such paths. Since $A\cap B$ has at most $2k-2$ vertices and L is $8c_1k\sqrt{\log k}$-connected, it follows that A contains L. By Claim **1**, $B-A-Z=\varnothing$. It follows from Theorem 6.3 that $A-B$ can be expressed as $L'\cup W_1\cup W_2\cup\cdots\cup W_m$, for some integer m with L' a bipartite graph containing L and each W_i a block. In addition, each W_i contains a cut-vertex v_i to L' so that W_i-v_i consists of vertices in Z, since otherwise there would be a separation (A',B') of order at most $2k$ and both $A'-B'-Z$ and $B'-A'-Z$ non-empty, contradicting Claim **1**. Since $|Z|\le 4k$, we have $m\le 4k$. Because $B-A-Z=\varnothing$, we need at most $4k$ colours for Z, at most $4k$ colours for the vertices $v_1,v_2,\dots,v_m$, two colours for L', and at most $2k-2$ colours for $A\cap B$ (since $A\cap B$ has order at most $2k-2$). These colours are enough to colour G, and we need only at most $10k\le 8c_1k\sqrt{\log k}+4k$ colours for the colouring of G. This completes the proof of Claim **3**.

The next claim is proved in [35], although the proof there is slightly different; its proof is inspired by proofs of Robertson and Seymour [57] and Bollobás and Thomason [11].

(4) *Let G be a graph, let $S=\{s_1,s_2,\dots,s_k\}$ be a set of k vertices, and suppose that G has a K_{2k}-minor and is k-connected. Then G has k vertex-disjoint non-empty connected subgraphs $C_1,C_2,\dots,C_k$ such that, for each $1\le i\le k$, the subgraph C_i contains s_i and is adjacent to all of the other subgraphs $C_1,C_2,\dots,C_{i-1},C_{i+1},\dots,C_k$.*

Let $P_1,P_2,\dots,P_k$ be the parity-breaking paths for L with ends p_i and p'_i, for $1\le i\le k$. By our choice of c, L has a K_{4k}-minor. So claims **3** and **4** imply the following result.

(5) *G has a K_{2k}-minor M consisting of $2k$ disjoint trees $N_1,N_2,\dots,N_{2k}$, each of which is contained in the bipartite graph L, and in which there are k disjoint parity-breaking paths $P_1,P_2,\dots,P_k$ in which P_i has one end p_i in N_{2i-1} and the other end p'_i in N_{2i}, but is otherwise disjoint from M.*

The induced subgraph consisting of trees $N_1,N_2,\dots,N_{2k}$ in L is clearly bipartite. We now use Claim **5** to construct an odd-K_k-minor. Suppose that we can 2-colour

$N_j \cup P_j \cup N_{j+k}$ so that the trees $N_i \cup P_i \cup N_{i+k}$ consist of an odd K_j-minor for $1 \le i \le j$. We claim that we can 2-colour $N_{j+1} \cup P_{j+1} \cup N_{j+1+k}$ so that the trees $N_i \cup P_i \cup N_{i+k}$ consist of an odd K_{j+1}-minor for $1 \le i \le j+1$. This is certainly possible since the 2-colouring of N_{j+1+k} can be interchanged with the 2-colouring of N_{j+1} in the bipartition of L. So, for any $i \le j$, either the edge between N_i and N_{j+1} or the edge between N_{i+k} and N_{j+1+k} has both vertices the same colour. In this way, we can construct the desired odd K_k-minor, and this completes the proof of the theorem. ■

7. Independent sets and Hadwiger's conjecture

There is another way to think about Hadwiger's conjecture, based on the relation between the independence number and the chromatic number of a graph. Recall that in a coloring of a graph, each color-class is a set of independent vertices, so that if G has independence number α, then $\chi(G) \ge n/\alpha$. Hadwiger's conjecture asserts that G has $K_{\chi(G)}$ as a minor, and so, if true, implies the following conjecture, which was first explicitly stated and put into context by Woodall [77]. Throughout this section, n denotes the order of the graph G, and α denotes its independence number.

Conjecture A Every graph G has a $K_{\lceil n/\alpha \rceil}$ minor.

This conjecture seems weaker than Hadwiger's conjecture; however, for $\alpha = 2$, the two are equivalent (see [54]). In 1982 Duchet and Meyniel [25] proved the following remarkable result.

Theorem 7.1 *Every graph G has a $K_{\lceil n/(2\alpha-1) \rceil}$ minor.*

This result was improved first by Plummer, Toft and Kawarabayashi (see [41]) and subsequently in [6], [27] and [48]. Currently, the strongest result of this type is due to Balogh and Kostochka [5], who showed that $\lceil n/(2\alpha-1) \rceil$ can be replaced by $\lceil n/(2-c)\alpha \rceil$, for some constant $c > \frac{1}{19.2} = 0.052\,0833....$

We note that it seems quite hard to improve the constant $\frac{1}{3}$ in Theorem 7.1 for $\alpha = 2$ (a question asked independently by Seymour and Mader). As far as we know, no progress has yet been made on this question, although some related results appear in [16].

Analogous questions for the odd Hadwiger conjecture have also been studied; one is the following, due to Komlós and Szemerédi [48].

Theorem 7.2 *Every graph G has an odd $K_{\lceil n/(2\alpha-1) \rceil}$ minor.*

A slightly stronger result is known when $\alpha(G) = 3$. This is an analogue of a result in [48].

Theorem 7.3 *Every graph G with $\alpha = 3$ has an odd $K_{\lceil n/4 \rceil}$ minor.*

A graph is called *claw-free* if it does not contain $K_{1,3}$ as an induced subgraph; clearly every graph G with $\alpha = 2$ has this property. It is therefore likely to be very hard to show Hadwiger's conjecture for claw-free graphs. One result on minors and claw-free graphs was proved by Fradkin [28].

Theorem 7.4 *Every claw-free graph with no K_{2k}-minor is $3k$-colourable.*

8. Other variants of the conjecture

In this section we consider some other concepts for which Hadwiger-like conjectures are quite natural; in particular, we consider subdivisions and immersions of graphs, and then move on to list-colourings.

Subdivisions

Although Kuratowski's well-known theorem can be stated in terms of both subdivisions and minors, similarities between the two concepts do not seem to carry much beyond that. For example, graphs without K_5-minors were already characterized by Wagner [74] in 1937, but graphs without K_5-subdivisions are still mysterious and a characterization is perhaps out of reach. Another difference between the two is that, as shown by Robertson and Seymour [60] in proving Wagner's famous conjecture, graphs are well-quasi-ordered by the minor relation, but not by the subdivision relation.

In connection with the four-colour problem, Hajós conjectured in the 1940s that, for all k, every graph without a subdivision of K_k is $(k-1)$-colourable. For $k \leq 3$ the result is obvious, while for $k = 4$ it was proved independently by Hadwiger [31] and Dirac [21]. However, it was disproved by Catlin [13] for $k \geq 7$. In fact, Erdős and Fajtlowicz [26] proved that the conjecture is false for almost all graphs (see also Bollobás and Catlin [10]). Recently, Thomassen [71] has given many families of graphs that are counter-examples to Hajós's conjecture. In fact, it is known that the order of magnitude for the chromatic number of graphs without a K_k-subdivision is $\Theta(k^2)$ (see [10], [11] and [45]). Thus, we have this modified version (which some might consider more of a question) of Hajós's original conjecture.

Conjecture B Every graph without a subdivision of K_4 is 3-colourable, and every graph without a subdivision of K_5 is 4-colourable.

The story does not end there however, as a restricted form of subdivision has led to some interesting questions. In a *totally odd subdivision* of a graph G, each edge of G has been replaced by a path of odd length. Totally odd subdivisions of K_4 play an important role in both graph theory and combinatorial optimization. In graph theory, their importance arises in graph colouring. Toft [72] conjectured that any graph with no totally odd K_4-subdivision is 3-colourable, a generalization of the $k = 4$ case of Hajós's original conjecture. This was finally settled by Zang [78] and Thomassen [70] independently. (This conjecture also relates to the so-called 't-perfect graphs' and 'strongly t-perfect graphs' from the theory of perfect graphs; for more details, see the book by Schrijver [63].)

A general result by Thomassen [68] asserts that there is a function $f : \mathbb{N} \to \mathbb{N}$ with the property that every $f(k)$-chromatic graph has a subdivision of K_k in which each edge corresponds to a path of prescribed parity in the subdivision. As pointed out by Thomassen [69], totally odd K_k-subdivisions are also interesting in this connection. Unfortunately, the bound on the chromatic number given by Thomassen [68] is far from best possible, even in the totally odd case. Here we give an essentially sharp bound.

Theorem 8.1 *For any set S of at most k^2 vertices in a graph G, either G has a totally odd K_k-subdivision or every precolouring of the induced subgraph $\langle S \rangle$ can be extended to a $\frac{79}{4}k^2$-colouring of G.*

A result in [11] and [45] asserts that there exists a constant c for which every graph with minimum degree ck^2 contains a subdivision of K_k. This implies that the chromatic number of any graph without a subdivision of K_k is at most $ck^2 - 1$; furthermore, this result is sharp (up to constants). Hence, our result is best possible in a sense, and generalizes the result in [11] and [45].

Because the proof of Theorem 8.1 is quite similar to that of Theorem 6.2, we do not include it here. We note, however, that the proof implies that the same conclusion holds if 'totally odd' is replaced by 'prescribed parity'.

Immersions

An *immersion* of one graph H into another graph G is a mapping η that is one-to-one from $V(H)$ into $V(G)$ and maps the edges in H to edge-disjoint paths in G in such a way that if $e = vw$ is an edge of H, then $\eta(e)$ is a path joining $\eta(v)$ and $\eta(w)$ in G.

Previous investigations on immersions were mainly conducted from an algorithmic standpoint (see [12]), but it would be interesting to consider structural issues, since the notions of an immersion and a minor seem to be quite similar, and a structural approach concerning graph minors has been extremely successful. In fact, Robertson and Seymour [61] extended their proof of Wagner's conjecture [60] to prove that graphs are well-quasi-ordered by the immersion relation, and this proves a conjecture of C. Nash-Williams. The proof is based on the entire series of graph minors papers, and thus we may expect that a structural approach to immersions is very difficult – perhaps as hard as structural results on minors.

Immersion containment is not quite the same as either subdivision or minor containment. Clearly, a subdivision of H in G implies the existence of an immersion of H, but the converse is apparently not true. Also, a subdivision of H implies a minor of H, but the converse is not true. So, the existence of a subdivision is stronger than either a minor or an immersion. On the other hand, a minor and an immersion do not seem to be comparable. It would therefore be interesting to investigate relations between the chromatic number of a graph G and the largest order of a complete graph immersed in G. In fact, Abu-Khzam and Langston [1] conjectured the following.

Conjecture C Every graph with no immersions of K_k is $(k-1)$-colourable.

This conjecture, like Hadwiger's conjecture and Hajós's conjecture, is true for $k \leq 4$. In fact, this follows from the truth of Hajós's conjecture for $k \leq 4$. Abu-Khzam and Langston [1] proved the weaker result that K_5 with one edge removed has an immersion in every 5-chromatic graph. They also pointed out that structural investigations have been extensively carried out for graphs without a K_4-immersion (see [12]).

Independently, DeVos *et al.* [19] and Lescure and Meyniel [51] have proved cases $k = 5, 6$ and 7 of the conjecture. Recently, DeVos *et al.* [18] proved a very general result that says that Conjecture **C** is true if k is replaced by $200k$ – in fact, they proved that the average degree $200k$ is enough to force a K_k-immersion. This actually differentiates between Hadwiger's conjecture and Conjecture **C** since the average degree $\Theta(k\sqrt{\log k})$ forces K_k as a minor (see Kostochka [50] and Thomason [66]), and this bound is sharp. On the other hand, only an average degree of $200k$ is needed to force K_k as an immersion.

List-colourings

The last variation that we consider in this section involves a different type of colouring rather than a different type of graph. Motivated by the weak Hadwiger conjecture, Kawarabayashi and Mohar [40] proved the following result. (A constant $f(k)$ is *computable* means that $f(k)$ can be expressed as a specific value, depending on k (see [9]).)

Theorem 8.2 *For each positive integer k, there exists a computable constant $f(k)$ such that every graph G with no K_k-minor admits a vertex-partition $\{V_1, V_2, \ldots, V_{\lceil 31k/2 \rceil}\}$ for which every component of each $G[V_i]$ has at most $f(k)$ vertices.*

When $f(k) = 1$, we get a colouring of G. Thus, the theorem gives a lessening of the requirements of a colouring, and hence a relaxation of Hadwiger's conjecture, and is the first result in this direction. In fact, since it is still not known whether there exists a constant c for which every ck-chromatic graph has a K_k-minor, this may be viewed as a first step in attacking that question.

Actually, the main result in [9] is the list-colouring version of Theorem 8.2 (see Chapter 6). When any of a set of ck colours is allowed at each vertex (called *ck-choosability*), we obtain this list-colouring version of Hadwiger's conjecture.

Conjecture D *There is a constant c for which every graph without a K_k-minor is ck-choosable.*

The list-colouring version of Hadwiger's conjecture itself is false; for example, there are planar graphs, which of course have no K_5 minors, that are not 4-choosable. Furthermore, Conjecture **D** is not true for $c = 1$; an example of Barat, Joret and Wood [7] shows that there is a K_{3k+2}-minor-free graph that is not $4k$-choosable.

Fractional colourings

Fractional colourings are closely related to the independent sets in a graph; in fact, the definition involves them, so we let $\mathcal{I}$ denote the family of independent sets in a graph G. For k a non-negative rational number, a *fractional k-colouring* of graph G is a map $q : \mathcal{I} \to \mathbb{Q}^+$ (the set of non-negative rationals) such that

- for each vertex v, $\sum q(S) = 1$, where the sum is over all $S \in \mathcal{I}$ that contain v
- $\sum q(S) \leq k$, where the sum is over all $S \in \mathcal{I}$.

Thus, when k is an integer, a graph G is k-colourable if and only if it has a fractional k-colouring q that has its values in $(0, 1)$. Consequently, if true, Hadwiger's conjecture would imply that every graph with no K_{t+1}-minor has a fractional t-colouring. In this direction, Reed and Seymour [55] proved the following result.

Theorem 8.3 *For each integer $t \geq 1$, every graph with no K_t-minor has a fractional $2t$-colouring.*

Motivated by the odd conjecture and Theorem 8.3, Kawarabayashi and Reed [43] deduced the same conclusion when only odd minors are excluded.

Theorem 8.4 *For each integer $t \geq 1$, every graph with no odd K_t-minor has a fractional $2t$-colouring.*

9. Open problems

We conclude the chapter with some open questions related to Hadwiger's conjecture. The first unsettled case of the conjecture is the following.

Conjecture E Every 7-chromatic graph has a K_7 minor.

Mader [52] proved that a minimal counter-example to Hadwiger's conjecture is 7-connected. Hence, Conjecture **E** is implied by the next one.

Conjecture F Every 7-connected graph with no K_7 minor has two vertices whose removal leaves a planar graph.

Motivated by the fact that a complete characterization for graphs without K_6 minors is not known, Jørgensen [34] made the following beautiful conjecture.

Conjecture G Every 6-connected graph G with no K_6 minors contains a vertex whose removal leaves a planar graph.

This conjecture was solved for large graphs in [41], but their method cannot give a complete solution. The next conjecture actually seems much harder than Conjecture **G**, and even out of reach for large graphs.

Conjecture H Every 7-connected graph has K_6 as a minor.

In fact, for a complete characterization of graphs with no K_6 minors, we may need to find a characterization of 5-connected graphs with no K_6 minor. The following weakening of Conjecture **H** would be a step in the right direction.

Conjecture I Every sufficiently large 7-connected graph G with no K_7-minors has two vertices v, w such that $G - v - w$ is planar.

As we have seen in this chapter, there are many weaker versions of Hadwiger's conjecture, all of which currently appear to be intractable. Let us state again the weak version of Hadwiger's conjecture.

Hadwiger's conjecture (weak version) There exists a constant c such that for $k \geq 1$, every ck-chromatic graph contains the complete graph K_k as a minor.

A stronger conjecture has also been made (see [40]).

Conjecture J There is a constant c for which every graph without a K_k-minor is ck-choosable.

Finally, we mention two other related conjectures. The first was made independently by Chartrand, Geller and Hedetniemi [14] and Woodall [77], while the second is due to Bojan Mohar.

Conjecture K For integers $1 \leq h \leq k$, every graph G without K_{k+1} or a $K_{\lfloor (k+2)/2 \rfloor, \lceil (k+2)/2 \rceil}$-minor, has a partition of G into $k-h+1$ parts, each part inducing a graph without K_{h+1} or a $K_{\lfloor (h+2)/2 \rfloor, \lceil (h+2)/2 \rceil}$-minor.

Conjecture L There are only finitely many 3-connected k-colour-critical graphs without K_k as a minor.

References

1. F. N. Abu-Khzam and M. A. Langston, Graph coloring and the immersion order, to appear.
2. K. Appel and W. Haken, Every planar map is four colorable, Part I. Discharging, *Illinois J. Math.* **21** (1977), 429–490.
3. K. Appel, W. Haken and J. Koch, Every planar map is four colorable, Part II. Reducibility, *Illinois J. Math.* **21** (1977), 491–567.
4. S. Arnborg and A. Proskurowski, Linear time algorithms for NP-hard problems restricted to partial k-trees, *Discrete Appl. Math.* **23** (1989), 11–24.
5. J. Balogh and A. Kostochka, Large minors in graphs with given independent number, *Discrete Math.* **311** (2011), 2204–2215.
6. J. Balogh, J. Lenz and H. Wu, Complete minors, independent sets, and chordal graphs, *Discuss. Math. Graph Theory*, to appear.

7. J. Barat, G. Joret and D. Wood, Disproof of the list Hadwiger conjecture, *Electron. J. Combin.* **18** (2011), P232.
8. H. L. Bodlaender, A linear-time algorithm for finding tree-decompositions of small treewidth, *SIAM J. Comput.* **25** (1996), 1305–1317.
9. T. Böhme, K. Kawarabayashi, J. Maharry and B. Mohar, Linear connectivity forces large complete bipartite minors, *J. Combin. Theory (B)* **99** (2009), 557–582.
10. B. Bollobás and P. A. Catlin, Topological cliques of random graphs, *J. Combin. Theory (B)* **30** (1981), 224–227.
11. B. Bollobás and A. Thomason, Highly linked graphs, *Combinatorica* **16** (1996), 313–320.
12. H. D. Booth, R. Govindan, M. A. Langston and S. Ramachandramurthis, Sequential and parallel algorithms for K_4-immersion testing, *J. Algorithms* **30** (1999), 344–378.
13. P. A. Catlin, A bound on the chromatic number of a graph, *Discrete Math.* **22** (1978), 81–83.
14. G. Chartrand, D. Geller and S. Hedetniemi, Graphs with forbidden subgraphs, *J. Combin. Theory (B)* **10** (1971), 12–41.
15. M. Chudnovsky and A. O. Fradkin, Hadwiger's conjecture for quasi-line graphs, *J. Graph Theory* **59** (2008), 17–33.
16. M. Chudnovsky and P. Seymour, Packing seagulls, *Combinatorica*, to appear.
17. V. Chvátal, On certain polytopes associated with graphs, *J. Combin. Theory (B)* **18** (1975), 138–154.
18. M. DeVos, Z. Dvořák, J. Fox, J. McDonald, B. Mohar and D. Scheide, A minimum degree condition forcing complete graph immersion, *Combinatorica* **34** (2014), 279–298.
19. M. DeVos, K. Kawarabayashi, B. Mohar and H. Okamura, Immersing small complete graphs, *Ars Math. Contemp.* **3** (2010), 139–146.
20. R. Diestel, *Graph Theory* (2nd edn.), Springer, 2000.
21. G. A. Dirac, A property of 4-chromatic graphs and some remarks on critical graphs, *J. London Math. Soc.* **27** (1952), 85–92.
22. G. A. Dirac, Trennende Knotenpunktmengen und Reduzibilität abstrakter Graphen mit Anwendung auf das Vierfarbenproblem, *J. Reine Angew. Math.* **204** (1960), 116–131.
23. G. A. Dirac, Homomorphism theorems for graphs, *J. Reine Angew. Math.* **153** (1964), 69–80.
24. G. A. Dirac, On the structure of 5- and 6-chromatic abstract graphs, *Math. Ann.* **214/215** (1964), 43–52.
25. P. Duchet and H. Meyniel, On Hadwiger's number and the stability number, *Graph Theory* (Cambridge, 1981), North-Holland (1982), 71–73.
26. P. Erdős and S. Fajtlowicz, On the conjecture of Hajós, *Combinatorica* **1** (1981), 141–143.
27. J. Fox, Complete minors and independence number, *SIAM J. Discrete Math.* **24** (2010) 1313–1321.
28. A. O. Fradkin, Clique minors in claw-free graphs, *J. Combin. Theory (B)* **102** (2012), 71–85.
29. J. Geelen, B. Gerards, B. Reed, P. Seymour and A. Vetta, On the odd variant of Hadwiger's conjecture, *J. Combin. Theory (B)* **99** (2009), 20–29.
30. B. Gerards, An extension of König's theorem to graphs with no odd-K_4, *J. Combin. Theory (B)* **47** (1989), 330–348.
31. H. Hadwiger, Über eine Klassifikation der Streckenkomplexe, *Vierteljschr. Naturforsch. Ges. Zürich* **88** (1943), 133–142.
32. M. R. Henzinger, S. Rao and H. N. Gabow, Computing vertex connectivity: new bounds from old techniques, *J. Algorithms* **34** (2000), 222–250.
33. T. R. Jensen and B. Toft, *Graph Coloring Problems*, Wiley–Interscience, 1995.
34. L. Jørgensen, Contraction to K_8, *J. Graph Theory* **18** (1994), 431–448.

35. K. Kawarabayashi, Rooted minors problem in highly connected graphs, *Discrete Math.* **287** (2004), 121–123.
36. K. Kawarabayashi, On the connectivity of minimal counterexamples to Hadwiger's conjecture, *J. Combin. Theory (B)* **97** (2007), 144–150.
37. K. Kawarabayashi, Note on coloring graphs with no odd-K_k-minors, *J. Combin. Theory (B)* **99** (2009), 738–741.
38. K. Kawarabayashi, Minors in 7-chromatic graphs, preprint.
39. K. Kawarabayashi, A. Kostochka and G. Yu, On sufficient degree conditions for a graph to be k-linked, *Combin. Probab. Comput.* **15** (2006), 685–694.
40. K. Kawarabayashi and B. Mohar, A relaxed Hadwiger's conjecture for list-colorings, *J. Combin. Theory (B)* **97** (2007), 647–651.
41. K. Kawarabayashi, S. Norine, R. Thomas and P. Wollan, K_6 minors in large 6-connected graphs, submitted.
42. K. Kawarabayashi, M. Plummer and B. Toft, Improvements of the theorem of Duchet and Meyniel on Hadwiger's conjecture, *J. Combin. Theory (B)* **96** (2005), 152–167.
43. K. Kawarabayashi and B. Reed, Fractional coloring and the odd Hadwiger's conjecture, *Europ. J. Combin.* **29** (2008), 411–417.
44. K. Kawarabayashi and B. Reed, Hadwiger's conjecture is decidable, *STOC'09 Proc. 2009 ACM International Symposium on Theory of Computing*, ACM (2009), 445–454.
45. K. Kawarabayashi and Z. Song, Some remarks on the odd Hadwiger's conjecture, *Combinatorica* **27** (2007), 429–438.
46. K. Kawarabayashi and B. Toft, Any 7-chromatic graph has K_7 or $K_{4,4}$ as a minor, *Combinatorica* **25** (2005), 327–353.
47. K. Kawarabayashi and G. Yu, Connectivities for k-knitted graphs and for minimal counterexamples to Hadwiger's conjecture, *J. Combin. Theory (B)* **103** (2013), 320–326.
48. J. Komlós and E. Szemerédi, Topological cliques in graphs II, *Combin. Prob. Comput.* **5** (1996), 79–80.
49. A. Kostochka, The minimum Hadwiger number for graphs with a given mean degree of vertices, *Metody Diskret. Analiz* **38** (1982), 37–58.
50. A. Kostochka, Lower bound of the Hadwiger number of graphs by their average degree, *Combinatorica* **4** (1984), 307–316.
51. F. Lescure and H. Meyniel, On a problem upon configurations contained in graphs with given chromatic number, *Graph Theory in Memory of G. A. Dirac* (Sandbjerg, 1985), Ann. Discrete Math. **41**, North-Holland (1989), 325–331.
52. W. Mader, Homomorphiesätze für Graphen, *Math. Ann.* **178** (1968), 154–168.
53. W. Mader, Existenz n-fach zusammenhängender Teilgraphen in Graphen genügend grosser Kantendichte, *Abh. Math. Sem. Univ. Hamburg* **37** (1972), 86–97.
54. M. Plummer, M. Stiebitz and B. Toft, On a special case of Hadwiger's conjecture, *Discuss. Math. Graph Theory* **23** (2003), 333–363.
55. B. Reed and P. D. Seymour, Fractional chromatic number and Hadwiger's conjecture, *J. Combin. Theory (B)* **74** (1998), 147–152.
56. N. Robertson, D. P. Sanders, P. D. Seymour and R. Thomas, The four-color theorem, *J. Combin. Theory (B)* **70** (1997), 2–44.
57. N. Robertson and P. D. Seymour, Graph minors XIII. The disjoint paths problem, *J. Combin. Theory (B)* **63** (1995), 65–110.
58. N. Robertson and P. D. Seymour, Graph minors XVII. Taming a vortex, *J. Combin. Theory (B)* **77** (1999), 162–210.
59. N. Robertson and P. D. Seymour, Graph minors XVI. Excluding a non-planar graph, *J. Combin. Theory (B)* **89** (2003), 43–76.

60. N. Robertson and P. D. Seymour, Graph minors XX. Wagner's conjecture, *J. Combin. Theory (B)* **92** (2004), 325–357.
61. N. Robertson and P. D. Seymour, Graph minors XXIII, Nash-Williams' immersion conjecture, *J. Combin. Theory (B)* **100** (2010), 181–205.
62. N. Robertson, P. D. Seymour and R. Thomas, Hadwiger's conjecture for K_6-free graphs, *Combinatorica* **13** (1993), 279–361.
63. A. Schrijver, *Combinatorial Optimization: Polyhedra and Efficiency*, Springer, 2003.
64. Z. Song and R. Thomas, The extremal function for K_9 minors, *J. Combin. Theory (B)* **96** (2006), 240–252.
65. R. Thomas and P. Wollan, An improved linear edge bound for graph linkages, *Europ. J. Combin.* **26** (2005), 309–324.
66. A. Thomason, An extremal function for contractions of graphs, *Math. Proc. Cambridge Philos. Soc.* **95** (1984), 261–265.
67. A. Thomason, The extremal function for complete minors, *J. Combin. Theory (B)* **81** (2001), 318–338.
68. C. Thomassen, Graph decomposition with applications to subdivisions and path systems modulo *k*, *J. Graph Theory* **7** (1983) 261–271.
69. C. Thomassen, Parity, cycle space, and K_4-subdivision in graphs, *Surveys in Combinatorics, 1999*, London Math. Soc. Lecture Notes **267**, Cambridge University Press (1999), 223–238.
70. C. Thomassen, Totally odd K_4-subdivisions in 4-chromatic graphs, *Combinatorica* **21** (2001), 417–443.
71. C. Thomassen, Some remarks on Hajós' conjecture, *J. Combin. Theory (B)* **93** (2005), 95–105.
72. B. Toft, Problem 11, *Recent Advances in Graph Theory* (ed. M. Fiedler), Academia Praha (1975), 543–544.
73. B. Toft, A survey of Hadwiger's conjecture, *Congr. Numer.* **115** (1996), 249–283.
74. K. Wagner, Über eine Eigenschaft der ebenen Komplexe, *Math. Ann.* **114** (1937), 570–590.
75. K. Wagner, Beweis einer Abschwächung der Hadwiger-Vermutung, *Math. Ann.* **153** (1964), 139–141.
76. D. Wood, Independent sets in graphs with an excluded clique minor, *Discrete Math. Theor. Comput. Sci.* **9** (2007), 171–175.
77. D. R. Woodall, Subcontraction-equivalence and Hadwiger's conjecture, *J. Graph Theory* **11** (1987), 197–204.
78. W. Zang, Proof of Toft's conjecture: every graph containing no fully odd K_4 is 3-colorable, *J. Combin. Optim.* **2** (1998), 117–188.

5
Edge-colourings

JESSICA MCDONALD

The focus of this chapter is on two cornerstones in modern edge-colouring: the famous Goldberg–Seymour conjecture, and ideas culminating in the method of Tashkinov trees. We also discuss extreme examples, the classification problem, and the computational complexity of edge-colouring.

1. Introduction

The central question concerning edge-colourings is: given a graph G, what is its *chromatic index* $\chi'(G)$ – that is, the minimum number of colours needed to edge-colour G? We explore this question in detail in this chapter.

The most obvious lower bound on the chromatic index of a graph G is $\chi'(G) \geq \Delta$ (where Δ is the maximum degree in G), and in 1916 König [39] proved that all bipartite graphs satisfy this bound with equality.

Theorem 1.1 *If G is a bipartite graph, then $\chi'(G) = \Delta$.*

The most famous upper bounds on the chromatic index are the following two results. The first of these was proved by Shannon [75] in 1949, and the second was proved by Vizing [81], and independently by Gupta [27], in the 1960s. The second result is widely known as *Vizing's theorem*, and it makes use of the maximum edge-multiplicity of a graph G, denoted by $\mu = \mu(G)$.

Theorem 1.2 *For every graph G, $\chi'(G) \le \frac{3}{2}\Delta$.*

Theorem 1.3 *For every graph G, $\chi'(G) \le \Delta + \mu$.*

In terms of exactly determining the chromatic index of a graph G, Theorem 1.2 gives a range of about $\frac{1}{2}\Delta$ possible values. When $\mu < \frac{1}{2}\Delta$, Vizing's theorem further narrows this window. In particular, when G is a simple graph ($\mu = 1$), Vizing's theorem says that there are only two possible values for the chromatic index: Δ and $\Delta + 1$. However, even in the case of simple graphs (and in fact even in the case of 3-regular simple graphs), it is NP-hard to determine the chromatic index exactly, as shown by Holyer [35].

Each colour class in an edge-colouring of a graph is a matching in that graph. Hence, given a set S of vertices in a graph G, where $|S|$ is odd and at least 3, the maximum size of a colour class in $G[S]$ is $\frac{1}{2}(|S| - 1)$. Thus the following is a natural lower bound for the chromatic index:

$$\rho = \rho(G) = \max\left\{\frac{2|E(G[S])|}{|S| - 1} : S \subseteq V(G), |S| \ge 3 \text{ and odd}\right\}.$$

In fact, we also get $\chi'(G) \ge \lceil \rho \rceil$, since the chromatic index is an integer. The famous *Goldberg–Seymour conjecture*, posed independently by Goldberg [18] and Seymour [73] in the 1970s, says that any graph with chromatic index higher than $\Delta + 1$ must have chromatic index $\lceil \rho \rceil$.

Conjecture A *For any graph G, $\chi'(G)$ is Δ, $\Delta + 1$ or $\lceil \rho \rceil$.*

Conjecture **A** is an amazing generalization of Vizing's theorem for simple graphs. Moreover, it suggests a dichotomy within edge-colouring. While Holyer's result tells us that it is NP-hard to distinguish between chromatic index Δ and $\Delta + 1$, the truth of the Goldberg–Seymour conjecture would imply the following for any graph G:

Determining whether or not $\chi'(G) > \Delta + 1$ can be done in polynomial time. Furthermore, the exact value of $\chi'(G)$ can be computed in polynomial time whenever $\chi'(G) > \Delta + 1$.

This implication is not immediate, but it can be deduced from well-known results in combinatorial optimization. In Section 7 we explain this in detail.

The focus of this chapter is on the Goldberg–Seymour conjecture and the techniques used in edge-colouring, particularly the 'method of Tashkinov trees'. In Section 2 we explain the core concepts of this method, taking as an example a proof of König's theorem. In Section 3 we discuss Tashkinov trees, and present easy proofs of Shannon and Vizing's theorems. We also present other important upper bounds for the chromatic index, all of which can be obtained in a similar manner. In Section 4 we survey the current status of work towards the Goldberg–Seymour conjecture, and show more advanced proof ideas that lead to approximations of the conjecture.

Given a known bound for the chromatic index – say, $\chi'(G) \leq k$ for all graphs G – can we characterize those G for which $\chi'(G) = k$? Our discussion of such extreme graphs begins in Section 5, where we also touch on graphs at the other end of the scale – those with chromatic index Δ. When we restrict the characterization of extreme graphs to simple graphs, we meet the so-called *classification problem*, the NP-hard problem of distinguishing simple graphs with chromatic index Δ or $\Delta + 1$. This is a very rich area, and we provide a brief survey in Section 6. Section 7 contains the explanation mentioned above, concerning the complexity implications of the Goldberg–Seymour conjecture and the dichotomy between low and high chromatic index. Our final section contains some final thoughts about this dichotomy, Conjecture **A** and the method of Tashkinov trees.

There are many well-studied variations and problems concerning edge-colourings that we do not attempt to cover here. The curious reader is referred to the thorough and thoughtful new book, *Graph Edge-colouring: Vizing's Theorem and Goldberg's Conjecture*, by Stiebitz, Scheide, Toft and Favrholt [77]. This book also contains more information on nearly everything we discuss here and we refer to it henceforth as Stiebitz *et al.* [77].

2. Elementary sets and Kempe changes

A *partial edge-colouring* φ of a graph G is an edge-colouring of some subgraph of G. The edges of this subgraph are referred to as the *domain* of φ and denoted by $\mathrm{dom}(\varphi)$. Suppose we want to find a k-edge-colouring of a graph G, or at least show that one exists. It is trivial to find a partial k-edge-colouring φ of G with $|\mathrm{dom}(\varphi)| = 1$, and we can view this as a starting point. The next step is to find a k-edge-colouring φ' of G for which $|\mathrm{dom}(\varphi')| = |\mathrm{dom}(\varphi)| + 1$, and then iterate this process. We can phrase such an argument in an algorithmic way (given any partial k-edge-colouring whose domain is not $E(G)$, we find one with a larger domain) or in an extremal way (given a partial k-edge-colouring with maximum domain, we show that its domain is $E(G)$). Each phrasing has its own merit and will suit us at different times in this chapter.

Given a graph G and a partial edge-colouring φ of G, let $\overline{\varphi}(v)$ be the set of colours in φ that are not used on any edge incident to v – that is, the set of colours *missing* at v. A set of vertices $W \subseteq V(G)$ is said to be *φ-elementary* if no pair of vertices in W have a common missing colour – that is, $\overline{\varphi}(w_1) \cap \overline{\varphi}(w_2) = \varnothing$ for each pair of distinct vertices $w_1, w_2 \in W$. If the ends of an uncoloured edge are *not* φ-elementary, then there is some colour missing at both ends, and that colour could be used on the edge in question. It is worth stating this formally, as follows.

Theorem 2.1 *Let G be a graph and let φ be a partial k-edge-colouring of G that leaves $e = vw$ uncoloured. If $\{v, w\}$ is not φ-elementary, then there is a k-edge-colouring of $\mathrm{dom}(\varphi) \cup \{e\}$.*

In the next section we discuss Tashkinov's Theorem (Theorem 3.1), which is a generalization of Theorem 2.1. A prerequisite to this topic is the fundamental concept of Kempe changes, which we describe now and use, along with Theorem 2.1, to prove König's theorem.

Given a partial k-edge-colouring of a graph G, any pair of colours $a, b \in \{1, 2, \ldots, k\}$ induces a subgraph $G(a, b)$ of G, where every component is a path or an even cycle. Note that switching a and b on any such component of $G(a, b)$ results in another k-edge-colouring of G. We call any such switch a *Kempe change*. First introduced by Kempe in his erroneous proof of the four-colour theorem, Kempe changes have been widely used in edge-colouring.

Given a partial k-edge-colouring φ of a graph G and a vertex v in G with $a \in \overline{\varphi}(v)$, $b \notin \overline{\varphi}(v)$, one can consider a maximal (a, b)*-alternating path* starting at v – that is, a path whose edges alternate in the colours a and b. Performing a Kempe change along this path, we obtain a k-edge-colouring with domain $\mathrm{dom}(\varphi)$, where $b \in \overline{\varphi}(v)$ and $a \notin \overline{\varphi}(v)$. This simple procedure is a staple in edge-colouring.

Let us now give a quick proof of König's theorem, that all bipartite graphs are Δ-edge-colourable. Let φ be a partial Δ-edge-colouring of a bipartite graph G with maximum domain, and suppose (for a contradiction) that some edge $e = vw$ is uncoloured. Then there exist some colour $a \in \overline{\varphi}(v)$ and some colour $b \in \overline{\varphi}(w)$. By Theorem 2.1, $\{v, w\}$ must be φ-elementary, so $a \neq b$ and we can make a Kempe change along the maximal (a, b)-alternating path beginning at w. The alternating path cannot end at v, since this would create an odd cycle, and so in the resulting colouring φ', $b \in \overline{\varphi'}(v) \cap \overline{\varphi'}(w)$. However, Theorem 2.1 then guarantees a Δ-edge-colouring of $\mathrm{dom}(\varphi') \cup \{e\}$, contradicting the fact that φ has maximum domain. König's theorem is thus proved.

The above proof of König's theorem can be thought of as a blueprint for the method of Tashkinov trees that we see next. We have already mentioned that the statement of Tashkinov's theorem (Theorem 3.1) is a generalization of Theorem 2.1. Moreover, the proof of Tashkinov's theorem is made up entirely of sophisticated sequences of Kempe changes.

3. Tashkinov trees and upper bounds

Let G be a graph and let φ be a partial edge-colouring of G in which edge e_1 is uncoloured. A φ*-Tashkinov tree* is a sequence $T = (v_1, e_1, v_2, e_2, \ldots, v_t)$ of distinct vertices $v_1, v_2, \ldots, v_t$ and edges $e_1, e_2, \ldots, e_{t-1}$ of G such that $e_1 = v_1v_2$ and, for each e_i with $i \geq 2$,

(T1) e_i has endpoints v_l, v_{i+1}, for some $l \leq i$
(T2) $\varphi(e_i) \in \overline{\varphi}(v_k)$, for some $k \leq i$.

We say that such a Tashkinov tree *starts* at its uncoloured edge e_1, and that it is φ*-elementary* if $V(T)$ is φ-elementary.

Given a partial edge-colouring φ of a graph G and an edge e not coloured by φ, e and its endpoints constitute a φ-Tashkinov tree. Hence the following result, proved by Tashkinov [78] in 2000 and known as *Tashkinov's theorem*, generalizes Theorem 2.1 when $k \geq \Delta + 1$.

Theorem 3.1 *Let G be a graph, let φ be a partial k-edge-colouring of G with $k \geq \Delta + 1$, and assume that e is uncoloured by φ. If there is a φ-Tashkinov tree starting at e that is not φ-elementary, then there is a k-edge-colouring of* $\mathrm{dom}(\varphi) \cup \{e\}$.

Theorem 3.1 actually generalizes Theorem 2.1, not only in the sense that its outcome is to guarantee the existence of a 'better' colouring, but in that it actually provides an algorithm to find such a colouring. This algorithm – that is, the proof of Theorem 3.1 – consists of a sophisticated sequence of Kempe changes. It provides the promised new colouring using at most a polynomial number (in n and Δ) of Kempe changes (see [45] or [77]).

There are two special cases of Tashkinov's theorem that are much easier to prove – that is, they require fewer Kempe changes. In (T1), if we can replace 'l' with 'i' then T is also called a *φ-Kierstead path*; if we can replace 'l' with '1' then T may be called a *φ-Vizing fan*.

Theorem 3.2 *Let G be a graph, let φ be a partial k-edge-colouring of G, and assume that e is uncoloured by φ. Assume that*
either
(1) $k \geq \Delta + 1$ *and there is a φ-Kierstead path T starting at e;*
or that
(2) $k \geq \Delta$ *and there is a φ-Vizing fan T starting at e.*
If T is not φ-elementary, then there is a k-edge-colouring of $dom(\varphi) \cup \{e\}$.

Case 2 of the above theorem was essentially proved by Vizing himself in his original argument for Vizing's theorem. In 1984 Kierstead [38] gave a new proof of Vizing's theorem, and on the way he proved case 1 of Theorem 3.2, a result which directly paved the way for Theorem 3.1. In fact, Tashkinov's proof of Theorem 3.1 is inductive and reduces to the case where the tree in question is a path – that is, a Kierstead path.

To apply Tashkinov's theorem, we typically want to consider a *maximal φ-Tashkinov tree* starting at e, where 'maximal' refers to the number of vertices in the tree. There may be a variety of such trees, with edge- and vertex-orderings that are distinct. However, it is easy to see that all maximal φ-Tashkinov trees starting at e must have the same vertex-set. We denote the vertex-set of any maximal φ-Tashkinov tree starting at e by $\mathcal{T}(e, \varphi)$. It is often helpful to consider a *maximal φ-Kierstead path* starting at e, or a *maximal φ-Vizing fan* starting at e, although the vertex-sets of such trees can vary considerably.

We now show how Theorem 3.1 implies both Shannon's and Vizing's theorems – in fact, we will see that these results follow from Case 1 of Theorem 3.2. To this end, let φ be a partial k-edge-colouring of a graph G, where $k \geq \Delta + 1$. (The value of k is either $\lfloor \frac{3}{2}\Delta \rfloor$ or $\Delta + \mu$ here, and we note that the case of $\lfloor \frac{3}{2}\Delta \rfloor < \Delta + 1$ is trivially resolved.) Suppose, for a contradiction, that φ has maximum domain but some edge e is left uncoloured. By Tashkinov's theorem, $\mathcal{T}(e,\varphi)$ is φ-elementary, so setting $|\mathcal{T}(e,\varphi)| = t$ we get

$$k \geq |\cup_{v\in\mathcal{T}(e,\varphi)} \overline{\varphi}(v)| \geq 2 + \sum_{v\in\mathcal{T}(e,\varphi)} (k - \deg(v)) \geq 2 + t(k - \Delta), \qquad (1)$$

where the '2' accounts for the edge e being uncoloured. Rearranging gives

$$k \leq \Delta + \frac{\Delta - 2}{t - 1}. \qquad (2)$$

We know that $t \geq 3$ since if $e = vw$, each colour in $\overline{\varphi}(v)$ appears on an edge incident with w, and we may pick any such edge as a second edge for a φ-Tashkinov tree. However, using $t \geq 3$ and $k = \lfloor \frac{3}{2}\Delta \rfloor$ in inequality (2) yields a contradiction. Thus we have proved Shannon's theorem.

To get Vizing's theorem, we go in a slightly different direction within the same argument, and pick some vertex $x \in \mathcal{T}(e,\varphi)$ that is not an endpoint of e. Since $\mathcal{T}(e,\varphi)$ is φ-elementary, x must be incident to an edge of every colour in $\bigcup_{v\in\mathcal{T}(e,\varphi),v\neq x} \overline{\varphi}(v)$, and by maximality, all these edges must be between x and $\mathcal{T}(e,\varphi) - x$. As there can only be at most $\mu(t-1)$ such edges, we get

$$\mu(t-1) \geq |\cup_{v\in\mathcal{T}(e,\varphi),v\neq x}\overline{\varphi}(v)| \geq 2 + (t-1)(k - \Delta), \qquad (3)$$

where the second inequality comes from inequality (1). Rearranging, we get

$$(t-1)(\Delta + \mu - k) \geq 2. \qquad (4)$$

Setting $k = \Delta + \mu$ in (4) yields a contradiction, and this proves Vizing's theorem.

The above proofs of Vizing's and Shannon's theorems are stated in the language of Tashkinov trees, but can be rephrased in terms of Kierstead paths, so that only Case 2 of Theorem 3.2 need be applied. To do so, simply consider a maximal φ-Kierstead path beginning at e and proceed as before, choosing x as the last vertex in this path in the proof of Vizing's theorem.

We discuss results that require the full strength of Tashkinov's theorem in the next section. First, however, we note that we can get even more from Theorem 3.2.

We can actually improve the condition $t \geq 3$ in the above argument to $t \geq g_o$, where g_o is the odd girth of the underlying simple graph. We see this by taking colours $a \in \overline{\varphi}(v)$ and $b \in \overline{\varphi}(w)$ and considering a maximal (a,b)-alternating path from w. (This path must end at v, since otherwise a single Kempe change causes $\{v,w\}$ to be non-elementary and hence contradict Theorem 2.1.) Putting $t \geq g_o$ in equation (2) yields a contradiction when $k = \Delta + 1 + \lfloor(\Delta - 2)/(g_o - 1)\rfloor$, and thus leads to a proof that of the following result, originally proved by Goldberg [21] in 1984.

Theorem 3.3 *For every non-bipartite graph G, $\chi'(G) \leq \Delta + 1 + \frac{\Delta-2}{g_o-1}$.*

Theorem 3.3 generalizes Shannon's theorem. A more careful analysis of the above argument for Vizing's theorem also yields a generalization. This generalization was proved by Steffen [76] in 2000 and uses the girth g of the underlying graph.

Theorem 3.4 *For every graph G containing a cycle, $\chi'(G) \leq \Delta + 1 + \frac{(\mu-1)}{\lfloor g/2 \rfloor}$.*

There is another well-known way of generalizing Vizing's theorem. Ore [53] proved the following local refinement in the 1960s, using the parameter $\mu(v)$ to denote the maximum multiplicity of an edge incident with v; we also use $\mu(v, w)$ to denote the number of edges between a specific pair of vertices v and w.

Theorem 3.5 *For every graph G, $\chi'(G) \leq \max\{\deg(v) + \mu(v) : v \in V(G)\}$.*

Another recent local refinement is due to Stiebitz *et al.* [77].

Theorem 3.6 *For every graph G with at least two vertices,*

$$\chi'(G) \leq \Delta + \min\{\mu(G - v) : v \in V(G)\}.$$

Theorem 3.6 generalizes an earlier result of Chetwynd and Hilton [4]. Theorem 3.5 is best proved using Case 2 of Theorem 3.2, as follows. We observe that if φ is a partial edge-colouring of a graph G, and $T = (v_1, e_1, v_2, \ldots, v_t)$ is a maximal φ-Vizing fan that is φ-elementary, then

$$\left|\bigcup_{i=2}^{t} \overline{\varphi}(v_i)\right| = \sum_{i=2}^{t} |\overline{\varphi}(v_i)| \leq \left(\sum_{i=2}^{t} \mu(v_1, v_i)\right) - 1, \tag{5}$$

since every colour that is missing at one of $v_2, v_3, \ldots, v_t$ must be present on an edge between v_1 and $v_2, v_3, \ldots, v_t$ (since otherwise the fan is not maximal), and e is uncoloured. On the other hand, as in (1), we get

$$\sum_{i=2}^{t} |\overline{\varphi}(v_i)| \geq 1 + \sum_{i=2}^{t} (k - \deg(v_i)). \tag{6}$$

Combining (5) and (6) and using Case 1 of Theorem 3.2, we get the following.

Theorem 3.7 *Let G be a graph and let φ be a partial k-edge-colouring of G with $k \geq \Delta$. Assume that φ has maximum domain but that $e = vw$ is uncoloured. Then there is a vertex-set $X \subseteq N(v)$ such that $|X| \geq 2$, $w \in X$, and*

$$\sum_{x \in X} (\deg(x) + \mu(v, x) - k) \geq 2.$$

Theorem 3.7 is commonly referred to as the *fan inequality*. If we set $k = \max\{\deg(v) + \mu(v) : v \in V(G)\}$ in Theorem 3.7 we see that for this value it is not possible to have a partial k-edge colouring of maximum domain that does not colour all of G, hence proving Ore's theorem. The proof of Theorem 3.6 also relies on the fan inequality, although we omit those details here.

The bounds mentioned in this section are arguably the most important in edge-colouring, and we discuss graphs that achieve them in Section 5. First however, we present more advanced results using Tashkinov trees, requiring the full generality of Theorem 3.1 – namely, we discuss approximations towards the Goldberg–Seymour conjecture.

4. Towards the Goldberg–Seymour conjecture

An equivalent way of stating the Goldberg–Seymour conjecture (Conjecture **A**), as given in our introductory section, is that for any graph G,

$$\chi'(G) \leq \max\{\lceil\rho\rceil, \Delta+1\}.$$

This conjecture is far from being established, but there is mounting evidence towards it, as we discuss in this chapter.

The Goldberg–Seymour conjecture is true asymptotically: Kahn [37] has shown that, for any graph G,

$$\chi'(G) \sim \max\{\rho, \Delta\} \quad \text{as} \quad \max\{\rho, \Delta\} \to \infty.$$

The conjecture was shown by Marcotte [43] to be true for all graphs that do not contain a K_5^--minor, where K_5^- is K_5 with an edge removed. Most other results towards Conjecture **A** take the form of an approximation – namely, that $\chi'(G) \leq \max\{\lceil\rho\rceil, \Delta+s\}$ for some specific $s > 1$.

There is a family of results, starting in the 1970s, which assert that

$$\chi' \leq \max\left\{\lceil\rho\rceil, \quad \Delta+1+\frac{\Delta-2}{m-1}\right\}, \tag{7}$$

for increasing values of m. Such results have been obtained by Goldberg [21], [20] for $m=9$, Nishizeki and Kashiwagi [51] and (independently) Tashkinov [78] for $m=11$, Favrholt, Stiebitz and Toft [16] for $m=13$, Scheide [69], [66] for $m=15$, and Kurt [41] for $m=25$. Since $(\Delta-2)/24 < 1$ when $\Delta \leq 25$, the best of these results show that the Goldberg–Seymour conjecture holds when the maximum degree is at most 25.

Scheide [69] (see also [67]) and Chen, Yu and Zang [3] independently proved a different sort of approximation to the Goldberg–Seymour conjecture. Building on earlier work by Favrholdt, Stiebitz and Toft [16], they proved the following.

Theorem 4.1 *For every graph G, $\chi'(G) \leq \max\left\{\lceil\rho\rceil, \quad \Delta+\sqrt{\Delta/2}\right\}$.*

Theorem 4.1 greatly improves the above-mentioned approximations when Δ is large. This result can be tightened slightly by replacing $\sqrt{\Delta/2}$ by $\sqrt{(\Delta-1)/2}$ (see [68], [45]), and Kurt [41] has announced that he can replace it with $\sqrt[3]{\Delta/2}$. We would like to replace this term by $\log\Delta$ as a next step, but this remains out of reach at this point. On the other hand, by allowing the parameter μ, Haxell and McDonald [28] proved the following.

Theorem 4.2 *For every graph G, $\chi'(G) \le \max\left\{\lceil\rho\rceil,\ \ \Delta + 2\sqrt{\mu\log\Delta}\right\}$.*

Another way to obtain approximations for the Goldberg–Seymour conjecture is to use the number of vertices n as a parameter. Plantholt [56] proved that every graph with even order $n \ge 572$ satisfies

$$\chi'(G) \le \max\{\Delta, \lceil\rho\rceil\} + 1 + \sqrt{n\log(n/6)},$$

and he recently made a further improvement that applies to all values of n (see [58]). In fact, the Goldberg–Seymour conjecture is known to be true for graphs for which n is very small. In 1997 Plantholt and Tipnis [59] completely determined the chromatic index of all graphs with at most 10 vertices, and showed that all graphs of this size satisfied the Goldberg–Seymour conjecture. Scheide's result of (7) (or rather, its proof) also implies the truth of the conjecture for graphs with at most 15 vertices. We explain this in more detail shortly.

The aforementioned work of Kahn, Marcotte and Plantholt *et al.* does not use the method of Tashkinov trees, but all the other results do, even those that predate Theorem 3.1. Tashkinov's association with (7) refers to his 2000 paper where Tashkinov trees were first introduced. However, as we saw in the last section, specialized versions of this method have long been used in edge-colouring.

In Section 3 we saw how to use Tashkinov trees (or even Kierstead paths) to k-edge-colour a graph G when $k = \lfloor\frac{3}{2}\Delta\rfloor$ (Shannon's theorem), $k = \Delta + \mu$ (Vizing's theorem) and $k = \Delta + 1 + \lfloor(\Delta-2)/(g_0-1)\rfloor$ (Goldberg's theorem). But how can we use this technique to prove k-edge-colourability when $k = \max\{\lceil\rho\rceil, \Delta + s\}$ for some $s \ge 1$? We now explore this question.

Let φ be a partial k-edge-colouring of a graph G with maximum domain, and let e be an edge uncoloured by φ. Then $\mathcal{T}(e,\varphi)$ is φ-elementary by Tashkinov's theorem. This means that if $|\mathcal{T}(e,\varphi)| = t$, then every colour that is missing at a vertex in $\mathcal{T}(e,\varphi)$ must occur on exactly $\frac{1}{2}(t-1)$ edges in $G[\mathcal{T}(e,\varphi)]$ (and hence t must be odd), as such a colour cannot be missing at any other vertex in $\mathcal{T}(e,\varphi)$ and cannot occur on any edge in $\delta(\mathcal{T}(e,\varphi))$ (since then this edge could be added to $\mathcal{T}(e,\varphi)$). By (1), the number of such missing colours is at least $t(k-\Delta)+2$. It follows from the definition of ρ that

$$\chi'(G) \ge \rho \ge \frac{2|E(G[\mathcal{T}(e,\varphi)])|}{t-1} > t(k-\Delta)+2, \tag{8}$$

where the strict inequality arises from the fact that the uncoloured edge e also lies in $G[\mathcal{T}(e,\varphi)]$. *If* we knew instead that *all* k colours occurred on exactly $\frac{1}{2}(t-1)$ edges of $G[\mathcal{T}(e,\varphi)]$, we could deduce (by using $k = \max\{\lceil\rho\rceil, \Delta + s\}$ in (8)) that

$$\chi'(G) \ge \rho > k \ge \rho,$$

which is a contradiction and hence proves the k-colourability of G. Thus, we may assume that there is some colour in $\{1, 2, \ldots, k\}$ that appears on fewer than $\frac{1}{2}(t-1)$ edges of $G[\mathcal{T}(e,\varphi)]$. Since this colour cannot be missing at any vertex in $G[\mathcal{T}(e,\varphi)]$, it must appear on multiple (in fact, at least 3) edges of $\delta(\mathcal{T}(e,\varphi))$; such a colour is called a *defective colour*. The existence of a defective colour yields further information about G: in particular, Favrholdt, Stiebitz and Toft [16] (see also [77]) have proved several such facts using Kempe changes, allowing us to improve the bound in inequality (2), and obtain the following result; similar statements are given in [77] and [45].

Theorem 4.3 *Let G be a graph, s be a positive integer, and $k = \max\{\Delta + s, \lceil\rho\rceil\}$. If there is a partial k-edge-colouring φ of G with maximum domain where e is uncoloured and $|\mathcal{T}(e,\varphi)| \geq t$, then*

$$s \leq \frac{\Delta - 3}{t+1}.$$

In the light of this result, if we wish to prove that

$$\chi'(G) \leq \max\left\{\lceil\rho\rceil, \Delta + 1 + \frac{\Delta - 3}{t+1}\right\}$$

for a given value of t, we need 'only' build a large enough Tashkinov tree ($|\mathcal{T}(e,\varphi)| \geq t$) and use Theorem 4.3 to get a contradiction. The results of form (7) are roughly obtained in this way by guaranteeing $t \geq m-2$, but as m increases this tends to require substantial case analysis. The bound of Theorem 4.1 requires showing that t is large compared to the number of colours k $(= \Delta + s)$ – in particular, $t \geq 2s+3$; this can be done easily by analyzing the number of times that a colour on an edge of a Tashkinov tree must appear in $\mathcal{T}(e,\varphi)$. Taking this idea further, and considering the parameter μ as well, one can show that $t > 1 + (1 + \frac{1}{2}s\mu)^s$, from which Theorem 4.2 follows.

If we set $s = 1$ in the above arguments and are able to show the existence of a Tashkinov tree of size t, then we can deduce that all graphs with fewer than t vertices satisfy the Goldberg–Seymour conjecture. In fact, using results from [16] we can tighten this further to prove the conjecture whenever $n < t + 3$; in particular, this yields the aforementioned result due to Scheide [67] for graphs with up to 15 vertices.

5. Extreme graphs

Which graphs have the highest (and lowest) chromatic indices? For example, if we know that all graphs G satisfy $\chi'(G) \leq k$, can we characterize those G for which $\chi'(G) = k$?

If H is a simple graph, then we denote by $H^{(\mu)}$ the *μ-multiple* of H – that is, the graph obtained from H by replacing each edge with μ parallel edges. Multiples of even cliques and multiples of even cycles are known to have chromatic index equal to Δ. On the other hand, by considering the vertex-set of an odd clique or an odd cycle, we can see that

$$\rho(K_{d+1}^{(\mu)}) \geq \Delta + \mu \quad \text{and} \quad \rho(C_{g_o}^{(\mu)}) \geq \Delta + 1 + \lfloor(\Delta - 2)/(g_o - 1)\rfloor,$$

for all even d and odd g_o. Since $\lceil \rho \rceil$ is a lower bound on the chromatic index and we can use the theorems of Vizing and Goldberg, this tells us that

$$\chi'(K_{d+1}^{(\mu)}) = \Delta + \mu \quad \text{and} \quad \chi'(C_k^{(\mu)}) = \Delta + 1 + \lfloor (\Delta - 2)/(g_o - 1) \rfloor$$

for all d even and k odd. We thus obtain a class of examples (multiples of odd cliques) that achieve Vizing's bound and a class of examples (multiples of odd cycles) that achieve Goldberg's bound; the latter can be specified to yield a class of examples (multiples of triangles) that achieve Shannon's bound. These actually turn out to be the only extreme graphs for Shannon's and Goldberg's bounds – namely, Vizing [83] proved that

$$\chi'(G) = \tfrac{3}{2}\Delta \quad \text{if and only if} \quad G = K_3^{(\mu)},$$

and McDonald [46] generalized this to

$$\chi'(G) = \Delta + 1 + \frac{\Delta - 2}{g_o - 1} \text{ if and only if } \; G = C_{g_o}^{(\mu)} \text{ and } (g_o - 1) \text{ divides } 2(\mu - 1).$$

These proofs can be deduced from the proof of Goldberg's bound presented in Section 3 by letting $k = \chi'(G) - 1$; instead of seeking a contradiction, we seek to understand the necessity of the last colour. This value of k in (2) implies that $t \leq g_o$, and hence inequalities (1) and (2) are satisfied with equality, leading to our desired characterization.

While all multiples of odd cliques achieve Vizing's bound of $\chi'(G) = \Delta + \mu$, these are not the only graphs to do so; for example, we can easily check that $\chi'(K_7^{(2)} - e) = \Delta + \mu$. Scheide and Stiebitz [70] listed pairs (Δ, μ) for which there exists a graph with maximum degree Δ and multiplicity μ achieving the bound, and proved that their list is complete provided that Conjecture **A** holds. Necessary structural conditions for graphs achieving Vizing's bound are also known (see [38], [46]), and if we restrict ourselves to multiples of simple graphs, a descriptive characterization of sorts is attainable; here it turns out that odd cliques do tell the full story (see [47]).

In the absence of a general descriptive characterization, however, simply characterizing non-simple graphs achieving Vizing's bound as having chromatic index $\lceil \rho \rceil$ would be highly desirable. This would mean proving that $\chi'(G) \leq \max\{\lceil \rho \rceil, \Delta + \mu - 1\}$ when $\mu \geq 2$, and partial results towards this have been obtained (see [28] and [16]). Without the condition on μ Holyer's result says that there would be no hope here, and perhaps this helps to explain why the problem is so difficult. In contrast, one can extend the characterization of Goldberg's bound even further to deduce that $\chi'(G) \leq \max\{\lceil \rho \rceil, \Delta + 1 + (\Delta - 3)/(g_o + 3)\}$ (see [48]).

We have already encountered two families of graphs that have the lowest possible chromatic index (Δ): bipartite graphs and multiples of even cliques. Because of Holyer's work we know that we cannot hope for a characterization of such graphs. However, Seymour [72] made the following conjecture, characterizing regular planar graphs of chromatic index Δ.

Conjecture B *Let G be a regular planar graph. Then $\chi'(G) = \Delta$ if and only if $\rho(G) \leq \Delta$.*

Note that this conjecture does not follow from the Goldberg–Seymour conjecture, which says that $\chi'(G) = \Delta$ or $\Delta + 1$ if $\rho(G) \leq \Delta$. The truth of Conjecture **B** is well known to be equivalent to the four-colour theorem when $\Delta = 3$, and has been established for all $\Delta \leq 8$ with a string of deep papers by Guenin [26], Dvořák, Kawarabayashi and Král [12], Kawarabayashi and Edwards (see [14], [9]) and Chudnovsky, Edwards and Seymour [10]. There are also significant results towards two generalizations of Conjecture **B**, one asserting that it holds even if 'planar graph' is replaced by 'graph containing no Petersen minor', and the other asserting that the chromatic index of any planar graph is equal to $\max\{\Delta, \lceil\rho\rceil\}$ (see [74], [44], [61], [63] and [62]). The methods in these papers vary greatly from what we have been discussing so far – in particular, we see induction, discharging and reducible configurations, as introduced in Chapter 1.

When we restrict our discussion of extreme graphs to simple graphs, the scope of our problem paradoxically seems to increase. Since simple graphs can take only one of two values for the chromatic index, every simple graph can be considered extreme, in the sense that it meets either the upper bound $\Delta + 1$ of Vizing's theorem or the lower bound Δ. Work on edge-colouring simple graphs has been prolific; Fiorini and Wilson's 1977 book *Edge-colourings of Graphs* [17] (see also the later survey by Hilton and Wilson [30]) focuses almost exclusively on simple graphs. We devote the next section to a short survey of this important topic.

6. The classification problem and critical graphs

Vizing's theorem divides all simple graphs into one of two classes: those with chromatic index Δ and those with chromatic index $\Delta + 1$. The former simple graphs are called *class 1* and the latter are called *class 2*, according to the convention introduced by Wilson. The problem of distinguishing between them, which we have already seen is NP-hard, is commonly called the *classification problem.*

To follow up on results from the previous section, the classification problem has been well studied for planar graphs. Vizing [82] conjectured that all simple planar graphs with $\Delta \geq 6$ are of class 1, and this has now been established save for the case $\Delta = 6$; the case $\Delta = 7$ was recently proved independently by Grünewald [23], Sanders and Zhao [64] and Zhang [88]. The classification problem has also been studied for graphs embedded on other surfaces by Sanders and Zhao [65] and Luo and Zhao [42].

Even when we step away from talking about planar (or embedded) graphs, graphs of class 1 are much more abundant than those of class 2. In fact, Erdős and Wilson [15] proved in 1977 that almost all simple graphs are of class 1.

One way to prove that a simple graph is of class 2 is to show that $\lceil\rho\rceil \geq \Delta + 1$; the easiest case is when we need only analyze the entire vertex-set $V(G)$ (as opposed to various odd subsets S of $V(G)$). With this in mind, Hilton defined a simple graph G to be *overfull* if $m \geq \lfloor \frac{1}{2}n \rfloor \Delta + 1$. Such an inequality is not possible unless n is odd and $n \geq 3$, so we see that every overfull graph has too many edges to be of class 1 – that is, for any overfull graph G,

$$\chi'(G) \geq \lceil \rho \rceil \geq \left\lceil \frac{2m}{n-1} \right\rceil \geq \left\lceil \frac{\Delta(n-1)+2}{n-1} \right\rceil = \Delta + 1.$$

Every simple regular graph of odd order is easily seen to be overfull, and is thus of class 2; additional examples can be found in the early paper of Beineke and Wilson [2]. Hoffman and Rodger [34] proved that a complete multipartite graph is of class 2 if and only if it is overfull. *Hilton's overfull conjecture*, posed by Hilton in 1985 and first appearing in [5] and [6], states that when $\Delta > \frac{1}{3}|V(G)|$, overfullness is the only possible reason for a simple graph G to be of class 2.

Conjecture C *Let G be a simple graph with $\Delta > \frac{1}{3}n$. If G is of class 2, then G contains an overfull subgraph H with $\Delta(H) = \Delta(G)$.*

Note that the $\frac{1}{3}n$ condition is best possible: consider $P - v$, where P is the Petersen graph. An earlier and weaker version of Conjecture **C**, where G is a regular simple graph with $2n$ vertices and $\Delta \geq n$, appeared in Chetwynd and Hilton [4] and is known as the 1-*factorization conjecture*. This conjecture is known to be true asymptotically, as shown by Perković and Reed [55], and there are other supporting results (see, for example, [7] and [52]). Little progress has been made on Conjecture **C** in general, save for work by Plantholt [57], although Hilton and Johnson explored consequences of the conjecture in [29]. Generalizations of the 1-factorization conjecture and the overfull conjecture for all (not necessarily simple) graphs have been proposed by Plantholt and Tipnis [60] and Stiebitz *et al.* [77], respectively; the first of these was proved asymptotically by Plantholt and Tipnis [60] and Vaughan [80].

Many results on the classification problem give structural properties of class 2 graphs. Thomassen [79] proved that every class 2 graph contains two vertices of degree Δ that are joined by Δ edge-disjoint paths. A number of authors (see [33], [32], [4] and [7]) have studied the subgraph induced by all vertices of maximum degree, called the *core* of a graph, and found conditions for a graph to be of class 1 or class 2 depending on its core. Of special interest are the *Hilton graphs*, connected simple graphs of class 2 with a core of $\Delta = 1$ or 2 (see [31]).

Results about class 2 graphs (in particular, those mentioned in the above paragraph) very often make use of the following result, known as *Vizing's adjacency lemma*. Here we encounter for the first time in this chapter the important notion of *critical* as it relates to edge-colouring: an edge e in a graph G is *critical* if $\chi'(G - e) < \chi'(G)$, and G is *critical* if every edge of G is a critical edge.

Theorem 6.1 *Let G be a simple graph of class 2 and let $e = vw$ be a critical edge of G. Then v is adjacent to at least $\Delta - \deg(w) + 1$ vertices of degree Δ.*

This lemma, originally proved by Vizing [82], can be seen as a consequence of the fan inequality (Theorem 3.7). Namely, if we apply Theorem 3.7 with $k = \Delta$ and $\mu = 1$ and let $Z = X \setminus w$, where X is the set of vertices guaranteed by Theorem 3.7, we deduce that $w \notin Z \subseteq N(v)$ and

$$\sum_{z \in Z} (\deg(z) + 1 - \Delta) \geq \Delta - \deg(w) + 1.$$

Hence the vertices of Z give the outcome of Vizing's adjacency lemma. With considerably more care, one can also prove a number of generalizations of Vizing's adjacency lemma (which provide additional structure), including those due to Goldberg [19] and Andersen [1], Choudum and Kayarthi [8] and Kostochka and Stiebitz [40].

Every simple graph G of class 2 contains as a subgraph a critical simple graph G' of class 2 with the same maximum degree (see Vizing [82]). Thus, a study of simple graphs of class 2 amounts to a study of *critical* simple graphs of class 2. Further, the only critical simple graphs that are of class 1 are stars (as these are the only graphs where deleting any edge causes the maximum degree to drop). Therefore, the classification problem can be seen as the problem of describing critical simple graphs of class 2.

There are constructions for critical simple graphs of class 2 by authors such as Jacobsen [36] (using Hajós's construction), Grünewald and Steffen [24] and Yap [87]. Vizing [83], [84] also made the following conjectures about critical simple graphs of class 2.

Conjecture D *If G is a simple critical n-vertex graph of class 2, then:*

- *the independence number of G is at most $\frac{1}{2}n$*
- *G contains a 2-regular spanning subgraph*
- *the number of edges in G is at least $\frac{1}{2}((\Delta - 1)n + 3)$.*

Note that the first conjecture above, commonly known as *Vizing's independence number conjecture*, is implied by the second conjecture, referred to as *Vizing's 2-factor conjecture*. The latter is known to be true for all overfull graphs (see Grünewald and Steffen [25]) and the former is additionally known to be true when when $\frac{1}{2}$ is replaced by $\frac{3}{5}$ (Woodall [86]). The last conjecture above, *Vizing's average degree conjecture*, is known to be true when $\frac{1}{2}((\Delta-1)n+3)$ is replaced by $\frac{1}{3}(\Delta+1)n$ (Woodall [85]), and there are a number of other supporting results; see [77] for a survey.

It is worth noting that nearly all of our earlier discussions can be phrased in terms of critical graphs. In particular, both of the following conjectures are equivalent to the Goldberg–Seymour conjecture; the first is known as the *critical multigraph conjecture* and the second is known as *Andersen's conjecture*. Of interest is the fact that Andersen proposed his conjecture in 1975 while an M.Sc. student at the University of Aarhus; it was later published in [1].

Conjecture E *Every critical graph G with $\chi'(G) \geq \Delta + 2$ has n odd and $m = \frac{1}{2}(\chi'(G) - 1)(n - 1) + 2$.*

Conjecture F *If e is an edge in a critical graph G with $\chi'(G) \geq \Delta + 2$, then there is a $(\chi'(G) - 1)$-edge-colouring of $G - e$ such that no colour is missing at more than one vertex.*

Unlike the situation for simple graphs, we have reason to hope that there may be a polynomial-time algorithm for $\chi'(G)$-edge-colouring *any* graph G with $\chi'(G) \geq \Delta + 2$. This is why we chose not to restrict our earlier discussions to critical graphs. We explain this hope in the following section.

7. The dichotomy of edge-colouring

Let G be a graph, and let $\mathcal{M}(G)$ be the set of matchings in G. As previously mentioned, each colour class in an edge-colouring is a matching. Hence, we can express the chromatic index of G as

$$\chi'(G) = \min\left\{ \sum_{M \in \mathcal{M}(G)} x_M : \sum_{M : e \in M} x_M = 1 \text{ for all } e \in E,\ \ x \in \mathbb{Z}^{\mathcal{M}(G)} \text{ and } x \geq 0 \right\},$$

where $x_M \in \{0, 1\}$ indicates whether or not the matching M is a colour class in the edge-colouring $x \in \{0, 1\}^{\mathcal{M}(G)}$. The linear programming relaxation gives rise to the *fractional chromatic index* of G, $\chi'_f(G)$ – namely,

$$\chi'_f(G) = \min\left\{ \sum_{M \in \mathcal{M}(G)} x_M : \sum_{M : e \in M} x_M = 1 \text{ for all } e \in E,\ \ x \in \mathbb{R}^{\mathcal{M}(G)} \text{ and } x \geq 0 \right\},$$

where we view $x_M \in [0, 1]$ as the weight with which we pick the matching M. The *matching polytope* $\mathcal{MP}(G)$ of G is defined to be the convex hull of the characteristic vectors of matchings in G, and this definition leads immediately to the following:

$$\chi'_f(G) = \min\{\beta : (1/\beta, 1/\beta, \ldots, 1/\beta) \in \mathcal{MP}(G)\}. \tag{9}$$

Edmonds [13] gave a combinatorial description of $\mathcal{MP}(G)$ that is now known as *Edmonds's matching polytope theorem*, a short proof of which appears in [71, pp. 109–111]. This theorem and equation (9) yield the following useful interpretation of the fractional chromatic index:

$$\chi'_f(G) = \max\{\rho, \Delta\}. \tag{10}$$

We can now explain why the Goldberg–Seymour conjecture implies what we claimed it did in Section 1 – namely the following.

Determining whether or not $\chi'(G) > \Delta + 1$ can be done in polynomial time. Moreover, the exact value of $\chi'(G)$ can be computed in polynomial time whenever $\chi'(G) > \Delta + 1$.

To this end, note that the Goldberg–Seymour conjecture and (10) imply that

$$\chi'(G) > \Delta + 1 \iff \chi'(G) = \lceil \rho \rceil > \Delta + 1 \iff \chi'(G) = \lceil \chi'_f(G) \rceil > \Delta + 1.$$

Hence our claim amounts to being able to compute $\chi'_f(G)$ in polynomial time.

Linear programs are known to be solvable in time that is polynomial in their number of constraints. The linear program describing the fractional chromatic index has an exponential number of constraints, however, as is the case with the linear program describing the fractional chromatic number. Despite this, although computing the fractional chromatic number is known to be NP-hard, there is a polynomial-time algorithm to compute the fractional chromatic index because of (9) and the work of Edmonds.

Equation (9) reduces the computation of $\chi'_f(G)$ to the following separation problem over the matching polytope.

Given a weighting of the edges in G, either identify a constraint of the matching polytope that is violated, or prove that no such constraint exists.

This is because we know, for example, that $(\frac{1}{\Delta-1}, \frac{1}{\Delta-1}, \ldots, \frac{1}{\Delta-1}) \notin \mathcal{MP}(G)$ and $(\frac{3}{2\Delta}, \frac{3}{2\Delta}, \ldots, \frac{3}{2\Delta}) \in \mathcal{MP}(G)$ (by Shannon's theorem), and we can use bisection to find the minimum as in (9).

There are two ways to show that we can solve this separation problem in polynomial time, both relying on the work of Edmonds. First, Padberg and Rao [54] solved this directly by making use of Edmonds's matching polytope theorem. Alternatively, we now know that for the matching polytope (or even more generally), a separation problem can be solved in polynomial time if and only if an optimization problem can be solved in polynomial time (see [22] and [11, p.238]). A polynomial-time algorithm for optimizing over the matching polytope was found by Edmonds in the 1960s; it is commonly called *Edmonds's blossom algorithm.*

8. Final thoughts

Our discussions have centred on the Goldberg–Seymour conjecture (Conjecture **A**) and the method of Tashkinov trees. This method consists of modifying a given partial edge-colouring with sequences of Kempe changes and resulting *extensions* (colouring an edge e with a colour a, provided that a is missing at both ends of e). Therefore, if we want to know whether the method of Tashkinov trees can be used to prove the Goldberg–Seymour conjecture, we need to ask the following question.

Can every partial $\chi'(G)$-edge-colouring of G be turned into a $\chi'(G)$-edge-colouring of G by a sequence of Kempe changes and extensions?

An affirmative answer to this question would settle positively the following conjecture of Vizing from 1965, now known as *Vizing's interchange conjecture.* Observe (for what follows) that a Kempe change may reduce the frequency of a colour, and when this change involves a single edge, the frequency may decrease to 0.

Conjecture G *Every edge-colouring of G can be turned into a $\chi'(G)$-edge-colouring of G by a sequence of Kempe changes.*

Two colourings are said to be *Kempe equivalent* if one can be obtained from the other by a sequence of Kempe changes. Vizing [83] noted that not all edge-colourings of a given graph are Kempe equivalent, but that Conjecture **G** holds when $\chi'(G) = \Delta + \mu$ or $\chi'(G) = \frac{3}{2}\Delta$, and we see that the same is true for all the bounds proved using the method of Tashkinov trees. More results on Kempe equivalence of edge-colourings have been found by Mohar [50] and McDonald, Mohar and Scheide [49].

In the previous section we saw that the Goldberg–Seymour conjecture implies that there exists a polynomial-time algorithm for determining $\chi'(G)$, whenever $\chi'(G) > \Delta + 1$. Is it possible that when $\chi'(G) > \Delta + 1$, every partial $\chi'(G)$-edge-colouring of G can be turned into a $\chi'(G)$-edge-colouring of G by a polynomial number of Kempe changes and extensions? We know that such a polynomial-time algorithm cannot exist for graphs with lower chromatic index (see Holyer [35]), but there may well be such a dichotomy in edge-colouring. In fact, graphs that have extremely high chromatic index appear to be among the easiest to edge-colour efficiently, perhaps because they are 'far away' from the classification problem. Accordingly, while one might hope to prove the Goldberg–Seymour conjecture first for graphs with $\mu \leq 2$, or even just for 2-multiples of simple graphs, these may be among the most difficult cases to settle.

References

1. L. Andersen, On edge-colourings of graphs, *Math. Scand.* **40** (1977), 161–175.
2. L. W. Beineke and R. J. Wilson, On the edge-chromatic number of a graph, *Discrete Math.* **5** (1973), 15–20.
3. G. Chen, X. Yu and W. Zang, Approximating the chromatic index of multigraphs, *J. Combin. Optim.* **21** (2011), 219–246.
4. A. G. Chetwynd and A. J. W. Hilton, Regular graphs of high degree are 1-factorizable, *Proc. London Math. Soc. (2)* **50** (1985), 193–206.
5. A. G. Chetwynd and A. J. W. Hilton, Star multigraphs with three vertices of maximum degree, *Math. Proc. Cambridge Philos. Soc.* **100** (1986), 303–317.
6. A. G. Chetwynd and A. J. W. Hilton, The edge chromatic class of graphs with maximum degree at least $|V| - 3$, *Graph Theory in Memory of G. A. Dirac* (eds. L. Andersen, I. Jakobsen, C. Thomassen, B. Toft and P. Vestergaard), Vol. 1, Ann. Discrete Math. **41**, North-Holland (1989), 91–110.
7. A. G. Chetwynd and A. J. W. Hilton, 1-factorizing regular graphs of high degree – an improved bound, *Discrete Math.* **75** (1989), 103–112.
8. S. Choudum and K. Kayathri, An extension of Vizing's adjacency lemma on edge chromatic critical graphs, *Discrete Math.* **206** (1999), 97–103.
9. M. Chudnovsky, K. Edwards, K. Kawarabayashi and P. Seymour, Edge-colouring seven-regular planar graphs, to appear.

10. M. Chudnovsky, K. Edwards and P. Seymour, Edge-colouring eight-regular planar graphs, to appear.
11. W. J. Cook, W. H. Cunningham, W. R. Pulleyblank and A. Schrijver, *Combinatorial Optimization*, Wiley, 1998.
12. Z. Dvořák, K. Kawarabayashi and D. Král, Packing six T-joins in plane graphs, manuscript.
13. J. Edmonds, Maximum matching and a polyhedron with 0,1-vertices, *J. Research Nat. Bureau of Standards (B)* **69** (1965), 125–130.
14. K. Edwards, *Optimization and Packings of T-joins and T-cuts*, M.Sc. thesis, McGill University, 2011.
15. P. Erdős and R. J. Wilson, On the chromatic index of almost all graphs, *J. Combin. Theory (B)* **23** (1977), 255–257.
16. L. M. Favrholdt, M. Stiebitz and B. Toft, *Graph Edge Colouring: Vizing's Theorem and Goldberg's Conjecture*, Preprints 2006, No. 20, IMADA, University of Southern Denmark, 91 pages.
17. S. Fiorini and R. J. Wilson, *Edge-colourings of Graphs*, Research Notes in Mathematics **16**, Pitman, 1977.
18. M. K. Goldberg, On multigraphs with almost-maximal chromatic class [in Russian], *Diskret. Analiz* **23** (1973), 3–7.
19. M. K. Goldberg, A remark on the chromatic class of a multigraph [in Russian], *Vyčisl. Mat. i. Vyčisl. Tehn. (Kharkow)* **5** (1974), 128–130.
20. M. K. Goldberg, Structure of multigraphs with restrictions on the chromatic class [in Russian], *Metody Diskret. Analiz* **30** (1977), 3–12.
21. M. K. Goldberg, Edge-coloring of multigraphs: recolouring technique, *J. Graph Theory* **8** (1984), 123–137.
22. M. Grötschel, L. Lovász and A. Schrijver, *Geometric Algorithms and Combinatorial Optimization*, Springer-Verlag, 1988.
23. S. Grünewald, *Chromatic Index Critical Graphs and Multigraphs*, Ph.D. thesis, Universität Bielefeld, 2000.
24. S. Grünewald and E. Steffen, Chromatic-index critical graphs of even order, *J. Graph Theory* **30** (1999), 27–36.
25. S. Grünewald and E. Steffen, Independent sets and sets and 2-factors in edge-chromatic-critical graphs, *J. Graph Theory* **45** (2004), 113–118.
26. B. Guenin, Packing T-joins and edge-colouring in planar graphs, manuscript.
27. R. P. Gupta, *Studies in the Theories of Graphs*, Ph.D. thesis, Tata Institute, Bombay, 1967.
28. P. Haxell and J. McDonald, On characterizing Vizing's edge-colouring bound, *J. Graph Theory* **69** (2012), 160–168.
29. A. J. W. Hilton and P. D. Johnson, Graphs which are vertex-critical with respect to the edge-chromatic number, *Math. Proc. Cambridge Philos. Soc.* **102** (1987), 211–221.
30. A. J. W. Hilton and R. J. Wilson, Edge-colorings of graphs: a progress report, *Graph Theory and its Applications: East and West (Jinan, 1986)* (eds. M. Capobianco *et al.*), Ann. New York Acad. Sci. **576** (1989), 241–249.
31. A. J. W. Hilton and Y. Zhao, One the edge-colouring of graphs whose core has maximum degree two, *J. Combin. Math. Combin. Comput.* **21** (1996), 97–108.
32. D. Hoffman, Cores of class II graphs, *J. Graph Theory* **20** (1995), 397–402.
33. D. Hoffman and C. Rodger, Class 1 graphs, *J. Combin. Theory (B)* **44** (1988), 372–376.
34. D. Hoffman and C. Rodger, The chromatic index of complete multipartite graphs, *J. Graph Theory* **16** (1992), 159–163.
35. I. Holyer, The NP-completeness of edge-colouring, *SIAM J. Comput.* **10** (1981), 718–720.

36. I. T. Jakobsen, Some remarks on the chromatic index of a graph, *Arch. Math. (Basel)* **24** (1973), 440–448.
37. J. Kahn, Asymptotics of the chromatic index for multigraphs, *J. Combin. Theory (B)* **68** (1996), 223–254.
38. H. A. Kierstead, On the chromatic index of multigraphs without large triangles, *J. Combin. Theory (B)* **36** (1984), 156–160.
39. D. König, Über Graphen und ihre Anwendung auf Determinantentheorie und Mengenlehre, *Math. Ann.* **77** (1916), 453–465.
40. A. Kostochka and M. Stiebitz, Edge-colouring: a new fan equation, manuscript, 2006.
41. O. Kurt, *On the Edge Coloring of Graphs*, Ph.D. thesis, Ohio State University, 2009.
42. R. Luo and Y. Zhao, Finding the exact bound of the maximum degrees of class two graphs embeddable in a surface of characteristic $\varepsilon \in \{-1, -2, -3\}$, *J. Combin. Theory (B)* **306** (2008), 1788–1790.
43. O. Marcotte, On the chromatic index of multigraphs and a conjecture of Seymour, II, *Proc. DIMACS Workshop, Morristown, N. J.* (eds. W. Cook and P. D. Seymour), American Math. Soc. (1990), 245–279.
44. O. Marcotte, Optimal edge-colourings for a class of planar multigraphs, *Combinatorics* **21** (2001), 361–394.
45. J. McDonald, *Multigraphs With High Chromatic Index*, Ph.D. thesis, University of Waterloo, 2009.
46. J. McDonald, Achieving maximum chromatic index in multigraphs, *Discrete Math.* **309**(8) (2009) 2077–2084.
47. J. McDonald, On multiples of simple graphs and Vizing's theorem, *Discrete Math.* **310** (2010), 2212–2214.
48. J. McDonald, On a theorem of Goldberg, *J. Graph Theory* **68**(1) (2011), 8–21.
49. J. McDonald, B. Mohar and D. Scheide, Kempe equivalence of edge-colourings in subcubic and subquartic graphs, *J. Graph Theory* **70**(2) (2012), 226–239.
50. B. Mohar, Kempe equivalence of colorings, *Graph Theory in Paris: Conf. in Honour of Claude Berge* (eds. J. A. Bondy *et al.*), Birkhäuser (2006), 287–297.
51. T. Nishizeki and K. Kashiwagi, On the 1.1 edge-coloring of multigraphs, *SIAM J. Discrete Math.* **3(3)** (1990), 391–410.
52. T. Nissen and L. Volkmann, Class 1 conditions depending on the minimum degree and the number of vertices of maximum degree, *J. Graph Theory* **14** (1990), 225–246.
53. O. Ore, *The Four-color Problem*, Academic Press, 1967.
54. M. W. Padberg and M. R. Rao, Odd minimum cut-sets and b-matchings, *Math. Oper. Res.* **7**(1) (1982), 67–80.
55. L. Perković and B. Reed, Edge coloring regular graphs of high degree, *Discrete Math.* **165/166** (1997), 567–578.
56. M. Plantholt, A sublinear bound on the chromatic index of multigraphs, *Discrete Math.* **202** (1999), 201–213.
57. M. Plantholt, Overfull conjecture for graphs with high minimum degree, *J. Graph Theory* **47** (2004), 73–80.
58. M. Plantholt, A combined logarithmic bound on the chromatic index of multigraphs, *J. Graph Theory* **73** (2013), 239–259.
59. M. Plantholt and S. Tipnis, The chromatic index of multigraphs of order at most 10, *Discrete Math.* **177** (1997), 185–193.
60. M. Plantholt and S. Tipnis, All regular multigraphs of even order and high degree are 1-factorable, *Electron. J. Combin.* **8**(1) (2001), R41.
61. N. Robertson, P. Seymour and R. Thomas, Tutte's edge-coloring conjecture, *J. Combin. Theory (B)* **70** (1997), 166–183.

62. D. Sanders, P. Seymour and R. Thomas, Edge three-coloring cubic doublecross graphs, in preparation.
63. D. Sanders and R. Thomas, Edge three-coloring cubic apex graphs, in preparation.
64. D. Sanders and Y. Zhao, On the size of edge chromatic critical graphs, *J. Combin. Theory (B)* **86** (2002), 408–412.
65. D. Sanders and Y. Zhao, Coloring edges of graphs embedded in a surface of characteristic zero, *J. Combin. Theory (B)* **87** (2003), 254–263.
66. D. Scheide, On the 15/14 edge-colouring of multigraphs, Preprints 2007 No. 11, IMADA, University of Southern Denmark, 49 pages.
67. D. Scheide, *Edge-colourings of Multigraphs*, Ph.D. thesis, TU Ilmenau, 2009.
68. D. Scheide, A polynomial-time $\Delta + \sqrt{(\Delta - 1)/2}$-edge-colouring algorithm, Preprints 2009 No. 4, IMADA, University of Southern Denmark, 15 pages.
69. D. Scheide, Graph edge-colouring: Tashkinov trees and Goldberg's conjecture, *J. Combin. Theory (B)* **100** (2010), 68–96.
70. D. Scheide and M. Stiebitz, On Vizing's bound for the chromatic index of a multigraph, *Discrete Math.* **309** (2009), 4920–4925.
71. A. Schrijver, *Theory of Linear and Integer Programming*, Wiley, 1986.
72. P. Seymour, *Matroids, Hypergraphs and the Max.-Flow Min.-Cut Theorem*, D.Phil. Thesis, Oxford University, 1975.
73. P. Seymour, On multi-colourings of cubic graphs, and conjectures of Fulkerson and Tutte, *Proc. London Math. Soc.* **3** (1979), 423–460.
74. P. Seymour, Colouring series-parallel graphs, *Combinatorica* **10** (1990), 379–392.
75. C. E. Shannon, A theorem on colouring the lines of a network, *J. Math. Phys.* **28** (1949), 148–151.
76. E. Steffen, A refinement of Vizing's theorem, *Discrete Math.* **218** (2000), 289–291.
77. M. Stiebitz, D. Scheide, B. Toft and L. Favrholt, *Graph Edge-colouring: Vizing's Theorem and Goldberg's Conjecture*, Wiley, 2012.
78. V. A. Tashkinov, On an algorithm to color the edges of a multigraph (in Russian), *Diskret. Analiz* **7** (2000), 72–85.
79. C. Thomassen, Hajós' conjecture for line graphs, *J. Combin. Theory (B)* **97** (2007), 156–157.
80. E. Vaughan, An asymptotic version of the multigraph 1-factorization conjecture, *J. Graph Theory* **72** (2013), 19–29.
81. V. G. Vizing, On an estimate of the chromatic class of a p-graph, *Diskret. Analiz* **3** (1964), 25–30.
82. V. G. Vizing, Critical graphs with a given chromatic class, *Diskret. Analiz* **5** (1965), 9–17.
83. V. G. Vizing, The chromatic class of a multigraph, *Kibernetika* (Kiev) **1** (1965) 29–39 [in Russian]; English transl.: *Cybernetics* **1** (1965), 32–41.
84. V. G. Vizing, Some unsolved problems in graph theory (in Russian), *Uspekhi Mat. Nauk.* **23** (1968), 117–134; English transl.: *Russian Math. Surveys* **23** (1968), 125–141.
85. D. R. Woodall, The average degree of an edge-chromatic critical graph II, *J. Graph Theory* **56** (2007), 194–218.
86. D. R. Woodall, The independence number of an edge-chromatic critical graph, *J. Graph Theory* **66** (2011), 98–103.
87. H. P. Yap, *Some Topics in Graph Theory*, London Math. Soc. Lecture Notes **108**, Cambridge Univ. Press, 1986.
88. L. Zhang, Every planar graph with maximum degree 7 is class 1, *Graphs Combin.* **16** (2000), 467–495.

6

List-colourings

MICHAEL STIEBITZ and MARGIT VOIGT

Over the years, many generalizations of the graph colouring problem have been introduced and investigated in the graph theory literature. Such generalizations are typically obtained by relaxation of the colouring condition that each colour class forms an independent set, or by imposing restrictions on the colours that may be used. A popular example of the second type is the list-colour problem, in which it is required to choose a colour for each vertex v of G from an individual list $L(v)$ of available colours. The list-chromatic number $\chi_\ell(G)$ is the smallest integer k such that, whenever we assign a list of k colours to each vertex of G, there is a colouring of the vertices of G for which each vertex receives a colour from its list and adjacent vertices receive different colours. In recent years the list-colouring problem has attracted much attention and has led to many interesting results and open problems.

1. Introduction

Let $G = (V, E)$ be a graph, let $f : V(G) \to \mathbb{N}$, and let $k \geq 0$ be an integer. A *list-assignment* L of G is a function that assigns to each vertex v of G a set (list) $L(v)$ of colours: usually each colour is a positive integer. We say that L is an *f-assignment* if $|L(v)| = f(v)$ for all $v \in V$, and a *k-assignment* if $|L(v)| = k$ for all $v \in V$. A *colouring* of G is a function φ that assigns a colour to each vertex of G so that $\varphi(v) \neq \varphi(w)$ whenever $vw \in E$. An *L-colouring* of G is a colouring φ of G such that

$\varphi(v) \in L(v)$ for all $v \in V$. If G admits an L-colouring, then G is *L-colourable*, or *list-colourable* if L is understood. When $L(v) = [1,k]$ for all $v \in V$ (where $[1,k]$ is the set $\{1,2,\ldots,k\}$), the corresponding terms become a *k-colouring* and *k-colourable*, respectively. The graph G is said to be *f-list-colourable* if G is L-colourable for every f-assignment L of G. When $f(v) = k$ for all $v \in V$, the corresponding terms become *k-list-colourable* and *k-choosable*. The *list-chromatic number* or *choice number* $\chi_\ell(G)$ is the least number k such that G is k-list-colourable.

The study of list-colouring problems for graphs was initiated in the 1970s independently by Vizing [65] and Erdős, Rubin and Taylor [18]. In both papers it was observed that every k-list-colourable graph is k-colourable, so that $\chi \leq \chi_\ell$, but not conversely (see Fig. 1). Moreover, the gap between χ and χ_ℓ can be arbitrarily large. To see this, consider the complete bipartite graph $G = K_{m,m}$, with $m = \binom{2r-1}{r}$ for some integer $r \geq 2$. If C is a set of $2r-1$ colours, then there is an r-assignment L of G such that every r-subset of C is assigned to exactly one vertex on each side of the bipartition (A,B) of G. We claim that G is not L-colourable, and hence $\chi_\ell(G) \geq r+1$. Suppose on the contrary that there is an L-colouring φ of G. Then $X = \{\varphi(a) : a \in A\}$ and $Y = \{\varphi(b) : b \in B\}$ are disjoint subsets of C and, since $|C| = 2r-1$, one of the two sets, say X, has at most $r-1$ elements. But then there is a vertex $a \in A$ for which $L(a) \subseteq C \setminus X$, contradicting $\varphi(a) \in L(a)$.

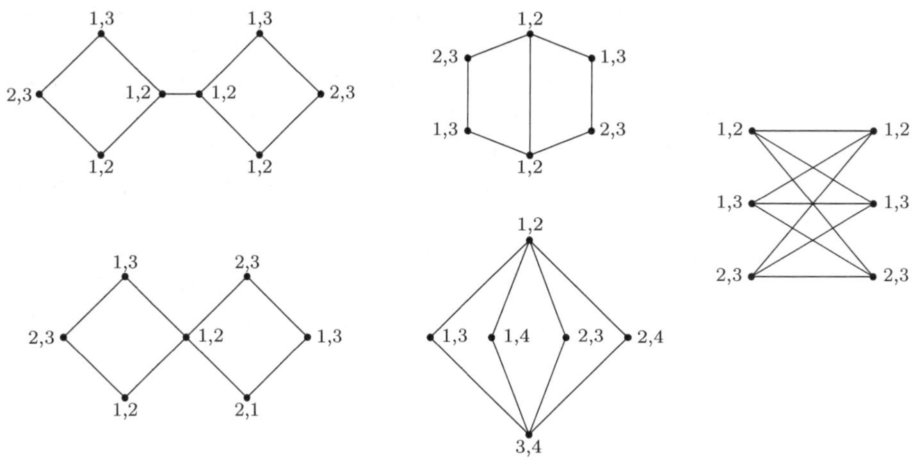

Fig. 1. Small bipartite graphs with uncolourable 2-assignments

Let $m(r,k)$ be the minimum number of edges in an r-uniform hypergraph with chromatic number greater than k, and denote by $N(k,r)$ the minimum number of vertices in a k-colourable graph with list-chromatic number greater than r. That the colouring of r-uniform hypergraphs relates to list-colourings of bipartite graphs was first observed by Erdős, Rubin and Taylor [18], who proved that

$$m(r,2) \leq N(r,2) \leq 2m(r,2), \text{ for } r \geq 2.$$

Erdős established the first non-trivial bounds on $m(r,2)$ – namely,

$$2^{r-1} \leq m(r,2) \leq r^2 2^r.$$

Further results concerning $m(r,k)$ are discussed in the survey paper [37].

For $m, r \geq 1$, let K_{m*r} be the complete r-partite graph with m vertices in each of the r parts. Then, as proved by Alon [3], there are positive constants c_1 and c_2 such that $c_1 r \log m \leq \chi_\ell(K_{m*r}) \leq c_2 r \log m$, where $m, r \geq 2$.

The *line graph* $L(G)$ of a graph G is the simple graph with vertex-set $E(G)$ in which $e, f \in E(G)$ are adjacent if they are adjacent as edges in G. A colouring of $L(G)$ is therefore a map $\varphi : E(G) \to C$ with $\varphi(e) \neq \varphi(f)$ for any adjacent edges e, f of G. A colouring of $L(G)$ is also called an *edge-colouring* of G, and $\chi'(G) = \chi(L(G))$ is called the *chromatic index* of G and $\chi'_\ell(G) = \chi_\ell(L(G))$ is called the *list-chromatic index* of G.

While the gap between $\chi(G)$ and $\chi_\ell(G)$ can be arbitrarily large, even for the class of bipartite graphs, it was conjectured independently by various researchers such as Vizing, Albertson, Collins, Tucker, Gupta, Bollobás and Harris that this cannot happen for the class of line graphs.

List-edge-colouring conjecture (LECC) *Every graph G satisfies $\chi'_\ell(G) = \chi'(G)$.*

The LECC seems difficult to attack, and might even be false for the class of all graphs. Hence it is reasonable to search for subclasses for which the conjecture can be proved. Dinitz suggested a candidate for such a subclass – the class of complete bipartite graphs $K_{m,m}$. One of the spectacular results about list-colourings is Galvin's 1995 proof [23] that the LECC holds for the class of bipartite graphs; the proof method is discussed in Section 2. That the LECC is true for the class of simple graphs, at least asymptotically, was proved by Kahn [35], who showed that any simple graph G satisfies $\chi'_\ell(G) = \Delta(G) + o(\Delta(G))$.

Another class of graphs for which the chromatic number and the list-chromatic number always coincide was recently recognized by Noel, Reed and Wu [50]. This remarkable result provides an affirmative answer to a conjecture proposed in 2002 by Ohba.

Theorem 1.1 *Every simple graph G with $|V(G)| \leq 2\chi(G) + 1$ satisfies $\chi_\ell(G) = \chi(G)$.*

There are several further classes of graphs for which it has been conjectured that $\chi_\ell = \chi$. These classes include claw-free graphs, total graphs and squares of graphs. A graph is *claw-free* if it contains no $K_{1,3}$ as an induced subgraph: line graphs are claw-free. The conjecture that every claw-free graph G satisfies $\chi_\ell(G) = \chi(G)$ is due to Gravier and Mafrey [24].

The *total graph* $T(G)$ of a graph G is the simple graph with vertex-set $V(G) \cup E(G)$ in which two vertices are adjacent whenever the corresponding elements in G are adjacent or incident. That every graph G satisfies $\chi_\ell(T(G)) = \chi(T(G))$ was conjectured about the same time (and independently of each other) by Borodin, Kostochka and Woodall [10], Juvan, Mohar and Škrekovski [32], and Hilton and

Johnson [29]. This conjecture became known as the *list-total-colouring conjecture* (LTCC).

The *square* G^2 of a simple graph G is the simple graph with the same vertex-set as G in which two vertices are adjacent if and only if their distance in G is at most 2. The *list-square-colouring conjecture* (LSCC), made by Kostochka and Woodall [38], says that every simple graph G satisfies $\chi_\ell(G^2) = \chi(G^2)$. If H is a graph and if G is the simple graph obtained from H by subdividing each edge of H, then $G^2 = T(H)$. Thus, the LSCC implies the LTCC.

One obvious way to find an L-colouring of a given graph G is the following sequential algorithm:

- choose a vertex-order $v_1, v_2, \ldots, v_n$ of G – either an arbitrary one, or one that satisfies a certain property
- consider the vertices in turn and colour each vertex v_i with a colour, (if any) from $L(v_i)$, different from all those colours already assigned to the neighbours of v_i among $v_1, v_2, \ldots, v_{i-1}$.

This procedure yields an L-colouring of G if $|L(v_i)| \geq d_{G_i}(v_i) + 1$ for all $i \in [1, n]$, where $G_i = \langle v_1, v_2, \ldots, v_i \rangle$. The least integer $p \geq 1$ for which G has a vertex-order in which each vertex is preceded by fewer than p of its neighbours is called the *colouring number* $\mathrm{col}(G)$ of G. Observe that $\mathrm{col}(G) \leq \Delta(G) + 1$. Furthermore, the *clique number* $\omega(G)$ (the largest number n for which K_n is a subgraph of G) is a lower bound for the chromatic number. So we have proved the following:

Every graph G is f-list-colourable, where $f(v) = d_G(v) + 1$ for all $v \in V(G)$, and

$$\omega(G) \leq \chi(G) \leq \chi_\ell(G) \leq \mathrm{col}(G) \leq \Delta(G) + 1. \tag{1}$$

It is well known that the colouring number of a non-empty graph G is equal to 1 + (the maximum minimum degree of the subgraphs of G). A graph G with $\mathrm{col}(G) \leq k + 1$ is *k-degenerate*, so a graph is 0-degenerate if and only if it is edgeless, and is 1-degenerate if and only if it is a forest.

Erdős, Rubin and Taylor [18] proved that a connected graph G is f-list-colourable, where $f(v) = d_G(v)$ for all $v \in V(G)$, unless each block of G is a complete graph or an odd cycle (see Chapter 2). From this result they deduced that a connected graph G satisfies $\chi_\ell(G) \leq \Delta(G)$, unless G is a complete graph or an odd cycle. This list-version of Brooks's theorem was independently obtained by Vizing [65].

It follows from (1) that if $\chi \to \infty$, then $\mathrm{col} \to \infty$. The complete bipartite graph $K_{m,m}$ with colouring number $m + 1$ shows that the converse statement is not true. However, the following result of Alon [4] shows that if $\mathrm{col} \to \infty$ then $\chi_\ell \to \infty$. Recall that $2|E(G)|/|V(G)|$ is the *average degree* of G.

Theorem 1.2 *Let G be a simple graph with average degree at least d. If s is an integer, and if*

$$d > 4\binom{s^4}{2} \log_2 \left(2\binom{s^4}{2}\right),$$

then $\chi_\ell(G) \geq s + 1$.

Let G be a simple graph, and let $f, g : V(G) \to \mathbb{N}$ be functions. We call G $(f:g)$-*list-colourable* if, given any f-assignment L of G, we can choose sets $S_v \subseteq L(v)$, with $|S_v| = g(v)$ for all $v \in V(G)$, so that $S_v \cap S_w = \varnothing$ whenever vw is an edge of G. When $f(v) = a$ and $g(v) = b$ for all $v \in V(G)$, the corresponding term becomes $(a:b)$-*list-colourable*. Erdős, Rubin and Taylor [18] asked whether every $(a:b)$-list-colourable graph is necessarily $(ma:mb)$-list-colourable? On the one hand, the only pair for which this is known to be true is $(a, b) = (2, 1)$ (see [63]). On the other hand, Kostochka and Woodall [38] proposed the following conjectures.

Conjecture A *The weak $(a:b)$-choosability conjecture:*
For all $a, b, m \in \mathbb{N}$, if a simple graph G is $(a:b)$-list-colourable, then G is $(ma:mb)$-list-colourable.

Conjecture B *The strong $(a:b)$-choosability conjecture:*
For all $a, b, m \in \mathbb{N}$, if a simple graph G is $(a:b)$-list-colourable, then $G[K_m]$ is $(ma:b)$-list-colourable.

Conjecture C *The $(a:b)$-choosability equivalence conjecture:*
For all $a, b \in \mathbb{N}$, a simple graph G is $(a:b)$-list-colourable if and only if $G[K_b]$ is a-list-colourable.

If G is a simple graph, then $G[K_m]$ is called a *uniform inflation* of G. The classes of line graphs, claw-free graphs and squares are closed under uniform inflations. The following result due to Kostochka and Woodall [38] thus shows that the $(a:b)$-choosability conjectures for these classes of graphs are implied by the LECC, the list-colouring conjecture for claw-free graphs, and the LSCC, respectively. Observe that the class of total graphs is not closed under uniform inflation.

Theorem 1.3 *If G is a simple graph satisfying $\chi_\ell(G[K_m]) = \chi(G[K_m])$ for all $m \in \mathbb{N}$, then all three $(a:b)$-choosability conjectures hold for G.*

2. Orientations and list-colourings

Let D be a digraph. Recall that, for a vertex $v \in V(D)$, the *out-degree* $d_D^+(v)$ is the number of edges in D that go out of v, and the *in-degree* $d_D^-(v)$ is the number of edges in D that go into v. The *maximum out-degree* is denoted by $\Delta^+(D)$. For a set $X \subseteq V(D)$, we denote by $D\langle X\rangle$ the *subdigraph induced* by X – that is, $V(D\langle X\rangle) = X$ and $E(D\langle X\rangle) = \{vw \in E(D) : v, w \in X\}$. A set $X \subseteq V(D)$ is a *kernel* if X is an independent set of D (that is, $D\langle X\rangle$ has no edges) and for each vertex $v \in V(D) \setminus X$ there is a vertex $w \in X$ for which $vw \in E(D)$. A digraph D is *kernel-perfect* if every induced subdigraph of D has a kernel. The *underlying graph* of D is the simple graph G with $V(G) = V(D)$ and $E(G) = \{vw : vw \text{ or } wv \in E(D)\}$.

Theorem 2.1 *Let D be a kernel-perfect digraph and let G be the underlying graph of D. If $f, g : V(D) \to \mathbb{N}$ are functions for which*

$$f(v) \geq g(v) + \sum_{vw \in E(D)} g(w)$$

whenever $g(v) \geq 1$, then G is $(f : g)$-list-colourable, and so $\chi_\ell(G) \leq \Delta^+(D) + 1$.

Theorem 2.1 was first used by Galvin [23] in order to show that the LECC holds for line graphs of bipartite (multi)graphs. The following extension of this landmark result was obtained by Borodin, Kostochka and Woodall [10].

Theorem 2.2 *Let G be a bipartite graph and let $f(e) = \max\{d_G(v), d_G(w)\}$ for each edge e joining v and w in G. Then the line graph of G is f-list-colourable. Consequently, $\chi'_\ell(G) = \chi'(G) = \Delta(G)$.*

Borodin, Kostochka and Woodall [10] used the above theorem to show that every graph G satisfies $\chi'_\ell(G) \leq \frac{3}{2}\Delta(G)$. This is analogous to Shannon's bound for the chromatic index (see Chapter 5).

Let D be a digraph with $n \geq 1$ vertices $v_1, v_2, \ldots, v_n$. To each vertex v_i we associate a variable x_i, and define the polynomial P_D by

$$P_D = \prod_{v_i v_j \in E(D)} (x_j - x_i). \tag{2}$$

Clearly, $P_D \in \mathbb{Z}[x_1, x_2, \ldots, x_n]$ is a polynomial over the integer ring $\mathbb{Z}$ in n variables; as usual, the product over the empty set is the unit of the underlying ring. The polynomial P_D can be written as a linear combination of monomials,

$$P_D = \sum_{d_1, d_2, \ldots, d_n} a_{d_1, d_2, \ldots, d_n} \prod_{i=1}^{n} x_i^{d_i}, \tag{3}$$

where the sum is over all possible tuples $(d_1, d_2, \ldots, d_n)$ of non-negative integers. The coefficient $a_{d_1, d_2, \ldots, d_n}$ belongs to the ring $\mathbb{Z}$ and is denoted by $\mathrm{koef}_D(d_1, d_2, \ldots, d_n)$. If G is the underlying graph of D and $L(v_i)$ is a non-empty list for $i \in [1, n]$, then G is non-L-colourable if and only if P_D vanishes over $L(v_1) \times L(v_2) \times \cdots \times L(v_n)$.

The following result of Alon and Tarsi [6] relates list-colourability to the coefficients of the polynomial P_D. This result is a consequence of the Combinatorial Nullstellensatz [5].

Theorem 2.3 *Let D be a digraph with vertex-set $V(D) = \{v_1, v_2, \ldots, v_n\}$ and let $m = |E(D)|$. If*

$$\mathrm{koef}_D(d_1, d_2, \ldots, d_n) \neq 0$$

for some tuple $(d_1, d_2, \ldots, d_n)$ of non-negative integers, then $d_1 + d_2 + \cdots + d_n = m$ and the underlying graph G of D is f-list-colourable, where $f(v_i) = d_i + 1$ for each $i \in [1, n]$.

Alon and Tarsi [6] established a combinatorial interpretation for the coefficients of the polynomial P_D, which makes the above theorem more applicable. A digraph D is *Eulerian* if the in-degree and out-degree are equal at each vertex. For a digraph D, let $\mathcal{E}(D)$ denote the set of all sets $F \subseteq E(D)$ for which $D - (E(D) \setminus F)$ is Eulerian. Furthermore, let $\mathcal{E}^o(D)$ be the set of all sets $F \in \mathcal{E}(D)$ for which $|F|$ is odd, and $\mathcal{E}^e(D)$ be the corresponding set when $|F|$ is even. Associate with these sets the following numbers:

$$\varepsilon(D) = |\mathcal{E}(D)|, \ \varepsilon_o(D) = |\mathcal{E}^o(D)| \ \text{ and } \varepsilon_e(D) = |\mathcal{E}^e(D)|.$$

As proved by Alon and Tarsi [6], every digraph D with vertex-set $V(D) = \{v_1, v_2, \ldots, v_n\}$ and with out-degrees $d_D^+(v_i) = d_i$, for $1 \leq i \leq n$, satisfies

$$\text{koef}_D(d_1, d_2, \ldots, d_n) = \varepsilon_e(D) - \varepsilon_o(D).$$

Together with Theorem 2.3 this implies the following result of Alon and Tarsi [6].

Theorem 2.4 *Let D be a digraph and let G be the underlying graph of D. If $\varepsilon_e(D) \neq \varepsilon_o(D)$, then G is f-list-colourable, where $f(v) = d_D^+(v) + 1$ for each vertex $v \in V(G)$. Moreover, $\chi_\ell(G) \leq \Delta^+(D) + 1$.*

Let $\mathcal{D}$ denote the class of digraphs D satisfying $\varepsilon_e(D) \neq \varepsilon_o(D)$. Then $\mathcal{D}$ contains acyclic digraphs, digraphs containing no directed cycles of odd order, and orientations of bipartite graphs. However, all these digraphs are also kernel-perfect. In general, it seems to be difficult to decide whether a digraph belongs to $\mathcal{D}$.

For Eulerian digraphs, however, the following observation due to Fleischner and Stiebitz [22] has proved useful. If D is an Eulerian digraph with an even number of edges and with $\varepsilon(D) \equiv 2 \pmod 4$, then $D \in \mathcal{D}$. Thus, the underlying graph G is f-list-colourable, where $f(v) = d_D^+(v) + 1$ for each vertex $v \in V(G)$. Consequently, if G is a simple graph with an even number of edges such that $d_G(v)$ is even for all $v \in V(G)$, and if D is an Eulerian orientation of G with $\varepsilon(D) \equiv 2 \pmod 4$, then $\chi_\ell(G) \leq \frac{1}{2}\Delta(G) + 1$. Fleischner and Stiebitz [22] used this result to show that if a 4-regular graph G on $3n$ vertices is the union of a Hamiltonian cycle and n disjoint triangles, then $\chi_\ell(G) = \chi(G) = 3$.

Ellingham and Goddyn [17] used Theorem 2.4 to show that any d-regular planar graph G satisfies $\chi'_\ell(G) = d$ if and only if $\chi'(G) = d$. By Theorem 1.3, this implies all the $(a:b)$-choosability conjectures for line graphs of planar d-regular graphs with chromatic index d.

Prowse and Woodall [51] used the Alon–Tarsi method to prove that $\chi_\ell(G) = \chi(G)$ if $G = C_n^p$ is the power of a cycle – that is, the vertices of G are $v_1, v_2, \ldots, v_n$ and $N(v_i) = \{v_{i-p}, \ldots, v_{i-1}, v_{i+1}, \ldots, v_{i+p}\}$ for $i = 1, 2, \ldots, n$, where subscripts are taken modulo n. This implies that all the $(a:b)$-choosability conjectures hold for powers of cycles.

If G is a graph whose underlying simple graph is a forest F, then an easy inductive argument shows that

$$\chi'_\ell(G) = \chi'(G) = \Delta(G), \chi_\ell(T(G)) = \chi(T(G)) = \Delta(G) + 1$$
$$\text{and } \chi_\ell(G^2) = \chi(G^2) = \Delta(F) + 1.$$

Hence G satisfies the LECC, LTCC and LSSC. Kostochka and Woodall [39] proved the LECC and LTCC for *ring graphs* – that is, graphs whose underlying simple graph is a cycle. The proof of the LECC for ring graphs is very short, while the proof of the LTCC for ring graphs is long and complicated; one part uses the kernel method and one part uses the Alon–Tarsi method.

3. Planar graphs

By Euler's formula, simple planar graphs are 5-degenerate and hence 6-list-colourable. On the other hand, the four-colour theorem states that planar graphs are 4-colourable. A comprehensive survey about colourings of planar graphs is given by Borodin [9]. Many references for the results in this section can be found there.

Planar bipartite graphs

In 1979 Erdős, Rubin and Taylor [18] asked whether there exists a planar bipartite graph that is not 3-list-colourable. That the answer is negative was proved by Alon and Tarsi [6]. A simple bipartite planar graph G has an orientation D satisfying $\Delta^+(D) \leq 2$; this follows from the fact that any subgraph H of G has at most $2|V(H)| - 4$ edges. Since G is bipartite, D satisfies $\varepsilon_e(D) = \varepsilon_o(D)$, and by Theorem 2.4 this implies that $\chi_\ell(G) \leq 3$. It is also easy to see that any orientation of a bipartite graph has a kernel and is therefore kernel-perfect. Theorem 2.1 thus implies that every planar bipartite graph is $(3m : m)$-list-colourable, for all $m \geq 1$.

For these graphs, the proof of Theorem 2.1 can be easily transformed into an algorithm. So let G be a simple planar bipartite graph, let D be an orientation of G with $\Delta^+(D) \leq 2$, and let L be a $3m$-assignment for G. The aim is to find a set $S_v \subseteq L(v)$ of m colours for each vertex v of G, such that $S_v \cap S_w = \varnothing$ whenever vw is an edge of G. Let α be a colour occurring in the lists and let $V_\alpha = \{v \in V(D) : \alpha \in L(v)\}$. Now construct a kernel K_α of $D_\alpha = D\langle V_\alpha \rangle$, as follows. If $v \in V_\alpha$ has no successor in D_α, then add v to K_α and remove v and all of its predecessors from D_α; continue in this way until the remaining digraph has no vertices or all vertices have successors in it; in the latter case, add one of the partition classes to K_α. For all vertices $v \in K_\alpha$ assign colour α to its colour set S_v and remove α from all lists. If $|S_v| = m$, remove v and its incident edges from the graph, and continue in this way until the graph has no vertices. Note that every vertex gets m colours, since a colour is removed from its list if and only if the vertex itself, or one of its two successors, gets this colour. Thus the algorithm terminates with the required colouring.

For a graph G, let $\mathrm{mad}(G) = \max_{\varnothing \neq H \subseteq G} 2|E(H)|/|V(H)|$ be the *maximum average degree* of G. As first observed by Alon and Tarsi [6], every simple bipartite graph G satisfies $\chi_\ell(G) \leq \lceil \frac{1}{2}\mathrm{mad}(G) \rceil + 1$.

5-list-colourability

In 1979 Erdős, Rubin and Taylor [18] conjectured that all planar graphs are 5-list-colourable, but that not all planar graphs are 4-list-colourable. The second part of this conjecture was verified in 1993 by Voigt [66], who constructed a non-4-list-colourable planar graph with 238 vertices. The smallest known example of such a graph has 63 vertices and was obtained by Mirzakhani [48].

How can we construct such a graph? Consider the basic graph W first used by Gutner [28] and the list-assignment L given by M. Voigt and B. Wirth (see Fig. 2).

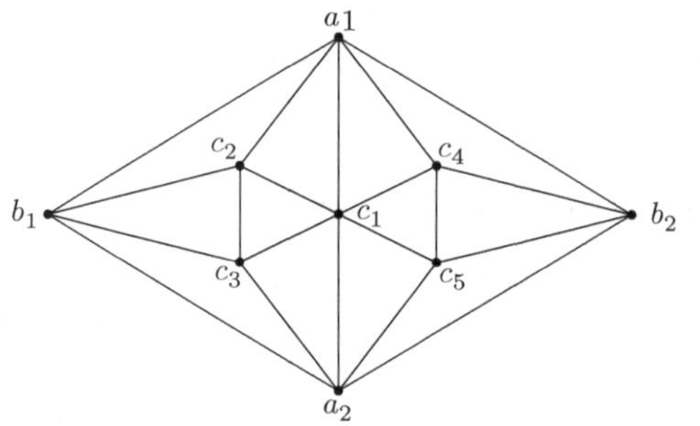

$L(a_1) = (\alpha)$
$L(a_2) = (\beta)$
$L(b_1) = (\alpha, \beta, \gamma, 5)$
$L(b_2) = (\alpha, \beta, \delta, 5)$
$L(c_1) = (\alpha, \beta, \gamma, \delta)$
$L(c_2) = (\alpha, \gamma, \delta, 5)$
$L(c_3) = (\beta, \gamma, \delta, 5)$
$L(c_4) = (\alpha, \gamma, \delta, 5)$
$L(c_5) = (\beta, \gamma, \delta, 5)$

Fig. 2. The graph W with a list-assignment

Then W is not L-colourable. To see this, assume first that c_1 is coloured δ; then the triangle $b_1c_2c_3$ cannot be properly coloured. Otherwise, c_1 is coloured γ and the triangle $b_2c_4c_5$ cannot be properly coloured. Now take twelve disjoint copies of the graph W – say, $W_1, W_2, \ldots, W_{12}$. For each vertex v of W, denote by $v(i)$ the copy of v in W_i. Identify all twelve vertices $a_1(i)$ to a new vertex p_1 and all twelve vertices $a_2(i)$ to a new vertex p_2, and add an edge joining p_1 and p_2. Identify $b_2(i)$ with $b_1(i+1)$ for $i = 1, 2, \ldots, 11$. The resulting graph G has $(12 \times 5) + 2 + 13 = 75$ vertices. For each vertex v of W, let $L(v(i))$ be the list obtained from $L(v)$ by replacing $\alpha, \beta, \gamma, \delta$ according to the following table. Finally, set $L(p_1) = L(p_2) = \{1, 2, 3, 4\}$.

i	1	2	3	4	5	6	7	8	9	10	11	12
α	1	1	4	4	4	3	3	3	1	2	2	2
β	2	3	3	2	1	1	2	4	4	4	3	1
γ	4	2	1	3	2	4	1	2	3	1	4	3
δ	3	4	2	1	3	2	4	1	2	3	1	4

Notice that $L(b_1) = \{\alpha, \beta, \gamma, 5\}$ and $L(b_2) = \{\alpha, \beta, \delta, 5\}$, implying that $L(b_2(i)) = L(b_1(i+1))$ for $i = 1, 2, \ldots, 11$. So G is a planar graph and L is a 4-assignment for G. Furthermore, for each colour pair $(\alpha, \beta) \in L(p_1) \times L(p_2)$ with $\alpha \neq \beta$, there is an $i \in [1, 12]$ for which $L(a_1(i)) = \{\alpha\}$ and $L(a_2(i)) = \{\beta\}$. This implies that G has no L-colouring, and so $\chi_\ell(G) \geq 5$.

The above graph G has list-chromatic number 5, but chromatic number 3. This answers in the negative a question of Jensen and Toft [31, Problem 2.13], who asked whether every 3-colourable planar graph is 4-list-colourable. Note that for the given list-assignment, the total set of colours has only 5 elements.

In 1994 Thomassen [57] proved a list version of Heawood's five-colour theorem, thus answering the conjecture of Erdős, Rubin and Taylor [18].

Theorem 3.1 *Every planar graph is 5-list-colourable.*

Thomassen's proof of the above result is surprisingly short and from first principles. In contrast to the usual proof of the five-colour theorem, Thomassen's proof does not use Euler's formula and involves no recolouring argument. The proof can be extended to show that every planar graph is $(5m : m)$-list-colourable for all $m \geq 1$ (see [63]); moreover, the proof yields a linear colouring algorithm. In 2007 Thomassen [60] proved that if G is a planar graph with n vertices and if L is a 5-assignment for G, then G has at least $2^{n/9}$ distinct L-colourings. Clearly, there are no more than 5^n distinct list-colourings. In 1998 Škrekovski [55] extended Thomassen's result and proved that every graph G not containing K_5 as a minor satisfies $\chi_\ell(G) \leq 5$. DeVos, Kawarabayashi and Mohar [13] proved that locally planar graphs are 5-list-colourable. If a graph G is embedded on a surface S, then the *edge-width* of G is the length of a shortest cycle of G that is non-contractible in S.

Theorem 3.2 *For each surface S there exists a constant w such that every graph that can be embedded in S with edge-width at least w is 5-list-colourable.*

In 2012 Kawarabayashi and Thomassen [36] proved the following extension of Theorem 3.1.

Theorem 3.3 *Let G be a connected graph embedded in a surface S of Euler characteristic ε. Then G has a vertex-set A with at most $1000(2-\varepsilon)$ vertices for which $G - A$ is 5-list-colourable.*

Forbidden cycles and 4-list-colourability

Many authors have investigated whether forbidden cycles of specific length in a planar graph ensure 4-list-colourability. It is easy to show that any simple planar graph without triangles is 3-degenerate, and hence 4-list-colourable. Lih and Wang [44] proved that if G is a planar graph in which any two distinct triangles are vertex-disjoint, then G is also 4-list-colourable. Xu considered the $(4m : m)$-list-colourability of planar graphs [75] and graphs embedded in the projective plane [76]. Xiaofang [74] generalized the result of Lih and Wang to toroidal graphs.

Theorem 3.4 *Let G be a graph in which any two distinct triangles are vertex-disjoint. If G can be embedded in the torus, then G is 4-list-colourable. If G can be embedded in the projective plane, then G is $(4m : m)$-list-colourable for all integers $m \geq 1$.*

Wang and Lih [45] proved that any simple planar graph without 5-cycles is 3-degenerate, and hence $(4m : m)$-list-colourable for all $m \geq 1$; the proof contains a small flaw, but it can easily be fixed. However, simple planar graphs without 4-cycles are not 3-degenerate in general. Based on the discharging method, Lam *et al.* [41] proved that any planar graph without 4-cycles is 4-list-colourable, and Lam, Shiu and Xu [42] showed that these graphs are even $(4m : m)$-list-colourable. In 2002 Fijavž *et al.* [21] managed to show that every simple planar graph without 6-cycles is 3-degenerate, and hence $(4m : m)$-list-colourable for all $m \geq 1$. Finally, every planar graph without 7-cycles is 4-list-colourable, as proved by Farzad [20]. In summary, we have the following result.

Theorem 3.5 *Let $k = 3, 4, 5$ or 6. If G is a planar graph with no cycle of length k, then G is $(4m : m)$-list-colourable for all $m \geq 1$. If G is a planar graph with no cycle of length 7, then G is 4-list-colourable.*

Gutner [28] constructed a non-4-list-colourable planar graph whose longest cycle has length 16. It is not known whether there exists a non-4-list-colourable planar graph without k-cycles, for $k \in \{8, 9, \ldots, 16\}$. Cai, Wang and Zhu [11] extended the above result for toroidal graphs, where the bounds for the list-chromatic number are tight.

Theorem 3.6 *Let G be a toroidal graph without cycles of length k. Then $\chi_\ell(G) \leq 4$ if $k = 3, 4$ or 5, $\chi_\ell(G) \leq 5$ if $k = 6$, and $\chi_\ell(G) \leq 6$ if $k = 7$.*

Triangle-free planar graphs

As mentioned above, triangle-free planar graphs are 4-list-colourable. A famous result of Grötzsch [25] says that triangle-free planar graphs are 3-colourable. However, as proved by Voigt [67], there are triangle-free planar graphs that are not 3-list-colourable. There are even examples of such graphs where the colours of a bad 3-assignment are taken from a set of five colours only. Whether 'five' can be replaced by 'four' is unknown.

The situation is different for planar graphs of higher girth. A planar graph of girth at least 6 is 2-degenerate, and hence 3-list-colourable. In 1995 Thomassen [58] extended this result to planar graphs of girth at least 5.

Theorem 3.7 *Every planar graph of girth at least 5 is 3-list-colourable.*

Moreover, Thomassen [59] proved that if G is a planar graph with girth at least 5, and if L is a 3-assignment for G, then G has at least $2^{n/10000}$ distinct L-colourings.

Several authors established conditions for the 3-list-colourability of planar graphs with girth at least 4 if some cycles of given length are missing. For details and references concerning the following result, see [15].

Theorem 3.8 *Let $k = 5, 6, 7$ or 8. Then a planar graph G is 3-list-colourable, unless G contains a cycle of length $3, k$ or $k + 1$.*

A further strengthening of Theorem 3.7 was given by Dvořák, Lidickí and Škrekovski [16] and (independently) by Guo and Wang [26]. They proved that if a triangle-free planar graph is not 3-list-colourable, then it contains a 4-cycle that intersects another 4-cycle or 5-cycle in exactly one edge.

The list version of Steinberg's conjecture

In 1976 Steinberg (see [31, Problem 2.9]) conjectured that every planar graph without 4-cycles and without 5-cycles is 3-colourable. In 1990 Erdős (see [9, ref. 43]) suggested the following relaxation of Steinberg's conjecture.

Problem *What is the smallest integer ℓ for which every planar graph without j-cycles for $4 \le j \le \ell$ is 3-colourable?*

The best partial result, due to Borodin *et al.* (see [9, ref. 43]), states that every planar graph G is 3-colourable, unless G contains a cycle of length $4, 5, 6$ or 7, and thus $\ell \le 7$. On the other hand, Abbott and Zhou (see [9, ref. 1]) constructed two infinite families of 4-critical planar graphs – one without cycles of length 4 and with six cycles of length 5, and the other without cycles of length 5 and with at most four cycles of length 4. For more references concerning this topic see [31]. Montassier suggested the corresponding question for list-colourings.

Problem *What is the smallest integer ℓ such that every planar graph without j-cycles for $4 \le j \le \ell$ is 3-list-colourable?*

Voigt [69] constructed a planar graph without 4-cycles or 5-cycles which is not 3-list-colourable. M. Montassier, A. Raspaud and W. Wang [49] found a non-3-list-colourable planar graph without 4-cycles, 5-cycles or intersecting triangles. Thus $\ell \ge 6$ in the above problem, and the list version of Steinberg's conjecture is false. On the other hand, $\ell \le 9$ (a consequence of the next theorem). For details and references concerning the proof of this theorem, see [71].

Theorem 3.9 *For every pair i, j of integers with $5 \le i < j \le 8$, a planar graph G is 3-list-colourable unless G contains a cycle of length $4, i, j$ or 9.*

Edge-list-colouring

In 1976 Vizing [65] conjectured that every simple graph G satisfies $\chi'_\ell(G) \le \Delta(G) + 1$. This conjecture is implied by the LECC and Vizing's bound for the chromatic index of simple graphs (see Chapter 5).

It follows from the inequality (1) that every simple graph G with $E(G) \neq \varnothing$ satisfies $\chi'_\ell(G) \leq 2\Delta(G) - 1$, since $\Delta(L(G)) \leq 2\Delta(G) - 2$. Moreover, we know from Brooks's theorem (see Chapter 2) that every simple graph G satisfies $\chi'_\ell(G) \leq 2\Delta(G) - 2$, unless G is an odd cycle; this implies Vizing's conjecture for $\Delta(G) = 3$. Juvan, Mohar and Škrekovski [33] confirmed Vizing's conjecture for graphs with maximum degree $\Delta = 4$. Borodin (see [9, ref. 27]) proved that every planar simple graph G with $\Delta(G) \geq 9$ satisfies $\chi'_\ell(G) \leq \Delta(G)+1$. Whether Vizing's conjecture holds for planar simple graphs G with $\Delta(G) = 5, 6, 7$ or 8 remains open.

Juvan, Mohar and Thomas [34] proved the LECC for series–parallel simple graphs (but not for series–parallel graphs in general), and hence for simple outerplanar graphs. The result for simple outerplanar graphs was independently obtained by Lih and Wang [43]. Lih and Wang also proved the LTCC for outerplanar graphs with maximum degree $\Delta \geq 4$. The following theorem combines results from [10], [54], [46] and [47].

Theorem 3.10 *If G is a planar graph, then $\chi'_\ell(G) = \Delta(G)$ if any of the following conditions holds:*
- $\Delta(G) \geq 12$
- $\Delta(G) \geq 8$ *and no two triangles have a vertex in common*
- $\Delta(G) \geq 8$ *and no triangle shares an edge with a* 4*-cycle*
- $\Delta(G) \geq 6$ *and G contains neither C_4 nor C_6, or G contains neither C_5 nor C_6.*

There are many results confirming Vizing's conjecture for planar graphs with special structures. Interestingly, planar graphs with maximum degree 5 seem to play a special role in the general case.

Theorem 3.11 *For $k = 3, 4, 5$ or 6, every planar simple graph G without a k-cycle satisfies $\chi'_\ell(G) \leq \Delta(G) + 1$.*

In 2004 Zhang and Wu [73] proved Theorem 3.11 for $k = 3$. In the same paper they settled the case $k = 4$ under the assumption that $\Delta(G) \neq 5$. The case $k = 4$ with $\Delta(G) = 5$ was proved independently by Wang [70] and Farzad in his Ph.D. thesis [19]. Lih and Wang [45] considered the case $k = 5$; their proof contains a small flaw, but it can be easily fixed. The first correct proof for $k = 6$ was given in 2005 by Farzad [19].

Richter's planar graph problem

Thomassen's five-colour theorem tells us that every planar graph G of order n is f-list-colourable if $f(v) = 5$ for all $v \in V(G)$; this gives $\sum_v f(v) = 5n$. On the other hand, if D is an acyclic orientation of a planar graph G with n vertices, then D is kernel-perfect and Theorem 2.1 implies that G is f-list-colourable if $f(v) = d^+_D(v) + 1$ for all $v \in V(G)$; this yields $\sum_v f(v) = |E(G)| + n \leq 4n - 6$.

In 2008 Hutchinson [30] published results on list-colourings of subclasses of planar graphs where, for each $v \in V(G), f(v) = \min\{d(v), k\}$ for some given integer k. She proved that a 2-connected outerplanar bipartite graph is f-list-colourable if $f(v) = \min\{d(v), 4\}$ for all $v \in V$, and a 2-connected outerplanar near-triangulation is f-list-colourable if $f(v) = \min\{d(v), 5\}$ for all $v \in V$ except when the graph is K_3 with identical 2-lists. In the same paper she mentioned the following problem of Bruce Richter.

Problem *Let G be a planar 3-connected graph that is not a complete graph. If $f(v) = \min\{d(v), 6\}$ for all $v \in V(G)$, is G f-list-colourable?*

From Thomassen's result on 5-list-colourability of planar graphs it follows immediately that the answer is 'yes' if G has minimum degree at least 5. Moreover, if G has maximum degree at most 6, then $f(v) = d(v)$ for all $v \in V(G)$ and G is f-list-colourable unless each block of G is a complete graph or an odd cycle [18]. In [12] it is observed that there are planar 2-connected non-complete graphs G with minimum degree $\delta(G) = 3$ and list-assignments L with $|L(v)| = \min\{d(v), 6\}$ for which G is not L-list-colourable. Thus, the answer is negative if the assumption of '3-connectivity' is relaxed to '2-connectivity' or $\delta(G) = 3$.

Hutchinson observed that the requirement of planarity cannot be weakened to having no K_5-minor. If we replace $f(v) = \min\{d(v), 6\}$ by $f(v) = \min\{d(v), 5\}$, then the answer is negative, even if the distance between vertices of degree smaller than 5 is at least 4, as shown by an example presented by Zs. Tuza and M. Voigt [64]. On the other hand, if $f(v) = \min\{d(v), 6\}$ and the distance between vertices of degree smaller than 6 is at least 3, then a planar graph is f-list-colourable, by a result of Albertson [1].

Theorem 3.12 *Suppose that G is r-list-colourable and $W \subseteq V(G)$ is such that the distance between any two vertices in W is at least 3. Let L be a list-assignment such that $|L(v)| \geq r+1$ for all $v \in V(G) \setminus W$. Then, any precolouring of W can be extended to an L-colouring of G.*

This result can be extended by some easy arguments. Let G be a planar graph with $f(v) = \min\{d(v), 6\}$ for all $v \in V$, and let H be the subgraph of G induced by the set of vertices of degree smaller than 6. If any two components of H have distance at least 5 in G, then G is L-colourable for every f-assignment L. To see this, choose for each component H_i of H a vertex $v_i \in V(G) \setminus V(H)$ for which v_i is adjacent to a vertex w_i in H_i. Colour v_i with a colour $\alpha \in L(v_i) \setminus L(w_i)$. Note that $|L(v_i)| > |L(w_i)|$, because of the choice of the vertices. By Theorem 3.12, we can colour the remaining vertices of $G - V(H)$. After that, every vertex w of H has still at least $d_H(w)$ available colours; moreover, for each component of H, the specified vertex w_i has at least $d_H(w_i) + 1$ available colours. By a result of Kostochka, Stiebitz and Wirth (see Chapter 2, ref. 30), the colouring can be completed. The general problem is still open. Some more partial results and references are given in [12].

4. Precolouring extensions

Extending colourings of an induced subgraph of a graph to the entire graph has become an important tool in graph colouring. A colouring of a complete subgraph can always be extended to an optimal colouring of the entire graph. For an edgeless subgraph, however, this need not to be the case. To see this, let G be the graph whose vertex-set consists of p pairwise disjoint sets $X_1, X_2, \ldots, X_p$, each with $k-1$ vertices, and $p+1$ additional vertices $v_1, v_2, \ldots, v_{p+1}$ for which $X_1, X_2, \ldots, X_p$ are cliques in G not joined by any edge, $N(v_1) = X_1, N(v_{p+1}) = X_p$ and $N(v_i) = X_{i-1} \cup X_i$ for $i = 2, 3, \ldots, p$. Then $\chi(G) = k$, and in any k-colouring of G the vertices v_1 and v_{p+1} receive the same colour. Note that G is planar if $k \leq 4$ and the distance between v_1 and v_{p+1} is $p+2$. The example is thus in sharp contrast to the following result of Albertson [1].

Theorem 4.1 *Let G be a planar graph, and let $W \subseteq V(G)$ such that the distance between any two vertices of W is at least* 4. *Then any* 5*-colouring of W can be extended to a* 5*-colouring of G.*

As pointed out by Albertson, the threshold for the distance is best possible. This result motivated Albertson [1] to propose the following question, which has received much attention.

Problem *Does there exist an integer d for which, if G is a planar graph and L is a* 5*-assignment for G, then any colouring of a vertex-set $W \subseteq V(G)$ can be extended to an L-colouring of G, provided that the distance between any two vertices of W is at least d?*

An example showing that $d \geq 5$ was presented by Tuza and Voigt [64]. Dvořák *et al.* [14] found an upper bound for the distance d, which is certainly not tight.

Theorem 4.3 *Let G be a planar graph, let L be a* 5*-assignment, and let $W \subseteq V(G)$ be such that the distance between any two vertices of W is at least* 19 828. *Then any colouring of W can be extended to an L-colouring of G.*

Let G be a graph and let L be a list-assignment for G. Furthermore, let H be an induced subgraph of G and let φ be an L-colouring of H. To decide whether this precolouring can be extended to an L-colouring of G, we define a list-assignment L' of $G' = G - V(H)$ by $L'(v) = L(v) \setminus \{\varphi(u) : u \in N(v) \cap V(H)\}$. Note that φ can be extended to an L-colouring of G if and only if G' admits an L'-colouring. This observation can be used to prove Theorem 3.12.

Axenovich [8] and (independently) Albertson, Kostochka and West [2] proved a precolouring extension of Brooks's theorem. In both papers it is proved that the threshold for the distance bound is sharp. However, Rackham [53] improved this bound to 4, provided that the graph is 3-edge-connected.

Theorem 4.4 *Let G be a graph with maximum degree $\Delta \geq 3$ containing no $K_{\Delta+1}$, and let $W \subseteq V(G)$ be such that the distance between any two vertices of W is at* *least* 8. *Furthermore, let L be a* Δ*-assignment for G. Then, every precolouring of W can be extended to an L-colouring of G.*

For the ordinary colouring problem there are many results where the set of precoloured vertices is not an independent set; for references, see [52]. The proof of the following list-colouring result about precolourings of outerplanar graphs can be found in [52].

Theorem 4.5 *Let G be an outerplanar graph, and let* $W \subseteq V(G)$ *be such that* $\langle W \rangle$ *is bipartite. Assume that the shortest distance between components of* $\langle W \rangle$ *is at least* 7. *If L is a* 4*-assignment for* $G - W$*, then any precolouring of* $\langle W \rangle$ *with two colours for each component can be extended to an L-colouring of G.*

5. Notes

The list-colouring concept was introduced in the 1970s, but in the following years only a few papers appeared on this subject. The situation changed in the early 1990s with the publication of the pioneering works of Alon and Tarsi [6], Alon [3], [4], Voigt [66], Thomassen [57] and Galvin [23]. Since then, the area has flourished and has attracted much attention, while the number of publications has increased immensely.

Several survey papers have helped readers to keep track of the subject. The first summary, published in 1997 by Tuza [61], contains more than eighty references on list-colourings. This survey was updated two years later by Kratochvíl, Tuza and Voigt [40], who discussed new trends in this area of research, including colouring extensions, set choosability, partial colourings, oriented colourings and the new concept of mixed hypergraphs.

The survey in 2001 by Woodall [72] covers more recent topics, such as defective or improper list-colourings. A typical result in this area says that if G is a planar graph and L is a 3-assignment, then there is an improper L-colouring of G for which each colour class induces a subgraph with maximum degree at most 2. This result was proved by Škrekovski and (independently) by Eaton and Hull (see [72, refs. 55 and 12]). It is not known whether, for every planar graph and every 4-assignment, there exists an improper list-colouring for which every colour class induces a subgraph with maximum degree at most 1.

As we saw in the Introduction, k-colourable graphs need not be k-list-colourable unless all lists are the same or the lists are pairwise different. A graph is called (k,p)-*list-colourable* if G is L-colourable for each k-assignment L for which $|L(v) \cap L(w)| \leq p$ whenever $vw \in E(G)$. As observed by Kratochvíl, Tuza and Voigt [40], every planar graph is (4,1)-list-colourable; this follows from the fact that if G is a simple planar graph, then G admits an orientation D for which $d^+(v) \leq 3$. Let L be a 4-assignment for G for which $|L(v) \cap L(w)| \leq 1$ for every edge vw of G. To find an L-colouring of G choose, for each vertex v of D, a colour from $L(v)$ that does not occur in the lists of its predecessors. While it is known that not every planar graph is (4,3)-list-colourable, it remains unknown whether every planar graph is (4,2)-list-colourable.

It is folklore that the chromatic number is a submultiplicative function with respect to the graph union – that is, $\chi(G \cup H) \le \chi(G) \times \chi(H)$. Assign to each vertex of $G \cup H$ a pair of colours, where the first colour refers to a colouring of G and the second colour refers to a colouring of H. This defines a colouring of $G \cup H$.

Conjecture D *For graphs G and H,* $\chi_\ell(G \cup H) \le \chi_\ell(G) \times \chi_\ell(H)$.

The truth of this conjecture is implied by the weak $(a\!:\!b)$-choosability conjecture (Conjecture **A**). To see this, suppose that $\chi_\ell(G) = k$ and $\chi_\ell(H) = h$. Let L be a kh-assignment for $G \cup H$, where we assume that $V(G) = V(H)$. The weak $(a\!:\!b)$-choosability conjecture implies that G is $(kh\!:\!h)$-list-colourable. Thus, for each vertex v of G, there is a set $L'(v) \subseteq L(v)$ with $|L'(v)| = h$ for which $L'(v) \cap L'(w) = \varnothing$ whenever vw is an edge of G. So L' is a h-assignment for H and so, since $\chi_\ell(H) = h$, there is an L'-colouring of H which is an L-colouring of $G \cup H$. Note that Conjecture **D** is true if G or H is a 2-list-colourable graph, since the weak $(a\!:\!b)$-choosability conjecture is proved for $a = 2, b = 1$. Moreover, if G or H is a k-degenerate graph with list-chromatic number $\chi_\ell = k + 1$, then it is $(\chi_\ell m\!:\!m)$-list-colourable for all m. Thus, Conjecture **D** is also true for this class of graphs which contains (for example) chordal graphs. If $G = F \cup H$ is the union of a forest F which is 2-list-colourable and a 2-degenerate graph H, then $\chi(G) \le \chi_\ell(G) \le 6$. M. Tarsi (see [31, Problem 4.2]) has asked whether $\chi(G) \le 5$.

As observed by N. Alon, it follows from Theorem 1.2 that there exists a function $f = f(k, \ell)$ for which $\chi_\ell(G \cup H) \le f(\chi_\ell(G), \chi_\ell(H))$.

The weak $(a\!:\!b)$-choosability conjecture says that any $(a\!:\!b)$-list-colourable graph is $(am\!:\!bm)$-list-colourable for all $m \ge 1$. Motivated by this conjecture, it is natural to investigate the set

$$CH(G) = \{a/b : G \text{ is } (a\!:\!b)\text{-list-colourable}\}.$$

Gutner proved in his thesis [27] that the elements of $CH(G)$ can be arbitrarily close to $\chi(G)$. The *choice ratio* of a graph G is defined by $\mathrm{chr}(G) = \inf CH(G)$. It is not hard to see that every odd cycle C_{2k+1} is $(2k + 1\!:\!h)$-list-colourable for all $h \le k$, and that the choice ratio of C_{2k+1} is $2 + (1/k)$ (see [7]), which is also the fractional chromatic number. That the equality between choice ratio and fractional chromatic number holds for all graphs was proved in 1997 by Alon, Tuza and Voigt [7]; in fact, there are two different proofs, one probabilistic and the other deterministic. Moreover, it follows that the infimum in the definition of the choice ratio can be replaced by the minimum. However, it is an open problem to find small integers a and b such that $\mathrm{chr}(G) = a/b$ and G is $(a : b)$-list-colourable. The proof yields only very huge integers with this property. Nevertheless, $\mathrm{chr}(G) \le \chi(G)$ for all G, in contrast to the list-chromatic number which can be arbitrarily large compared with the chromatic number.

The $(f:g)$-choosability concept is closely related to list-set-colourings or colourings of weighted graphs. A *weighted graph* is a pair (G, g) consisting of a simple graph G and a function $g : V(G) \to \mathbb{N}$. A colouring of (G, g) with colour-set C is a map φ that assigns to each vertex v a set $\varphi(v) \subseteq C$ of $g(v)$ colours for which $\varphi(v) \cap \varphi(w) = \varnothing$ whenever $vw \in E(G)$. If L is an list-assignment for G, then a colouring φ of (G, g) is called an *L-colouring* if $\varphi(v) \subseteq L(v)$ for all $v \in V(G)$. The weighted graph (G, g) is called *f-list-colourable* if (G, g) admits an L-colouring for every f-assignment L of G. If $f(v) = k$ for all $v \in V(G)$, the corresponding term is *k-list-colourable*. Note that (G, g) is f-list-colourable if and only if G is $(f:g)$-list-colourable. The *chromatic number* $\chi(G, g)$ is the least integer k for which (G, g) has a colouring with a set of k colours. The *list-chromatic number* $\chi_\ell(G, g)$ is the least integer k for which (G, g) is k-list-colourable.

It is also well known that the colouring problem for a weighted graph can be reduced to an ordinary colouring problem. For a weighted graph (G, g), let $G[g]$ denote the graph obtained from G by replacing each vertex v of G by a complete graph K^v of order $g(v)$, where all these complete graphs are vertex-disjoint, and by joining K^v and K^w by all possible edges whenever $vw \in E(G)$; the resulting graph $G[g]$ is called an *inflation* of G. If $g(v) = n$ for all $v \in V(G)$, then $G[g] = G[K_n]$ is *a uniform inflation* of G.

It is also well known that the fractional chromatic number $\chi^*(G)$ of a simple graph G satisfies $\chi^*(G) = \lim_{n\to\infty} \chi(G[K_n])/n$. It is easy to show that $\chi(G, g) = \chi(G[g])$. On the other hand, for the list-chromatic number we have only that $\chi_\ell(G, g) \le \chi_\ell(G[g])$, and it is unknown whether equality always holds. For a weighted graph (G, g),

$$\omega(G[g]) \le \chi(G, g) \le \chi_\ell(G, g) \le \chi_\ell(G[g]) \le \mathrm{col}(G[g]) \le \Delta(G[g]) + 1, \qquad (4)$$

where

$$\omega(G[g]) = \max\left\{\sum_{v\in X} g(v) \ : \ X \subseteq V(G) \text{is a clique}\right\}$$

and

$$\Delta(G[g]) = \max\left\{g(v) + \sum_{u\in N(v)} g(u) \ : \ v \in V(G)\right\} - 1.$$

The list version of Brooks's theorem implies that every weighted connected graph (G, g) with $g(v) \ge 1$ for all $v \in V(G)$ satisfies $\chi_\ell(G, g) \le \Delta(G[g]) + 1$, where equality holds if and only if G is a complete graph, or $g = 1$ and G is an odd cycle. As proved by Stiebitz, Tuza and Voigt [56], this result can be improved as follows. If (G, g) is a weighted connected graph such that $g(v) \ge 1$ for all $v \in V(G)$, then

$$\chi_\ell(G, g) \le \Delta(G[g]) + 1 - \min_{v\in V(G)} g(v),$$

unless G is a complete graph or an odd cycle. The proof uses Galvin's kernel method and the following result of Stiebitz, Tuza and Voigt. If G is a simple connected graph, then there is a kernel-perfect digraph D such that G is the underlying graph of D and $d^+(v) \leq d(v) - 1$ for all $v \in V(G)$, unless each block of G is a complete graph or an odd cycle.

In the study of colourings of graphs, critical graphs play an important role. A graph G is said to be *L-critical*, where L is a list-assignment for G, if every proper subgraph of G is L-colourable, but G itself is not L-colourable. If $L(v) = \{1, 2, \ldots, k-1\}$ for all $V \in V(G)$, where $k \geq 1$, we also use the term *k-critical.* The graph G is called *k-list-critical* if G is L-critical for some $(k-1)$-assignment L for G. Clearly, every k-critical graph is k-list-critical, but not conversely. A graph G is k-critical if and only if $\chi(H) < \chi(G) = k$ for every proper subgraph H of G. Thus, a k-critical graph cannot contain another k-critical graph as a proper subgraph. However, this may happen for k-list-critical graphs. To see this, let G be the graph obtained from two disjoint copies K, K' of K_k by adding an edge xx' with $x \in V(K)$ and $x' \in V(K')$. Let L be the $(k-1)$-assignment with $L(v) = \{1, 2, \ldots, k-1\}$ if $v \in V(G) \setminus \{x, x'\}$, and $L(x) = L(x') = \{2, 3, \ldots, k\}$. Then G is L-critical, and hence k-list-critical, but not k-critical. Clearly, K_k is a k-list-critical graph, and for $k = 1$ or 2 it is the only k-list-critical graph. It follows from König's characterization of bipartite graphs that the only 3-critical graphs are the odd cycles.

By a *minimal k-list-critical* graph we mean a k-list-critical graph that contains no other k-list-critical graph as a proper subgraph. It is easy to show that a graph G is minimal k-list-critical if and only if $\chi_\ell(H) < \chi_\ell(G) = k$ for every proper subgraph H of G. Thus, a graph G satisfies $\chi_\ell(G) \leq k-1$ if and only if G contains no minimal k-list-critical subgraph. Erdős, Rubin and Taylor [18] provided a characterization of 2-list-colourable graphs in terms of forbidden subgraphs. A graph G is minimal 3-list-critical if and only if G belongs to one of the following six types:

- an odd cycle
- two vertex-disjoint even cycles joined by a path
- two even cycles with exactly one vertex in common
- two vertices joined by three internally-disjoint even paths, where the length of at least two of the paths is not exactly 2
- two vertices joined by three internally disjoint odd paths
- two vertices joined by four internally disjoint even paths, where the length of at least three of the paths is exactly 2.

It seems hopeless to characterize the class of simple graphs G satisfying $\chi_\ell(G) = \chi(G)$. However, the situation changes if we require that the condition holds also for every subgraph. A graph G is said to be *strong choice-perfect* if $\chi_\ell(H) = \chi(H)$ for every subgraph H of G. The following result was obtained by P. Mihok and M. Voigt (unpublished).

Theorem 5.1 *A graph G is strong choice-perfect if and only if it contains no bipartite minimal* 3*-list-critical graph as a subgraph.*

Clearly, a strong choice-perfect graph cannot contain a bipartite minimal 3-list-critical subgraph. But suppose that G does not contain a bipartite minimal 3-list-critical graph as a subgraph, and let H be a subgraph of G. We show that $\chi_\ell(H) = \chi(H)$. The proof is by induction on $k = \chi(H)$. For $k = 2$, this follows from the assumption, since G contains no bipartite non-2-list-colourable subgraph. So assume $k \geq 3$. Consider a proper k-colouring of H and let $V_1, V_2, \ldots, V_k$ denote the k colour classes. Let $H_1 = H\langle V_1 \cup V_2 \cup \cdots \cup V_{k-1}\rangle$ and let $H_2 = (V, E_2)$ be the bipartite graph with vertex-set $V = V(H)$ and edge-set $E_2 = \{e \in E(H) : e \text{ has an endpoint in } V_k\}$. Clearly, $\chi(H_1) = k-1$ and $\chi(H_2) = 2$. It then follows from the induction hypothesis that $\chi_\ell(H_1) = k - 1$ and $\chi_\ell(H_2) = 2$. Let L be an arbitrary k-assignment for $H = H_1 \cup H_2$. By a result from [64], there exists a colour $c_v \in L(v)$ for each vertex $v \in V_k$ such that, for every vertex $w \in V(H_1)$, the reduced list $L'(w) = L(w) \setminus \{c_v : wv \in E_2\}$ satisfies $|L'(w)| \geq k-1$. Since $\chi_\ell(H_1) = k-1$, it follows that H_1 has an L'-colouring. Clearly, this leads to an L-colouring of H, as required.

We define a graph G to be *choice-perfect* if $\chi_\ell(H) = \chi(H)$ for every induced subgraph H of G. To characterize choice-perfect graphs it is not sufficient to exclude bipartite minimal 3-list-critical graphs. So we have $\chi(K_{3,3}) = 2$ and $\chi_\ell(K_{3,3}) = 3$ and $K_{3,3}$ contains no induced bipartite minimal 3-list-critical subgraph. The recent state of research concerning choice-perfect graphs is discussed in [62].

A conjecture related to this topic was made by Voigt in 1994. Let $G = (V, E)$ be a simple graph with $\chi(G) = k$. If every subgraph of G induced by two colour classes is 2-list-colourable, then G is k-list-colourable. This conjecture was disproved by Glenn Chappell (personal communication) in 2004 by the following example. Let G be the Cartesian product of K_3 and $K_{1,6}$: this graph can be drawn nicely as seven triangles – one in the centre, and the rest in a ring around it. Each triangle in the ring is then joined to the one in the centre via a matching. Every 3-colouring of this graph has the property that the union of any pair of colour classes induces a forest. However, the graph is not 3-list-colourable. To see this, give each vertex in the centre triangle the list {1,2,3}. Pick one of the ring triangles and give its vertices the lists $\{1,4,5\}$, $\{2,4,5\}$ and $\{3,4,5\}$. Now we cannot colour the vertices in the centre triangle 1, 2 and 3. There are six possible colourings of the centre triangle. We can use each triangle of the ring to forbid one of them, as above.

References

1. M. O. Albertson, You can't paint yourself into a corner, *J. Combin. Theory (B)* **73** (1998), 189–194.
2. M. O. Albertson, A. V. Kostochka and D. B. West, Precoloring extension of Brooks' theorem, *SIAM J. Discrete Math.* **18** (2004), 542–553.
3. N. Alon, Choice numbers of graphs; a probabilistic approach, *Combin. Probab. Comput.* **1** (1992), 107–114.
4. N. Alon, Restricted colorings of graphs, *Surveys in Combinatorics, 1993* (ed. K. Walker), Cambridge University Press (1993), 1–33.
5. N. Alon, The combinatorial Nullstellensatz, *Combin. Probab. Comput.* **8** (1999), 7–29.

6. N. Alon and M. Tarsi, Colorings and orientations of graphs, *Combinatorica* **12** (1992), 125–134.
7. N. Alon, Zs. Tuza and M. Voigt, Choosability and fractional chromatic number, *Discrete Math.* **165/166** (1997), 31–38.
8. M. Axenovich, A note on graph coloring extensions and list colorings, *Electronic J. Combin.* **10** (2003), N1.
9. O. V. Borodin, Colorings of planar graphs: a survey, *Discrete Math.* **313** (2013), 517–539.
10. O. V. Borodin, A. V. Kostochka and D. R. Woodall, List edge and list total colourings of multigraphs, *J. Combin. Theory (B)* **71** (1997), 184–204.
11. L. Cai, W. Wang and X. Zhu, Choosability of toroidal graphs without short cycles, *J. Graph Theory* **65** (2010), 1–15.
12. D. W. Cranston, A. Pruchnewski, Zs. Tuza and M. Voigt, List colorings of K_5-minor-free graphs with special list assignments, *J. Graph Theory* **71** (2012), 18–30.
13. M. DeVos, K. Kawarabayashi and B. Mohar, Locally planar graphs are 5-choosable, *J. Combin. Theory (B)* **98** (2008), 1215–1232.
14. Z. Dvořák, B. Lidický, B. Mohar and L. Postle, 5-list-coloring planar graphs with distant precolored vertices, *arXiv:1209.0366 [math.CO].*
15. Z. Dvořák, B. Lidický and R. Škrekovski, Planar graphs without 3-, 7-, and 8-cycles are 3-choosable, *Discrete Math.* **309** (2009), 5899–5904.
16. Z. Dvořák, B. Lidický and R. Škrekovski, 3-choosability of triangle-free planar graphs with constraints on 4-cycles, *SIAM J. Discrete Math.* **24** (2010), 934–945.
17. M. N. Ellingham and L. Goddyn, List edge colourings of some 1-factorizable multigraphs, *Combinatorica* **16** (1996), 343–352.
18. P. Erdős, A. L. Rubin and H. Taylor, Choosability in graphs, *Congr. Numer.* **XXVI** (1979), 125–157.
19. B. Farzad, *On Four Colourings of Graphs*, Ph.D. thesis, University of Toronto, 2005.
20. B. Farzad, Planar graphs without 7-cycles are 4-choosable, *SIAM J. Discrete Math.* **23** (2009), 1179–1199.
21. G. Fijavž, M. Juvan, B. Mohar and R. Škrekovski, Planar graphs without cycles of specific length, *European J. Combin.* **23** (2002), 377–388.
22. H. Fleischner and M. Stiebitz, A solution to a colouring problem of P. Erdős, *Discrete Math.* **101** (1992), 39–48.
23. F. Galvin, The list chromatic index of a bipartite multigraph, *J. Combin. Theory (B)* **63** (1995), 153–158.
24. S. Gravier and F. Maffrey, Choice number of 3-colorable elementary graphs, *Discrete Math.* **165/166** (1997), 353–358.
25. H. Grötzsch, Ein Dreifarbensatz für dreikreisfreie Netze auf der Kugel, *Wiss. Zeitschrift der Universität Halle, Math.-Nat.* **VIII/1** (1958), 109–119.
26. J.-L. Guo and Y.-L. Wang, 3-list-coloring planar graphs of girth 4, *Discrete Math.* **311** (2011), 413–417.
27. S. Gutner, *Choice Numbers of Graphs*, M.Sc. thesis, Tel Aviv University, 1992.
28. S. Gutner, The complexity of planar graph choosability, *Discrete Math.* **159** (1996), 119–130.
29. A. J. W. Hilton and P. D. Johnson, The Hall number, the Hall index, and the total Hall number of a graph, *Discrete Appl. Math.* **94** (1999), 227–245.
30. J. P. Hutchinson, On list-coloring outerplanar graphs, *J. Graph Theory* **59** (2008), 59–74.
31. T. R. Jensen and B. Toft, *Graph Coloring Problems,* Wiley Interscience, 1995.
32. M. Juvan, B. Mohar and R. Škrekovski, List-total colorings of graphs, *Combin. Probab. Comput.* **7** (1998), 181–188.

33. M. Juvan, B. Mohar and R. Škrekovski, Graphs of degree 4 are 5-edge-choosable, *J. Graph Theory* **32** (1999), 250–262.
34. M. Juvan, B. Mohar and R. Thomas, List edge-coloring of series-parallel graphs, *Electronic J. Combin.* **6(1)** (1999), R42.
35. J. Kahn, Asymtotically good list colourings, *J. Combin. Theory (A)* **73** (1996), 1–59.
36. K. Kawarabayashi and C. Thomassen, From the plane to higher surfaces, *J. Combin. Theory (B)* **102** (2012), 852–868.
37. A. V. Kostochka, Color-critical graphs and hypergraphs with few edges: a survey, *Bolyai Society Mathematical Studies* **15** (2006), 175–197.
38. A. V. Kostochka and D. R. Woodall, Choosability conjecture and multicircuits, *Discrete Math.* **240** (2001), 123–143.
39. A. V. Kostochka and D. R. Woodall, Total choosability of multicircuits I, II, *J. Graph Theory* **40** (2002), 26–43, 44–67.
40. J. Kratochvíl, Zs. Tuza and M. Voigt, New trends in the theory of graph colorings: Choosability and list colorings, *DIMACS, Series in Discrete Mathematics and Theoretical Computer Science* **49** (1999), 183–198.
41. P. C. B. Lam, J. Liu and B. Xu, The 4-choosability of plane graphs without 4-cycles, *J. Combin. Theory (B)* **76** (1999), 117–126.
42. P. C. B. Lam, W. C. Shiu and B. Xu, On structure of some plane graphs with application to choosability, *J. Combin. Theory (B)* **82** (2001), 285–296.
43. K. Lih and W. Wang, Choosability, edge-choosability and total choosability of outerplane graphs, *Europ. J. Combin.* **22** (2001), 71–78.
44. K. Lih and W. Wang, Choosability and edge choosability of planar graphs without intersecting triangles, *SIAM J. Discrete Math.* **15** (2002), 538–545.
45. K. Lih and W. Wang, Choosability and edge choosability of planar graphs without five cycles, *Appl. Math. Lett.* **15** (2002), 561–565.
46. R. Li and B. Xu, Edge choosability and total choosability of planar graphs with no 3-cycles adjacent to 4-cycles, *Discrete Math.* **311** (2011), 2158–2163.
47. B. Liu, J. Hou and G. Liu, List edge and list total colorings of planar graphs without short cycles, *Inform. Process. Lett.* **108** (2008), 347–351.
48. M. Mirzakhani, A small non-4-choosable planar graph, *Bull. Inst. Combin. Appl.* **17** (1996), 15–18.
49. M. Montassier, A. Raspaud and W. Wang, Bordeaux 3-color conjecture and 3-choosability, *Discrete Math.* **306** (2006), 573–579.
50. J. A. Noel, B. A. Reed and H. Wu, A proof of a conjecture of Ohba, manuscript, (2012).
51. A. Prowse and D. R. Woodall, Choosability and powers of circuits, *Graphs Combin.* **19** (2003), 137–147.
52. A. Pruchnewski and M. Voigt, Precoloring extension for K_4-minor-free graphs, *J. Graph Theory* **60** (2009), 284–294.
53. T. Rackham, A note on $K^-_{\Delta+1}$-free precolouring with Δ colours, *Electronic J. Combin.* **16** (2009), N28.
54. H. Sheng and Y. Wang, A structural theorem for planar graphs with some applications, *Discrete Appl. Math.* **159** (2011), 1183–1187.
55. R. Škrekovski, Choosability of K_5-minor-free graphs, *Discrete Math.* **190** (1998), 223–226.
56. M. Stiebitz, Zs. Tuza and M. Voigt, Orientations of graphs with prescribed weighted out-degrees, *Graphs Combin.* (2013).
57. C. Thomassen, Every planar graph is 5-choosable, *J. Combin. Theory (B)* **62** (1994), 180–181.
58. C. Thomassen, 3-list-coloring planar graphs of girth 5, *J. Combin. Theory (B)* **64** (1995), 101–107.

59. C. Thomassen, Many 3-colorings of triangle-free planar graphs, *J. Combin. Theory (B)* **97** (2007), 334–349.
60. C. Thomassen, Exponentially many 5-colorings of planar graphs, *J. Combin. Theory (B)* **97** (2007), 571–583.
61. Zs. Tuza, Graph colorings with local constraints – A survey, *Discuss. Math.– Graph Theory* **17** (1997), 161–228.
62. Zs. Tuza, Choice-perfect graphs, *Discuss. Math.– Graph Theory* **33** (2013), 231–242.
63. Zs. Tuza and M. Voigt, On a conjecture of Erdős, Rubin and Taylor, *Discuss. Math.– Graph Theory* **9** (1996), 69–82.
64. Zs. Tuza and M. Voigt, List colorings and reducibility, *Discrete Appl. Math.* **79** (1997), 247–256.
65. V. G. Vizing, Coloring the vertices of a graph in prescribed colors, (in Russian), *Metody Diskret. Anal. v Teorii Kodov i Schem* **29** (1976), 3–10.
66. M. Voigt, List colorings of planar graphs, *Discrete Math.* **120** (1993), 215–219.
67. M. Voigt, A not 3-choosable planar graph without 3-cycles, *Discrete Math.* **146** (1995), 325–328.
68. M. Voigt, Choosability of planar graphs, *Discrete Math.* **150** (1996), 457–460.
69. M. Voigt, A non-3-choosable planar graph without cycles of length 4 and 5, *Discrete Math.* **294** (2005), 297–301.
70. W. Wang, Edge choosability of planar graphs without short cycles, *Science in China (A)– Mathematics* **48** (2005), 1531–1544.
71. Y. Wang, Q. Wu and L. Shen, Planar graphs without cycles of length 4, 7, 8 or 9 are 3-choosable, *Discrete Appl. Math.* **159** (2011), 232–239.
72. D. R. Woodall, List colourings of graphs, *Surveys in Combinatorics, 2001* (ed. J.W.P. Hirschfeld), London Math. Soc. Lecture Notes **288**, Cambridge University Press (2001), 269–301.
73. L. Zhang and B. Wu, Edge choosability of planar graphs without small cycles, *Discrete Math.* **283** (2004), 289–293.
74. L. Xiaofang, The 4-choosability of toroidal graphs without intersecting triangles, *Inform. Process. Lett.* **102** (2007), 85–91.
75. B. Xu, $(4m,m)$-choosability of plane graphs, *J. Systems Sci. Complexity* **14** (2001), 174–178.
76. B. Xu, On structure of graphs embedded on surfaces of nonnegative characteristic with application to choosability, *Discrete Math.* **248** (2002), 283–291.

7

Perfect graphs

NICOLAS TROTIGNON

Perfect graphs were defined by Claude Berge in the 1960s. They are important objects for graph theory, linear programming and combinatorial optimization. Berge made a conjecture about them (now called the strong perfect graph theorem or SPGT) which was proved by Chudnovsky, Robertson, Seymour and Thomas in 2002. This survey about perfect graphs mostly focuses on the SPGT.

1. Introduction

Every graph G clearly satisfies $\chi(G) \geq \omega(G)$, where $\omega(G)$ is the clique number of G, because the vertices of a clique must receive different colours. A graph G is *perfect* if every induced subgraph H of G satisfies $\chi(H) = \omega(H)$. A chordless cycle of length $2k + 1$, for $k \geq 2$, satisfies $3 = \chi > \omega = 2$, and its complement satisfies $k + 1 = \chi > \omega = k$; these graphs are therefore *imperfect*. Since perfect graphs are closed under taking induced subgraphs, they must be defined by excluding a family $\mathcal{F}$ of graphs as induced subgraphs. The *strong perfect graph theorem* (SPGT for short) states that the two examples just given are the only members of $\mathcal{F}$.

Let us make this more formal. A *hole* in a graph G is an induced cycle of length at least 4, and an *antihole* is a hole of $\overline{G}$. A graph is *Berge* if it does not contain an odd hole or an odd antihole. The following was conjectured by Berge [3] in the 1960s and was the object of much research until it was finally proved in 2002 by Chudnovsky, Robertson, Seymour and Thomas [13].

Theorem 1.1 (Strong perfect graph theorem) *A graph is perfect if and only if it is Berge.*

One direction of the proof is easy: every perfect graph is Berge since, as we observed above, odd holes and antiholes satisfy $\chi = \omega + 1$. The proof of the converse statement is very long and relies on structural graph theory. The main step is a *decomposition theorem* (Theorem 5.1), which states that every Berge graph is either in a well-understood *basic* class of perfect graphs (see Section 3), or has some *decomposition* (see Section 4). The goal of this survey is to make this result and its meaning understandable to a non-specialist mathematician. For this purpose, not only the proof is surveyed, but also some results that were seminal to it, and some that were proved after it. Another goal is to present perfect graphs as a lively subject for researchers, by mentioning several attractive open questions.

We start with an open question that has the same flavour as the SPGT. In 1987 Gyárfás [35] proposed the following generalization of a perfect graph: a graph G is *χ-bounded by a function f* if every induced subgraph H of G satisfies $\chi(H) \leq f(\omega(H))$; thus, perfect graphs are χ-bounded by the identity function. There might exist a short proof that, for some function f (possibly fastly increasing), all Berge graphs are χ-bounded by f, but so far the proof of the SPGT is the only known proof of this fact. In [35] the following conjecture is stated.

Conjecture A *There exists a function f such that, for all graphs G, if G has no odd hole, then G is χ-bounded by f.*

The SPGT can be rephrased as follows: perfect graphs and Berge graphs form the same class. We still keep two names for the same class, and use *Berge* when we concentrate on excluding holes and antiholes, and *perfect* when we focus on colourings. Note that the distinction between *Berge* and *perfect* has to be kept when we sketch the proof of the SPGT, since in the proof the two classes are supposed to be potentially different. A *minimally imperfect graph* is an imperfect graph, every proper induced subgraph of which is perfect. A restatement of the SPGT is thus that minimally imperfect graphs are precisely the odd holes and antiholes. Many statements about minimal imperfect graphs are therefore trivial to check by using the SPGT, but when proving the SPGT it is essential to prove them by other means. Observe that a minimal (and therefore minimum) counter-example to the SPGT has to be a Berge minimally imperfect graph.

We mostly follow the terminology from [13], which is not fully standard. Since the whole theory deals with induced subgraphs, we write *G contains H* to mean that H is an induced subgraph of G and write *path* instead of chordless path or induced path.

When a and b are vertices of a path P, we denote by aPb the subpath of P whose ends are a and b. A subset A of $V(G)$ is *complete* to a subset B of $V(G)$ if A and B are disjoint and every vertex of A is adjacent to every vertex of B; we also say that B is *A-complete*.

We use the prefix *anti* to mean a property or a structure of the complement (as for holes and antiholes). For instance, an *antipath* in G is a path in $\overline{G}$, A is *anticomplete* to B means that no edge of G has an end in A and the other one in B, and G is *anticonnected* if its complement is connected. By $C(A)$ we mean the set of all A-complete vertices in G, and by $\overline{C}(A)$ the set of all A-anticomplete vertices. By a *colouring* of a graph, we mean an optimal colouring of the vertices.

Since we focus on the SPGT and its proof, most of this survey is on the structure of Berge graphs, and we omit many other important aspects of the theory, such as the ellipsoid method used by Gröstchel, Lovász and Schrijver [34] to colour any input perfect graph in polynomial time. The theory of semi-definite programming started there.

Several earlier surveys exist. The survey of Lovász [45] from the 1980s is still good, while two books are completely devoted to perfect graphs [4], [50], and contain a lot of material cited below. Part VI of Schrijver [55] is the most comprehensive survey on perfect graphs. Chudnovsky, Robertson, Seymour and Thomas [12] wrote a good survey just after their proof, while Roussel, Rusu and Thuillier [53] wrote a good survey about the long sequence of results, attempts and conjectures that finally led to the successful approach.

For more historical aspects, see Section 67.4g of Schrijver [55]. Berge and Ramírez Alfonsín ([50, pp.1–12]) wrote an article about the origin of perfect graphs, and Seymour [56] wrote an article, interesting even to a non-mathematician, telling the story of how the SPGT was proved.

2. Lovász's perfect graph theorem

As pointed out by Preissmann and Sebő (see [50, pp. 185–214]), the following theorem conveniently gives a weak (1) and a strong (2) characterization of perfect graphs. Hence, to prove perfection checking the weak one is enough, while to use perfection one may rely on the strong one. This allows a kind of leverage that is used often in what follows.

Theorem 2.1 *Let G be a graph. The following three statements are equivalent.*

(1) *G is perfect.*
(2) *For each induced subgraph H of G and each $v \in V(H)$, H contains a stable set that contains v and intersects all maximum cliques of G.*
(3) *For each induced subgraph H of G, H contains a stable set that intersects all maximum cliques of G.*

Proof To prove that (1) implies (2), consider an optimal colouring of H and the colour class S that contains v. Then S is a stable set that intersects all maximum cliques of H, for otherwise $\chi(H \setminus S) \geq \omega(H)$, which is a contradiction.

Trivially, (2) implies (3).

To prove that (3) implies (1), consider the following greedy colouring algorithm:

Step 1. Set $i \leftarrow 1$ and $H \leftarrow G$.
Step 2. While H is non-empty, consider a stable set of H, as in (3), give it colour i, and set $H \leftarrow H \setminus S$ and $i \leftarrow i + 1$.

Since ω decreases at each iteration in Step 2, this algorithm produces a colouring of G with $\omega(G)$ colours, and it can be processed for any induced subgraph of G. Thus, G is perfect. ■

Replicating a vertex v of a graph G means adding a new vertex v' adjacent to v and to all neighbours of v. Note that the property $\chi = \omega$ is not preserved by replication. As an example, consider the non-perfect graph G obtained by replicating one vertex of C_5. Clearly, $\chi(G) = \omega(G) = 3$. However, by replicating any vertex of degree 2 in G, a graph G' such that $\chi(G') = 4 > \omega(G') = 3$ is obtained. Thus, the following lemma due to Lovász [44] is more surprising than it may look at first glance.

Theorem 2.2 (Replication lemma) *Perfection is preserved by replication.*

Proof Let G be a perfect graph and let G' be obtained from G by replicating a vertex v. We show that G' satisfies Theorem 2.1(3). Let H be an induced subgraph of G'. We look for a stable set S that intersects all maximum cliques of H. If H contains at most one vertex from $\{v, v'\}$, then it is isomorphic to an induced subgraph of G, so clearly S exists. Otherwise, $H \setminus v'$ is perfect, so, by Theorem 2.1(2), there exists a stable set S that contains v and intersect all maximum cliques of $H \setminus v'$. In fact, S intersects all maximum cliques of H, because a maximum clique of H contains v if and only if it contains v'. ■

The following result was conjectured by Berge as the *weak perfect graph conjecture* and is now called the *perfect graph theorem*. We give the proof that, among the available ones, is most closely related to the rest of this chapter.

Theorem 2.3 *If a graph is perfect, then its complement is perfect.*

Proof Let G be a perfect graph. Construct a graph G' as follows: start from G and replicate every vertex v $\alpha_v - 1$ times, where α_v is the number of maximum stable sets of G that contain v; note that replicating -1 times means deleting the vertex, and replicating 0 times means doing nothing. From its construction, G' can be covered by k disjoint maximum stable sets that therefore form an optimal colouring of G'. Since G' is perfect, by Theorem 2.2, it follows that G' has a clique K' of size k. Since a clique and a stable set intersect in at most one vertex, K' intersects all maximum stable sets of G'. Now construct a clique K of G by taking, for each vertex of K', the vertex of G from which it is replicated. The clique K that we obtain intersects all maximum stable sets of G. By the same argument, a clique intersecting all maximum stable sets can be found in any induced subgraph of G. Thus the complement of G satisfies Theorem 2.1(3) and is therefore perfect. ■

In [35], the question of generalizing Theorem 2.3 is discussed, but since then, little progress has occurred in this direction (see [36]). In particular, the following neat generalization of Theorem 2.3 (due to Gyárfás) is still open.

Conjecture B *There exists a function f such that, for all graphs G, if G is χ-bounded by $x \mapsto x + 1$, then $\overline{G}$ is χ-bounded by f.*

The perfect graph theorem has a polyhedral proof found by Fulkerson [30], related to polyhedral characterizations of perfect graphs discovered by him and Chvátal [18]. Lovász [43] proved a deep characterization of perfect graphs suggested by Hajnal.

Theorem 2.4 *A graph G is perfect if and only if every induced subgraph H of G satisfies $\alpha(H)\,\omega(H) \geq |V(H)|$.*

This characterization implies that deciding the perfection of an input graph is a CoNP problem (see [50, pp. 185–214]). It can be proved by a simple argument, relying on linear algebra, discovered by Gasparian [32]. Since the characterization is self-complementary, it gives another proof of the perfect graph theorem. This characterization is the starting point of many developments of great signifiance, such as the theory of partitionable graphs surveyed by Preissmann and Sebő (see [50, pp. 185–214]). It has deep connections with combinatorial optimization, as explained in Cornuéjols [25].

3. Basic graphs

In this section, we survey the five basic classes that are used in the proof of the SPGT: bipartite graphs, their complements, line graphs of bipartite graphs, their complements, and double split graphs. Bipartite graphs are easily checked to be perfect, and so, by Theorem 2.3, their complements are perfect; this can be restated as: 'every bipartite graph G satisfies $\theta(G) = \alpha(G)$'. Since $|V(G)| = \theta(G) + \nu(G) = \alpha(G) + \tau(G)$ for any triangle-free graph G, we deduce that every bipartite graph G satisfies $\nu(G) = \tau(G)$; this can be rephrased as: 'the complements of line graphs of bipartite graphs are perfect'. Again applying Theorem 2.3, we deduce that line graphs of bipartite graphs are perfect; this can be restated as: 'every bipartite graph G satisfies $\Delta(G) = \chi'(G)$'. Theorem 2.3 thus implies the perfection of three of the four *historical basic classes* of perfect graphs: bipartite graphs, their complements, their line graphs and the complements of their line graphs.

We now turn our attention to the fifth class that is first presented in the proof of the SPGT: *double split graphs*. As presented in [13], this class is not closed under taking induced subgraphs, so we prefer to use *doubled graphs*, that are easily seen to form the class of induced subgraphs of double split graphs.

A *good partition* of a graph G is a partition (X, Y) of $V(G)$ (possibly, X or $Y = \varnothing$) such that:

- Every component of $G[X]$ has at most two vertices, and every anticomponent of $G[Y]$ has at most two vertices.
- For every component C_X of $G[X]$, every anticomponent C_Y of $G[Y]$, and every vertex v in $C_X \cup C_Y$, there exist at most one edge and at most one antiedge between C_X and C_Y that are incident to v.

A graph is *doubled* if it has a good partition. Doubled graphs are easily seen to be perfect, by a direct colouring argument. They are also closed under taking induced subgraphs and complements. A graph is *basic* if it belongs to at least one of the five classes defined here. For each basic class, the characterization by excluding induced subgraphs is known (see Beineke [2] for line graphs, and Alexeev, Fradkin and Kim [1] for doubled graphs); the recognition can be performed in polynomial time (see Lehot [40] or Roussopoulos [54] for line graphs and Maffray [46] for doubled graphs). Also the colouring and maximum clique problems can be solved in polynomial time (see Schrijver [55] for the historical classes and Maffray [46] for doubled graphs).

In this section, we have focused on the basic graphs that play an important role in the proof of the SPGT. But any class of graphs whose perfection is simple to prove can potentially serve as a basic class of a decomposition theorem, so all classes are potentially of interest. The book of Brandstädt, Le and Spinrad [6] is on general graph classes, but contains a lot of material on perfect graphs. The most complete catalogue of classes of perfect graphs seems to be that written by Hougardy [38], which describes 120 classes. Chapter 66 in Schrijver [55] also contains an extensive survey about classes of perfect graphs. Golumbic's book [33] surveys algorithmic aspects of several classes of perfect graphs.

The most important class that we omit from here is the seminal class of hole-free graphs; it is the first class with a decomposition theorem, has some connections with graph minor theory and tree-width, and is also important in fast graph-searching algorithms (see Sections 9.7–9.8 in Bondy and Murty [5]). Another important class that has connections with ordered sets is the class of comparability graphs, introduced by Gallai [31]; the English translation by Maffray and Preissmann ([50, pp. 25–66]) contains a short survey.

4. Decompositions

By a *decomposition* of a graph we mean a way to partition its vertices with some prescribed adjacencies. A decomposition is *useful* if it can be proved that a minimum counter-example to the SPGT cannot admit the decomposition. Indeed, suppose that we can prove a statement such as: 'every Berge graph is either basic or has a useful decomposition'. The SPGT can then be proved as follows: consider a minimum counter-example, which is thus a Berge graph, imperfect, and of minimum size. It does not admit a decomposition (because the decomposition is useful), and since it is imperfect, it cannot be basic, contradicting the statement.

A *2-join* of a graph G, first defined by Cornuéjols and Cunningham [26], is a partition (X_1, X_2) of $V(G)$ for which there exist disjoint non-empty sets $A_1, B_1 \subseteq X_1$, $A_2, B_2 \subseteq X_2$ satisfying:

- A_1 is complete to A_2, B_1 is complete to B_2, and these edges are the only ones between X_1 and X_2.
- $|X_1| \geq 3$ and $|X_2| \geq 3$.
- every component of $G[X_i]$ intersects A_i and B_i, for $i = 1$ and 2.
- if $|A_i| = |B_i| = 1$, then $G[X_i]$ is not a path of length 2 joining the members of A_i and B_i, for $i = 1$ and 2.

Cornuéjols and Cunningham [26] proved that a minimally imperfect graph admitting a 2-join must be an odd hole (so 2-joins are useful). Note that here, *path 2-joins* are allowed: these are 2-joins such that, for $i = 1$ and 2, $|A_i| = |B_i| = 1$ and $G[X_i]$ is a path from the unique vertex in A_i to the unique vertex in B_i. Some papers (mostly those involving Conforti, Cornuéjols or Vušković) restrict the notion of 2-joins to *non-path 2-joins*. Section 8 discusses the relevance of excluding path 2-joins. When X_i, A_i and B_i are as above, it is customary to set $C_i = X_i \setminus (A_i \cup B_i)$. It is easy to prove that in a Berge graph, all paths from A_i to B_i with interior in C_i have the same parity, since otherwise an odd hole exists. There are thus two kinds of 2-joins: *odd* 2-joins and *even* ones. If (X_1, X_2) is a 2-join of $\overline{G}$, then it is a *complement 2-join* of G.

When G is a graph and $A \subseteq V(G)$, we denote by $C(A)$ the set of vertices of G complete to A, and by $\overline{C}(A)$ the set of vertices of G anticomplete to A. A *homogeneous pair* (first defined by Chvátal and Sbihi [20], who proved that they are useful) is a pair of disjoint sets $A, B \subseteq V(G)$ for which $|A|, |B| \geq 2$, every vertex of A has a neighbour and a non-neighbour in B, every vertex of B has a neighbour and a non-neighbour in A, and the sets

$$A,\ B,\ C(A) \cap \overline{C}(B),\ \overline{C}(A) \cap C(B),\ C(A) \cap C(B) \text{ and } \overline{C}(A) \cap \overline{C}(B)$$

are all non-empty and partition $V(G)$.

The decompositions presented so far are nice in the following sense. When applied recursively, they yield decomposition trees of polynomial size that allow the solution of several problems; the machinery is too heavy to be presented here (see Section 10).

We now turn our attention to other cutsets that do not have this nice property. A *star cutset* in a graph G (first defined by Chvátal [19]) is a set S of vertices for which $G \setminus S$ is disconnected and S contains a vertex v called the *centre*, complete to $S \setminus v$. The following is a result of Chvátal [19].

Theorem 4.1 (Star cutset lemma) *A minimally imperfect graph has no star cutset.*

Proof Let G be a minimally imperfect graph, and suppose for a contradiction that G has a star cutset S centred at v. Let (X,Y) be a partition of $V(G \setminus S)$ for which $|X|$ and $|Y| \geq 1$ and X is anticomplete to Y. We now prove that G satisfies Theorem 2.1(3); this implies that G is perfect, giving the required contradiction. Since every proper induced subgraph of G is perfect, it remains to find the desired stable set in G. By Theorem 2.1(2), there exists a stable set A_X in $G[S \cup X]$ that contains v and intersects all maximum cliques of $G[S \cup X]$. A similar stable set A_Y exists in $G[S \cup Y]$. We see that $A_X \cup A_Y$ is then a stable set of G that intersects all maximum cliques of G. ■

It is quite easy to turn the above proof into a colouring algorithm that would, for instance, colour any perfect graph every induced subgraph of which is either basic or decomposable by a star cutset. The algorithm would output a stable set that intersects all maximum cliques and, following the lines of the proof of Theorem 2.1, would yield a colouring algorithm. Unfortunately, this algorithm does not run in polynomial time, because the star cutset can be very big (for instance, all the graph except for two vertices). In this case, the complexity analysis of the recursive calls leads to an exponential number of calls. There is a similar problem with the generalizations that we consider next. This is the main reason why the decomposition of Berge graphs does not lead to a polynomial-time colouring algorithm.

A *skew partition* of a graph G (first defined by Chvátal [19]) is a partition (A,B) of $V(G)$ for which $G[A]$ is not connected and $G[B]$ is not anticonnected; in this case, we say that B is a *skew cutset*. Following a prophetic insight that some *self-complementary* decomposition generalizing the star cutset should play a role, Chvátal conjectured that a minimally imperfect graph has no skew partition and (a less formal statement) that skew partitions should appear in the decomposition of Berge graphs. Observe that if a graph with at least five vertices and at least one edge has a star cutset, then it has a skew partition.

The proof of Chv́atal's conjectures escaped their investigators, but several fruitful attempts were made. In particular, many special kinds of skew partitions were proved not to be in minimal imperfect graphs (see Reed [51] for a survey). In the opposite direction, a generalization of skew partitions was proved to decompose all Berge graphs in the following theorem of Conforti, Cornuéjols and Vušković [23]. Here, a *double star cutset* in a graph G is a set $S \subseteq V(G)$ such that $G \setminus S$ is disconnected and S contains an edge vw for which every vertex of S is adjacent to at least one of v and w. Observe that antiholes of length at least 6 have double star cutsets. Note that, for a Berge graph G, the following yields information for both G and $\overline{G}$.

Theorem 4.2 *A graph with no odd hole is either basic, or has a 2-join or a double star cutset.*

One of the breakthroughs made in the proof of the SPGT is the concept of a *balanced skew partition*. For a graph G, a partition (skew or not) (A,B) of $V(G)$ is *balanced* if every path of length at least 3 with ends in B and interior in A, and every antipath of length at least 3 with ends in A and interior in B have even length. It is straightforward to check that a partition (A,B) of a Berge graph is balanced if and only if adding a vertex complete to B and anticomplete to A yields a Berge graph.

As we will see, the notion of a balanced skew partition is sufficiently particular to allow a short proof that a minimum counter-example to the SPGT cannot contain it, and is sufficiently general to be found in all non-basic Berge graphs that cannot be decomposed otherwise.

Theorem 4.3 *If (A,B) is a balanced partition of a perfect graph G, and if every Berge graph of order at most $|V(G)|+1$ is perfect, then $G[B]$ admits a colouring that can be extended to a colouring of G.*

Proof Consider the graph G' that is obtained by adding to G a clique of size $k = \omega(G) - \omega(G[B])$ complete to B and anticomplete to A. It is Berge because (A,B) is balanced. So it is perfect when $k \leq 1$ from our assumption, and it is also perfect when $k \geq 2$ by several applications of Theorem 2.2. Observe that $\omega(G') = \omega(G)$. An $\omega(G)$-colouring of G' then yields a colouring of $G[B]$ that extends to a colouring of G. ■

The following theorem is due to Chudnovsky, Robertson, Seymour and Thomas [13].

Theorem 4.4 *A minimum imperfect Berge graph admits no balanced skew partition.*

Proof Let G be a minimum imperfect Berge graph, so $\chi(G) > \omega(G)$. By Theorem 2.3, $\overline{G}$ is also a minimum imperfect Berge graph. Let (A,B) be a balanced skew partition in G, so A partitions into two sets A_1 and A_2 anticomplete to one another, and B partitions into two sets X and Y complete to one another. By Theorem 4.1 $|A_1|$ and $|A_2| \geq 2$, for otherwise the vertex in A_1 or A_2 would be the centre of a star cutset in $\overline{G}$. It follows from the minimality of G that every Berge graph of size $|B \cup A_i|+1$ is perfect for $i = 1$ and 2. By Theorem 4.3, consider an $\omega(G[B])$ colouring C_i of $G[B]$ that extends to a colouring of $G[B \cup A_i]$. Let X_i be the set of vertices of $G[B \cup A_i]$ whose colour in the colouring C_i is present in X, and let $Y_i = (B \cup A_i) \setminus X_i$. Because of the colouring C_i, $\omega(G[X_i]) = \omega(G[X])$, so $\omega(G[X_1 \cup X_2]) = \omega(G[X])$. By the minimality of G, it follows that $G[X_1 \cup X_2]$ has an $\omega(G[X])$-colouring. Because of the colouring C_i, $\omega(G[Y_i]) = \omega(G) - \omega(G[X_i]) = \omega(G) - \omega(G[X])$, so $\omega(G[Y_1 \cup Y_2] = \omega(G) - \omega(G[X])$. By the minimality of G, it follows that $G[Y_1 \cup Y_2]$ has an $(\omega(G) - \omega(G[X]))$-colouring, and hence that G has an $\omega(G)$-colouring, which is a contradiction. ■

The following question may be hopeless to answer, because a proof would imply a direct argument for the skew partition conjecture (no such argument exists at present). Also, a proof of the following, together with Theorem 4.2, would yield a new proof of the SPGT.

Problem *Find a direct proof of the following: if G is a minimum Berge imperfect graph, then at least one of G or $\overline{G}$ admits no double star cutset.*

Rusu (see [50, pp. 167–183]) wrote a survey about cutsets in perfect graphs, and Reed [51] wrote a survey about skew partitions (on which this section is mostly based).

An important question about decompositions is their detection in polynomial time. For a graph with n vertices and m edges, the following decompositions can be detected in polynomial time: 2-join (in time $O(n^2m)$, see Charbit, Habib, Trotignon and Vušković [8]), homogeneous pair (in time $O(n^2m)$, see Habib, Mamcarz and de Montgolfier [37]) and skew partitions (in time $O(n^4m)$, see Kennedy and Reed [39]). Trotignon [57] showed that balanced skew partitions are NP-hard to detect, but devised an $O(n^9)$-time non-constructive algorithm that certifies whether an input Berge graph has a balanced skew partition.

5. The strategy of the proof

The main result in [13] is Theorem 5.1 below and, as we know from the earlier sections, it implies the SPGT. Its statement is the result of a long sequence of attempts by many researchers, as explained in the introduction of [13] or in [53]. A slight variant on what seems now to be the right statement was first conjectured by Conforti, Cornuéjols and Vušković [22]. They proved it in the square-free case, and some of the arguments that they discovered are essential in the strategy described below – in particular, the attachments to prisms, and the use of Truemper configurations.

Theorem 5.1 *Every Berge graph is basic, or has a 2-join, a complement 2-join, a homogeneous pair or a balanced skew partition.*

The strategy used by Chudnovsky, Robertson, Seymour and Thomas to prove Theorem 5.1 is classical in structural graph theory. It consists of identifying a 'dense' basic class and a 'sparse' basic class, as we now explain. The 'dense' basic class does not contain the obstruction (here an odd hole or antihole), but 'almost' contains it, so that if a graph G contains an induced dense subgraph H, then any vertex exterior to H must attach in a very specific way to H, either enlarging the basic graph to a bigger basic graph, or entailing a decomposition. For the sake of proving the decomposition theorem, it can therefore be assumed that this particular class of basic graphs is excluded. The process can then be iterated with a new kind of dense basic graphs. The sparse class is what remains when all dense substructures are excluded. Observe that this method is in some sense safer than proofs by induction. Finding the right induction hypothesis is time-consuming because, for each failure, one has to restart from scratch, whereas a result asserting that some dense substructure entails a decomposition is a true statement that can be used in the future, even if the strategy of the proof changes.

For the proof of Theorem 5.1 the sparse class is formed by bipartite graphs and their complements. The dense class is more complicated. At the beginning of the proof it is formed by 'sufficiently' connected line graphs, their complements and doubled graphs. The simplest line graphs in this context are the prisms, where a *prism* is a graph made of three vertex-disjoint paths $P_1 = a_1 \dots b_1$, $P_2 = a_2 \dots b_2$ and $P_3 = a_3 \dots b_3$ of length 1 or more, for which $a_1a_2a_3$ and $b_1b_2b_3$ are triangles and no edges exist between the paths except those of the two triangles. Note that for a prism in a Berge graph, the lengths of the three paths must have the same parity. The prism is *odd* or *even* according to this parity.

To understand why prisms and their generalizations are 'dense', one can check that a Berge graph formed of a prism and one vertex not in the prism is either a larger line graph, or has a 2-join or some kind of skew partition such as a star cutset. For this purpose, it is convenient to know that Berge graphs have no pyramids, where a *pyramid* is a graph made of three paths $P_1 = a \dots b_1$, $P_2 = a \dots b_2$ and $P_3 = a \dots b_3$ of length at least 1, two of which have length at least 2, are vertex-disjoint except at a, and are such that $b_1b_2b_3$ is a triangle and no edges exist between the paths except those of the triangle and the three edges incident with a. What makes this work is that prisms are 'close to' containing pyramids, and a pyramid always contains an odd hole. One can also try to prove a similar statement for the line graph of the 2-subdivision of K_4, or similar statements with the vertex outside the prism replaced by some path with neighbours in exactly two paths of the structure, for instance. All this should lead to variants of Theorem 10.1 from [13], whose proof is easy to read since it does not rely on any technical lemmas. More on attachments to prisms is explained in Section 6 below.

In fact, a dozen dense basic graphs are considered: first, several kinds of line graphs of bipartite subdivisions of K_4; even prisms; then long prisms (*long* means that at least one of the paths has length at least 2); the *double diamond* (the graph obtained from a cycle $abcd$ by adding an edge vw such that v is adjacent to a and b and w to c, d, and a second edge $v'w'$ with the same attachment to $abcd$); various kinds of wheels (where a *wheel* is a graph formed by a hole H together with a vertex with at least three neighbours in the hole); and antiholes of length at least 6. For each of these dense basic classes, it was proved that containing it leads to being basic or having some decomposition; the Berge graphs handled next are thus supposed not to contain this kind of induced subgraph. At the end of this process, so many induced subgraphs are excluded that the graph under consideration, or its complement, is bipartite. Needless to say, identifying this long sequence of 'dense' graphs is a *tour de force*, especially since for each of them, the technicalities are really involved.

Despite all these technicalities, the objects considered in the proof of Theorem 5.1 are very combinatorial. This leads to the following problem.

Problem *Can the above strategy be transformed into a polynomial-time algorithm whose input is an arbitrary graph G and whose output is either an odd hole, an odd antihole, or a partition of the vertices of G certifying one of the outcomes of Theorem 5.1?*

In [13] the global strategy of the proof is well explained at the beginning, and more about the strategy can be found in [56] and [12]. How structural methods can be used generally for classes closed under taking induced subgraphs is discussed by Chudnovsky and Seymour [14]. Prisms, pyramids and wheels are 'Truemper configurations', graphs that play an important role in many decompostion theorems (see the survey of Vušković [60]). To make a start on the above problem, the first step is the detection in polynomial time of the structures that are used in the proof (line graphs of a bipartite subdivision of K_4, even prisms, odd prisms and wheels). Apart from wheels they all can be detected in polynomial time in a Berge graph (see Maffray and Trotignon [47]). Diot, Tavenas and Trotignon proved that detecting a wheel in a Berge graph is an NP-complete problem, but for the sake of the problem above, this problem might be by-passed.

6. Book from the Proof

We are now ready to investigate some technicalities of the proof of the SPGT. A lemma due to Roussel and Rubio [52] is used at many steps in [13]; in fact, the authors of [13] rediscovered it (in joint work with Thomassen) and initially named it the *Wonderful lemma* because of its many applications.

The Roussel–Rubio lemma states that, in a sense, any anticonnected set of vertices of a Berge graph behaves like a single vertex. How does a vertex v 'behave' in a Berge graph? If a path of odd length (at least 3) has both ends adjacent to v, then v must have other neighbours in the path, for otherwise there is an odd hole. The lemma states roughly that an anticonnected set T of vertices behaves similarly: if a path of odd length (at least 3) has both ends complete to T, then at least one internal vertex of the path is also complete to T (there are exceptions, see the statement below). When $P = xx' \dots y'y$ is a path of length at least 3, a *leap* for P is a pair of non-adjacent vertices v, w such that $N(v) \cap V(P) = \{x, x', y\}$ and $N(w) \cap V(P) = \{x, y', y\}$. We denote by P^* the interior of a path P. The result of Roussel and Rubio is as follows.

Theorem 6.1 *Let T be an anticonnected set of vertices in a Berge graph G. If P is a path of odd length at least 3, vertex-disjoint from T, whose ends are T-complete, then one of the following holds:*

(1) *an internal vertex of P is T-complete;*
(2) *there is a leap for P in T;*
(3) *P has length 3, and its two internal vertices are the ends of an antipath of odd length whose interior is in T.*

We now give a corollary of Theorem 6.1 that is used constantly in [13].

Corollary 6.2 *Let T be an anticonnected set of vertices in a Berge graph G. If P is a path with odd length at least 3, vertex-disjoint from T, whose ends are T-complete and such that no internal vertex of P is T-complete, then every T-complete vertex has a neighbour in P^*.*

Proof Let v be a T-complete vertex, and suppose that v has no neighbour in P^* (so v is not in $V(P) \cup T$). Apply Theorem 6.1 to T and P. If conclusion (2) holds, then a path of odd length with same interior as P joins the members of the leap. Together with v it forms an odd hole, giving a contradiction. If conclusion (3) holds, then the antipath of odd length can be completed to an odd antihole through v, giving a contradiction. ■

We now present a self-contained lemma from [13] in order to give some taste of the technicalities. We place this lemma in context by stating informally how line graphs are handled.

A line graph K of a bipartite graph is formed from a collection of cliques and a collection of paths linking them. If K is contained in a Berge graph G, a vertex $v \in V(G \setminus K)$ is *major* if it has many (at least 2) neighbours in each clique, and *minor* if it has neighbours in at most one of the paths or in at most one of the cliques. An important result is that every vertex is major or minor, or allows one to obtain a larger line graph. (This is not fully true, but vertices that are neither major nor minor can be considered as part of some of the paths in what is called a generalized line graph; therefore, when the generalized line graphs are properly defined, every vertex is major or minor with respect to a maximal generalized line graph.) A next step is to prove that connected (or sometimes anticonnected) components of vertices of $G \setminus K$ behave as a vertex. For more details of these components, see the proof of 10.1 in [13]; we deal with anticomponents below.

If there are no major vertices, then the graph is formed of the line graph and possibly a collection of minor components. If some of these components are attached to a clique then there is a balanced skew partition, and if some are attached to a path there is a 2-join. If there are major vertices, then it can be proved that an anticomponent of these is formed of vertices that are all major *in the same way*, meaning that they are all attached to exactly the same vertices of the cliques. An anticomponent of major vertices is therefore complete to its attachment and forms a skew cutset separating parts of the line graph. The next lemma (a variant of 7.3 of [13]) proves a statement of this form. It illustrates another breakthrough made in [13]: showing how the Roussel–Rubio lemma can be used to find skew partitions.

Theorem 6.3 *In a Berge graph G, let K be a prism with triangles $a_1a_2a_3$ and $b_1b_2b_3$ and paths $P_i = a_i \dots b_i$ for $i = 1, 2, 3$. Suppose that, for each $i = 1, 2, 3$, P_i has length at least* 2. *Let Y be an anticonnected set of vertices each of which has at least two neighbours in $\{a_1, a_2, a_3\}$ and at least two neighbours in $\{b_1, b_2, b_3\}$ (so Y is disjoint from K). Then at least two members of $\{a_1, a_2, a_3\}$ and at least two members of $\{b_1, b_2, b_3\}$ are Y-complete.*

Proof Suppose not; then there is an antipath with interior in Y joining two vertices both of which are in $\{a_1, a_2, a_3\}$ or in $\{b_1, b_2, b_3\}$. Let Q be a shortest such antipath and suppose up to symmetry that Q is from a_1 to a_2. Every vertex in Y is adjacent to either a_1 or a_2, so Q has length at least 3. From the minimality of Q, a_3 is

Q^*-complete, and so is at least one of $\{b_1, b_2, b_3\}$ (say b_i). Since Q can be completed to an antihole via $a_1b_ia_2$, it follows that Q has even length, at least 4. Let $Q = a_1q_1 \dots q_na_2$, let a_i' be the neighbour of a_i in P_i for $i = 1, 2$, and $P = a_1'P_1b_1b_2P_2a_2'$. We now prove two claims.

(1) *At least one internal vertex of P is $(Q^* \setminus q_1)$-complete, and at least one internal vertex of P is $(Q^* \setminus q_n)$-complete.*

If one of b_1 or b_2 is Q^*-complete, the claim is obviously true. Otherwise, none of b_1, b_2 is Q^*-complete, so there exists a antipath from b_1 to b_2 whose interior is in Q^*, and from the minimality of Q this antipath has the same interior as Q. It follows that one of b_1, b_2 is complete to $Q^* \setminus q_1$ and the other is complete to $Q^* \setminus q_n$. This proves (**1**).

(2) *If an internal vertex of P is $(Q^* \setminus q_1)$-complete or $(Q^* \setminus q_n)$-complete, then it is Q^*-complete. If a_1' is $(Q^* \setminus q_1)$-complete, then it is Q^*-complete.*

If an internal vertex v of P is $(Q^* \setminus q_n)$-complete then it is Q^*-complete, since otherwise, vq_nQa_1v is an odd antihole. If v is a_1' is an internal vertex of P, and if v is $(Q^* \setminus q_1)$-complete, then v is Q^*-complete, for otherwise, vq_1Qa_2v is an odd antihole. This proves (**2**).

If both a_1', a_2' are Q^*-complete, then Q can be completed to an odd antihole via $a_1a_2'a_1'a_2$, giving a contradiction. So, up to symmetry, we may now suppose from here on that a_1' is not Q^*-complete. It follows by (**2**) that a_1' is not $(Q^* \setminus q_1)$-complete.

Let x be the $(Q \setminus q_1)$-complete vertex of P closest to a_1' along P. By (**1**), x exists and is an internal vertex of P. By (**2**), x is in fact Q^*-complete.

If $a_1'a_1Px$ has odd length, then by Corollary 6.2 applied to $a_1a_1'Px$ and $Q^* \setminus q_1$, there is a contradiction because a_3 is $(Q^* \setminus q_1)$-complete and has no neighbour in the interior a_1Px. So a_1Px has even length. It follows that $xPa_2'a_2$ has odd length. By Corollary 6.2 applied to $xPa_2'a_2$ and $Q^* \setminus q_n$, and because of a_3, there must be an internal vertex v of $xPa_2'a_2$ that is $(Q^* \setminus q_n)$-complete. By (**2**), v is in fact Q^*-complete. But then, $a_3a_1a_1'Px$ is odd and its ends are Q^*-complete, but none of its internal vertex is Q^*-complete. This contradicts Corollary 6.2 because of v. ■

The above lemma shows how the Roussel–Rubio lemma 'generates' antiholes to provide contradictions. Maybe it can be used to investigate the structure of several subclasses of Berge graphs, such as in the next problem.

Problem *Is there a structural characterization of Berge graphs with no antihole of length at least* 6*?*

A proof of the Roussel–Rubio lemma is given in Maffray and Trotignon [48], where some applications are also presented. A shorter proof in the case when T is a stable set is due to Kapoor, Vušković and Zambelli (see [58] where several very simple applications are presented). For applications of the Roussel–Rubio lemma, see [13, 3.1].

About technicalities, the best source is of course Chudnovsky, Robertson, Seymour and Thomas [13]. This paper is well organized: the first four sections are devoted to technical lemmas that are used intensively in the sequel, but as a result of this organization, it is hard to extract a meaningful chunk, and it is difficult to appreciate how useful the technical lemmas are without knowing the sequel, and the sequel is difficult to understand without mastering the technicalities of the lemmas. The easiest parts of [13] are perhaps 15.1, Section 15 and Section 16. Section 10 is devoted to even prisms and is a self-contained chunk that has fewer technicalities than the rest of the paper, but still relies greatly on technical lemmas from the previous sections.

7. Recognizing perfect graphs

From here on we investigate results on perfect graphs that were obtained after the proof of the SPGT. Soon after this proof, in 2002, another open question was solved, when Chudnovsky, Cornuéjols, Liu, Seymour and Vušković [11] found a polynomial-time algorithm that decides whether an input graph is Berge. Note that this algorithm is independent of the SPGT: it takes an arbitrary graph G as input and in time $O(n^9)$ outputs an odd hole or antihole of G, if there is one. We give here a brief outline, from the introduction to [11]. In what follows, G is the input graph.

> ...we would like to decide either that G is not Berge, or that G contains no odd hole. (To test Bergeness, we just run this algorithm on G and then again on the complement of G.) If there is an odd hole in G, then there is a shortest one, say C. A vertex of the remainder of G is C-major if its set of neighbours in C is not a subset of the vertex set of any 3-vertex path of C; and C is *clean* (in G) if there are no C-major vertices in G. If there happens to be a clean shortest odd hole in G, then it stands out and can be detected relatively easily; and that essentially is the first step of our algorithm, a routine to test whether there is a clean shortest odd hole. The remainder of the algorithm consists of reducing the general problem to the 'clean' case that was just handled. If C is a shortest odd hole in G, let us say a subset X of $V(G)$ is a *cleaner* for C if $X \cap V(C) = \varnothing$ and every C-major vertex belongs to X. Thus if X is a cleaner for C then C is a clean hole in $G \setminus X$. The idea of the remainder of the algorithm is to generate polynomially many subsets of $V(G)$, such that if there is a shortest odd hole C in G, then one of the subsets will be a cleaner for C. If we can do that, then we delete each of these subsets in turn, thereby generating polynomially many induced subgraphs; and we know that there is an odd hole in G if and only if in one of these subgraphs there is a clean shortest odd hole. Thus we can decide whether G has an odd hole by testing whether any of these subgraphs has a clean shortest odd hole.

Theorem 7.1 *There exists an algorithm that decides whether an input graph is Berge in time $O(n^9)$.*

Let us now explain briefly the two steps from the sketch above.

The first step relies on the *shortest-path detector*, a method designed by Chudnovsky and Seymour [11]. The shortest-path detector enumerates all possible subsets of fixed size of vertices that are potentially particular vertices of some structure to be detected, and computes shortest paths between them to complete the structure until the desired structure is detected. At first glance this should fail because between the paths chords could exist (the paths might fail to be disjoint also). But a clever argument shows that this failure leads to a smaller structure similar to the one to be detected, so that when the enumeration suddenly matches the particular vertices of a *smallest* structure in the graph, the problem does not arise, and the structure is detected. This method is used in [11] to decide in time $O(n^9)$ whether a graph contains a pyramid.

The second step relies on a powerful technique discovered by Conforti and Rao [24], called the *cleaning*, which consists in 'guessing' a cleaner. Here, the way to guess the cleaner comes from some Roussel–Rubio-like lemmas, asserting that the set X of vertices to be 'guessed' that are all major vertices of some possible smallest odd hole, are all common neighbours of some restricted set S of neighbours. One can therefore enumerate all possible sets S by brute force, and for each of them nominate X as the set of S-complete vertices.

The following problem is still open.

Problem *Is there a polynomial-time algorithm to decide whether an input graph has an odd hole?*

In [11], another algorithm for recognizing Berge graphs is given. It relies on decompositions and uses Theorem 4.2. Detecting odd holes can be solved in polynomial time under the assumption that the largest size of a clique in the input graph is bounded by some constant. This was proved by Conforti, Cornuéjols, Liu, Vušković and Zambelli [21], a good source for understanding the main ideas of the recognition of Berge graphs. Simple examples of shortest path detectors can be found in Maffray and Trotignon [47].

8. Berge trigraphs

An obvious question arising from Theorem 5.1 is whether it is best possible. Is each outcome necessary? Are there outcomes that can be made more precise? A careful reader of [13], noticing that the outcome 'homogeneous pair' is obtained only once in the whole proof, might therefore wonder whether it is really necessary. Chudnovsky proved that it is not. Let us explain how.

A way of proceeding is to consider a smallest Berge graph G for which the homogeneous pair is the only outcome satisfied by G in Theorem 5.1, and to look for a contradiction. A natural idea is then to 'contract' a homogeneous pair (A, B)

so as to find a smaller Berge graph G', and to apply Theorem 5.1 to G' and prove that any outcome of Theorem 5.1 in G' yields an outcome in G (because G' and G are very similar). From our initial assumption, G' therefore satisfies no outcome of Theorem 5.1, and this provides a contradiction. We call this a *bootstrap method*, because it improves structural theorems somehow 'for free'. The natural way to 'contract' a homogeneous pair (A, B) of G is to replace A by a vertex a complete to $C(A)$ and anticomplete to $\overline{C}(A)$, and to replace B by a vertex b complete to $C(B)$ and anticomplete to $\overline{C}(B)$.

The bootstrap method is hard to implement in this context. The problem is with ab: should it be an edge or an antiedge of G'? Both choices lead to difficult technicalities: if ab is chosen to be a non-edge, then some skew cutset may separate a from b in G' while A and B are linked in G. If a is chosen to be adjacent to b, then the same phenomenon may happen in the complement. No neat example of these bad phenomena can be given, because (as we will see) the outcome 'homogeneous pair' is not necessary in Theorem 5.1. But any attempt of proof must face this issue and is likely to fail.

Chudnovsky's idea was to leave undecided the adjacency between a and b in G'. For this purpose, she defined *trigraphs* as graphs with edges, anti-edges, and a third kind of adjacency: a *switchable pair*. For each pair of distinct vertices v and w, vw is an edge, an antiedge or a switchable pair. A *realization* G of a trigraph T is any graph on $V(T)$ such that all edges of T are edges of G and all antiedges of T are antiedges of G (so every switchable pair is transformed into an edge or an antiedge). A trigraph is *Berge* if every realization is Berge.

We do not give the long list of definitions that extends the vocabulary of graphs to trigraphs. The key point is in the definitions of the decompositions: all the 'important' edges in these definitions are not allowed to be switchable pairs. For instance, if (X, Y) is a 2-join of a trigraph T, no switchable pair of T is from X to Y. If (X, Y) is a skew partition of a trigraph T, then X must be partitioned into sets X_1 and X_2 for which no edge and no switchable pair exist between X_1 and X_2, and Y must be partitioned into sets Y_1 and Y_2 for which no antiedge and no switchable pair exist between Y_1 and Y_2. It is easy to guess how useful this requirement is: for instance, when building G' from G as in the paragraph above, ab is defined to be a switchable pair, and so the problem that we mentioned with the skew cutset separating a from b does not arise any more.

One slight problem with this notion of trigraph is that Theorem 5.1 has to be proved again for Berge trigraphs. Chudnovsky proved several decomposition theorems similar to Theorem 5.1 for Berge trigraphs. The proof of her main one is self-contained (that is, it does not rely on the proof of Theorem 5.1) and runs to more than two hundred pages. We do not give the precise statements here since we do not give the precise definitions of decompositions and basic classes of trigraph. The precise statements of the theorems are in [10], where only parts of the proofs are given, and the complete proof is in [9]. With trigraphs, the bootstrap method works smoothly and yields the following result of Chudnovsky. We translate back to graphs, since a graph is a particular trigraph.

Theorem 8.1 *Every Berge graph is basic, or has a* 2*-join, a complement* 2*-join or a balanced skew partition.*

Another potential improvement of Theorem 5.1 concerns the technical requirements in the definition of a 2-join. Some authors consider only *non-path 2-joins*, and in some applications it is essential to use these, because one needs to replace one side of a 2-join by a long path in proofs by induction or in recursive algorithms, and this obviously fails if the side is already a long path. Trotignon [57] investigated this question and obtained the following result starting from Theorem 8.1 and with the bootstrap method applied to graphs. It easily implies the different variants of Theorem 5.1. *Path cobipartite graphs* and *path double-split graphs* are Berge graphs obtained by subdividing edges in complements of bipartite graphs and in double split graphs respectively (a more precise definition can be given, but this one is enough here). Note that subdividing an edge in any graph creates a path 2-join. In [57], several graphs that yield a unique outcome in the following theorem of Trotignon are given.

Theorem 8.2 *If G is a Berge graph, then G is basic or has a balanced skew partition, or one of G, $\overline{G}$ is a path-cobipartite graph, or is a path-double split graph, or has a non-path* 2*-join, or has a homogeneous pair and a path* 2*-join.*

Little work has been devoted to the algorithmic aspects of trigraphs. The following problem is still open.

Problem *What is the complexity of recognizing Berge trigraphs?*

It seems now that trigraphs are an important tool for graph decompositions, as suggested by their use in claw-free and bull-free graphs (see [14]). For algorithms for Berge trigraphs, Chudnovsky, Trotignon, Trunck and Vušković [17] seems to be the only available reference.

9. Even pairs: a shorter proof of the SPGT

An *even pair* in a graph is a pair of distinct vertices such that every path between them has even length. This notion is involved in algorithmic aspects of perfect graphs and in a significantly shorter proof of the SPGT due to Chudnovsky and Seymour [15]. Meyniel [49] proved the following result.

Theorem 9.1 *A minimally imperfect graph has no even pair.*

It follows that even pairs may be an ingredient of a useful decomposition theorem for Berge graphs. Every known example of a Berge graph with no even pair contains either an odd prism or an antihole. *Artemis graphs* are such graphs with no odd holes, no antiholes of length at least 5, and no prisms. The following result was proved by Maffray and Trotignon [48].

Theorem 9.2 *Every Artemis graph is a clique or has an even pair.*

Recall that given two vertices v and w in a graph G, the operation of *contracting* them means removing v and w and adding one vertex with edges to every vertex of $G \setminus \{v, w\}$ that is adjacent in G to at least one of v or w; we denote by G/vw the graph that results from this operation. Fonlupt and Uhry [29] proved that *if G is a perfect graph and if $\{v, w\}$ is an even pair in G, then G/vw is perfect and has the same chromatic number as G.* Using this idea, it is easy to turn the proof of Theorem 9.2 into a polynomial-time algorithm that colours every Artemis graph (see [42] for a faster algorithm).

The proof of the next theorem uses several ideas from the proof of Theorem 9.2, is very readable, and (as we will see) saves about fifty pages in the proof of the SPGT. An *odd wheel* (C, T) in a graph G consists of a hole C of length at least 6, and a non-empty anticonnected subset $T \subseteq V(G) \setminus V(C)$ for which at least three vertices of C are T-complete and there is a path P of C with odd length at least 3, whose ends are not T-complete and whose internal vertices are all T-complete. We say that G is *impoverished* if G is Berge and if G and $\overline{G}$ contain no odd wheel, long prism or double diamond. A *dominant pair* in G is a pair $\{v, w\}$ of non-adjacent vertices for which every other vertex of G is adjacent to at least one of v, w. The following theorem is due to Chudnovsky and Seymour [15].

Theorem 9.3 *If G is impoverished, then either G admits a star cutset, an even pair or a dominant pair, or G is a complete graph.*

It is not difficult to prove that a minimally imperfect graph has no dominant pair (see [15]), and it follows easily by Theorem 9.1 and Theorem 4.1 that every impoverished graph is perfect. Since the last fifty-five pages of the proof of the SPGT are devoted to finding a skew partition in an impoverished graph, they are no longer necessary and can be replaced by the nine pages needed to prove Theorem 9.3.

A good survey (with many conjectures) on even pairs is by Everett, de Figueiredo, Linhares Sales, Maffray Porto and Reed (see [50, pp. 67–92]). It seems that even pairs provide a cornucopia of conjectures: other conjectures on even pairs can be found in Burlet, Maffray and Trotignon [7] and Lévêque and de Werra [41]. Short proofs of the existence of even pairs in classical classes of perfect graphs (Meyniel graphs and weakly chordal graphs) can be found in Trotignon and Vušković [58].

10. Colouring perfect graphs

In the 1980s Gröstchel, Lovász and Schrijver [34] devised a polynomial-time algorithm that colours any input perfect graph. This algorithm relies on the ellipsoid method, and one may wonder whether a more combinatorial algorithm exists. Even if there is no formal definition of what a *combinatorial* algorithm should be, most mathematicians agree that any algorithm that relies on graph searches and decompositions, or even on classical linear programming, can be called 'combinatorial' and that the ellipsoid method cannot be, so this question is considered as open. A more precise question is whether a fast colouring algorithm can be derived from Theorem 5.1. We describe here recent progress in this direction.

Gröstchel, Lovász and Schrijver [34] devised a polynomial-time combinatorial algorithm that colours every Berge graph, under the assumption that a polynomial-time algorithm is available for the maximum stable set (this is also explained in [17]). So our question now is whether Theorem 5.1 (or one of its variants from Section 8) helps us to find a maximum weighted stable set. It turns out that the question is easy for all basic classes – for the historical ones see [55], and for doubled graphs see [17]. Also homogeneous pairs and complement 2-joins are not hard to handle (see [17]).

For 2-joins, the situation is more complicated because a maximum stable set may overlap a 2-join in many ways. In fact, Naves (see [59]) observed that for some class of graphs decomposable by 2-joins into very simple graphs, it is NP-hard to compute a maximum stable set. We now explain how 2-joins can actually help to find stable sets *in Berge graphs*. If (X_1, X_2) is a 2-join and if A_1, B_1, C_1, A_2, B_2, C_2 are as in the definition of a 2-join, we set $\alpha_{AC} = \alpha(G[A_1 \cup C_1])$, $\alpha_{BC} = \alpha(G[B_1 \cup C_1])$, $\alpha_C = \alpha(G[C_1])$ and $\alpha_X = \alpha(G[X_1])$. The next inequalities from Trotignon and Vušković [59] tell us how stable sets and 2-joins overlap in Berge graphs.

Theorem 10.1 *If (X_1, X_2) is an odd 2-join of G, then $\alpha_C + \alpha_X \leq \alpha_{AC} + \alpha_{BC}$, and if (X_1, X_2) is an even 2-join of G, then $\alpha_{AC} + \alpha_{BC} \leq \alpha_C + \alpha_X$.*

The above lemma allows to construct blocks of decomposition of a 2-join that preserve being Berge and allow us to keep track of α (see [17] for a precise definition of the blocks). Interestingly, 2-joins are used to compute α in other classes of graphs, while they seem to be hard to use in general: for claw-free graphs see Faenza, Oriolo and Stauffer [28], and for even-hole-free graphs with no star cutsets see Trotignon and Vušković [59].

In [17], Theorem 10.1 is used to prove the following result. Many technicalities are needed, some of which come from the fact that the decomposition blocks for 2-joins that keep track of α do not preserve having no balanced skew partition. It is also proved in [17] that for Berge graphs with no balanced skew partition there exist *extreme* decompositions, in which one of the blocks is basic; these are very convenient for proofs by induction. To handle all these technicalities, it is convenient (if not essential) to work with trigraphs, but here we state the result for graphs. The following result is due to Chudnovsky, Trotignon, Trunck and Vušković [17].

Theorem 10.2 *There exists an $O(n^7)$-time algorithm whose input is a Berge graph with no balanced skew partition and whose output is a maximum weighted stable set of G and a colouring of G.*

Up to now, no-one has discovered how skew partitions could be handled so as to provide a polynomial colouring algorithm. One might think that a Berge graph that is uniquely decomposable with a balanced skew partition must have an even pair. To support this idea, Chudnovsky and Seymour [16] studied the structure of Berge graphs with no K_4 and no even pair. They describe them quite precisely in the next result.

Theorem 10.3 *If G is a 3-connected K_4-free Berge graph with no even pair and with no clique cutset, then one of G and $\overline{G}$ is the line graph of a bipartite graph.*

Unfortunately, it seems that this theorem does not generalize to larger values of ω, as shown by the *WKBGSF*, a graph G (discovered by Chudnovsky and Seymour, unpublished) shown in Fig. 1. Every edge of G is the middle edge of a P_4 so that, in the complement, each pair of non-adjacent vertices can be linked by a P_4, so $\overline{G}$ has no even pairs. However, G is perfect, and the balanced skew partition is the only outcome of Theorem 5.1 satisfied by G.

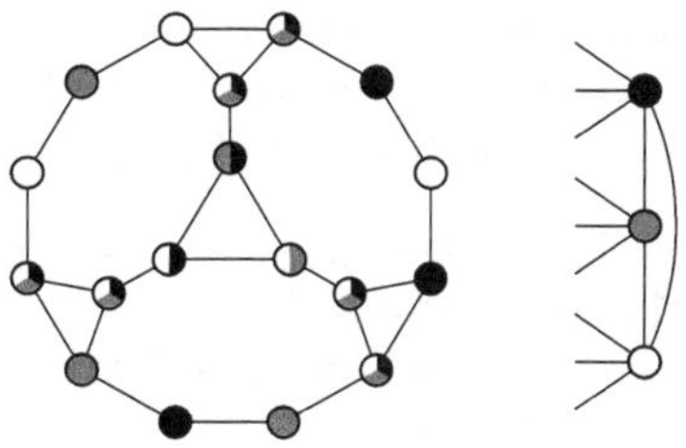

Fig. 1. The WKBGSF (Worst Known Berge Graph So Far): a vertex in the left part is adjacent to one of the three vertices in the right part if they share one of the colours black, white or grey

The following two problems may be the most important ones about perfect graphs.

Problem *Describe the structure of Berge graphs with no even pairs.*

Problem *Find a combinatorial polynomial-time algorithm that colours every Berge graph.*

Acknowledgement We thank Maria Chudnovsky, Frédéric Maffray and Irena Penev for their help.

References

1. N. Alexeev, A. Fradkin and I. Kim, Forbidden induced subgraphs of double-split graphs, *SIAM J. Discrete Math.* **26** (2012), 1–14.
2. L. W. Beineke, Characterizations of derived graphs, *J. Combin. Theory* **9** (1970), 129–135.
3. C. Berge, Färbung von Graphen, deren sämtliche bzw. deren ungerade Kreise starr sind, Technical report, *Wiss. Z. der Martin-Luther-Univ. Halle-Wittenberg, Math.-Natur. Reihe* **10**, 1961.

4. C. Berge and V. Chvátal (eds.), *Topics on Perfect Graphs*, Ann. Discrete Math. **21**, North-Holland, 1984.
5. J. A. Bondy and U. S. R. Murty, *Graph Theory*, Graduate Texts in Math. **244**, Springer, 2008.
6. A. Brandstädt, V. B. Le and J. P. Spinrad, *Graph Classes: A Survey*, SIAM, 1999.
7. M. Burlet, F. Maffray and N. Trotignon, Odd pairs of cliques, *Graph Theory in Paris, Proc. Conf. in Memory of Claude Berge* (eds. A. Bondy *et al.*), Birkhäuser (2007), 85–95.
8. P. Charbit, M. Habib, N. Trotignon and K. Vušković, Detecting 2-joins faster, *J. Discrete Algorithms* **17** (2012), 60–66.
9. M. Chudnovsky, *Berge Trigraphs and their Applications*, Ph.D. thesis, Princeton University, 2003.
10. M. Chudnovsky, Berge trigraphs, *J. Graph Theory* **53** (2006), 1–55.
11. M. Chudnovsky, G. Cornuéjols, X. Liu, P. Seymour and K. Vušković, Recognizing Berge graphs, *Combinatorica* **25** (2005), 143–186.
12. M. Chudnovsky, N. Robertson, P. Seymour and R. Thomas, Progress on perfect graphs, *Math. Programming (B)* **97** (2003), 405–422.
13. M. Chudnovsky, N. Robertson, P. Seymour and R. Thomas, The strong perfect graph theorem, *Ann. of Math.* **164** (2006), 51–229.
14. M. Chudnovsky and P. Seymour, Excluding induced subgraphs, *Surveys in Combinatorics, 2007*, London Math. Soc. Lecture Notes **346** (2007), 99–119.
15. M. Chudnovsky and P. Seymour, Even pairs in Berge graphs, *J. Combin. Theory (B)* **99** (2009), 370–377.
16. M. Chudnovsky and P. Seymour, Three-colourable perfect graphs without even pairs, *J. Combin. Theory (B)* **102** (2012), 363–394.
17. M. Chudnovsky, N. Trotignon, T. Trunck and K. Vušković, Coloring perfect graphs with no balanced skew-partitions, submitted.
18. V. Chvátal, On certain polytopes associated with graphs, *J. Combin. Theory (B)* **18** (1975), 138–154.
19. V. Chvátal, Star-cutsets and perfect graphs, *J. Combin. Theory (B)* **39** (1985), 189–199.
20. V. Chvátal and N. Sbihi, Bull-free Berge graphs are perfect, *Graphs Combin.* **3** (1987), 127–139.
21. M. Conforti, G. Cornuéjols, X. Liu, K. Vušković and G. Zambelli, Odd hole recognition in graphs of bounded clique size, *SIAM J. Discrete Math.* **20** (2006), 42–48.
22. M. Conforti, G. Cornuéjols and K. Vušković, Square-free perfect graphs, *J. Combin. Theory (B)* **90** (2004), 257–307.
23. M. Conforti, G. Cornuéjols and K. Vušković, Decomposition of odd-hole-free graphs by double star cutsets and 2-joins, *Discrete Appl. Math.* **141** (2004), 41–91.
24. M. Conforti and M. R. Rao, Testing balancedness and perfection of linear matrices, *Math. Programming* **61** (1993), 1–18.
25. G. Cornuéjols, *Combinatorial Optimization: Packing and Covering*, CBMS-NSF Regional Conf. Series in Applied Math. **74**, SIAM, 2001.
26. G. Cornuéjols and W. H. Cunningham, Composition for perfect graphs, *Discrete Math.* **55** (1985), 245–254.
27. E. Diot, S. Tavenas and N. Trotignon, Detecting wheels, *Analysis Discrete Math.* **8** (2014), 111–122.
28. Y. Faenza, G. Oriolo and G. Stauffer, An algorithmic decomposition of claw-free graphs leading to an $O(n^3)$-algorithm for the weighted stable set problem, *SODA* (2011), 630–646.

29. J. Fonlupt and J. P. Uhry, Transformations which preserve perfectness and h-perfectness of graphs, *Bonn Workshop on Combinatorial Optimization* (eds. A. Bachem, M. Grötschel and B. Korte), *Ann. Discrete Math.* **16** (1982), 83–85.
30. D. R. Fulkerson, Anti-blocking polyhedra, *J. Combin. Theory (B)* **12** (1972), 50–71.
31. T. Gallai, Transitiv orientierbare Graphen, *Acta Math. Acad. Sci. Hungar.* **18** (1967), 25–66.
32. G. S. Gasparian, Minimal imperfect graphs: a simple approach, *Combinatorica* **16** (1996), 209–212.
33. M. C. Golumbic, *Algorithmic Graph Theory and Perfect Graphs*, Elsevier, 2004.
34. M. Gröstchel, L. Lovász and A. Schrijver, *Geometric Algorithms and Combinatorial Optimization*, Springer-Verlag, 1988.
35. A. Gyárfás, Problems from the world surrounding perfect graphs, *Zastowania Mat. Appl. Math.* **19** (1987), 413–441.
36. A. Gyárfás, Z. Li, R. C. S. Machado, A. Sebő, S. Thomassé and N. Trotignon, Complements of nearly perfect graphs, *J. Combin.* **4** (2013), 299–310.
37. M. Habib, A. Mamcarz and F. de Montgolfier, Algorithms for some H-join decompositions, *LATIN 2012*, 446–457.
38. S. Hougardy, Classes of perfect graphs, *Discrete Math.* **306** (2006), 2529–2571.
39. W. S. Kennedy and B. Reed, *Fast Skew Partition Recognition*, Lecture Notes in Computer Science **4535** (2008), 101–107.
40. P. G. H. Lehot, An optimal algorithm to detect a line graph and output its root graph, *J. Assoc. Comp. Mach.* **21** (1974), 569–575.
41. B. Lévêque and D. de Werra, Graph transformations preserving the stability number, *Discrete Appl. Math.* **160** (2012), 2752–2759.
42. B. Lévêque, F. Maffray, B. Reed and N. Trotignon, Coloring Artemis graphs, *Theor. Comp. Sci.* **410** (2009), 2234–2240.
43. L. Lovász, A characterization of perfect graphs, *J. Combin. Theory (B)* **13** (1972), 95–98.
44. L. Lovász, Normal hypergraphs and the perfect graph conjecture, *Discrete Math.* **2** (1972), 253–267.
45. L. Lovász, Perfect graphs, *Selected Topics in Graph Theory* (eds. L. W. Beineke and R. J. Wilson), Academic Press (1983), 55–87.
46. F. Maffray, *Fast recognition of doubled graphs*, Technical Report, *Les Cahiers Leibniz* **202**, 2013.
47. F. Maffray and N. Trotignon, Algorithms for perfectly contractile graphs, *SIAM J. Discrete Math.* **19** (2005), 553–574.
48. F. Maffray and N. Trotignon, A class of perfectly contractile graphs, *J. Combin. Theory (B)* **96** (2006), 1–19.
49. H. Meyniel, A new property of critical imperfect graphs and some consequences, *Europ. J. Combin.* **8** (1987), 313–316.
50. J. L. Ramírez Alfonsín and B. A. Reed (eds.), *Perfect Graphs*, Wiley Interscience, 2001.
51. B. A. Reed, Skew partitions in perfect graphs, *Discrete Appl. Math.* **156** (2008), 1150–1156.
52. F. Roussel and P. Rubio, About skew partitions in minimal imperfect graphs, *J. Combin. Theory (B)* **83** (2001), 171–190.
53. F. Roussel, I. Rusu and H. Thuillier, The Strong Perfect Graph Conjecture: 40 years of attempts and its resolution, *Discrete Math.* **309** (2009), 6092–6113.
54. N. D. Roussopoulos, A max $\{m,n\}$ algorithm for determining the graph H from its line graph G, *Inform. Proc. Letters* **2** (1973), 108–112.
55. A. Schrijver, *Combinatorial Optimization, Polyhedra and Efficiency*, Springer, 2003.

56. P. Seymour, How the proof of the strong perfect graph conjecture was found, *Gazette des Math.* **109** (2006), 69–83.
57. N. Trotignon, Decomposing Berge graphs and detecting balanced skew partitions, *J. Combin. Theory (B)* **98** (2008), 173–225.
58. N. Trotignon and K. Vušković, On Roussel–Rubio-type lemmas and their consequences, *Discrete Math.* **311** (2011), 684–687.
59. N. Trotignon and K. Vušković, Combinatorial optimization with 2-joins, *J. Combin. Theory (B)* **102** (2012), 153–185.
60. K. Vušković, The world of hereditary graph classes viewed through Truemper configurations, *Surveys in Combinatorics 2013*, London Math. Soc. Lecture Notes **409** (2013), 265–326.

8
Geometric graphs

ALEXANDER SOIFER

In this chapter we address a colourful area of geometric graph theory: finding the chromatic number of the plane and related problems. In addition to results, we present open problems of a classical kind that are easy to understand, but hard to solve.

1. The chromatic number of the plane

In the definition of a graph, edges symbolize the adjacency of points and nothing else: we ignore their shape and length. However, the past century has witnessed a great deal of interest in *geometric graphs*, where geometrical considerations such as distance define the adjacency. The wealth of material related to geometric graphs is so vast that one can easily imagine a book written on this topic alone. In view of space and time limitations and our emphasis on chromatic graph theory,we have chosen to go deep and address a small but colourful area of geometric graphs: the problem of finding the chromatic number of the plane and related problems. My monograph [45] contains further material on this topic, although there is some new material here that did not appear in that book.

We can create a graph G from the Euclidean plane E^2 by taking all of its points as vertices, and joining two vertices by an edge if and only if they are at distance 1 apart. More generally, we call a graph *unit-distance* when any two vertices are adjacent if and only if they are at distance 1 apart. The main open problem in the subject is as follows.

Problem *Find the chromatic number of the above graph G.*

This number is called the *chromatic number of the plane* (CNP) and is denoted by $\chi(E^2)$. As outlined in [45], this problem was created in late 1950 by the 18-year-old Edward Nelson, who determined a lower bound; his 20-year-old friend John Isbell found an upper bound:

$$4 \leq \chi(E^2) \leq 7 \text{ – that is, } \chi(E^2) = 4, 5, 6 \text{ or } 7.$$

There are similarities between the problem of finding the chromatic number of the plane (CNP) and the four-colour problem (4CP); both refer to plane colourings and both withstood all assaults for a long time, but whereas the latter problem was eventually solved, the former one remains open. Both problems had their distinguished promoters who kept the problems alive in their infancy: in the case of the 4CP it was Augustus De Morgan, and in the case of the CNP problem it was Paul Erdős. Yet, the problems differ vastly in their history. In 1890 P. J. Heawood proved the five-colour theorem, showing that the answer to the four-colour problem is 4 or 5. After over sixty years of intensive work, using tools from geometry, graph theory, abstract algebra, topology and measure theory, we still today have no improvement in the general case on the above bounds for $\chi(E^2)$, a surprisingly wide spread.

In a 2002 lecture, Ronald L. Graham offered a reward for the first solution to a special case that he called 'another four-colour conjecture'.

Problem *Is it possible to* 4*-colour the plane so as to forbid a monochromatic distance* 1*?*

Graham now believes that the chromatic number of the plane is 5 or 6 (see [19] and [20]). He cited a theorem of O'Donnell [31] showing the existence of 4-chromatic unit-distance graphs of arbitrarily large girth (see Theorem 3.2 below) as 'perhaps, the evidence that χ is at least 5'. Erdős [13], [14] believed that the chromatic number of the plane is 5, 6 or 7. Somewhat disjointly from Graham and Erdős, I predict that the answer is either 4 or 7.

2. The polychromatic number: lower bounds

Numerous problems related to the chromatic number of the plane have been posed. Let us look at one of them here. We say that a point set S *realizes the distance* d if S contains two points at distance d apart. In 1958 Erdős posed the following question.

Problem *What is the smallest number of colours needed to colour the plane so that no colour realizes all distances?*

In 1992 Soifer [40] named this invariant *the polychromatic number of the plane*, and denoted it by $\chi_p(E^2)$. Clearly $\chi_p(E^2) \le \chi(E^2) \le 7$. In 1970 upper and lower bounds for $\chi_p(E^2)$ were published by a Russian high-school student, Dmitry Raiskii [37]. We look at the lower bound here and return to the upper bound in Section 4.

In 1961 Leo and Willie Moser produced the seven-point plane configuration that we now call the *Mosers spindle* (see Fig. 1): each edge in the spindle has length 1. In [29] they proved the following result.

Theorem 2.1 *Any three points of the Mosers spindle contain two points at distance* 1 *apart. Consequently, in a colouring of the Mosers spindle that forbids monochromatic distance* 1*, at most two points can be of the same colour.*

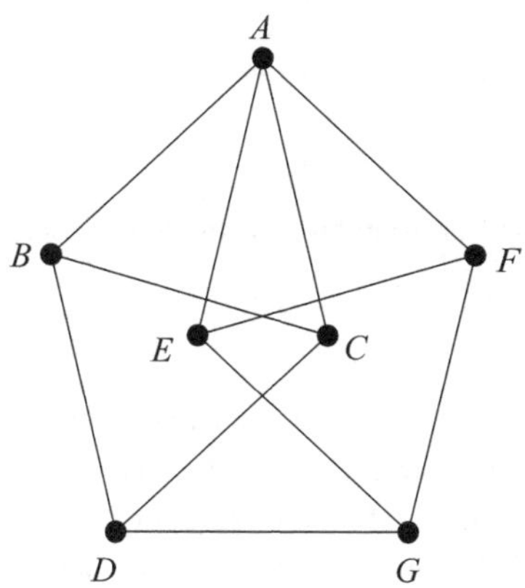

Fig. 1. The Mosers spindle

Raiskii's lower bound was as follows. A striking proof was found in 1997 by Alexei Merkov, another Russian high-school student, and appeared in an obscure brochure [34]. The following proof is due to Merkov, with modifications by the author.

Theorem 2.2 $\chi_p(E^2) \ge 4$.

Proof We assume that the plane is coloured in three colours, red, white and blue, and that each colour forbids a distance r, w and b, respectively. Equip the 3-coloured plane with Cartesian coordinates with origin O, and construct three 7-point sets S_r, S_w and S_b, each being the Mosers spindle (Fig. 2), in such a way that all three spindles share O as one of their seven vertices and have edges coloured r, w and b, respectively. This construction defines eighteen vectors: six 'red' ones, $v_1, v_2, \ldots, v_6$ from the origin O to each remaining point of S_r, six 'white' ones, $v_7, v_8, \ldots, v_{12}$ from O to the points of S_w, and six 'blue' ones $v_{13}, v_{14}, \ldots, v_{18}$ from O to the points of S_b.

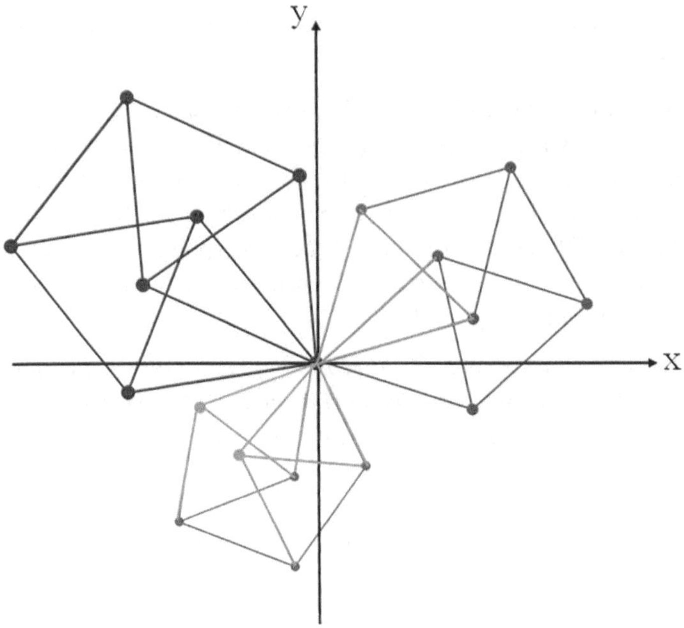

Fig. 2. The Cartesian plane with three Mosers spindles

We now introduce 18-dimensional Euclidean space E^{18} and a function M: $E^{18} \to E^2$ that maps each vector $(a_1, a_2, \ldots, a_{18})$ to $a_1v_1 + a_2v_2 + \ldots + a_{18}v_{18}$. This function induces a 3-colouring of E^{18} by assigning to a point of E^{18} the colour of the corresponding point of the plane. In accordance with the colours of the vectors v_i,we call the first six axes of E^{18} 'red', the next six axes 'white', and the last six axes 'blue'.

Let W be the subset of E^{18} consisting of all points whose coordinates include at most one equal to 1 for each of the three colours of the axes, and the remaining (fifteen or more) coordinates equal to 0; it is easy to verify that W has $7^3 = 343$ points. For any fixed array of coordinates allowable in W on white and blue axes, we get the 7-element set A of points in W with fixed coordinates on white and blue axes. The image $M(A)$ forms in the plane a translate of the original 7-point set S_r. If we fix another array of white and blue coordinates we obtain another 7-element set in E^{18}, whose image under M in the plane would form another translate of S_r. Thus, the set W gets partitioned into 49 subsets, each of which maps into a translate of S_r.

By Theorem 2.1, any translate of S_r has at most two red points among its 7 points. Since W has been *partitioned* into translates of S_r, at most $^2/_7$ of the points of Warered. We can now start all over again, and show similarly that at most $^2/_7$ of the points of W are white, and at most $^2/_7$ of the points are blue. But $^2/_7 + ^2/_7 + ^2/_7 < 1$. This contradiction implies that at least one of the colours realizes all distances, as required. ■

3. The de Bruijn–Erdős reduction to finite sets

We can expand the notion of the chromatic number to any subset S of the plane. The *chromatic number* $\chi(S)$ is the smallest number of colours sufficient to colour the points of S in such a way that forbids monochromatic pairs of points at distance 1 apart.

In 1951 de Bruijn and Erdős [11] published a powerful lemma that implies the following 'compactness theorem'; it assumes the axiom of choice.

Theorem 3.1 *The chromatic number of the plane is equal to the maximum chromatic number of its finite subsets.*

Accordingly, Erdős used to claim that the problem of finding the chromatic number of the plane is a problem about finite sets in the plane. In 1975 he posed the following problem [12].

Problem *Let S be a subset of the plane which contains no equilateral triangle of side length* 1. *Join two points of S if their distance is* 1. *Does this graph have chromatic number at most* 3*? If the answer is no, assume that the graph defined by S contains no cycles C_l of length l, for $3 \leq l \leq t$, and ask the same question.*

Erdős was unsure of the outcome: he expected triangle-free unit distance graphs to have chromatic number at most 3, or else that a chromatic number of 3 can be forced by prohibiting all small cycles up to C_t, for sufficiently large t.

In 1979 Wormald [51] disproved the first (easier) triangle-free conjecture above, and Erdős promptly reported the result in a lecture and later in print (see [15]):

In a recent paper (still unpublished), Wormald found a set S for which the unit distance graph $G_1(S)$ has girth 5 *and chromatic number* 4. *His construction involved elaborate computations and is fairly complicated. Indeed, aided by a computer, he had proved in* [51] *the existence of a set S of* 6448 *points with chromatic number* 4 *and without triangles or quadrilaterals with all sides of length* 1.

In a talk in 1992 the author shared this problem of Erdős, in the form of a competition:

Problem *Find the smallest number σ_4 of points in a plane set with chromatic number* 4 *without unit equilateral triangles, and classify all such sets S of σ_4 points.*

A number of young mathematicians entered the race, and the graphs obtained by the record setters were as mathematically significant as they were beautiful (see [45]). The two record-holding graphs, created by Hochberg and O'Donnell [22], are shown in Figs. 3 and 4; both graphs are 4-chromatic unit-distance graphs.

The problem of finding a 5-chromatic unit distance graph – or proving that none exists – remains open. However, much has been learned about 4-chromatic unit-distance graphs. The best of these results appeared in O'Donnell's Ph.D. thesis [31], where he answered Erdős's problem above in the negative (see [45, Theorem 48.4]). O'Donnell's result can be stated as follows.

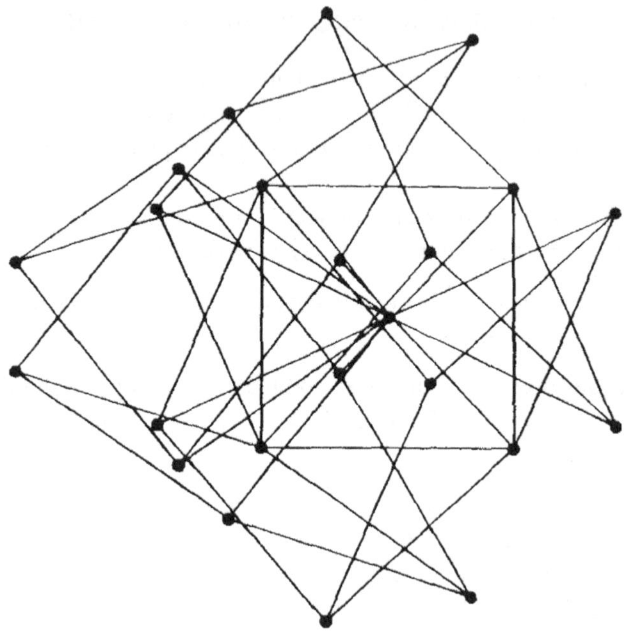

Fig. 3. The Hochberg-O'Donnell fish graph, of girth 4 and order 23

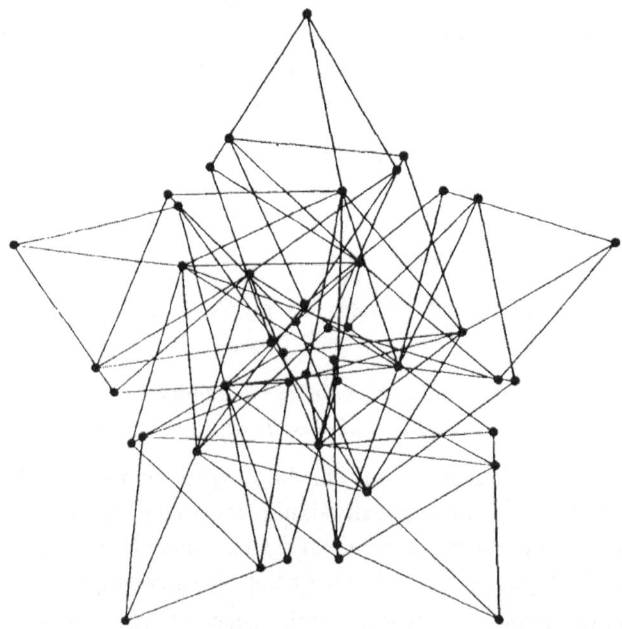

Fig. 4. The Hochberg-O'Donnell star graph, of girth 5 and order 45

Theorem 3.2 *There exist 4-chromatic unit-distance graphs with arbitrarily large finite girth.*

4. The polychromatic number: upper bounds

Raiskii's paper [37] also contained the upper bound $\chi_p(E^2) \leq 6$. An example proving this bound was found by Stechkin and published by Raiskii. We present Stechkin's construction (see Fig. 5).

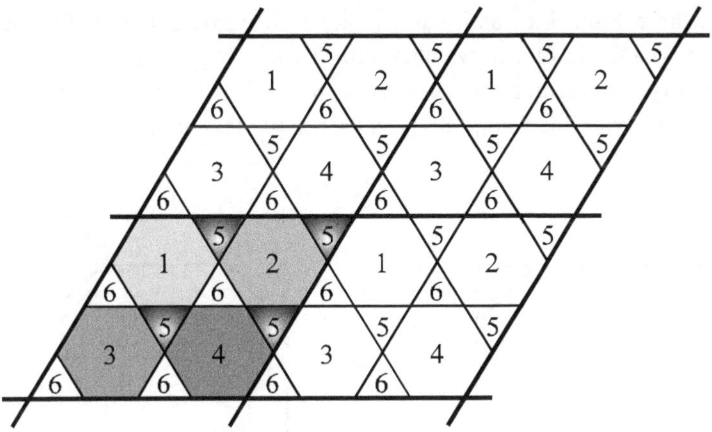

Fig. 5. Stechkin's 6-colouring of the plane

The 'unit of construction' is a parallelogram consisting of four regular hexagons and eight equilateral triangles, all of side-length 1. We colour the hexagons with colours 1, 2, 3 and 4. We then partition the triangles into two types: assign colour 5 to the triangles with a vertex below their horizontal base, and colour 6 to those with a vertex above their horizontal base. While colouring, assume that every hexagon includes its entire boundary, except for its one right-most and two lowest vertices, and that every triangle includes none of its boundary points. We can now tile the entire plane with translates of the 'unit of construction'.

If our ultimate goal is to find the chromatic number of the plane, or at least to improve the known bounds of 4 and 7, it might be worthwhile to 'measure' how close a given colouring of the plane is to achieving this goal. In 1992 such a measurement was introduced by the author and named the *colouring type* (see Soifer [40], [41]): given an n-colouring of the plane for which colour i does not realize the distance d_i (for $1 \leq i \leq n$), we say that this colouring is of *type* $(d_1, d_2, \ldots, d_n)$.

It would greatly improve our search for the chromatic number of the plane if we could find a 6-colouring of type (1,1,1,1,1,1), or show that one does not exist. With an appropriate choice of unit, Stechkin's colouring in Fig. 5 has type $(1,1,1,1,{}^1/_2, {}^1/_2)$. In 1973 Woodall [50] found a second 6-colouring of the plane with no colour realizing all distances; his colouring had the special property that each of the six monochromatic sets is closed. His example, however, had three 'missing distances': it had type $(1,1,1,\ 1/\sqrt{3},\ 1/\sqrt{3},\ 1/2\sqrt{3})$.

In 1991 a new 6-colouring was found by the author [51], using a tiling of the plane by squares and non-regular octagons; it had type $(1,1,1,1,1,\ 1/\sqrt{5})$. To construct it, we start with two squares, one with side 2 and the other with diagonal 1 (see Fig. 6), and use them to create a tiling of the plane with squares and non-regular octagons (see Fig. 8). We use colours 1–5 for the octagons and colour 6 for all the squares. With each octagon and each square we include half of its boundary (bold lines in Fig. 7) without the endpoints of that half. It is easy to verify that $\sqrt{5}$ is not realized by any of the colours 1–5, and that 1 is not realized by colour 6. By shrinking all linear sizes by a factor of $\sqrt{5}$, we obtain the 6-colouring of type $(1,1,1,1,1,\ 1/\sqrt{5})$.

To simplify the verification, we define the unit of construction as the region bounded by the bold line in Fig. 8; its translates tile the plane.

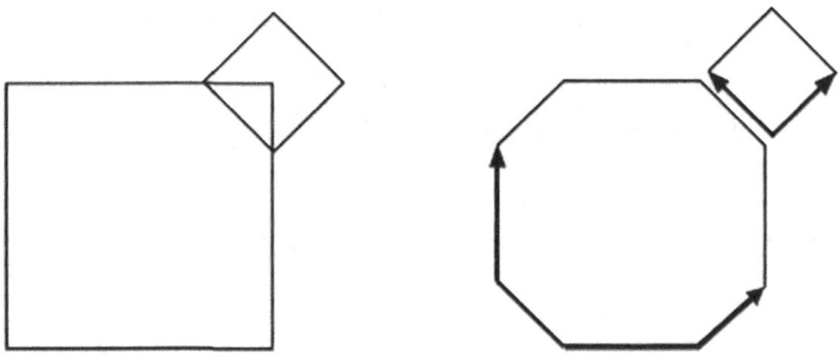

Figs. 6 and 7

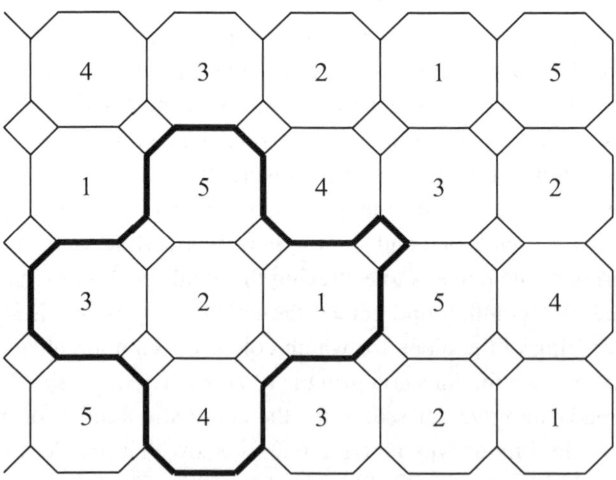

Fig. 8. Soifer's 6-colouring of the plane

The above 6-colouring gave birth to a new definition (see [23]). The *almost chromatic number* $\chi_a(E^2)$ of the plane is the minimum number of colours that are required to colour the plane so that all but one of the colours forbid distance 1, and the remaining colour forbids a distance that is not necessarily 1. Note that $\chi_a(E^2) = 4$, 5 or 6; the lower bound follows from Raiskii [37], and the upper bound follows from the above 6-colouring. The problem of determining $\chi_a(E^2)$ is still open (see [45]).

5. The continuum of 6-colourings

In 1993 Hoffman and the author [23], [24] found another 6-colouring, with type $(1, 1, 1, 1, 1, \sqrt{2} - 1)$; the story of its discovery can be found in [45]. To construct it we first tile the plane with squares of diagonals 1 and $\sqrt{2} - 1$(see Fig. 9). We use colours 1–5 for the larger squares and colour 6 for all the small squares. With each square we include the left and lower sides of its boundary without the endpoints of this half (see Fig. 10). To verify that this colouring does the job, define the unit of construction that is bounded by the bold line in Fig. 9; its translates tile the plane.

In 1993 the above two examples prompted the introduction of the new terminology and the translation of earlier results and problems into this new language (see [42],

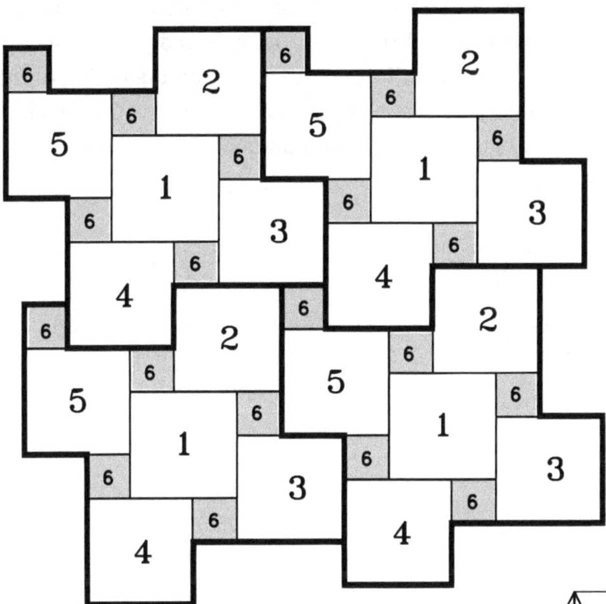

Fig. 9. Hoffman–Soifer's 6-colouring of the plane

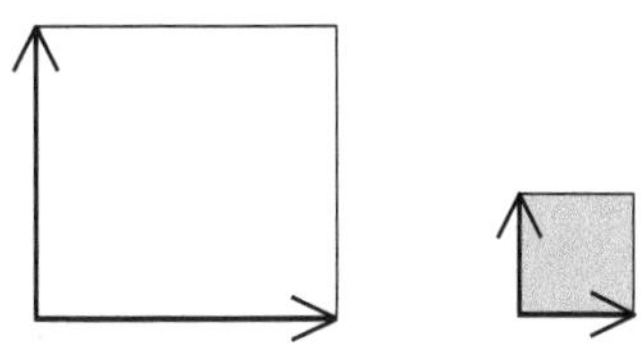

Fig. 10.

[43]). We let X_6 denote the 6-*realizable set* of all positive numbers α for which there exists a 6-colouring of the plane of type (1, 1, 1, 1, 1, α).The problem, which is still open and extremely difficult, is to find X_6.

We know from our above discussion that $1/\sqrt{5}$ and $\sqrt{2} - 1$ both lie in X_6. As shown in [42] and [43](see also [45]), these are extreme examples of the general case, which includes a continuum of 'working' 6-colourings: for each α between $\sqrt{2} - 1$ and $1/\sqrt{5}$, there is a 6-colouring of type (1, 1, 1, 1, 1, α).

Theorem 5.1 *X_6 contains the closed interval* $[\sqrt{2} - 1, 1/\sqrt{5}]$.

Outline of proof Partly cover a unit square by a smaller square that cuts off the unit square's vertical and horizontal segments (say, of lengths x and y, respectively) and forms with it an angle ω (see Fig. 11). These squares induce a tiling of the plane that consists of congruent non-regular octagons and 'small' squares (see Fig. 12).

We now colour this tiling in six colours. Denote by F the unit of our construction, bounded by a bold line and consisting of five octagons and four 'small' squares. Use colours 1–5 for the octagons inside F and colour 6 for all 'small' squares, and include in the colours of octagons and 'small' squares those parts of their boundaries that are shown in bold in Fig. 13. Translates of F tile the plane and thus determine the 6-colouring of the plane.

We now wish to select parameters to guarantee that each colour forbids a distance. In fact, a stronger result was proved (see [42], [43]) – namely that, for each angle ω between the small and the large squares in Fig. 11, there are uniquely determined sizes of the two squares, and uniquely determined parameters x and y, for which the constructed 6-colouring has type (1, 1, 1, 1, 1, α) for a uniquely determined α.

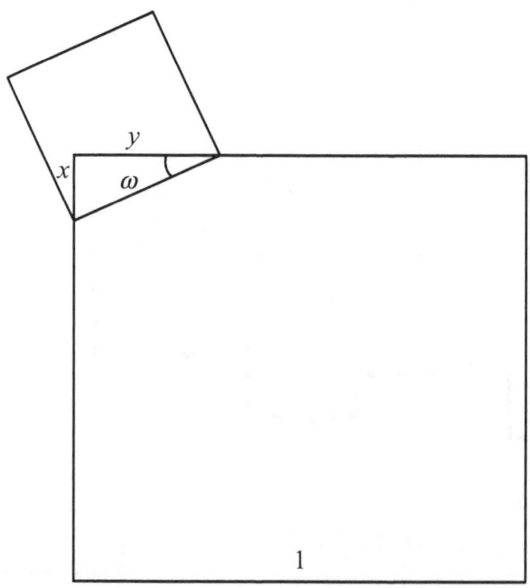

Fig. 11.

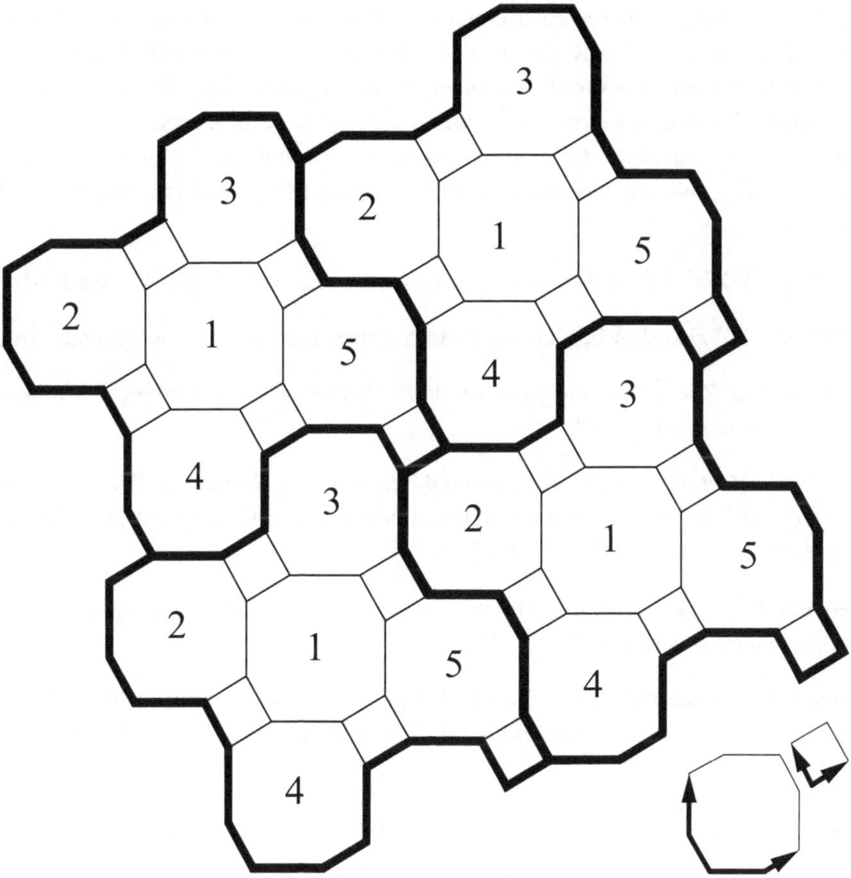

Fig. 12. Continuum of 6-colorings of the plane

Fig. 13.

Note that the 'working' solutions barely exist – they comprise something of a curve in three-dimensional space of the angle ω and two linear variables x and y. We have thus found a continuum of permissible values for α and a continuum of 'working' 6-colourings of the plane.

We remark that the problem of finding the 6-realizable set X_6 is closely related to the problem of finding the chromatic number $\chi(E^2)$ of the plane: finding X_6 would shed light on – if not solve – the latter problem, for if 1 is in X_6 then $\chi(E^2) \leq 6$, and if 1 is not in X_6 then $\chi(E^2) = 7$.

6. Special circumstances

In 1973 Woodall [50] attempted to prove a lower bound for the chromatic number of the plane for the special case of 'map-type colourings' of the plane. However, in 1979

Stephen Townsend constructed a counter-example that showed that one essential idea of Woodall's proof was incorrect. By that time Townsend had already proved the same result, and his proof was much more elaborate than Woodall's attempt. For decades Townsend's proof was unavailable until he produced a clear version of it where the complete definition of the map-type colouring can also be found (see [45]). The following result was thus conjectured by Woodall and proved by Townsend.

Theorem 6.1 *The chromatic number of the plane under map-type colouring is* 6 *or* 7.

Woodall [50] also showed that this result implies another one worth mentioning.

Theorem 6.2 *The chromatic number of the plane under colouring with closed monochromatic sets is* 6 *or* 7.

In 1993–94 three American undergraduate students, Nathanial Brown, Nathan Dunfield and Greg Perry, proved that a similar result is true for colouring with open monochromatic sets (see [3], [4] and [5]).

Theorem 6.3 *The chromatic number of the plane under colouring with open monochromatic sets is* 6 *or* 7.

Meanwhile, Falconer [16], while still a graduate student, proved the following important result, that the *measurable chromatic number* $\chi_m(E^2)$ of the plane is 5, 6 or 7.

Theorem 6.4 *If* $E^2 = \bigcup_{i=1}^{4} A_i$ *is a covering of the Euclidean plane* E^2 *by four disjoint measurable sets, then one of the sets* A_i *realizes distance* 1.

Recently, Falconer wrote a more detailed and self-contained exposition (see [45]).

Theorem 6.4 implies that if a 4-colouring of the plane forbids monochromatic distance 1, then one of the classes will not be 'nice' (it will be non-measurable). There is a supplementary result by Thomassen [49], who proved that if the colour classes are sufficiently 'nice' and forbid monochromatic distance 1, then seven colours are needed.

7. Space explorations

Around 1961 Erdős generalized the problem of finding the chromatic number of the plane to n-dimensional Euclidean space E^n. He was interested both in asymptotic behaviour as n increases, and in exact values of the chromatic number $\chi(E^n)$ for small n (especially $n = 2$ and 3).

In 1970 Raiskii [37] proved that $\chi(E^n) \geq n + 2$, for all $n > 1$; thus, for $n = 3$ we have $\chi(E^3) \geq 5$. This lower bound for E^3 lasted until 2000, when Oren Nechushtan proved that $\chi(E^3) \geq 6$ (see [30]). For upper bounds Coulson [9], [10] proved that $\chi(E^3) \leq 15$, by using a face-centred cubic lattice (see Conway and Sloane [8] for more on cubic lattices). Coulson informally conjectured that his upper bound of 15 is best possible for lattice-based colouring. For higher dimensions, Cantwell [6] proved in 1996 that $\chi(E^4) \geq 7$ and $\chi(E^5) \geq 9$; these remain the best results known. Then, in 2008, Josef Cibulka [7] proved that $\chi(E^6) \geq 11$.

Many years ago, Erdős conjectured that the chromatic number $\chi(E^n)$ increases exponentially with n. This conjecture was settled in the affirmative by two results, an exponential upper bound, found in 1972 by Larman and Rogers [26], and an exponential lower bound obtained in 1981 by Frankl and Wilson [18]:

$$\text{for all } n, \ (1 + o(1))1.2^n \ \leq \chi(E^n) \leq (3 + o(1))^n.$$

Asymptotically, Larman and Rogers's upper bound remains the best possible today. In 2000 Frankl and Wilson's asymptotic lower bound was improved by Raigorodskii [36] to $(1.239\ldots + o(1))^n$. It would be desirable to narrow the gap between these bounds.

The polychromatic number $\chi_p(E^2)$ also generalizes to higher dimensions. Raiskii [37] was the first to obtain a result, $\chi_p(E^n) \geq n + 2$, for all $n > 1$. Larman and Rogers's upper bound implies that $\chi_p(E^n) \leq (3 + o(1))^n$. They conjectured that $\chi_p(E^n)$ grows exponentially with n, and this was proved by Frankl and Wilson [18], who showed that $(1 + o(1))1.2^n \leq \chi_p(E^n)$.

8. Rational spaces

Another approach to the chromatic number of the Euclidean plane E^2 is to use Cartesian coordinates. As usual, E^2 is the set of all ordered pairs (x, y) with real coordinates x and y, and the distance between two points is defined in the usual Euclidean way.

By de Bruijn and Erdős's Theorem 3.1 it suffices to deal with finite subsets of E^2, and so we can restrict the coordinates to some subset C of E. The problem is: which subset should we choose?

Problem *Find a countable subset C of the set of real numbers E for which the chromatic number $\chi(C^2)$ equals that of the plane.*

The set $\mathbb{Q}$ of all rational numbers does not work, as shown by Woodall [50] in 1973.

Theorem 8.1 $\chi(\mathbb{Q}^2) = 2$.

Proof We need to colour the points of the rational plane $\mathbb{Q}^2$ – that is, the set of ordered pairs (r_1, r_2), where r_1 and r_2 are rational numbers. We first partition $\mathbb{Q}^2$ into disjoint classes by putting two pairs (r_1, r_2), and (q_1, q_2) into the same class if and only if $r_1 - q_1$ and $r_2 - q_2$ both have odd denominators when written in their lowest terms.

This partition of $\mathbb{Q}^2$ into subsets has an important property: if the distance between two points of $\mathbb{Q}^2$ is 1, then both points belong to the same subset of the partition. Indeed, let the distance between (r_1, r_2), and (q_1, q_2) be equal to 1. Then

$$(r_1 - q_1)^2 + (r_2 - q_2)^2 = 1.$$

Let $r_1 - q_1 = a/b$ and $r_2 - q_2 = c/d$ be these differences, written in their lowest terms. Then

$$(a/b)^2 + (c/d)^2 = 1 \text{ – that is, } a^2d^2 + b^2c^2 = b^2d^2.$$

Thus b and d must both be odd and so, by our definition above, (r_1, r_2) and (q_1, q_2) must belong to the same class.

Since any class of our partition can be obtained from any other class by a translation, it suffices to colour just one class and extend the colouring to the whole of $\mathbb{Q}^2$ by translations. Let us colour the class that contains the point (0, 0). This class consists of the points (r_1, r_2), where (in their lowest terms) the denominators of r_1 and r_2 are both odd. We colour red the points of the form $(o/o,\ o/o)$ and $(e/o,\ e/o)$, and colour blue the points of the form $(o/o,\ e/o)$ and $(e/o,\ o/o)$, where o stands for an odd number and e for an even number. In this colouring, two points of the same colour cannot be at distance 1 apart. ■

In 1975 there appeared a 'legendary unpublished manuscript', as P. D. Johnson, Jr. referred to a manuscript by Miro Benda and Micha Perles. This admired and widely circulated manuscript was called *Colorings of Metric Spaces*; Johnson tells its story in *Geombinatorics*, where in January 2000 the Benda–Perles paper was finally published [2]. Their results include the following.

Theorem 8.2 $\chi(\mathbb{Q}^3) = 2$ *and* $\chi(\mathbb{Q}^4) = 4$.

Benda and Perles [2] then posed some important open problems.

Problem *Find* $\chi(\mathbb{Q}^5)$ *and, in general,* $\chi(\mathbb{Q}^n)$.

Problem *Find the chromatic number of* $\mathbb{Q}(\sqrt{2})^2$ *and, in general, of any algebraic extension of* $\mathbb{Q}^2$.

This direction was developed by P. D. Johnson, Jr., Joseph Zaks, Klaus Fischer, Kiran B. Chilakamarri, Michael Reid, Douglas Jungreis, David Witte and Timothy Chow. In 2006 Johnson [25] published in *Geombinatorics* 'A tentative history and compendium' of this direction of inquiry.

More recently Matthias Mann [27] proved that $\chi(\mathbb{Q}^5) \geq 7$. This jump from $\chi(\mathbb{Q}^4) = 4$ explains the difficulty of finding $\chi(\mathbb{Q}^5)$, whose exact value is still unknown. Mann [28] then obtained further lower bounds: $\chi(\mathbb{Q}^6) \geq 10$, $\chi(\mathbb{Q}^7) \geq 13$ and $\chi(\mathbb{Q}^8) \geq 16$. In 2008 Cibulka [7] obtained new lower bounds for these chromatic numbers, improving some of Mann's results: $\chi(\mathbb{Q}^5) \geq 8$ and $\chi(\mathbb{Q}^7) \geq 15$.

9. One odd graph

In 1994 Moshe Rosenfeld [38] defined the odd-distance graph E_{odd} to be the graph with vertex-set E^2 in which two vertices are adjacent whenever the distance between them is an odd integer. He showed that E_{odd} does not have a subgraph K_4, and asked whether the chromatic number of E_{odd} is finite. In fact, while the problem was new, the absence of K_4-subgraphs was not, following from the following much more general result of Graham, Rothschild and Straus [21].

Theorem 9.1 *In E^n there exist $n + 2$ points, the distance between any two of which is an odd integer if and only if $n \equiv 14 \pmod{16}$.*

In the necessary part of the proof, the authors used an old result about determinants by Arthur Cayley. The main problem of this section remains wide open.

Problem *Find* $\chi(E_{\text{odd}})$.

We do not even know whether $\chi(E_{\text{odd}})$ is finite. In 2009 Ardal, Manuch, Rosenfeld, Shelah and Stacho [1] improved the lower bound to $\chi(E_{\text{odd}}) \geq 5$.

We denote the measurable chromatic number of the odd-distance graph by $\chi_m(E_{\text{odd}})$. In 1986 Falconer and Marstrand [17] proved that plane sets with positive density at infinity contain all large distances. This implies that $\chi_m(E_{\text{odd}}) \geq \aleph_0$. In 2009 the MIT undergraduate Jacob Steinhardt [48] found an alternative proof of this result using tools of spectral graph theory, which may be beneficial in solving other colouring problems.

10. Influence of set theory axioms

In this section, we need some ideas from set theory. We denote the standard Zermelo–Fraenkel choice system of axioms for set theory by **ZFC**, the countable axiom of choice by $\mathbf{AC}_{\aleph_0}$ and the principle of dependent choices by **DC**. We will use one further axiom, **LM**: every set of real numbers is Lebesgue measurable.

Our first task is to extend the definition of the chromatic number of a graph. Without the axiom of choice, the chromatic number of a graph may not exist. When allowing a system of axioms for set theory to exclude the axiom of choice, we need to create a much broader definition of the chromatic number than the usual one, if we want it to exist. In fact, instead of the chromatic number we ought to talk about *the set of chromatic cardinalities*. There are several meaningful ways to define this. Here is one that I chose in [45].

Let G be a graph and let A be a system of axioms for set theory. The *set of chromatic cardinalities* $\chi^A(G)$ of G is the set of all cardinal numbers $\tau \leq |V(G)|$ for which there is a proper colouring of the vertices of G in τ colours, and τ is minimum with respect to this property. As can be seen, the set of chromatic cardinalities need not have just one element as was the case with $A = \mathbf{ZFC}$. It can also be empty. The advantage of this definition is its simplicity. Best of all, we can use inequalities on sets of chromatic cardinalities as follows.

Let τ be a cardinal number. The inequality $\chi^A(G) > \tau$ means that $\sigma > \tau$ for every $\sigma \in \chi^A(G)$; the inequalities $<$, $\leq$ and $\geq$ are defined analogously. We also agree that the empty set is greater than or equal to any other set of cardinal numbers. Finally, if τ is a cardinal number and $\chi^A(G) = \{\tau\}$ is a one-element set of chromatic cardinalities (as is the case with the chromatic number when $A = \mathbf{ZFC}$), then we will simplify our notation by omitting parentheses and writing $\chi^A(G) = \tau$.

An infinite cardinal $\aleph_\alpha$ is *regular* if cf $\omega_\alpha = \omega_\alpha$, and κ is a *strong limit cardinal* if, for every cardinal λ, $\lambda < \kappa$ implies that $2^\lambda < \kappa$. A cardinal κ is *inaccessible* if $\kappa > \aleph_0$ and κ is regular and a strong limit cardinal. Assuming the existence of an inaccessible cardinal, and using Paul Cohen's forcing, Solovay [47] constructed in 1964 (and published in 1970) a model that proved a remarkable theorem. In his honour the author [45] defined the *Zermelo–Fraenkel–Solovay system of axioms* **ZFS** for set theory by

$$\mathbf{ZFS} = \mathbf{ZF} + \mathbf{AC}_{\aleph_0} + \mathbf{LM} \text{ and } \mathbf{ZFS+} = \mathbf{ZF} + \mathbf{DC} + \mathbf{LM}.$$

Solovay's theorem can now be formulated very concisely.

Theorem 10.1 **ZFS+** *is consistent.*

Shelah and Soifer [39] constructed the following example. Define a graph G as follows: the vertex-set is the set of real numbers, and the set of edges is $\{(s, t) : s - t - \sqrt{2} \in \mathbb{Q}\}$. They then proved that, for this graph, $\chi^{\mathbf{ZFC}}(G) = 2$, while $\chi^{\mathbf{ZFS}}(G) > \aleph_0$.

Similar examples with the plane E^2 and (in general) E^n as the vertex-set were constructed in [46] and [44], respectively. These examples illuminate the influence of the system of axioms for set theory on combinatorial results. They also suggest that the chromatic number of E^n may not exist 'in the absolute' (that is, in **ZF**), but may depend upon the system of axioms chosen for set theory. These examples naturally prompt the following open problem.

Problem *For which values of n is the chromatic number $\chi(E^n)$ uniquely defined 'in the absolute' – that is, in* **ZF***– regardless of the addition of the axiom of choice or its relative?*

An important example came from the Australian student Michael Payne [32], who started with the unit-distance graph G_1 with vertex-set $\mathbb{Q}^2$, where two vertices are adjacent if and only if they are at distance 1 apart. He then showed that the desired unit-distance graph G on the vertex-set E^2 is obtained by tiling the plane by translates of G_1 – that is, its edge-set is $\{(p_1, p_2) : p_1, p_2 \in E^2, p_1 - p_2 \in \mathbb{Q}^2$ and $|p_1 - p_2| = 1\}$, and proved that $\chi^{\mathbf{ZFC}}(G) = 2$ and $\chi^{\mathbf{ZFS}}(G) = 3, 4, 5, 6$ or 7.

Payne proved first that any measurable set S of positive Lebesgue measure contains the endpoints of a path of length 3 in G; of course, this ruled out a 2-colouring of S. Payne continued: 'We can then proceed in a similar fashion to Shelah and Soifer's proof' [39]. In 2009 he constructed a new class of unit distance graphs on the vertex-set E^n whose chromatic number depends upon the system of axioms for set theory [33].

11. Predicting the future

In 2003 Shelah and Soifer [39] obtained the following surprising result.

Theorem 11.1 *Assume that any finite unit-distance plane graph has chromatic number not exceeding 4. Then* $\chi^{\mathbf{ZFC}}(E^2) = 4$, *but* $\chi^{\mathbf{ZFS+}}(E^2) \geq 5$.

Can we obtain any results unconditionally? Yes, we can (see [45]), but not yet in **ZFC**.

Theorem 11.2 $\chi^{\mathbf{ZFS+}}(E^2) \geq 5$.

We conclude with the author's conjectures [45] of the expected value of the chromatic number of the plane, and more generally of E^n, in **ZFC**.

Conjecture $\chi(E^2) = 4$ *or* 7.

If the chromatic number of the plane were 4, then Theorem 11.1 would imply that the chromatic number of the plane does not exist in absolute, but depends upon the choice of the system of axioms for set theory. However, if I were limited to conjecturing a single value for the chromatic number of the plane, I would choose the value 7.

If the last conjecture is true, then a unit-distance 7-chromatic finite graph must exist in the plane. In 1998 Pritikin [35] published a lower bound for the order of such a graph.

Theorem 11.3 *Any unit-distance* 7*-chromatic graph G has at least* 6198 *vertices.*

In fact, the order of the smallest such graph may have to be much larger.

For 3-dimensional space the author [45] conjectured that $\chi(E^3) = 15$. My main conjecture is as follows.

Main conjecture For any positive integer $n > 1$, $\chi(E^n) = 2^{n+1} - 1$.

To paraphrase Paul Erdős's words about some of his conjectures, we can say that this conjecture is likely to withstand centuries, but we shall see!

Acknowledgement This chapter was written in 2013 and celebrates the centenary of the birth of Paul Erdős.

References

1. H. Ardal, J. Manuch, M. Rosenfeld, S. Shelah and L. Stacho, The odd-distance plane graph, *Discrete Comput. Geom.* **42** (2009), 132–141.
2. M. Benda and M. Perles, Colorings of metric spaces, *Geombinatorics* **IX** (3) (2000), 113–126.
3. N. Brown, N. Dunfield and G. Perry, Colorings of the plane I, *Geombinatorics* **III** (2) (1993), 24–31.
4. N. Brown, N. Dunfield and G. Perry, Colorings of the plane II, *Geombinatorics* **III** (3) (1993), 64–74.
5. N. Brown, N. Dunfield and G. Perry, Colorings of the plane III, *Geombinatorics* **III** (4) (1993), 110–114.
6. K. Cantwell, All regular polytopes are Ramsey, *J. Combin. Theory (A)* **114** (2007), 555–562.
7. J. Cibulka, On the chromatic number of real and rational spaces, *Geombinatorics* **XVIII** (2) (2008), 53–65.
8. J. H. Conway and N. J. A. Sloane, *Sphere Packings, Lattices and Groups*, 3rd edn., Springer-Verlag, 1999.
9. D. Coulson, A 15-colouring of 3-space omitting distance one, *Discrete Math.* **256** (2002), 83–90.
10. D. Coulson, Tilings and colourings of 3-space, *Geombinatorics* **XII** (3) (2003), 102–116.
11. N. G. de Bruijn and P. Erdős, A color problem for infinite graphs and a problem in the theory of relations, *Indag. Math.* **13** (1951), 369–373.
12. P. Erdős, Problem (p. 681); Unsolved problems, *Proceedings of the Fifth British Combinatorial Conference 1975, University of Aberdeen, July 1975* (eds. C. St.J. A. Nash-Williams and J. Sheehan), *Congr. Numer.* **XV**, Utilitas Mathematica, 1976.
13. P. Erdős, Combinatorial problems in geometry and number theory, *Relations between combinatorics and other parts of mathematics*, *Proc. Sympos. Pure Math.* (Ohio State Univ., 1978), Proc. Sympos. Pure Math. **XXXIV**, Amer. Math. Soc. (1979), 149–162.
14. P. Erdős, Some combinatorial problems in geometry, *Geom. & Diff. Geom.* (Proc. Haifa, 1979), Lecture Notes in Math. **792**, Springer (1980), 46–53.
15. P. Erdős, Some new problems and results in graph theory and other branches of combinatorial mathematics, *Combinatorics and Graph Theory* (Proc. Symp. Calcutta 1980), Lecture Notes in Math. **885**, Springer (1981), 9–17.
16. K. J. Falconer, The realization of distances in measurable subsets covering R^n, *J. Combin. Theory (A)* **31** (1981), 187–189.
17. K. J. Falconer, and J. M. Marstrand, Plane sets with positive density at infinity contain all large distances, *Bull. London Math. Soc.* **18** (1986), 471–474.
18. P. Frankl and R. M. Wilson, Intersection theorems with geometrical consequences, *Combinatorica* **1** (1981), 357–368.
19. R. L. Graham, Some of my favourite problems in Ramsey theory, *Proceedings of the 'Integers Conference 2005' in Celebration of the 70th Birthday of Ronald Graham,*

Carrolton, Georgia, USA, 2005 (eds. B. Landman *et al.*), Walter de Gruyter (2007), 229–236.

20. R. L. Graham, Old and new problems and results in Ramsey theory, *Horizons of Combinatorics* (Conference and EMS Summer School, Hungary, 2006) (eds. E. Győri, G. Katona and L. Lovász), Bolyai Society Mathematical Studies (2008), 105–118.
21. R. L. Graham, B. L. Rothschild and E. G. Straus, Are there $n + 2$ points in E^n with odd integral distances?, *Amer. Math. Monthly* **81** (1974), 21–25.
22. R. Hochberg and P. O'Donnell, Some 4-chromatic unit-distance graphs without small cycles, *Geombinatorics* **V**(4) (1996), 137–141.
23. I. Hoffman and A. Soifer, Almost chromatic number of the plane, *Geombinatorics* **III** (2) (1993), 38–40.
24. I. Hoffman and A. Soifer, Another six-colouring of the plane, *Discrete Math.* **150** (1996), 427–429.
25. P. D. Johnson, Jr., Coloring the rational points to forbid the distance one – A tentative history and compendium, *Geombinatorics* **XVI** (1) (2006), 209–218.
26. D. G. Larman and C. A. Rogers, The realization of distances within sets in Euclidean space, *Mathematika* **19** (1972), 1–24.
27. M. Mann, A new bound for the chromatic number of the rational five-space, *Geombinatorics* **XI** (2) (2001), 49–53.
28. M. Mann, Hunting unit-distance graphs in rational n-spaces, *Geombinatorics* **XIII** (2) (2003), 86–97.
29. L. Moser and W. Moser, Solution to problem 10, *Canad. Math. Bull.* **4** (1961), 187–189.
30. O. Nechushtan, On the space chromatic number, *Discrete Math.* **256** (2002), 499–507.
31. P. O'Donnell, *High Girth Unit-distance Graphs*, Ph.D. thesis, Rutgers University, 1999.
32. M. S. Payne, A unit distance graph with ambiguous chromatic number, *arXiv*: 0707.1177v1 [math.CO], 9 July 2007.
33. M. S. Payne, Unit distance graphs with ambiguous chromatic number, *Electronic J. Combin.* **16** (2009).
34. Materialy Konferenzii 'Poisk-97', Moscow, 1997 (in Russian).
35. D. Pritikin, All unit-distance graphs of order 6197 are 6-colourable, *J. Combin. Theory (B)* (1998),159–163.
36. A. M. Raigorodskii, On the chromatic number of a space, *Russian Math. Surveys* **55** (2000), 351–352.
37. D. E. Raiskii, Realizing of all distances in a decomposition of the space R^n into $n + 1$ parts, *Mat. Zametki* **7** (1970), 319–323 (in Russian); English transl., *Math. Notes* **7** (1970), 194–196.
38. M. Rosenfeld, Odd integral distances among points in the plane, *Geombinatorics* **V** (4) (1996), 156–159.
39. S. Shelah and A. Soifer, Axiom of choice and chromatic number of the plane, *J. Combin. Theory (A)* **103** (2003), 387–391.
40. A. Soifer, Relatives of chromatic number of the plane I, *Geombinatorics* **I** (4) (1992), 13–15.
41. A. Soifer, A six-colouring of the plane, *J. Combin. Theory (A)* **61** (1992), 292–294.
42. A. Soifer, Six-realizable set X_6, *Geombinatorics* **III** (4) (1994), 140–145.
43. A. Soifer, An infinite class of 6-colorings of the plane, *Congr. Numer.* **101** (1994), 83–86.
44. A. Soifer, Axiom of choice and chromatic number of R^n, *J. Combin. Theory (A)* **110** (2005), 169–173.
45. A. Soifer, *The Mathematical Coloring Book: Mathematics of Coloring and the Colorful Life of its Creators*, Springer, 2009.
46. A. Soifer and S. Shelah, Axiom of choice and chromatic number: an example on the plane, *J. Combin. Theory (A)* **105** (2004), 359–364.

47. R. M. Solovay, A model of set theory in which every set of reals is Lebesgue measurable, *Ann. of Math.* **92** (1970), 1–56.
48. J. Steinhardt, On coloring the odd-distance graph, *Electronic J. Combin.* **16** (12) (2009), 1–7.
49. C. Thomassen, On the Nelson unit distance coloring problem, *Amer. Math. Monthly* **106** (1999), 850–853
50. D. R. Woodall, Distances realized by sets covering the plane, *J. Combin. Theory (A)* **14** (1973), 187–200.
51. N. C. Wormald, A 4-chromatic graph with a special plane drawing, *J. Austral. Math. Soc. (A)* **28** (1979), 1–8.

9

Integer flows and orientations

HONGJIAN LAI, RONG LUO and CUN-QUAN ZHANG

The theory of integer flows was introduced by Tutte as a generalization of map-colouring problems. This chapter is a brief survey of integer flows, including their extensions: circular flows, modulo orientations, group connectivity, and an update of recent progress on Tutte's flow conjectures.

1. Introduction

The concept of an integer flow was introduced by Tutte [55], [56] as a generalization of map-colouring problems (see Theorem 1.1). This chapter is a brief survey of integer flow theory; for further study in this area, see Zhang [63].

Let $G = (V, E)$ be a graph. Given an orientation D of $E(G)$, we denote the resulting directed graph by $D(G)$, and for each vertex $v \in V(G)$, let $E^+(v)$ and $E^-(v)$ be the sets of arcs of $D(G)$ with their tails and heads (respectively) at v.

Let G be a graph, let D be an orientation of G, let Γ be an abelian group (an additive group with 0 as the identity) and let $f : E(G) \to \Gamma$ be a mapping. Then the ordered pair (D, f) is called a *flow* (or *group* Γ*-flow*) of G if

$$\sum_{e \in E^+(v)} f(e) = \sum_{e \in E^-(v)} f(e),$$

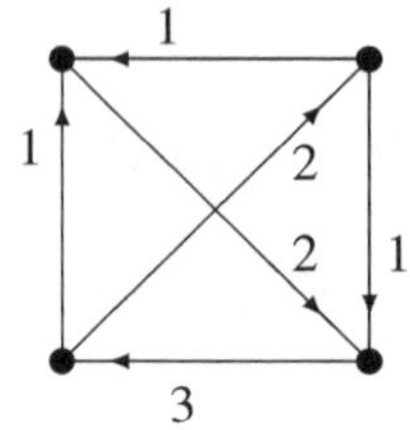

Fig. 1. A nowhere-zero 4-flow of the complete graph K_4.

for each vertex $v \in V(G)$. In this chapter, we are interested in finite abelian groups, infinite groups $\mathbb{Z}$ (the set of integers), $\mathbb{Q}$ (the set of rational numbers), $\mathbb{R}$ (the set of real numbers) and $\mathbb{Z}_k$ (the cyclic group of order k).

Let (D, f) be a Γ-flow of a graph G and let k be an integer. Then (D, f) is called an *integer flow* if $\Gamma = \mathbb{Z}$, and an integer flow is a *k-flow* if $|f(e)| < k$ for each edge $e \in E(G)$. (D, f) is a *mod-k-flow* if $f : E(G) \to \mathbb{Z}$ is such that

$$\sum_{e \in E^+(v)} f(e) \equiv \sum_{e \in E^-(v)} f(e) (\text{mod } k),$$

for each $v \in V(G)$ – that is, (D, f) is a group $\mathbb{Z}_k$-flow.

The *support* supp (f) of a Γ-flow (D, f) is the set of all edges of G with $f(e) \neq 0$. A flow (D, f) is a *nowhere-zero flow* if supp $(f) = E(G)$. An example of a nowhere-zero 4-flow is given in Fig. 1.

Flow-colouring duality

The following theorem of Tutte [56] indicates the important relation between map colouring and integer flows, and motivates the study of the theory of integer flows.

Theorem 1.1 *Let G be a planar bridgeless graph. Then G is k-face-colourable if and only if G admits a nowhere-zero k-flow.*

Note that the 'only if' part of Theorem 1.1 holds not only for planar graphs, but also for all graphs embeddable on some orientable surface. Tutte [56] also proved the following result.

Theorem 1.2 *Let G be a bridgeless graph with a closed 2-cell embedding on some orientable surface. If G is k-face-colourable, then G admits a nowhere-zero k-flow.*

Tutte's conjectures

The following conjectures are the most famous in the theory of integer flows. They were proposed by Tutte ([56], [59] and Problem 48 in [7]).

Conjecture A (5-flow conjecture) *Every bridgeless graph admits a nowhere-zero 5-flow.*

Conjecture B (4-flow conjecture) *Every bridgeless graph containing no subdivision of the Petersen graph admits a nowhere-zero 4-flow.*

Conjecture C (3-flow conjecture) *Every bridgeless graph containing no 3-edge-cut admits a nowhere-zero 3-flow.*

These well-known conjectures are motivated by the following map-colouring theorems of Heawood [23], Appel and Haken [2], [3] and Grötzsch [21].

Theorem 1.3 (The five-colour theorem) *Every bridgeless planar graph is 5-face-colourable.*

Theorem 1.4 (The four-colour theorem) *Every bridgeless planar graph is 4-face colourable.*

Theorem 1.5 (The three-colour theorem) *Every bridgeless planar graph without a 3-edge-cut is 3-face colourable.*

Although six decades have passed and some significant and important approaches have been made toward these conjectures, they remain essentially open.

2. Basic properties

In this section, we introduce some basic definitions and properties concerning integer flows.

Equivalence of k-flows

The following fundamental theorem is due to Tutte [55], [56].

Theorem 2.1 *Let G be a graph, let k be a positive integer, and let Γ be an abelian group of order k. Then the following statements are equivalent:*

(1) *G admits a nowhere-zero integer k-flow*
(2) *G admits a nowhere-zero mod-k-flow*
(3) *G admits a nowhere-zero group Γ-flow.*

From Theorem 2.1, all nowhere-zero flows are equivalent, so whenever we say that 'a graph G admits a nowhere-zero k-flow', it always means that G admits a nowhere-zero integer k-flow, a nowhere-zero group Γ-flow with $|\Gamma| = k$, or a nowhere-zero mod-k-flow. Note that each definition of a nowhere-zero flow has its own special advantages, depending on the topic being studied.

The equivalence of (1) and (2) is strengthened by a useful technical result (Theorem 2.3). The equivalence of (2) and (3) was originally proved by Tutte by using the flow polynomial technique.

From the definition of an integer flow, the following observations of Tutte [56], [57] are straightforward.

Theorem 2.2 *If a graph G admits a nowhere-zero integer k-flow, then G admits a nowhere-zero integer h-flow, for each $h \geq k$.*
A graph G admits a nowhere-zero 2-flow if and only if the degree of every vertex is even.

Mod-k-flows

In Theorem 2.1, the proof of '(a) $\Rightarrow$ (b)' is trivial, since a nowhere-zero integer k-flow is also a nowhere-zero mod-k-flow. The proof of '(b) $\Rightarrow$ (a)' follows from the following stronger result (see [55]).

Theorem 2.3 *If a graph G admits a mod-k-flow (D, f_a), then G admits an integer k-flow (D, f_b) for which $f_a(e) \equiv f_b(e)$ (mod k) for each edge $e \in E(G)$.*

Note that both flows in Theorem 2.3 correspond to the same orientation D.

Products of flows

The following result has been used in the proofs of some landmark theorems (see Jaeger [26] and Seymour [47]).

Theorem 2.4 *Let G be a graph and let k_1 and k_2 be integers. If G admits a k_1-flow (D, f_1) and a k_2-flow (D, f_2), and if* $\text{supp}\,(f_1) \cup \text{supp}\,(f_2) = E(G)$, *then $(D, k_2 f_1 + f_2)$ and $(D, f_1 + k_1 f_2)$ are nowhere-zero (k_1k_2)-flows of G.*

This result can be generalized to an 'if and only if' result (see Zhang [63]).

Theorem 2.5 *Let G be a graph and let k_1 and k_2 be integers. Then G admits a nowhere-zero (k_1k_2)-flow if and only if G admits a k_1-flow (D, f_1) and a k_2-flow (D, f_2) with* $\text{supp}\,(f_1) \cup \text{supp}\,(f_2) = E(G)$.

Sums of flows

Results on the sum of flows have been obtained by Little, Tutte and Younger [40].

Theorem 2.6 *For each non-negative integer k-flow (D, f) of a graph G, G admits $k-1$ non-negative 2-flows (D, f_r) $(r = 1, 2, \ldots, k-1)$ with $f = \sum_{r=1}^{k-1} f_r$.*

A directed graph is *even* if the in-degree of each vertex equals its out-degree. Since the support of a non-negative 2-flow is a directed even subgraph with orientation D, the following theorem on directed even subgraph covering is equivalent to Theorem 2.6.

Corollary 2.7 *Let G be a graph and let D be an orientation of G. Then the graph G admits a positive k-flow (D, f) if and only if $D(G)$ contains $k-1$ directed even subgraphs such that each arc of $D(G)$ appears in at least one of them.*

3. 4-flows

In this section we study properties, open problems, and some partial results related to the following major open problem in flow theory.

Recall Tutte's 4-flow conjecture (Conjecture **B**), that every bridgeless graph containing no subdivision of the Petersen graph admits a nowhere-zero 4-flow, and (from Theorem 1.4) that every bridgeless planar graph admits a nowhere-zero 4-flow.

It is well known that, for a cubic graph G, the graph G admits a nowhere-zero 4-flow if and only if G is 3-edge-colourable. Tutte [60] also conjectured. that every bridgeless cubic graph containing no subdivision of the Petersen graph admits a nowhere-zero
4-flow.

Jaeger [28] asked whether the latter conjecture is equivalent to Conjecture **B**. It was eventually proved by Robertson, Sanders, Seymour and Thomas [52], [48], while Conjecture **B** remains open.

Theorem 3.1 *Every bridgeless cubic graph containing no subdivision of the Petersen graph is 3-edge-colourable, and thus admits a nowhere-zero 4-flow.*

The proof of this will consist of a series of papers (see [52]).

The following theorem for highly connected graphs was proved by Jaeger [26].

Theorem 3.2 *Every 4-edge-connected graph admits a nowhere-zero 4-flow.*

4. 3-flows

Recall that a major open problem in integer flow theory is Conjecture **C** (Tutte's 3-flow conjecture), which is a generalization of Grötzsch's 3-colouring theorem (Theorem 1.5) for planar graphs.

The following result was observed by Tutte [55].

Theorem 4.1 *Let G be a cubic graph. Then G admits a nowhere-zero 3-flow if and only if G is bipartite.*

A weak version of Conjecture **C** was proposed by Jaeger [26], that there is an integer h for which every h-edge-connected graph admits a nowhere-zero 3-flow. Some early partial results on this weak conjecture can be found in Lai and Zhang [39] and Alon, Linial and Meshulam [1]. It was recently proved by Thomassen [54] with $h = 8$. Thomassen's theorem [54] was further improved by Lovász, Thomassen, Wu and Zhang [41].

Theorem 4.2 *Every* 6*-edge-connected graph admits a nowhere-zero* 3*-flow.*

For embedded graphs, we recall Grötzsch's theorem (Theorem 1.5) that every 4-edge-connected planar graph is 3-face colourable, and so admits a nowhere-zero 3-flow. The following generalizations can be viewed as partial results for the 3-flow conjecture; they are due, respectively, to Grünbaum [22], Steinberg and Younger [51] and Thomassen [53].

Theorem 4.3 *Every bridgeless planar graph containing at most three* 3*-edge-cuts is* 3*-face-colourable and so admits a nowhere-zero* 3*-flow.*

Theorem 4.4 *Every* 2*-edge-connected graph with at most one* 3*-edge-cut that can be embedded in the projective plane admits a nowhere-zero* 3*-flow.*

Theorem 4.5 *Let G be a graph embedded in the torus such that all contractible cycles are of length at least* 5*. Then G is* 3*-vertex-colourable.*

Tutte's 3-flow conjecture was originally proposed for graphs with no 1-edge-cut and no 3-edge-cut. It was pointed out in [26], [28] and [47] that no 2-edge-cut exists in any smallest counter-example to some well-known flow conjectures, including Conjecture **C**.

A graph G is *λ_o-odd edge-connected* if every odd edge-cut of G has at least λ_o edges. Theorem 4.2 has a stronger version for odd edge-connectivity, as proved by Lovász, Thomassen, Wu and Zhang [41].

Theorem 4.6 *Every* 7*-odd edge-connected graph admits a nowhere-zero* 3*-flow.*

For any smallest counter-example to Conjecture **C**, the following proposition was obtained by applying a vertex-splitting lemma in [64].

Theorem 4.7 *Any smallest counter-example to the* 3*-flow conjecture is* 5*-regular and* 5*-odd edge-connected.*

Kochol [32] also proved that it suffices to prove the 3-flow conjecture for 5-edge-connected graphs.

Theorem 4.8 *The following statements are equivalent.*

- *Every* 4*-edge-connected graph admits a nowhere-zero* 3*-flow.*
- *Every* 5*-edge-connected graph admits a nowhere-zero* 3*-flow.*

Unlike k-flows with $k > 3$, nowhere-zero 3-flows can be viewed as a modulo orientation problem. An orientation D of a graph G is called a *modulo* 3*-orientation* if $|E^+(v)| \equiv |E^-(v)| \pmod 3$, for each $v \in V(G)$.

The following observation appeared in [62] and [51].

Theorem 4.9 *A bridgeless graph G admits a nowhere-zero* 3*-flow if and only if G has a modulo* 3*-orientation D.*

This observation is further generalized in Section 7.

5. 5-flows

The main topic of this section is the 5-flow conjecture (Conjecture **A**) of Tutte [56]. When Tutte introduced the concept of an integer flow, he first conjectured that there is an integer k for which every bridgeless graph admits a nowhere-zero k-flow, and pointed out that $k \geq 5$, since the Petersen graph does not admit a nowhere-zero 4-flow. This conjecture was proved independently by Jaeger [26] and Kilpatrick [30] with $k = 8$ (the 8-flow theorem). The best approach to it is currently the 6-flow theorem of Seymour [47].

Theorem 5.1 *Every bridgeless graph admits a nowhere-zero* 6*-flow.*

The approaches in the proofs of the 8-flow and 6-flow theorems were different. The 8-flow theorem was proved by applying a theorem of Tutte and Nash-Williams [58], [43] and showing the existence of three edge-disjoint spanning trees in $2G$ for a 3-edge-connected graph G. The 6-flow theorem was proved by showing that every bridgeless graph admits two flows (D, f_1) and (D, f_2), where one is a 2-flow and the other is a 3-flow with $\text{supp}(f_1) \cup \text{supp}(f_2) = E(G)$. The 5-flow conjecture is still open.

For graphs embedded on some surface, the dual of vertex-colouring is the flow problem. Thus, Heawood's colouring result (Theorem 1.3) implies the 5-flow conjecture for certain families of embedded graphs.

The following result of Möller, Carstens and Brinkmann [42] and Fouquet [18] is related to the 5-colour theorem and the 5-flow conjecture.

Theorem 5.2 *Every bridgeless graph embeddable in an orientable surface of genus $g \leq 2$, or in a non-orientable surface of genus $g \leq 4$, admits a nowhere-zero* 5*-flow.*

The 5-flow conjecture has been proved for some special families of graphs. The following result was proved by Jaeger [25].

Theorem 5.3 *Let e be an edge of a graph G. If G admits a nowhere-zero* 4*-flow and $G - e$ is bridgeless, then $G - e$ admits a nowhere-zero* 5*-flow.*

Let H be a spanning even subgraph of a graph G. The *oddness of H* is the number of components of H containing an odd number of odd-degree vertices of G. The *oddness of G* is the minimum of the oddnesses of all spanning even subgraphs of G. It is straightforward to show that G admits a nowhere-zero 4-flow if and only if the oddness of G is 0.

Special cases of such graphs $G - e$ in Theorem 5.3 are graphs containing a Hamiltonian path, and (in general) graphs of oddness at most 2. On the other hand, Celmins [11] observed a similar result in the opposite direction.

Theorem 5.4 *If a bridgeless graph G has an edge e for which $G - e$ admits a nowhere-zero* 4*-flow, then G admits a nowhere-zero* 5*-flow.*

Both results have been further extended to the deletion of more than one edge by Steffen [49], [50], under some conditions of cyclic edge-connectivity.

For vertex-deletions, Gerards and Seymour (personal communication, 1995) proved the 5-flow conjecture for apex graphs. (An *apex graph* is one in which at least one vertex is adjacent to all the others.)

Theorem 5.5 *Every apex graph admits a nowhere-zero* 5*-flow.*

By applying a lemma in [46], that every cubic graph with girth at least 6 must have a Petersen minor, Kochol [31] further generalized the above theorem.

Theorem 5.6 *Every Petersen minor-free cubic graph admits a nowhere-zero* 5*-flow.*

A smallest counter-example to the 5-flow conjecture must be cubic and not 3-edge-colourable. The following further result was proved by Kochol [34].

Theorem 5.7 *A smallest counter-example G to the* 5*-flow conjecture has girth at least* 11 *and cyclic edge-connectivity at least* 6.

The first result was proved by using a computer-aided search, which extended the girth results in some earlier articles.

Modulo 5-orientations

An 8-flow is obtained as the product of three 2-flows (see Jaeger [26]), while a 6-flow is obtained as the product of a 2-flow and a 3-flow (see Seymour [47]). But 8 and 6 are composite numbers, while 5 is a prime number. So what can we do for 5-flows? Certainly they cannot be the product of smaller flows. Various approaches have been proposed, such as orientable 5-even subgraph double covers (see Archdeacon [4] and Jaeger [28]), modulo 5-orientations (see Jaeger [28]) and bipartizing matching (see Fleischner [17]).

The modulo 5-orientation was proposed by Jaeger [28] as an approach to the 5-flow conjecture (see Section 7 for a detailed discussion on modulo orientation).

Jaeger proved [28] that the following conjecture implies the 5-flow conjecture.

Conjecture D *Every* 9*-edge-connected graph has a modulo* 5*-orientation.*

A partial result concerning this conjecture is a result of Lovász, Thomassen, Wu and Zhang [41].

Theorem 5.8 *Every* 12*-edge-connected graph has a modulo* 5*-orientation.*

6. Bounded orientations and circular flows

Let $\{A, B\}$ be a partition of the vertex-set $V(G)$, and let D be an orientation of G. The set of arcs of $D(G)$ with tails in A and heads in B is denoted by $[A, B]_D$, or simply by $[A, B]$ if no confusion arises.

The following is a revised version of a result by Hoffman (see [6]).

Theorem 6.1 *Let G be a bridgeless graph, let D be an orientation of G and let a and b be two positive integers with $a \leq b$. Then the following statements are equivalent:*

(a) $a/b \leq |[A,B]_D|/|[B,A]_D| \leq b/a$, *for every edge-cut* (A,B) *of* G*;*

(b) *G admits a nowhere-zero integer flow (D, f_1) such that $a \leq f_1(e) \leq b$ for each $e \in E(G)$;*

(c) *G admits a nowhere-zero real-valued flow (D, f_2) such that $a \leq f_2(e) \leq b$ for each $e \in E(G)$.*

Corollary 6.2 *A graph G admits a nowhere-zero k-flow if and only if G has an orientation D for which*

$$\frac{1}{k-1} \leq \frac{|[A,B]_D|}{|[B,A]_D|} \leq k-1,$$

for every edge-cut (A,B) of G.

Let k and d be two integers with $0 < d \leq \frac{1}{2}k$. An integer flow (D,f) of a graph G is *a circular k/d-flow* if $f : E(G) \to \{\pm d, \pm(d+1), \ldots, \pm(k-d)\} \cup \{0\}$. The concept of a circular flow, introduced by Goddyn, Tarsi and Zhang [20], is a generalization of integer flows and is a dual version of the circular colouring problem (see Zhu [65] for a comprehensive survey of this area).

Theorem 6.3 *Let G be a bridgeless graph, let D be an orientation of G and let $k, d \in \mathbb{Z}^+$ and $q \in \mathbb{Q}^+$ be such that $q = k/d \geq 2$. Then the following statements are equivalent:*

(a) *G admits a positive circular k/d-flow (D, f_1);*

(b) *G admits a rational-valued flow (D, f_2) such that $f_2 : E(G) \to [1, q-1]$;*

(c)
$$q-1 = \frac{k-d}{d} \geq \frac{|[U,\ V(G)-U]_D|}{|[V(G)-U,\ U]_D|} \geq \frac{d}{k-d} = \frac{1}{q-1},$$

for each non-empty proper subset $U \subset V(G)$.

An immediate corollary of Theorem 6.3, proved in [20], is the following result, analogous to Theorem 2.2.

Theorem 6.4 *Let G be a graph and let $p \in \mathbb{Q}^+$. If G admits a nowhere-zero circular p-flow, then G admits a nowhere-zero circular q-flow for every $q \in \mathbb{Q}^+$ with $q \geq p$.*

For a bridgeless graph G, the *flow index* $\varphi(G)$ is the smallest rational number ℓ for which G admits a nowhere-zero circular ℓ-flow. It is natural to ask, for any given rational number ℓ, whether there is a graph G for which the flow index $\varphi(G)$ of the graph is precisely ℓ. The following result of Pan and Zhu [44] answers this question.

Theorem 6.5 *For every rational number ℓ in the interval $[2,5]$, there is a graph G for which $\varphi(G) = \ell$.*

Tutte's Theorem 1.1 states the relation between integer flows and face-colouring for planar graphs. The following result of DeVos, Goddyn, Mohar, Vertigan and Zhu [13] extends this theorem to locally planar graphs.

Theorem 6.6 *For a given orientable surface Σ and positive number ε, there is a function $f(\Sigma, \varepsilon)$ such that, for every graph G embedded on Σ with edge-width at least $f(\Sigma, \varepsilon)$, if the dual graph G^* admits a nowhere-zero circular ℓ-flow, then the graph G is circular $(\ell + \varepsilon)$-vertex-colourable.*

Together with Theorem 1.2, Theorem 6.6 provides a close relationship between face-colouring and flow index for 'locally planar' graphs (graphs with sufficiently large edge-width).

7. Modulo orientations and (2 + 1/*t*)-flows

As with 3-flows and mod-3-orientations (Theorem 4.9), a circular $(2 + 1/t)$-flow can also be considered as a modulo $(2t + 1)$-orientation for each positive integer t.

Let k be an odd integer. An orientation D of a graph G is called a *modulo k-orientation* if $d_D^+(v) \equiv d_D^-(v) \pmod k$, for every $v \in V(G)$. The following result is due to Jaeger [27].

Theorem 7.1 *Let G be a graph and let t be a positive integer. Then G has a modulo $(2t + 1)$-orientation if and only if G admits a nowhere-zero circular $(2 + 1/t)$-flow.*

The following conjecture was proposed by Jaeger [27] (see also [28] and [63]).

Conjecture E *Let G be a graph and let k (≥ 3) be an odd integer. If G is $(2k - 2)$-edge-connected, then G has a modulo k-orientation.*

The 3-flow conjecture (by Theorem 4.9) and Conjecture **D** are special cases of Conjecture **E** (for $k = 3$ and 5).

A weak version of this conjecture, the $(2 + \varepsilon)$-flow conjecture, was proposed by Seymour (personal communication 1999) and Galluccio, Goddyn and Hell [19] as an analogue to Jaeger's weak-3-flow conjecture [26]. It was proved recently by Thomassen [54] and improved by Lovász, Thomassen, Wu and Zhang [41].

Theorem 7.2 *Let G be a graph and let $k = 2t + 1 \geq 3$ be an odd integer. Then G has a modulo k-orientation if G is $(3k-3)$-edge-connected, and so admits a nowhere-zero circular $(2 + 1/t)$-flow.*

As we discussed in Section 4, odd edge-connectivity plays an important role for flows and modulo orientations. The following conjecture was proposed in [64] as a refinement of Conjecture **E** and Theorem 7.2 for $(2 + 1/t)$-flows (modulo $(2t + 1)$-orientations).

Conjecture F *For each positive integer t, every graph with odd edge-connectivity at least $4t + 1$ admits a nowhere-zero circular $(2 + 1/t)$-flow.*

A weak version [64] of Conjecture **F** has now been solved by Lovász, Thomassen, Wu and Zhang [41].

Theorem 7.3 *For each positive integer t, every* $(6t+1)$*-odd edge-connected graph admits a nowhere-zero circular* $(2+1/t)$*-flow.*

This result relaxes the edge-connectivity requirements in Theorem 7.2. Theorem 4.6 is a special case of this general result.

8. Contractible configurations

Contraction is one of the most useful and powerful operations in the inductive study of graph theory, if the resulting graph preserves a given graph property. In this section, we introduce *contractible configurations* and *collapsible graphs*; contractions of such graphs preserve some properties, such as integer flows.

Let $\mathcal{P}$ be a graph-theoretic property. A graph H is a *contractible configuration of* $\mathcal{P}$ if, for each supergraph G of H, G/H has property $\mathcal{P}$ if and only if G does.

Group connectivity

Group connectivity was introduced by Jaeger, Linial, Payan and Tarsi [29] as a generalization of integer flows. Let G be a graph, let Γ be an abelian group, and let $\beta : V(G) \to \Gamma$. Then β is called a *boundary* if it has *zero-sum* – that is, if $\sum_{v \in V(G)} \beta(v) = 0$. The graph G is Γ*-connected* if, for each boundary β, there are an orientation D_β and a nowhere-zero weight f_β of $E(G)$ such that

$$\sum_{e \in E^+_{D_\beta}(v)} f_\beta(e) - \sum_{e \in E^-_{D_\beta}(v)} f_\beta(e) = \beta(v), \qquad (1)$$

for each vertex $v \in V(G)$.

By the definition of group connectivity, we easily deduce the following result.

Theorem 8.1 *If H is Γ-connected, then H is a contractible configuration for Γ-flow.*

As for Tutte's flow conjectures, several open problems were proposed by Jaeger, Linial, Payan and Tarsi [29].

Conjecture G
Every 5-edge connected graph is $\mathbb{Z}_3$*-connected.*
Every 3-edge connected graph is $\mathbb{Z}_5$*-connected.*

Note that the 5-edge-connectivity is sharp for the first of these conjectures, since some 4-edge-connected counter-examples were discovered in [29] and [37]. The following theorem of Jaeger, Linial, Payan and Tarsi [29] gives some partial results.

Theorem 8.2
Every 3-edge-connected graph is Γ-connected for every group Γ of order at least 6. Every 4-edge-connected graph is Γ-connected for every group Γ of order at least 4.

The first part of Conjecture **G** was verified by Lai and Li [38] for planar graphs, and by Lovász, Thomassen, Wu and Zhang [41] for 6-edge-connected graphs.

Short cycles

Analogous to the girth studies of small counter-examples to Tutte's conjectures, shorter cycles are Γ-connected for larger groups Γ. The following observation was due to Jaeger, Linial, Payan and Tarsi [29].

Theorem 8.3 *Every cycle of length at most $r-1$ is Γ-connected for every abelian group Γ of order at least r.*

Note that a cycle C of length r is not Γ-connected if $|\Gamma| \leq r$: one can easily see that a constant boundary $\beta = 1$ is a 'bad' boundary – that is, there is no f_β satisfying equation (1).

From Theorem 8.3, we know that a digon is a contractible configuration for the 3-flow problem, and we also see that a triangle is not a contractible configuration for 3-flows. However, some graphs with many triangles (for example, even wheels) can be contractible configurations for the 3-flow problem.

A graph G is *triangularly connected* if, for each pair of edges e and f of G, there are triangles $T_1, T_2, \ldots, T_r$ such that $e \in E(T_1), f \in E(T_r)$, and $E(T_i) \cap E(T_{i+1}) \neq \emptyset$, for each $i = 1, 2, \ldots, r-1$.

Let $\mathcal{WF}$ be the subfamily of triangularly connected graphs constructed (recursively) as follows:

- the triangle is a member of $\mathcal{WF}$
- odd wheels are members of $\mathcal{WF}$
- for a pair of $\mathcal{WF}$-graphs G_1 and G_2, let $e_i \in E(G_i)$; a new $\mathcal{WF}$-graph G is constructed from G_1 and G_2 by merging e_1 and e_2 into a single edge e (see Fig. 2)

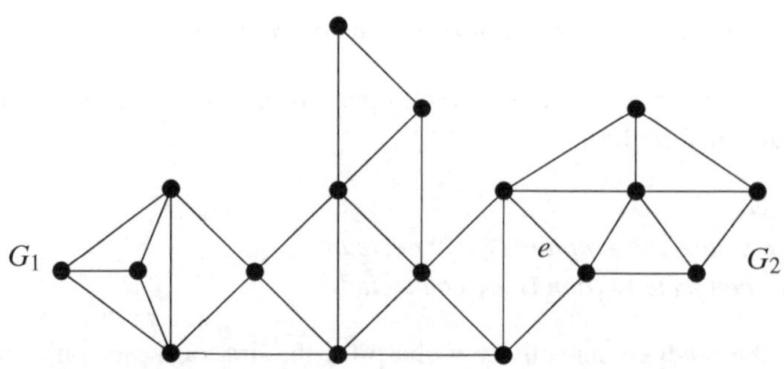

Fig. 2. A $\mathcal{WF}$-graph.

The following result of Fan, Lai, Xu, Zhang and Zhou [16] characterizes all triangularly connected graphs that are $\mathbb{Z}_3$-connected.

Theorem 8.4 *A triangularly connected graph is $\mathbb{Z}_3$-connected if and only if it is not a member of $\mathcal{WF}$.*

This result yields 3-flows for many families of dense graphs, such as locally connected graphs [36], squares of graphs [15], [61], triangulations of embedded graphs [5] and certain products of graphs [24].

A noticeable feature for 4-flows or their contractible configurations is that, although a 4-cycle is not $\mathbb{Z}_4$-connected, it is a contractible configuration for 4-flows. Catlin [9] proved the following result.

Theorem 8.5 *Let G be a graph and let C be a cycle of length* 4. *Then G admits a nowhere-zero* 4*-flow if and only if G/C admits a nowhere-zero* 4*-flow – that is,* 4*-cycles are contractible configurations for the* 4*-flow problem.*

Collapsible graphs

'Collapsible graphs', introduced by Catlin [8], are also contractible configurations for the 4-flow problem, due to the close relationship between supereulerian graphs and 4-flows.

A graph H is *collapsible* if, for each $X \subseteq V(H)$ of even order, H has a connected spanning subgraph H_X for which $X = O(H_X)$, the set of all odd-degree vertices in H_X.

The following result appeared in Lai [35].

Theorem 8.6 *Collapsible graphs are Γ-connected for every abelian group Γ of order* 4.

For more information on collapsible graphs, see Catlin's survey [10] and its supplement [12].

Group structure

For group-flow problems, the structure of the group makes no difference to the existence of nowhere-zero k-flows (see Theorem 2.1) if k, the order of the group, is fixed. However, the situation seems different for group connectivity problems. For example, it remains open as to whether $\mathbb{Z}_4$-connectivity is equivalent to $(\mathbb{Z}_2 \times \mathbb{Z}_2)$-connectivity. The following conjecture is due to Jaeger, Linial, Payan and Tarsi [29].

Conjecture H *A graph G is $\mathbb{Z}_4$-connected if and only if it is $(\mathbb{Z}_2 \times \mathbb{Z}_2)$-connected.*

Modulo orientations with boundaries

Beyond group connectivity, which mainly targets *integer-valued* k-flow problems for $k = 3, 4, \cdots$, the investigation of contractible configurations has been extended to circular $(2 + 1/t)$-flows with $t \geq 1$ (see [64], [37] and [54]).

Let G be a graph and let k be an odd integer. A mapping $\beta : V(G) \to \mathbb{Z}_k$ is called a *$\mathbb{Z}_k$-boundary* of G if $\sum_{v \in V(G)} \beta(v) \equiv 0 \pmod{k}$. An orientation D of G is called a *β-orientation* of G if $d_D^+(v) - d_D^-(v) \equiv \beta(v) \pmod{k}$.

Theorem 8.7 *Let G be a graph and let $k = 2t + 1$ be an odd integer. Then the following statements are equivalent:*

- *G has a β-orientation for each $\mathbb{Z}_k$-boundary β of G*
- *G is a contractible configuration for modulo k-orientation*
- *G is a contractible configuration for a circular $(2 + 1/t)$-flow.*

Just as short cycles are contractible configurations for integer flows, so parallel edges are contractible configurations for modulo orientation (see [64]).

Theorem 8.8 *Let $k \geq 3$ be an odd integer. Then the parallel edge $(k-1)K_2$ is a contractible configuration for modulo k-orientation.*

The following conjecture was proposed by Lai [37].

Conjecture I *If G is a $(2k-1)$-edge-connected graph, $k \geq 3$ is an odd integer and β is a $\mathbb{Z}_k$-boundary of G, then G has a β-orientation.*

A weak version [37] of Conjecture **I** was proved recently by Thomassen [54] and further improved by Lovász, Thomassen, Wu and Zhang [41].

Theorem 8.9 *Let G be a graph and let $k \geq 3$ be an odd integer. If G is $(3k-3)$-edge-connected, then G has a β-orientation for every $\mathbb{Z}_k$-boundary β, so every $(3k-3)$-connected graph is a contractible configuration for modulo k-orientation.*

9. Related problems

Together with the 5-flow conjecture, the cycle double cover conjecture and the Berge–Fulkerson conjecture are major open problems for snark graphs (cubic graphs with chromatic index 4). These conjectures have some important common properties: they all hold for 3-edge-colourable cubic graphs, but remain open for snarks, and investigations of one problem may be related to others, but little is known about how closely they are related.

Cycle double cover

In this section, we present some relations between flow problems and cycle double cover problems. In [55], Tutte proved the following results for cubic graphs. They were generalized and reformulated by Jaeger and Seymour.

Theorem 9.1 *For a graph G, the following statements are equivalent:*

- *G admits a nowhere-zero 4-flow*

- *G has a* 3*-even subgraph double cover*
- *G has a* 4*-even subgraph double cover.*

Preissmann [45] and Celmins [11] proposed the following conjecture.

Conjecture J *Every bridgeless graph has a* 5*-even-subgraph double cover.*

However, we have no knowledge yet about the relationship between 5-even-subgraph double covers and 5-flows.

An early result of Tutte [55] about orientable cycle double covers is closely related to integer flows.

Theorem 9.2 *Let* $r = 3$ *or* 4 *and let* G *be a graph. Then* G *admits a nowhere-zero* r*-flow if and only if* G *has an orientable* r*-even-subgraph double cover.*

A conjecture of Archdeacon [4] and Jaeger [28] analogous to Conjecture **J** is as follows.

Conjecture K *Every bridgeless graph has an orientable* 5*-even-subgraph double cover.*

It is evident that Conjecture **K** is stronger than Conjecture **J**, and they both seem to have some relationship with the 5-flow conjecture. We know that Conjecture **K** implies the 5-flow conjecture, but we do not know whether they are equivalent.

Flow double covering

The following conjecture was made by Zhu (unpublished notes, 2005).

Conjecture L *Let* G *be a bridgeless graph with an orientation* D*, and let* $r = 2, 3, 4$ *or* 5*. The graph* G *admits a set of* r*-flows* (D, f_1), $(D, f_2), \ldots, (D, f_{7-r})$ *for which each edge is covered by the supports of precisely two of these* r*-flows.*

Note that, for $r = 2$, Conjecture **L** is equivalent to Conjecture **J**, and for $r = 5$, Conjecture **L** is equivalent to the 5-flow conjecture.

Cycle space minors

The following concepts and conjecture were introduced by Jaeger [28].

Let G_1 and G_2 be bridgeless graphs. We write $G_1 \leq_C G_2$ if G_1 has a subdivision H and there is a bijection $\varphi : E(H) \to E(G_2)$ such that, for each even subgraph C of H, $\varphi(C)$ is an even subgraph in G_2.

In the following, the graph with one vertex and one loop is denoted by L, the graph with two vertices and three parallel edges is denoted by $3K_2$, and the Petersen graph is denoted by P_{10}. Jaeger [28] observed that $3K_2 \leq_C G$ if and only if G admits a nowhere-zero 4-flow.

Let $\mathcal{G}$ be the set of all bridgeless graphs. A graph G is called a *cycle space minor* if G is a minimal member of $\mathcal{G}$ under the order $\leq_C$. The cycle space of a graph is denoted by $\mathcal{C}(G)$ and the rank of $\mathcal{C}(G)$ is denoted by $r_C(G)$; the ranks of the cycle spaces of L, $3K_2$ and P_{10} are 1, 2 and 6, respectively. It is not hard to prove that there is no cycle space minor M with $r_C(M) = 3, 4$ or 5. Is there any cycle space minor with rank higher than 6? The following conjecture of Jaeger [28] implies many cycle cover conjectures.

Conjecture M *The only three cycle space minors are L, $3K_2$ and P_{10}.*

Conjecture **M** implies the cycle double cover conjecture, the 5-even subgraph double cover conjecture and the Berge–Fulkerson conjecture. Since the mapping φ preserves even subgraphs but not orientations, the cycle space minor problem is mainly related only to 2^t-flows, where t is a positive integer.

The concept of cycle space minor was further extended by DeVos, Nešetřil and Raspaud [14], to the case where a generalized graph mapping preserves both flow-values and orientations.

References

1. N. Alon, N. Linial and R. Meshulam, Additive bases of vector spaces over prime fields, *J. Combin. Theory (A)* **57** (1991), 203–210.
2. K. Appel and W. Haken, Every map is four colorable, Part I: Discharging, *Illinois J. Math.* **21** (1977), 429–490.
3. K. Appel, W. Haken and J. Koch, Every map is four colorable, Part II: Reducibility, *Illinois J. Math.* **21** (1977), 491–567.
4. D. Archdeacon, Face coloring of embedded graphs, *J. Graph Theory* **8** (1984), 387–398.
5. J. Barát and C. Thomassen, Claw-decompositions and Tutte-orientations, *J. Graph Theory* **52** (2006), 135–146.
6. C. Berge, *Graphs and Hypergraphs* (transl. E. Minieka), North-Holland, 1973.
7. J. A. Bondy and U. S. R. Murty, *Graph Theory with Applications*, Macmillan, 1976.
8. P. A. Catlin, A reduction method to find spanning eulerian subgraph, *J. Graph Theory* **12** (1988), 29–45.
9. P. A. Catlin, Double cycle covers and the Petersen graph, *J. Graph Theory* **13** (1989), 465–483.
10. P. A. Catlin, Supereulerian graphs: a survey, *J. Graph Theory* **16** (1992), 177–196.
11. U. A. Celmins, *On Cubic Graphs that do not have an Edge 3-coloring*, Ph.D. thesis, University of Waterloo (1984).
12. Z.-H. Chen and H.-J. Lai, Reductions techniques for supereulerian graphs and related topics – a survey, *Combinatorics and Graph Theory 95*, Vol. 1 (ed. Ku Tung-Hsin), World Scientific (1995), 53–69.
13. M. DeVos, L. Goddyn, B. Mohar, D. Vertigan and X.-D. Zhu, Coloring-flow duality of embedded graphs, *Trans. Amer. Math. Soc.* **357** (2005), 3993–4016.
14. M. DeVos, J. Nešetřil and A. Raspaud, On edge-maps whose inverse preserves flows and tensions, *Graph Theory, Trends in Mathematics* (eds. J. A. Bondy, J. Fonlupt, J.-L. Fouquet, J.-C. Fournier and J. L. R. Alfonsin), Birkhäuser (2006), 109–138.
15. M. DeVos, R. Xu and G. Yu, Nowhere-zero Z_3-connectivity, *Discrete Math.* **306** (2006), 26–30.

16. G.-H. Fan, H.-J. Lai, R. Xu, C.-Q. Zhang and C. Zhou, Nowhere-zero 3-flows in triangularly connected graphs, *J. Combin. Theory (B)* **98** (2008), 1325–1336.
17. H. Fleischner, Bipartizing matchings and Sabidussi's compatibility conjecture, *Discrete Math.* **244** (2002), 77–82.
18. J. L. Fouquet, Conjecture du 5-flot pour les graphes presque planaires, *Sèminaire de Mathématiques Discrètes et Applications,* Grenoble, 7 November 1985.
19. A. Galluccio, L. A. Goddyn and P. Hell, Large girth graphs avoiding a fixed minor are nearly bipartite, *J. Combin. Theory (B)* **83** (2001), 1–14.
20. L. A. Goddyn, M. Tarsi and C.-Q. Zhang, On (k,d)-colorings and fractional nowhere zero flows, *J. Graph Theory* **28** (1998), 155–161.
21. H. Grötzsch, Ein Dreifarbensatz für dreikreisfreie Netze auf der Kugel, *Wiss. Z. der Martin-Luther-Universität Halle-Wittenberg, Math.-Naturwissenschaftliche Reihe* **8** (1958/59), 109–120.
22. B. Grünbaum, Grötzsch's theorem on 3-colorings, *Michigan Math. J.* **10** (1963), 303–310.
23. P. J. Heawood, Map color theorem, *Quart. J. Pure and Applied Math.* **24** (1890), 332–338.
24. W. Imrich, I. Peterin, S. Špacapan and C.-Q. Zhang, NZ-flows in strong products of graphs, *J. Graph Theory* **64** (2010), 267–276.
25. F. Jaeger, On interval hypergraphs and nowhere-zero flow in graphs, *Research Report of Mathematics Application and Information*, Université Scientifique et Medicale et Institut National Polytechnique de Grenoble **126**, 1978.
26. F. Jaeger, Flows and generalized coloring theorems in graphs, *J. Combin. Theory (B)* **26** (1979), 205–216.
27. F. Jaeger, On circular flows in graphs, *Finite and Infinite Sets* (Eger, 1981), Colloq. Math. Soc. János Bolyai **37**, North-Holland (1984), 391–402.
28. F. Jaeger, Nowhere-zero flow problems, *Selected Topics in Graph Theory 3*, (eds. L. W. Beineke and R. J. Wilson), Academic Press (1988), 71–95.
29. F. Jaeger, N. Linial, C. Payan and M. Tarsi, Group connectivity of graphs – A nonhomogeneous analogue of nowhere-zero flow properties, *J. Combin. Theory (B)* **56** (1992), 165–182.
30. P. A. Kilpatrick, *Tutte's First Colour-Cycle Conjecture*, Ph.D. thesis, Cape Town, 1975.
31. M. Kochol, Cubic graphs without a Petersen minor have nowhere-zero 5-flows, *Acta Math. Univ. Comenianae* **LXVIII** (1999), 249–252.
32. M. Kochol, An equivalent version of the 3-flow conjecture, *Discrete Math.* **83** (2001), 258–261.
33. M. Kochol, Reduction of the 5-flow conjecture to cyclically 6-edge-connected snarks, *J. Combin. Theory (B)* **90** (2004), 139–145.
34. M. Kochol, Smallest counterexample to the 5-flow conjecture has girth at least eleven, *J. Combin. Theory (B)* **100** (2010), 381–389.
35. H.-J. Lai, Extending a partial nowhere zero 4-flow, *J. Graph Theory* **30** (1999), 277–288.
36. H.-J. Lai, Nowhere-zero 3-flows in locally connected graphs. *J. Graph Theory* **42** (2003), 211–219.
37. H.-J. Lai, Mod $(2p+1)$-orientations and $K_{1,2p+1}$-decompositions, *SIAM J. Discrete Math.* **21** (2007), 844–850.
38. H.-J., Lai and X.-W. Li, Group Chromatic number of planar graphs with girth at least 4, *J. Graph Theory* **52** (2006), 51–72.
39. H.-J. Lai and C.-Q. Zhang, Nowhere-zero 3-flows of highly connected graphs, *Discrete Math.* **110** (1992), 179–183.
40. C. H. C. Little, W. T. Tutte and D. H. Younger, A theorem on integer flows, *Ars Combin.* **26A** (1988), 109–112.

41. L. M. Lovász, C. Thomassen, Y.-Z. Wu and C.-Q. Zhang, Nowhere-zero 3-flows and modulo k-orientations, *J. Combin. Theory (B)* **103** (2013), 587–598.
42. M. Möller, H. G. Carstens and G. Brinkmann, Nowhere-zero flows in low genus graphs, *J. Graph Theory* **12** (1988), 183–190.
43. C. St.J. A. Nash-Williams, Edge-disjoint spanning trees of finite graphs, *J. London Math. Soc.* **36** (1961), 445–450.
44. Z. Pan and X.-D. Zhu, Construction of graphs with given circular flow numbers. *J. Graph Theory* **43** (2003), 304–318.
45. M. Preissmann, *Sur les Colorations des Arêtes des Graphes Cubiques*, Thèse de Doctorat de 3eme, Université de Grenoble, 1981.
46. N. Robertson, P. D. Seymour and R. Thomas, Girth six cubic graphs have Petersen minors, *arXiv:1405.0533*.
47. P. D. Seymour, Nowhere-zero 6-flows, *J. Combin. Theory (B)* **30** (1981), 130–135.
48. P. D. Seymour, Tutte's three-edge-colouring conjecture, *Proceedings of Graph Theory@Georgia Tech*, a conference honouring the 50th birthday of Robin Thomas, May 2012, *https://smartech.gatech.edu/handle/1853/44224.*
49. E. Steffen, Tutte's 5-flow conjecture for highly cyclically connected cubic graphs, *Discrete Math.* **310** (2010), 385–389.
50. E. Steffen, Intersecting 1-factors and nowhere-zero 5-flows, preprint, 2012.
51. R. Steinberg and D. H. Younger, Grötzsch's theorem for the projective plane, *Ars Combin.* **28** (1989), 15–31.
52. R. Thomas, Generalizations of the four color theorem, *http://people.math.gatech.edu/~thomas/FC/generalize.html.*
53. C. Thomassen, Grötzsch's 3-color theorem and its counterparts for the torus and the projective plane, *J. Combin. Theory (B)* **62** (1994), 268–279.
54. C. Thomassen, The weak 3-flow conjecture and the weak circular flow conjecture, *J. Combin. Theory (B)* **102** (2012), 521–529.
55. W. T. Tutte, On the imbedding of linear graphs in surfaces, *Proc. London Math. Soc. (2)* **51** (1949), 474–483.
56. W. T. Tutte, A contribution on the theory of chromatic polynomial, *Canad. J. Math.* **6** (1954), 80–91.
57. W. T. Tutte, A class of Abelian groups, *Canad. J. Math.* **8** (1956), 13–28.
58. W. T. Tutte, On the problem of decompositing a graph into n connected factors, *J. London Math. Soc.* **36** (1961), 221–230.
59. W. T. Tutte, On the algebraic theory of graph colourings, *J. Combin. Theory* **1** (1966), 15–50.
60. W. T. Tutte, A geometrical version of the four color problem, *Combinatorial Mathematics and its Applications* (eds. R. C. Bose and T. A. Dowling), Univ. North Carolina Press, 1967.
61. R. Xu and C.-Q. Zhang, Nowhere-zero 3-flows in squares of graphs, *Electron. J. Combin.* **10** (2003), R5.
62. D. H. Younger, Integer flows, *J. Graph Theory* **7** (1983), 349–357.
63. C.-Q. Zhang, *Integer Flows and Cycle Covers of Graphs*, Marcel Dekker, 1997.
64. C.-Q. Zhang, Circular flows of nearly eulerian graphs and vertex-splitting, *J. Graph Theory* **40** (2002), 147–161.
65. X. D. Zhu, Recent developments in circular colouring of graphs, *Topics in Discrete Mathematics* (eds. M. Klazar *et al.*), Springer (2006), 497–550.

10

Colouring random graphs

ROSS J. KANG and COLIN MCDIARMID

Typically how many colours are required to colour a graph? In other words, given a graph chosen randomly, what can we expect its chromatic number to be? We survey the classic interpretation of this question, with the binomial or Erdős–Rényi random graph and the usual chromatic number. We also treat a few variations, not only of the random graph model, but also of the chromatic parameter.

1. Introduction

How many colours are typically necessary to colour a graph?

We survey a number of perspectives on this natural question, which is central to random graph theory and to probabilistic and extremal combinatorics. It has stimulated a vibrant area of research, with a rich history extending back through more than half a century.

Erdős and Rényi [36] asked a form of this question in a celebrated early paper on random graphs in 1960. Let $G_{n,m}$ be a graph chosen uniformly at random from the set of graphs with vertex-set $[n] = \{1, 2, \ldots, n\}$ and m edges. In this probabilistic model, we cannot rule out the possibility that $G_{n,m}$ is, for example, the disjoint union of one large clique and some isolated vertices, or perhaps one Turán graph (a balanced complete multipartite graph) and some isolated vertices. The resulting range is large: in the former situation the chromatic number could be about $\sqrt{2m}$, while in the latter it could be 2 if $m \leq n^2/4$. These outcomes are unlikely, however, and we are interested in the most probable ones. To state the question properly, we say that an event A_n (which here always describes a property of a random graph on the vertex-set $[n]$) holds *asymptotically almost surely* (*a.a.s.*) if the probability that A_n holds satisfies $\mathbb{P}(A_n) \to 1$ as $n \to \infty$. Erdős and Rényi asked the following question.

Suppose that $m \sim cn$ for some $c \geq \frac{1}{2}$. Is there a positive integer function $f = f_c(n)$ for which $\chi(G_{n,m}) = f$ a.a.s., and if so how large is it?

They already noted that $f = 3$ if $c = \frac{1}{2}$, which tells us that $\mathbb{P}(\chi(G_{n,n/2}) = 3) \to 1$ as $n \to \infty$; however, as we will see, handling larger fixed values of c is a very challenging task.

If, however, we instead consider $G_{n,m}$ with the symmetric choice $m = \frac{1}{2}\binom{n}{2}$, then this is nearly the same as the basic model $G_{n,1/2}$, chosen uniformly at random from graphs with vertex-set $[n]$. Equivalently, form $G_{n,1/2}$ by including as an edge each of the $\binom{n}{2}$ possible pairs from $\binom{[n]}{2}$ independently at random with probability $\frac{1}{2}$. Even the following seemingly innocent question was open for decades in spite of serious and sustained efforts, thus taking the pattern of other hard problems in combinatorics and graph theory.

What is the a.a.s. first-order approximate behaviour of the chromatic number of a uniformly chosen graph with vertex-set $[n]$?

In other words, this question asks for an elementary function $f(n)$ (if there is one) that satisfies $\chi(G_{n,1/2}) \sim f(n)$ a.a.s. The typical value of the chromatic number is an attractive concept in its own right, but there is also a strong interplay with other research areas. This provides some contrasting viewpoints we now briefly discuss.

Researchers in 'deterministic' (chromatic) graph theory frequently look to random graphs for examples and counter-examples. In 1959 Erdős [34] elegantly proved the existence of graphs with both girth and chromatic number arbitrarily large – a clean constructive proof came somewhat later. Also, Erdős and Fajtlowicz [35] showed that 'almost all' graphs are counter-examples to Hajós's conjecture, which, as an over-strengthened form of Hadwiger's conjecture, claimed that every k-chromatic graph contains a subdivision of K_k. Both of these well-known applications of the probabilistic method used lower bounds for the chromatic number of a random graph.

In theoretical computer science, determining the chromatic number of a given graph is an archetypal NP-hard optimization problem. One might wonder whether the situation is simpler in a probabilistic setting. The chromatic number of random graphs (as well as the performance of algorithms that compute or approximate it) may be viewed as a crude indication of 'average-case' computational behaviour.

Surprisingly, the study of the chromatic number of random graphs is equivalent to an important model of spin glasses studied in statistical physics. It has been referred to variously as the 'diluted mean-field antiferromagnetic Potts model' and the 'zero-temperature Curie–Weiss–Potts antiferromagnet'. Spin glass models are insightful representations for the physical phenomenon of a *phase transition*, such as the change between water and ice.

This topic thus provides common ground for combinatorialists, probabilists, theoretical computer scientists and statistical physicists. Later we will see examples where the mixing of fields in this area has been stimulating and fruitful.

A main focus of ours is to address the two questions in italics above, which correspond to standard density regimes in the theory of classical Erdős–Rényi random graphs. The latter question, treated in Section 2, is representative of the 'dense' regime of random graphs, since the expected number of edges in $G_{n,1/2}$ is $\Theta(n^2)$. The former question (discussed in Section 3) belongs to the 'sparse' regime, since there are only a linear number of edges. Random graphs have very interesting properties and go through a number of well-defined phases according to the graph's density. For more on the theory and evolution of Erdős–Rényi random graphs, see the standard texts by Bollobás [15] and Janson, Łuczak and Ruciński [49].

The overarching question posed at the very beginning can be ascribed a broader meaning. It is well separated into its constituent parts, one of which concerns 'random graphs', the other of which concerns 'colouring'. The second half of this chapter is devoted to selectively developing these parts independently.

Random graph theory continues to develop rapidly and steadily, partly through the influence of networks. Many probability spaces for graphs can be interpreted as models for real networks such as the internet, protein–protein interaction, mathematical collaboration and telecommunication. We treat a selection of random graph models (apart from Erdős–Rényi) for which the chromatic number is important. Random regular graphs are discussed in Section 4; random geometric graphs, which are considered a standard model for frequency allocation in ad hoc communication networks, are treated in Section 5; and random graph models related to the colouring of planar graphs are considered in Section 6.

It is more than evident from the chapters in this book that chromatic graph theory is a broad and mature field. Chromatic *random* graph theory is also well developed. In the last section, Section 7, we present results on a few specific chromatic parameters, concentrating for brevity on the dense Erdős–Rényi random graph. First we discuss strengthenings of the chromatic number, then we cover edge and total colourings, and last we consider generalizations of the chromatic number.

We must mention at least two other major research areas that combine probability with graph colouring. Although they are closely related to this topic, they are outside the scope of this chapter. The generation of a uniformly random colouring of a graph, often with the Markov chain Monte Carlo method, has been extensively studied (see [44]). The application of the probabilistic method in graph colouring, using the Lovász local lemma, for instance, has been very successful and is covered in depth in the monograph of Molloy and Reed [75].

In this chapter, the symbols $\mathbb{P}$ and $\mathbb{E}$ denote probability and expectation, respectively. The level of probabilistic expertise needed for this material is not burdensome, but for more background see Alon and Spencer [9] or one of the random graph texts [15], [49]. Mostly we just use the first moment (Markov's inequality) and second moment (Chebyshev's or the Paley–Zygmund inequality) methods. Given a non-negative random variable X, Markov's inequality asserts that $\mathbb{P}(X \geq a) \leq \mathbb{E}(X)/a$ for any $a > 0$, while the Paley–Zygmund inequality implies that $\mathbb{P}(X > 0) \geq (\mathbb{E}(X))^2/\mathbb{E}(X^2)$. We refer to other inequalities for the establishment of concentrated probability measure – Chernoff, Azuma–Hoeffding, Janson's and Talagrand's inequalities – all of which are covered in the above-mentioned texts. We also use the approximation $(n/k)^k \leq \binom{n}{k} \leq (en/k)^k$, which is a consequence of the fact that $e^x \geq x^k/k!$ for $x \geq 0$, and the basic inequality $(1-a)^n \leq \exp(-an)$ for $a \in (0,1)$. When we write an unadorned logarithm symbol log, it is assumed to have base 2; ln has the natural base e. We omit symbols for floors and ceilings, unless they are important. We have adopted standard asymptotic (Landau) notation: O, Ω, Θ, $\sim$, o and ω. Unless otherwise specified, the asymptotics are as $n \to \infty$. Recall that, for functions $f(n)$ and $g(n) > 0$, we write $f = O(g)$ if $f(n) \leq cg(n)$ for some constant c, $f = \Theta(g)$ if both $f = O(g)$ and $g = O(f)$, $f \sim g$ if $f(n)/g(n) \to 1$, $f = o(g)$ if $f(n)/g(n) \to 0$, and $f = \omega(g)$ if $g = o(f)$.

2. Dense random graphs

By the dense regime of Erdős–Rényi random graphs, we mean $G_{n,p}$ for fixed $p \in (0,1)$. For clarity, we have chosen to restrict our attention to $p = \frac{1}{2}$ – that is, 'almost all graphs', though all results in this section extend naturally across the entire dense regime, to other fixed $p \in (0,1)$.

Upon encountering the problem for the first time, one might observe that the graph $G_{n,1/2}$ has 'global' rather than 'local' structure. In particular, any given vertex has a reasonable chance to be adjacent (or not) to any other vertex in the graph. This suggests that its chromatic number might not necessarily be determined by a local quantity, such as a large clique or a small maximum degree. As it happens, both the clique number of $G_{n,1/2}$ (which is about $2\log n$ a.a.s., as we shall see) and the maximum degree of $G_{n,1/2}$ (which is concentrated around $\frac{1}{2}n$ a.a.s.) are quite far away from $\chi(G_{n,1/2})$.

Instead, it turns out that the value of $\chi(G_{n,1/2})$ strongly depends upon large stable sets in $G_{n,1/2}$. (Throughout this chapter, we exclusively use *stable set*, rather than *independent set*, to refer to a vertex subset that induces a subgraph having no edges.) Recall that in any proper colouring of $G_{n,1/2}$, each colour class has order at most $\alpha(G_{n,1/2})$ (where $\alpha(G)$ denotes the number of vertices in a largest stable set of G), and thus $\chi(G_{n,1/2}) \geq n/\alpha(G_{n,1/2})$ always. Fortunately, as we see in the next subsection, $\alpha(G_{n,1/2})$ can be precisely determined without much difficulty. On the other hand, a thorough understanding of the 'global' behaviour of large stable sets in $G_{n,1/2}$ is necessary for a good upper bound on $\chi(G_{n,1/2})$. This vague narrative should become clearer later.

The stability number

Let $\mathcal{S}_k$ be the collection of stable sets of order k in $G_{n,1/2}$. To determine $\alpha(G_{n,1/2})$, we first study the expectation of $|\mathcal{S}_k|$. There are $\binom{n}{k}$ sets of order k, and the probability that such a set is stable (that is, each of its $\binom{k}{2}$ edges is absent) is $2^{-\binom{k}{2}}$. We thus have

$$\mathbb{E}(|\mathcal{S}_k|) = \binom{n}{k} 2^{-\binom{k}{2}}.$$

For relevant values of k, this expression for the first moment of $|S_k|$ is decreasing in k. We are interested in the values of k over which it changes from a large quantity (much greater than 1) to a small quantity (somewhat less than 1). In the latter case, there is positive probability that there are no stable sets of order k (that is, $\mathbb{P}(|\mathcal{S}_k| = 0) > 0$), whereas in the former, if the expectation is large enough, we would hope that the distribution of such sets is sufficiently well behaved that we could then pin down such a set. As we now show, the change in $\mathbb{E}(|\mathcal{S}_k|)$ is large and the range of k over which it happens is small.

With foresight, we consider, for some fixed $\varepsilon \in (0, 1)$, the choice

$$k = k_{+\varepsilon} = \lfloor k_0 + \varepsilon \rfloor, \quad \text{where } k_0 = 2\log n - 2\log\log n + 2\log e - 1.$$

Now, by an approximation of the binomial coefficient,

$$\mathbb{E}(|\mathcal{S}_{k_{+\varepsilon}+1}|) \le \left(\frac{en}{k_{+\varepsilon}+1}\right)^{k_{+\varepsilon}+1} 2^{-\binom{k_{+\varepsilon}+1}{2}},$$

and so, since $k_{+\varepsilon} + 1 \ge k_0 + \varepsilon$,

$$\frac{2\log \mathbb{E}(|\mathcal{S}_{k_{+\varepsilon}+1}|)}{k_{+\varepsilon}+1} \le 2\log e + 2\log n - 2\log k_0 - k_0 - \varepsilon + 1 = -\varepsilon + o(1),$$

since $\log k_0 = \log\log n + 1 + o(1)$. As we are interested in an asymptotic statement, we may allow an arbitrarily large choice of n. Then this last expression becomes at most $-\frac{1}{2}\varepsilon$, and $k_{+\varepsilon} \ge \log n$. It follows that, for sufficiently large n, $\mathbb{E}(|\mathcal{S}_{k_{+\varepsilon}+1}|) \le \exp\left(-\frac{1}{4}\varepsilon \log n\right)$, a vanishing quantity. We immediately deduce from an application of Markov's inequality that

$$\mathbb{P}(\alpha(G_{n,1/2}) > k_{+\varepsilon}) = \mathbb{P}(|\mathcal{S}_{k_{+\varepsilon}+1}| \ge 1) \le \mathbb{E}(|\mathcal{S}_{k_{+\varepsilon}+1}|) \to 0 \text{ as } n \to \infty.$$

In other words, $\alpha(G_{n,1/2}) \le k_{+\varepsilon}$ a.a.s.

Next let us consider, for some fixed $\varepsilon > 0$, the choice $k = k_{-\varepsilon} = \lfloor k_0 - \varepsilon \rfloor$, where k_0 is as above. Using another approximation of the binomial coefficient (in the other direction), we obtain by almost identical manipulations that, for n large enough,

$\mathbb{E}(|\mathcal{S}_{k_{-\varepsilon}}|) \geq \exp\left(\frac{1}{4}\varepsilon \log n\right)$, an unbounded quantity. So we see that as a result of a small decrease in k (from $k_{+\varepsilon} + 1$ to $k_{-\varepsilon}$) there is a large increase in $\mathbb{E}(|\mathcal{S}_k|)$. Around 1970 Matula [62], [63] showed that moreover $|\mathcal{S}_{k_{-\varepsilon}}|$ is concentrated around its mean. In particular, it is possible to show that the second moment $\mathbb{E}(|\mathcal{S}_{k_{-\varepsilon}}|^2)$ of $|\mathcal{S}_{k_{-\varepsilon}}|$ is very small compared to $(\mathbb{E}(|\mathcal{S}_{k_{-\varepsilon}}|))^2$, by using calculations similar to (but more involved than) those just given above for the first moment. Consequently, an application of Chebyshev's or the Paley–Zygmund inequality yields $\alpha(G_{n,1/2}) \geq k_{-\varepsilon}$ a.a.s. Thus the value of the stability number of $G_{n,1/2}$ is very precisely determined a.a.s.

Theorem 2.1 *For fixed* $\varepsilon \in (0,1)$, $\alpha(G_{n,1/2}) \in [k_{-\varepsilon}, k_{+\varepsilon}]$ *a.a.s.*

A slightly sharper form of this was proved independently by Bollobás and Erdős [17] in 1976.

A greedy algorithm

In the last subsection, we saw with a first moment argument that $\alpha(G_{n,1/2}) \leq k_{+\varepsilon}$ and so the lower bound $\chi(G_{n,1/2}) \geq n/k_{+\varepsilon} \sim n/(2\log n)$ holds a.a.s. We now show how another first moment argument, due to Grimmett and McDiarmid [48] in 1975, also gives a reasonable upper bound on $\chi(G_{n,1/2})$, one that is only twice as large as the lower bound. This argument uses a simple colouring algorithm. It relies on the following observation for maximal stable sets – that is, stable sets that cannot be enlarged by adding a vertex.

Theorem 2.2 *With probability at least* $1 - \exp(-\Omega(\log^3 n))$, *every maximal stable set of* $G_{n,1/2}$ *has order greater than* $\log n - 3\log\log n$.

Proof Given a stable set in $G_{n,1/2}$ of order k, the probability that it is maximal is $(1 - 2^{-k})^{n-k}$, this being the probability that each vertex outside the stable set is adjacent to at least one of the vertices within the set. As there are at most $\binom{n}{k}$ stable sets of order k, and so fewer than n^k, the expected number of maximal stable sets of order at most $\log n - 3\log\log n$ is at most

$$\begin{aligned}\sum_{k=1}^{\log n - 3\log\log n} \binom{n}{k}(1-2^{-k})^{n-k} &\leq \sum_{k=1}^{\log n - 3\log\log n} n^k \exp(-(n-k)2^{-k})\\ &\leq (\log n) n^{\log n}\exp(-(1-k/n)\log^3 n)\\ &= \exp(-\Omega(\log^3 n)),\end{aligned}$$

where we have used the fact that $\exp(-(n-k)2^{-k})$ is maximized over $k \in [1, \log n - 3\log\log n]$ when $k = \log n - 3\log\log n$. So, by Markov's inequality, the probability of there being a maximal stable set of order at most $\log n - 3\log\log n$ is also at most $\exp(-\Omega(\log^3 n))$. ∎

We see now that the following greedy procedure quickly and reliably produces a stable set S whose order is almost half of $\alpha(G_{n,1/2})$: initialize S to be the empty set; at each step, choose a previously unchosen vertex of $[n] \setminus S$ and add it to S if it is adjacent to none of the vertices in S. The greedy colouring algorithm borrows this as a subprocedure to generate new colour classes iteratively. The validity of the analysis relies on the observation that the generation of S relies only on the examination (or 'exposure') of candidate edges that have an endpoint in S, and thus the graph obtained by the removal of S from some $G_{n',1/2}$ may be viewed as an independent copy of the random graph $G_{n'-|S|,1/2}$, upon which we may iterate. We continue until fewer than $n/\log^2 n$ vertices remain and we give each of these vertices its own colour. Thus the total number of colours used is less than

$$\frac{n}{\log n - 5\log\log n} + \frac{n}{\log^2 n} \sim \frac{n}{\log n},$$

assuming that each iteration successfully generates a colour class of order $\log n - 5\log n$. The probability that any iteration of the algorithm fails to extract such a colour class from $G_{n',1/2}$, with $n' \geq n/\log^2 n$, is at most $\exp(-\Omega(\log^3 n))$, by Theorem 2.2. Trivially, there are at most n stages, and thus by the union bound (Boole's inequality) the algorithm succeeds with probability at least $1 - n\exp(-\Omega(\log^3 n)) \to 1$ as $n \to \infty$. It is easily checked that this algorithm runs in polynomial time, and so we have the following result.

Theorem 2.3 *There is a polynomial-time algorithm that a.a.s. produces a proper colouring of $G_{n,1/2}$ with at most $(1 + o(1))n/\log n$ colours.*

In fact, McDiarmid [66] showed that many of the most common polynomial-time colouring heuristics a.a.s. produce colourings of $G_{n,1/2}$ with at most $(1+o(1))n/\log n$ colours.

Concentration and the chromatic number

We have seen that $(1 + o(1))n/(2\log n) \leq \chi(G_{n,1/2}) \leq (1 + o(1))n/\log n$ a.a.s. Grimmett and McDiarmid also showed that the expected number of proper $((1 + \varepsilon)n/(2\log n))$-colourings of $G_{n,1/2}$ tends to infinity, and then conjectured that the lower bound in the above range should be the correct first-order behaviour of $\chi(G_{n,1/2})$ a.a.s. This remained one of the most important open problems in random graph theory for well over a decade.

A turning point was the introduction of simple yet powerful tools from probability theory for proving concentration of measure. We owe this important insight to Shamir and Spencer [82] in 1987. With a direct application of the Azuma–Hoeffding martingale inequality (to the 'vertex exposure' martingale), they showed that the value of $\chi(G_{n,1/2})$ is a.a.s. concentrated within an interval roughly of length $O(\sqrt{n})$. Frustratingly, however, this method gave no clue as to the location of this interval!

In spite of this, the discovery of this connection proved decisive and the problem was resolved soon after. Several methods were developed, independently or in quick succession. In the next section, we discuss the 'expose-and-merge' algorithm of Matula [64]. The elegant method of Bollobás [14] applied martingale concentration to prove that a non-algorithmic form of the greedy process described in the last subsection can properly colour $G_{n,1/2}$ with large colour classes. This approach depends on the following strong upper bound on the probability of $G_{n,1/2}$ to have all stable sets slightly smaller than the values given by Theorem 2.1.

Theorem 2.4 $\mathbb{P}(\alpha(G_{n,1/2}) < 2\log n - 3\log\log n) < \exp(-\Omega(n^2/\log^5 n))$.

Before discussing this bound further, let us first see how it is used to determine the a.a.s. first-order behaviour of $\chi(G_{n,1/2})$.

Theorem 2.5 $\chi(G_{n,1/2}) \sim n/(2\log n)$ *a.a.s.*

Proof As the lower bound follows from the upper bound on $\alpha(G_{n,1/2})$ of Theorem 2.1, it suffices to prove the upper bound. To this end, let $\mathcal{A}_n$ be the set of graphs G on $[n]$ with $\alpha(G[S]) \geq 2\log n - 7\log\log n$ for all $S \subseteq [n]$ with $|S| \geq n/\log^2 n$. Then Theorem 2.4 yields, as $n \to \infty$, that

$$\begin{aligned}\mathbb{P}(G_{n,1/2} \notin \mathcal{A}_n) &\leq 2^n\, \mathbb{P}(\alpha(G_{n/\log^2 n,1/2}) < 2\log n - 7\log\log n) \\ &\leq 2^n \exp(-\Omega(n^2/\log^9 n)) \to 0.\end{aligned}$$

Thus $G_{n,1/2} \in \mathcal{A}_n$ a.a.s.

For any graph in $\mathcal{A}_n$, an iterative process similar to the greedy algorithm yields a good colouring. We start with $T = [n]$. As long as $|T| \geq n/\log^2 n$, the condition for membership in $\mathcal{A}_n$ guarantees that we can extract a new colour class S from T with at least $2\log n - 7\log\log n$ vertices and then set $T = T \setminus S$. After stopping, we give each vertex left in T its own colour, so fewer than $n/(2\log n - 7\log\log n) + n/\log^2 n$ $\sim n/(2\log n)$ colours are used. ∎

By our earlier use of the term 'non-algorithmic' (which was not strictly correct), we meant that, while Theorem 2.4 ensures that there is a large stable set, it does not provide any efficient method for producing one. Certainly, a brute-force test of every subset with the appropriate number of vertices is far too slow. In 1976 Karp [52] asked whether there is a polynomial-time algorithm that produces from $G_{n,1/2}$ a stable set of size $(1+\varepsilon)\log n$ a.a.s. This is open, even if (for example, by Theorem 2.2) there are fast algorithms to produce stable sets of size $(1-\varepsilon)\log n$.

As mentioned above, tools for proving strong concentration of measure were crucial. By now, there are a few ways to prove Theorem 2.4. Bollobás's proof made a clever choice of two random variables that track large stable sets. One of these never exceeds the other, but they serve different purposes: the former has expectation easily bounded from below, while the latter is a martingale amenable to the Azuma–Hoeffding inequality. More direct proofs use more sophisticated concentration bounds, such as Talagrand's inequality. We have chosen to outline the use of Janson's inequality – a powerful probabilistic tool partly inspired by Bollobás's method – to prove Theorem 2.4. The reader may choose to skip the following outline, and take away the rough message that, for sets that are large but not too large, Janson's inequality provides the required probability bound if we can control the pairwise dependence (measured by a parameter Δ below) among the events of being stable.

Outline of proof of Theorem 2.4 As before, let $\mathcal{S}_k$ be the collection of stable sets of order k in $G_{n,1/2}$. By Janson's inequality (Theorem 2.18(ii) in [49]),

$$\mathbb{P}(\alpha(G_{n,1/2}) < k) = \mathbb{P}(|\mathcal{S}_k| = 0) \le \exp\left(-\frac{\mu^2}{\mu+\Delta}\right),$$

where $\mu = \mathbb{E}(|\mathcal{S}_k|)$. On summing over ordered pairs $A, B \subseteq [n]$ with $2 \le |A \cap B| \le k-1$,

$$\Delta = \sum \mathbb{P}(A, B \in \mathcal{S}_k).$$

Recall that $\mu = \binom{n}{k}2^{-\binom{k}{2}}$, and observe that $2^{-2\binom{k}{2}+\binom{\ell}{2}}$ is the probability that two given subsets of $[n]$ of order k that overlap in exactly ℓ vertices are both in $\mathcal{S}_k$. We may then write

$$\Delta = \sum_{\ell=2}^{k-1}\binom{n}{k}\binom{k}{\ell}\binom{n-k}{k-\ell}2^{-2\binom{k}{2}+\binom{\ell}{2}} = \mu\sum_{\ell=2}^{k-1}\binom{k}{\ell}\binom{n-k}{k-\ell}2^{-\binom{k}{2}+\binom{\ell}{2}}.$$

It is a routine exercise to check, under the assumption that $k \le 2\log n - 3\log\log n$, that the first term is the largest in the sum, and therefore

$$\Delta \le \mu(k-2)\binom{k}{2}\binom{n-k}{k-2}2^{-\binom{k}{2}+1} \le \mu^2 O\left(\frac{k^5}{n^2}\right) = O\left(\frac{\mu^2\log^5 n}{n^2}\right).$$

The choice of k also implies that $\mu = \exp(\Omega((\log\log n)\log n))$, so $\mu^2/(\mu+\Delta) = \Omega(n^2/\log^5 n)$, as required. ■

Colouring rate and anti-concentration

Theorem 2.5 answers the latter question in the Introduction, but the story does not end here. We mentioned that Shamir and Spencer proved that $\chi(G_{n,1/2})$ is contained in an interval of length roughly $O(\sqrt{n})$ a.a.s., but it remains unclear precisely where the interval could be situated. (Note that this interval can be narrowed by a $\log n$ factor, see [9, Ex. 7.9].) Theorem 2.5 has been tightened recently, as we see in the following theorem, but the resulting explicit interval is still comparatively large, having length $\Theta(n/\log^2 n)$.

Theorem 2.6 *Let* $k_1 = 2\log n - 2\log\log n - 2$ $(= k_0 - 2\log e - 1)$. *Then*

$$\frac{n}{k_1+o(1)} \le \chi(G_{n,1/2}) \le \frac{n}{k_1-1-o(1)} \quad a.a.s.$$

The lower bound, due to Panagiotou and Steger [78] in 2009, and the upper bound, due to Fountoulakis, Kang and McDiarmid [38] a year later, improved upon the best previously known range (in [67]) from some two decades earlier. We prefer to restate the result in a modified form. We define the *colouring rate* $\overline{\alpha}(G)$ of a graph G to be $|V(G)|/\chi(G)$, which is the maximum average size of a colour class in a proper colouring of G. The above theorem says that the colouring rate of $G_{n,1/2}$ a.a.s. lies in the following interval of length $1 + o(1)$:

$$k_1 - 1 - o(1) \leq \overline{\alpha}(G_{n,1/2}) \leq k_1 + o(1).$$

Notice in particular that $\overline{\alpha}(G_{n,1/2})$ is strictly smaller than $\alpha(G_{n,1/2})$ a.a.s. A natural problem is to determine the a.a.s. behaviour of $\overline{\alpha}(G_{n,1/2})$ up to a $o(1)$ additive error.

What is the length of a smallest interval I for which $\mathbb{P}(\chi(G_{n,1/2}) \in I) \geq \varepsilon$? Lower bounds on this quantity are called 'anti-concentration' results. Various behaviours for anti-concentration of the chromatic number have been proposed (see [7], [16]). This problem is likely to be far from easy: while the length of the interval is at most about $O(\sqrt{n}/\log n)$, there are currently no results that show that $\chi(G_{n,1/2})$ is not contained a.a.s. in an interval of *constant* length. The possibility has been suggested [7] ('conjecture' would be too strong a word) that there may even be non-monotonicity in n – that is, it might be the case that very sharp concentration holds for infinitely many values of n, while for infinitely many other values it does not.

3. Sparse random graphs

In this section we consider the chromatic number for $G_{n,p}$ when $p = p(n)$ satisfies $p \to 0$ as $n \to \infty$. First we see how the chromatic number becomes more sharply concentrated for sparser graphs – that is, for $p(n)$ tending to 0 more quickly. In the second subsection we consider the specific case $p = d/n$, for some fixed choice of $d > 1$. By a standard equivalence, this amounts to the first displayed question in the Introduction. This question has recently benefited from the influence of analysis and statistical physics, leading to dramatic progress.

Expose-and-merge and concentration

A subtle colouring method of Matula [64], dubbed the 'expose-and-merge' algorithm, was used to prove Theorem 2.5 without the need for martingale- or Talagrand-type concentration, but rather with the more elementary inequalities of Chebyshev and Chernoff, as shown by Matula and Kučera [65]. For sparse $G_{n,p}$, Bollobás's method of extracting the colour classes one by one could not be extended further, though it worked down to edge probability about $p = n^{-1/3}$ (that is, expected average degree about $n^{2/3}$). However, Łuczak [59] realized that Matula's algorithm and the concentration bounds key to Bollobás's approach could be combined effectively to prove the following stronger result. This significantly extended the range of $p = p(n)$ for the known a.a.s. first-order term of $\chi(G_{n,p})$.

Theorem 3.1 *There is a constant d_0 such that, if $p = p(n)$ satisfies $np > d_0$ for n sufficiently large and $p = o(1)$, then a.a.s.*

$$\chi(G_{n,p}) = \frac{np}{2\ln(np)}\left(1 + \Theta\left(\frac{\ln\ln(np)}{\ln(np)}\right)\right).$$

Outline of proof For simplicity, we assume throughout that $p < \ln^{-7} n$. The lower bound follows by noticing that the first moment argument used for the upper bound in Theorem 2.1 extends routinely to the sparse regime:

$$\alpha(G_{n,p}) \le (2\ln(np) - 2\ln\ln(np) + 2 - 2\ln 2 + \varepsilon)/p \text{ a.a.s.}$$

For the upper bound, we employ the expose-and-merge method. We are somewhat rough with our description of this; for a more precise and detailed account, see [49, Sec. 7.5]. The engine behind this technique is a result of the following type, which a.a.s. guarantees the existence of many disjoint stable sets of nearly maximum order.

Theorem 3.2 *Fix $\varepsilon > 0$. There exists C_ε such that if $C_\varepsilon/n \le p \le \ln^{-7} n$ then, with probability at least $1 - 1/n$, $G_{n,p}$ contains a disjoint collection of y stable sets, each of order at least k, where*

$$k = k(n) = (2\ln(np) - 2\ln\ln(np) + 2 - 2\ln 2 - \varepsilon)/p \quad \text{and} \quad y = y(n) = \frac{n}{k\ln^5(np)}.$$

Matula and Kučera proved a dense version of this result by a lengthy though uncomplicated second moment computation. For the form stated here, specific to sparse random graphs, the second moment is not as well behaved. Luckily, it does not grow too quickly, so the standard argument succeeds when it is strengthened by the appropriate application of a martingale concentration inequality. This neat trick was first employed by Frieze [41] to determine the stability number of sparse $G_{n,p}$.

With Theorem 3.2 in hand, we may attempt to extract from $G_{n,p}$ several large stable sets at a time. Set $\hat{n} = n/\ln^C(np)$ for a fixed positive constant C to be determined and, with a view to applying Theorem 3.2 repeatedly, set

$$\hat{k} = k(\hat{n}) \sim (2\ln(np) - 2(C+1)\ln\ln(np))/p \text{ and } \hat{y} = y(\hat{n}) \sim \frac{n}{\hat{k}\ln^{C+5}(np)},$$

where the asymptotics implied in $\sim$ are as $np \to \infty$.

To begin, choose, or rather *expose*, some subset A_1 of $[n]$ uniformly at random among all subsets of order $\hat{n}$. The induced subgraph $H_1 = G_{n,p}[A_1]$ may be viewed as $G_{\hat{n},p}$, so by Theorem 3.2, with probability at least $1 - 1/\hat{n}$, it contains a collection of

$\hat{y}$ disjoint stable sets each of order $\hat{k}$. Let us choose one such collection, $S_1^1, S_1^2, \ldots, S_1^{\hat{y}}$, uniformly at random (giving each subset of A_1 of order $\hat{k}$ a fair chance to be part of this collection). We label all vertices in $\bigcup_j S_1^j$ as *used* and all pairs of vertices in A_1 as *exposed*. Due to our extraction of a particular collection of stable sets from H_1, we can no longer view the exposed pairs as random edges.

Next we choose a new subset A_2 of order $\hat{n}$ uniformly at random from only the unused vertices. Unfortunately, we may not view the induced subgraph $G_{n,p}[A_2]$ as $G_{\hat{n},p}$ because its structure depends possibly on some exposed pairs. The bald but critical idea now is simply to forget about this problem for a while. More precisely, for each exposed pair from A_2, we resample by performing a new independent random experiment to include that pair as an edge with probability p. Let H_2 denote the graph on A_2 formed in this way. The key property of H_2 is that it may now be viewed as an independent copy of $G_{\hat{n},p}$ and, using Theorem 3.2 as above, we may choose a collection, $S_2^1, S_2^2, \ldots, S_2^{\hat{y}}$, of disjoint stable sets each of order $\hat{k}$. We now label all vertices in $\bigcup_j S_2^j$ as *used* and all pairs of vertices in A_2 as *exposed*.

Over $\hat{x} = \ln^{C+5}(np) - \ln^{C+2}(np)$ iterations, the probability that some iteration fails to generate a collection of $\hat{y}$ disjoint stable sets, each of order $\hat{k}$, is at most $\hat{x}/\hat{n} = o(1)$. With all but negligible probability, we have at the end a collection of $\hat{x}\hat{y} \sim np/(2\ln(np))$ disjoint sets S_i^j such that $S_i^1, S_i^2, \ldots, S_i^{\hat{y}}$ are all stable in H_i. It now remains to show how to *merge* these sets into a proper colouring of nearly all of $G_{n,p}$. In particular, we sketch how the sets S_i^j can be pruned slightly to form a proper colouring of all but at most $O(n/\ln^3(np))$ vertices. (It can then be checked that the graph induced by the remaining vertices has small enough degeneracy that it can be polished off greedily, using only a few extra colours — recall that *degeneracy* is defined as the maximum over all induced subgraphs of the minimum degree.) The crux here is that, due to our temporary amnesia, although S_i^j is a stable set in H_i, it might not be stable in $G_{n,p}$. We are in trouble if S_i^j contains some pair of adjacent vertices that was exposed in a previous iteration but resampled as a non-edge in the ith iteration. With an appropriate choice of C and a first moment estimation, it is indeed possible (although we omit the details) to show that the number Y of such troublesome pairs is small, satisfying $\mathbb{P}(Y > n/\ln^3(np)) = O(1/\ln^2(np))$. Thus a.a.s. we may safely remove Y vertices from $\bigcup_{i,j} S_i^j$, one from each troublesome pair, to obtain the required proper colouring of nearly all of $G_{n,p}$. ∎

As in the dense regime, in the sparse case one can prove strong concentration results for $\chi(G_{n,p})$ without knowing an asymptotic value. In addition to the aforementioned bound for the dense case, Shamir and Spencer showed with martingales that, for $p = n^{-\alpha}$ with $\alpha > \frac{1}{2}$, $\chi(G_{n,p})$ is a.a.s. confined to an interval whose width is a constant dependent upon α. A few years later, their proof was sharpened by Łuczak [60] to demonstrate *two*-point concentration when $\alpha > \frac{5}{6}$. (By continuity considerations, there must exist edge probability functions $p(n)$ that evenly balance the probability mass between the events that the chromatic number evaluates to one of two consecutive integers; so two-point concentration is best possible in general.)

Another few years later, Alon and Krivelevich [7] extended the two-point result to hold for all (fixed) $\alpha > \frac{1}{2}$, using recolouring via the Lovász local lemma.

This brings us close to the situation that we encountered in the dense case. Martingale techniques show that there is a small range of values that the chromatic number can take, but they do not give any hint of where it is located. The explicit interval implied by Theorem 3.1 is of width $\Theta(np \ln\ln(np)/\ln^2(np))$. This is unbounded (and certainly larger than 2) if np is unbounded. On the other hand, if np is bounded, then we have the intuition (or at least some hope) that it would be possible to identify (nearly) exact values. We pursue this next.

Sharp thresholds for k-colourability

Random graphs exhibit behaviours that might seem counter-intuitive at first sight. Many properties or parameters of $G_{n,p}$, such as the stability number or the number of proper k-colourings, experience a sudden change across a small interval of p. This is sometimes referred to as a *threshold* (or *phase transition*, as mentioned earlier). It is well known that thresholds arise in wider contexts in mathematics, physics, economics, sociology and elsewhere. In $G_{n,p}$, their study is of great importance. The demonstration by Bollobás and Thomason [18] that thresholds exist in a general sense for graph properties (families of graphs that are closed under isomorphism) that are monotone (closed under edge-removal) is one of the fundamental results in random graphs.

In 1999 Friedgut [40] investigated how suddenly a given phase transition takes place. Imprecisely, given a property of $G_{n,p}$ that undergoes a transition, consider the ratio of the following two quantities:

- the length of the window of p over which the transition takes place
- the critical probability p_c for which G_{n,p_c} possesses the property with probability exactly $\frac{1}{2}$ (which is guaranteed to exist).

The property's threshold is *sharp* if as $n \to \infty$ this ratio tends to 0, whereas if the ratio is bounded away from 0 it is a called a *coarse* threshold. (The result of Bollobás and Thomason showed the ratio to be bounded above for any monotone property.) The foundational example of a sharp threshold is the critical probability for connectivity of $G_{n,p}$: it is located at around $p = (\ln n)/n$, but the length of the interval over which $G_{n,p}$ changes from being disconnected to connected is of order $1/n$. The seminal work of Friedgut used Fourier analytic techniques to give a deep criterion for monotone graph properties in $G_{n,p}$ to have sharp thresholds. In rough terms, a monotone graph property with a coarse threshold can be approximated by the property that it contains one of a fixed collection of graphs as a subgraph. With this characterization, Achlioptas and Friedgut [1] showed that, for $k \geq 3$, the property of k-colourability (that is, of having chromatic number at most k) has a sharp threshold sequence.

Theorem 3.3 *Let $k \geq 3$ be an integer. There exists $d_k = d_k(n)$ such that, for any $\varepsilon > 0$,*

$$\lim_{n\to\infty} \mathbb{P}\left(\chi\left(G_{n,(d_k-\varepsilon)/n}\right) \leq k\right) = 1 \quad \text{and} \quad \lim_{n\to\infty} \mathbb{P}\left(\chi\left(G_{n,(d_k+\varepsilon)/n}\right) \leq k\right) = 0.$$

(It is known that the threshold for 2-colourability is coarse, since the probabilities of $G_{n,d/n}$ having an odd cycle and not having an odd cycle are both bounded away from 0 for $d \in (0, 1)$.) From Theorem 3.3, we deduce that $\chi(G_{n,d/n})$ would be one-point concentrated for all but a vanishing proportion of $d > 1$ (in Lebesgue measure) if it could be shown that $d_k(n)$ converges as $n \to \infty$, or even that $d_k(n)$ is constrained to an interval of length independent of k for all large enough n. Unfortunately, Achlioptas and Friedgut's result does not directly yield anything on the actual value of the chromatic number.

The question of convergence of $d_k(n)$ as $n \to \infty$ is open, but in a breakthrough in 2005 Achlioptas and Naor [3] found an explicit interval within which the limit must lie if it exists.

Theorem 3.4 *Let $k \geq 3$ be an integer. For any $\varepsilon > 0$,*

$$\lim_{n\to\infty} \mathbb{P}\left(\chi\left(G_{n,(2(k-1)\ln(k-1)-\varepsilon)/n}\right) \leq k\right) = 1$$
$$\text{and} \quad \lim_{n\to\infty} \mathbb{P}\left(\chi\left(G_{n,((2k-1)\ln k)/n}\right) \leq k\right) = 0.$$

Thus the sharp threshold sequence $d_k = d_k(n)$ for k-colourability (as guaranteed by Theorem 3.3) satisfies, for all n,

$$2(k-1)\ln(k-1) \leq d_k \leq (2k-1)\ln k. \tag{1}$$

An immediate corollary of Theorem 3.4 is that a.a.s. the chromatic number of $G_{n,d/n}$ must take either one explicit value or one of two values.

Corollary 3.5 *Given $d > 0$, let k_d be the least integer k for which $d < 2(k-1)\ln(k-1)$. Then $\chi(G_{n,d/n})$ is $k_d - 1$ or k_d a.a.s. If $d \geq (2k_d - 3)\ln(k_d - 1)$, then $\chi(G_{n,d/n}) = k_d$ a.a.s.*

Building upon this work and combining it carefully with some other ideas used for existential sharp concentration of the chromatic number, Coja-Oghlan, Panagiotou and Steger [26] have obtained an explicit two- or three-point concentration result for denser $G_{n,p}$, specifically for $p = n^{-\alpha}$ with $\alpha > \frac{3}{4}$. Recently, Dyer, Frieze and Greenhill [33] have obtained an analogue of Theorem 3.4 for the weak chromatic number of binomial random k-uniform hypergraphs.

We briefly outline some of the ideas of the proof of Theorem 3.4. First of all, rather than $G_{n,d/n}$, it is easier, and also does not hurt, to work with the random *multigraph* $\widetilde{G}_{n,m}$ with $m = \frac{1}{2}dn$, where we throw in m edges independently and uniformly at random, allowing repetitions. Counting the number of proper k-colourings of $\widetilde{G}_{n,m}$ is equivalent to summing, over all possible partitions of $[n]$

into k parts, the indicator variable that the given partition is a proper colouring. The right-hand side of (1), as first observed by Devroye (see [23]), is a simple first moment estimate. It follows from the fact that a given partition into k parts is a proper colouring with probability at most $(1-1/k)^m$, and thus the expected number of proper k-colourings of $\widetilde{G}_{n,m}$ is at most $k^n(1-1/k)^m$.

For the much more difficult part, the lower bound on d_k, Achlioptas and Naor made a consummate application of the second moment method, which they reduced to a high-dimensional optimization problem. A powerful simplification is to restrict attention to partitions of $[n]$ that are *balanced* – that is, each part has n/k vertices (where we have also assumed that k divides n). These account for most proper k-colourings of $\widetilde{G}_{n,m}$. The number of balanced proper k-colourings of $\widetilde{G}_{n,m}$ is a sum of $n!/((n/k)!)^k$ indicators, and it follows that its first moment is

$$\frac{n!}{((n/k)!)^k}\left(1-\frac{1}{k}\right)^m.$$

To bound the second moment of this random variable, we bound the probability, for a given pair of balanced partitions, that both are proper k-colourings of $\widetilde{G}_{n,m}$. Using the property that the partitions are balanced, we may see that this probability is

$$\left(1-\frac{2}{k}+\sum_{i,j}\frac{{\ell_{ij}}^2}{n^2}\right)^m,$$

where ℓ_{ij} is the number of vertices of $[n]$ with colour i in one of the balanced partitions and colour j in the other. So the correlation between two variables that indicate whether a given balanced partition is proper is governed by k^2 parameters. Consider the set $\mathcal{L}$ of all $k \times k$ matrices $\mathbf{L} = (\ell_{ij})$ of non-negative integers for which the sum of each row and each column is n/k. The second moment of the number of balanced proper k-colourings can be expressed as

$$\sum_{\mathbf{L}\in\mathcal{L}}\frac{n!}{\prod_{i,j}\ell_{ij}!}\left(1-\frac{2}{k}+\sum_{i,j}\frac{{\ell_{ij}}^2}{n^2}\right)^{dn/2}.$$

Optimization over $\mathcal{L}$ can be translated into optimization over the Birkhoff polytope $\mathcal{B}_k$, the set of $k \times k$ doubly stochastic matrices. Thus, to bound the second moment appropriately, it suffices to show a particular entropy–energy inequality over $\mathcal{B}_k$. Achlioptas and Naor did so by first relaxing to singly stochastic matrices and then using sophisticated geometric and analytic ideas. Their argument then shows that the probability of being k-colourable is bounded above 0, whereupon an application of Theorem 3.3 implies the theorem.

Recent work of Coja-Oghlan and Vilenchik [27] and of Coja-Oghlan [24] has markedly improved both ends of the interval (1) (which we remark has length $(1+o(1))\ln k$ as $k \to \infty$). Guided by important developments in statistical physics, which used applications of the so-called 'cavity method' and 'belief/survey propagation' algorithms [21], [76], they have obtained a range of length that is independent of k:

$$(2k-1)\ln k - 2\ln 2 - o(1) \le d_k \le (2k-1)\ln k - 1 + o(1), \qquad (2)$$

where the $o(1)$ asymptotics are as $k \to \infty$. The above range for the k-colourability sharp threshold sequence (particularly the improved lower bound) is quite close to completely settling the original question of Erdős and Rényi from 1960. It shows that $\chi(G_{n,d/n})$ is concentrated on an explicit value – that is, the least integer k for which $d < (2k-1)\ln k - 2\ln 2$ holds – for all but a vanishing proportion of $d > 1$.

4. Random regular graphs

Given an integer $r \ge 1$, let $G_{n,r}$ be a graph chosen uniformly at random from the set of r-regular graphs with vertex-set $[n]$ (where rn is even, for feasibility). This model of random graphs has received much attention since the late 1970s, when Bollobás [13] probabilistically reinterpreted an enumerative formula of Bender and Canfield [11] using the now-standard 'configuration' or 'pairing' model. In some aspects, $G_{n,r}$ is qualitatively quite distinct from $G_{n,p}$; for example, $G_{n,r}$ has stronger connectivity properties at lower edge densities: notably, $G_{n,3}$ is Hamiltonian a.a.s. However, for larger r, $G_{n,r}$ seems to behave more like $G_{n,p}$; this was partially formalized by Kim and Vu [54]. (The rough intuition behind this is that, above a certain density, all the degrees of $G_{n,p}$ are highly concentrated around np.) Techniques for proving properties of $G_{n,r}$ are usually more difficult though, and are close in spirit to those used in analytic and enumerative combinatorics. The refinement of these techniques is useful in the study of degree-constrained random graphs, with relevance to real-world network modelling. Wormald [88] has a broad account of the theory of random regular graphs. It is important to note that, unlike $G_{n,p}$, the model $G_{n,r}$ is not generally one that is monotone in density. In other words, it is not true in general that, if a monotone property holds in $G_{n,r}$ a.a.s., then it also holds in $G_{n,r+1}$ a.a.s. (as can be seen by considering the property of containment of a perfect matching); however, for *fixed* $r \ge 2$, this statement is true as a consequence of 'contiguity' (see, for example, [49] or [88]).

Two of the first results on the chromatic number of $G_{n,r}$ were given in 1992. Molloy and Reed bounded the expected number of proper k-colourings in the pairing model of $G_{n,r}$, with r fixed, to conclude that, if $k(1-1/k)^{r/2} < 1$, then $\chi(G_{n,r}) > k$ a.a.s. This argument was given in Molloy's University of Waterloo Master's thesis, but it has been reiterated in the more easily obtained paper [84]. Frieze and Łuczak [43] further analyzed the pairing model to show that, if $r = r(n) < n^{1/3-\varepsilon}$ for some $\varepsilon > 0$ and $r \ge r_0$ for some constant r_0, then a.a.s.

$$\chi(G_{n,r}) = \frac{r}{2\ln r} + \Theta\left(\frac{r\ln\ln r}{\ln^2 r}\right).$$

About a decade later, with an 'edge-switching' strategy, Cooper, Frieze, Reed and Riordan [28] extended this asymptotic statement to hold if $r < n^{1-\varepsilon}$ for some $\varepsilon > 0$ (and $r \geq r_0$ for some constant r_0). In an independent and complementary project, Krivelevich, Sudakov, Vu and Wormald [57] used an expression for the probability that $G_{n,p}$ is r-regular to show in the range $n^{6/7+\varepsilon} < r < 0.9n$ that $\chi(G_{n,r}) \sim n/(2\log_b r)$, where $b = n/(n-r)$. An easy calculation checks that these estimates of $\chi(G_{n,r})$ agree asymptotically as $r \to \infty$ with those of $\chi(G_{n,p})$ with $p = r/n$.

As it did for $G_{n,p}$, significant progress on the determination of $\chi(G_{n,r})$ in the sparse case (when r is constant) came later. The values $r \leq 3$ are easy: trivially, $\chi(G_{n,1}) = 2$ a.a.s. For $r \geq 2$, $G_{n,r}$ contains an odd cycle a.a.s. and so $\chi(G_{n,2}) = 3$ a.a.s.; also a.a.s. $\chi(G_{n,3}) = 3$, since $G_{n,3}$ contains no K_4-subgraph and the statement follows from Brooks's theorem. The next values are more involved. For $4 \leq r \leq 10$, Shi and Wormald [83], [84] provided a fine analysis in 2007, using the *differential equations method* [87], of a tailored colouring algorithm, to obtain explicit one- or two-point determinations. (The idea behind this general analytic method is to show that a random variable of interest can be tracked accurately by some function expressed as the solution to a set of differential equations.) In particular, Shi and Wormald proved that $\chi(G_{n,4}) = 3$, $\chi(G_{n,5})$ is 3 or 4, and $\chi(G_{n,6}) = 4$ a.a.s.

A further advance was obtained a few years later. Improving on a previously-announced bound of Achlioptas and Moore [2], Kemkes, Pérez-Giménez and Wormald [53] used *subgraph conditioning* (see [88]) to prove the following theorem.

Theorem 4.1 *Given $r > 0$, let k_r be the least integer k for which $r < 2(k-1)\ln(k-1)$. Then $\chi(G_{n,r})$ is $k_r - 1$ or k_r a.a.s. If $r \geq (2k_r - 3)\ln(k_r - 1)$, then $\chi(G_{n,r}) = k_r$ a.a.s.*

The lower bound is just the aforementioned bound of Molloy and Reed. As we have often seen, the most difficult part in obtaining the upper bound is controlling the second moment. Even after an adaptation of the sophisticated analysis of Achlioptas and Naor, the second moment is too large for the application of either Chebyshev's inequality or the Paley–Zygmund inequality directly, and no sharp threshold-type result is available. In particular, the second moment is some constant times the square of the first moment. The broad intuition for the subgraph conditioning method is that much of the magnitude of the second moment may depend on the presence of some specific small but not too common subgraphs, such as short cycles. When the method applies, conditioning on counts of enough of these small subgraphs does not change the first moment by too much, but it reduces the second moment to any desired fraction of the original. Kemkes, Pérez-Giménez and Wormald showed that this method is applicable to the number of balanced k-colourings of $G_{n,r}$, in order to obtain the upper bound on $\chi(G_{n,r})$ in this theorem.

A comparison of Theorem 4.1 and Corollary 3.5 suggests that $G_{n,r}$ and $G_{n,r/n}$ have a close affinity with respect to k-colourability, which contrasts with the broader qualitative differences we pointed out earlier. In a recent manuscript, Coja-Oghlan,

Efthymiou and Hetterich [25] have proved a more direct connection between these parameters, and used that link to tighten the bounds on $\chi(G_{n,r})$ to match the bounds on $\chi(G_{n,p})$ in (2). In particular, there is some ε_k with $\varepsilon_k \to 0$ as $k \to \infty$ such that if $r \leq (2k-1)\ln k - 2\ln 2 - \varepsilon_k$, then $\chi(G_{n,r}) \leq k$ a.a.s., whereas if $r \geq (2k-1)\ln k - 1 + \varepsilon_k$, then $\chi(G_{n,r}) > k$ a.a.s.

For the upper bound, they combined the second moment calculations in [27] with the subgraph conditioning methodology in [53]. The lower bound demands more technical care, and for this they study the geometry of the set of proper k-colourings and use large deviations analysis.

Note that the result that $\chi(G_{n,4}) = 3$ a.a.s. is not captured by Theorem 4.1 or the more recent work [25]. The conjecture that $\chi(G_{n,5}) = 3$ a.a.s. remains open, but there is some positive evidence for it in [30]. Ben-Shimon and Krivelevich [10] have used the edge-switching method to prove existential two-point concentration of $\chi(G_{n,r})$ for $r = o(n^{1/5})$, though since the continuity argument fails for $G_{n,r}$ it is unclear even whether two-point concentration is best possible. For fixed r, it seems very likely that $\chi(G_{n,r})$ is always uniquely determined a.a.s. This is true with an explicit determination for 'most' values of r, by the new results in [25].

Random Cayley graphs

So far in this section, we have focused on the uniform model of a regular graph. We now mention briefly a very different algebraically defined family of random regular graph models that has recently garnered heightened interest, including in regard to the chromatic number. Let B be a finite group of order n and let $k \in [\lfloor \frac{1}{2}n \rfloor]$. Choose a subset $A \subseteq B$ by first choosing k elements of B independently and uniformly at random (with repetitions), and then letting A be these elements together with their inverses, without the identity. The *random Cayley graph* with respect to B and k is the graph with vertex-set B, where b_1 and b_2 are adjacent when $b_1 \cdot {b_2}^{-1} \in A$; this graph is regular of degree $|A| \leq 2k$. The study of such random graphs is important in many settings, such as information theory, computational complexity and additive combinatorics.

The behaviour of the chromatic number of random Cayley graphs is currently an active research area; we refer the interested reader to Alon [6] for a systematic study, where, for instance, it is shown that the chromatic number of the random Cayley graph with respect to any B and k is $O(k/\ln k)$. (Alon furthermore gives general lower bounds that nearly match the upper bound in the dense case $k = \Theta(n)$ but worsen for smaller k.) We also point to a recent manuscript of Green and Morris [47] on *random Cayley sum graphs* with respect to $\mathbb{Z}/n\mathbb{Z}$. In this model, $A \subseteq \mathbb{Z}/n\mathbb{Z}$ is chosen by including each element independently at random with probability $\frac{1}{2}$; then $x, y \in \mathbb{Z}/n\mathbb{Z}$ are adjacent in the graph when $x+y \in A$. With methods from additive combinatorics, such as an 'arithmetic regularity lemma', they prove that such graphs a.a.s. have no clique or stable set of order more than $(2 + o(1))\log n$. A matching upper bound on the chromatic number, although with a technical restriction that $\gcd(n, 6) = 1$, was recently proved by Green [46].

5. Random geometric graphs

Suppose that we wish to assign bands of radio frequencies to a collection of active transmitters scattered in some region. Minimizing the amount of spectrum allocated can be modelled (albeit with many simplifying assumptions) as follows. Given a collection $\mathcal{X} = \mathcal{X}_n$ of independent random points $X_1, X_2, \ldots, X_n \in \mathbb{R}^d$ with common probability distribution ν and a positive distance parameter $r = r(n) > 0$, we first construct a *random geometric graph* $G_n(\mathcal{X}, r)$ with vertex-set $[n]$, where i and j are adjacent when $\|X_i - X_j\| < r$. Then, taking $\|\cdot\|$ as the ordinary Euclidean norm and ν as a uniform distribution over $[0, 1]^2$, we may imagine the points X_i as transmitters in the plane and r as their common transmission (and interference) range. We are interested in the asymptotic behaviour of $G_n(\mathcal{X}, r)$ as $n \to \infty$ and assume that $r = r(n) = o(n)$. (In fact, the norm $\|\cdot\|$ may be any norm and ν may be any probability distribution with a bounded density function over $\mathbb{R}^d$.) We may thus consider a proper colouring as a good spectrum allocation, so that $\chi(G_n(\mathcal{X}, r))$ indicates the minimum amount needed.

The random geometric graph is a much-studied model because of its simple and natural definition, its history going back more than half a century, and its well-established links to areas such as statistics, probability theory and electrical engineering. Background and further properties of $G_n(\mathcal{X}, r)$ are covered in detail in the monograph of Penrose [79]. Intuitively, the parameter $r(n)$ plays the same role that the edge probability parameter $p(n)$ does in $G_{n,p}$. It controls density: the average degree of $G_n(\mathcal{X}, r)$ is $\Theta(r^d n)$ in general, and is about $\pi r^2 n$ for the Euclidean norm and a uniform distribution over $[0, 1]^2$. As we might expect, $G_n(\mathcal{X}, r)$ goes through a number of phases as r increases. We remark here that most of the results in this section are given with stronger convergence than a.a.s.: a property holds *almost surely (a.s.)* if it holds with probability 1.

We will see that there is a threshold in the a.s. behaviour of $\chi(G_n(\mathcal{X}, r))$. Loosely speaking, the chromatic number behaves very 'locally' in the sparse regime when the average degree is $o(\ln n)$, whereas it behaves essentially as if the points are uniformly spread in the dense regime when the average degree is $\omega(\ln n)$; the behaviour at the threshold $\Theta(\ln n)$ is most interesting. Moreover, the chromatic number of $G_n(\mathcal{X}, r)$ is quite close to both the clique number and the maximum degree, in contrast with $G_{n,p}$. This bolsters our natural intuition from the fact that the structure of $G_n(\mathcal{X}, r)$ is locally determined. (We comment here that, for a 'deterministic' geometric graph G in the plane with Euclidean norm, it is known that $\chi(G) \leq 3\omega(G) - 2$ and $\Delta(G) \leq 6\omega(G) - 7$.)

Before stating the main result, we define the parameters required of the common probability measure ν and the norm $\|\cdot\|$. The *maximum density* σ of ν is the essential supremum of the probability density function f of ν – that is to say, $\sigma = \sup\{t : \mathrm{vol}\,(\{x : f(x) > t\}) > 0\}$, where vol denotes the (Lebesgue measure) volume. For $K > 0$, let $N(K)$ be the maximum cardinality of a collection of pairwise disjoint translates of the unit ball $B = \{x \in \mathbb{R}^d : \|x\| < 1\}$ with centres in $(0, K)^d$. The (*translational*) *packing density* δ of the unit ball B with respect to $\|\cdot\|$ is defined as

$$\delta = \lim_{K\to\infty} \frac{N(K)\,\mathrm{vol}\,(B)}{K^d}.$$

This limit always exists and satisfies $0 < \delta \leq 1$. For the Euclidean norm in $\mathbb{R}^2$, it is known that $\delta = \pi/(2\sqrt{3}) \approx 0.907$. Informally, δ is the greatest proportion of $\mathbb{R}^d$ that can be filled with disjoint translates of B. We are now ready to see the full picture of phases for $\chi(G_n(\mathcal{X}, r))$.

Theorem 5.1 *For the random geometric graph $G_n(\mathcal{X}, r)$, the following hold.*

Very sparse regime: *Suppose that $r^d n \leq n^{-\alpha}$ for some fixed $\alpha > 0$, and let $k = k(n) = \left\lfloor |\ln n/\ln(r^d n)| + \frac{1}{2} \right\rfloor$. Then a.s. $\chi(G_n(\mathcal{X}, r)) = \omega(G_n(\mathcal{X}, r))$ is k or $k+1$ for all but finitely many n.*
Sparse regime: *Suppose that $r^d n = \omega(n^{-\varepsilon})$ for all $\varepsilon > 0$ but $r^d n = o(\ln n)$. Then a.s. both $\chi(G_n(\mathcal{X}, r))$ and $\omega(G_n(\mathcal{X}, r))$ are asymptotic to $\ln n/\ln(\ln n/(r^d n))$.*
Intermediate regime: *Suppose that $\sigma r^d n/\ln n \to t \in (0, \infty)$. Then a.s.*

$$\chi(G_n(\mathcal{X}, r)) \sim f_\chi(t)\cdot\sigma r^d n \ \text{ and } \ \omega(G_n(\mathcal{X}, r)) \sim f_\omega(t)\cdot\sigma r^d n$$

where the functions f_χ and f_ω are given explicitly in [72]. *They depend only on d and $\|\cdot\|$, and are continuous; f_χ is non-increasing, and satisfies $f_\chi(t) \to \mathrm{vol}\,(B)/(2^d\delta)$ as $t \to \infty$, and $f_\chi(t) \to \infty$ as $t \downarrow 0$; and f_ω is strictly decreasing, and satisfies $f_\omega(t) \to \mathrm{vol}\,(B)/2^d$ as $t \to \infty$, and $f_\omega(t) \to \infty$ as $t \downarrow 0$.*
Dense regime: *Suppose that $r^d n = \omega(\ln n)$ but $r \to 0$. Then a.s.*

$$\chi(G_n(\mathcal{X}, r)) \sim \frac{\mathrm{vol}\,(B)}{2^d\delta}\cdot\sigma r^d n \ \text{ and } \ \omega(G_n(\mathcal{X}, r)) \sim \frac{\mathrm{vol}\,(B)}{2^d}\cdot\sigma r^d n.$$

The results on $\omega(G_n(\mathcal{X}, r))$ essentially already appeared in Penrose's book [79] of 2003, except that an assumption has been dropped and some results amplified. Both Penrose [79] and McDiarmid [68] studied the chromatic number in the sparse and dense regimes, but prior to the 2011 work of McDiarmid and Müller [72] little was known about the chromatic number in the most difficult intermediate regime, $r^d n = \Theta(\ln n)$. The proof of Theorem 5.1 took an optimization viewpoint. In particular, McDiarmid and Müller thoroughly investigated the fractional chromatic number, which is the solution to the linear programming relaxation of the natural integer linear programme for the chromatic number. This could be reduced to analyzing weighted integrals that correspond to feasible solutions of the dual linear programme for the fractional chromatic number. The result then follows on showing that the chromatic and fractional chromatic numbers are close.

The most striking consequence of Theorem 5.1 is that $\chi(G_n(\mathcal{X},r))$ and $\omega(G_n(\mathcal{X},r))$ have the same asymptotic value when $r^d n = o(\ln n)$, but the two quantities differ by a multiplicative factor (if the packing density δ is less than 1) when $r^d n = \omega(\ln n)$. McDiarmid and Müller also found a sharp threshold between these behaviours (again, if $\delta < 1$).

Theorem 5.2 *If $\delta < 1$, there is a critical parameter t_0, depending only on d and $\|\cdot\|$, for which the following hold for the random geometric graph $G_n(\mathcal{X},r)$:*

- *if* $\limsup_{n\to\infty} \sigma r^d n/\ln n \le t_0$, *then* $\chi(G_n(\mathcal{X},r))/\omega(G_n(\mathcal{X},r)) \to 1$ *a.s.*
- *if* $\liminf_{n\to\infty} \sigma r^d n/\ln n > t_0$, *then* $\liminf_{n\to\infty} \chi(G_n(\mathcal{X},r))/\omega(G_n(\mathcal{X},r)) > 1$ *a.s.*

Müller [77] established a general principle of existential concentration for parameters of $G_n(\mathcal{X},r)$. Specific consequences of this principle are that the clique number $\omega(G_n(\mathcal{X},r))$, the chromatic number $\chi(G_n(\mathcal{X},r))$ and the degeneracy $\delta^*(G_n(\mathcal{X},r))$ of the random geometric graph are each a.a.s. two-point concentrated if $r^d n = o(\ln n)$. The two consecutive values upon which the chromatic number is concentrated remain unknown except in the very sparse regime as specified in Theorem 5.1. Furthermore, in the light of Theorem 5.2, Müller suggested that it would be interesting to determine the location of the threshold (if one exists) for the property that the difference $\chi(G_n(\mathcal{X},r)) - \omega(G_n(\mathcal{X},r))$ is bounded.

6. Random planar graphs and related classes

Given the storied history of chromatic graph theory, it would seem unjust to omit mention of planar and other embeddable graphs in a colouring survey. A considerable amount of work has been done on random planar graphs and related models, much of it in recent years. Let $\mathcal{P}_n$ be the set of planar graphs with vertex-set $[n]$ and let P_n be a graph chosen uniformly at random from $\mathcal{P}_n$. We already see the difficulty of this model in its definition: to get a good grip on this random object, we essentially need to enumerate planar graphs as combinatorial structures, regardless of any particular topological embedding. An advance in our understanding of P_n came about a decade ago, when McDiarmid, Steger and Welsh [74] used a superadditivity argument to show the existence of the *(labelled) planar graph growth constant* γ, satisfying

$$\lim_{n\to\infty}\left(\frac{|\mathcal{P}_n|}{n!}\right)^{1/n} = \gamma.$$

More recently, Giménez and Noy [45] built on work of Bender, Gao and Wormald [12] in a masterful application of analytic combinatorics. They applied classical graph-theoretic decompositions according to connectivity, showed that these translate into basic functional relationships among the corresponding generating functions, and then applied singularity analysis to these functions to determine γ exactly. The mere existence of γ, though, was already enough to derive a range of properties of P_n, including results on the numbers of vertices of a given degree, of faces of a given size in any embedding, and of automorphisms. These results were given as consequences of what is now known as the 'appearances theorem' [74, Theorem 4.1], which says roughly that a.a.s. any connected planar graph appears in P_n as a pendant copy a linear number of times. Another consequence of the appearances theorem is that P_n fails to contain a K_4 subgraph with probability $e^{-\Omega(n)}$. This implies the following, the first part of which is immediate.

Theorem 6.1 *A.a.s.* $\chi(P_n) = 4$. *Furthermore, there is an expected quadratic-time algorithm to provide an optimal colouring of* P_n.

The second part of this theorem is an observation of Taraz and Krivelevich (see [74]). In linear time, we can distinguish between two cases. The first is that there is a K_4 subgraph, in which case we apply the quadratic-time 4-colouring algorithm for planar graphs. In the second case, which occurs with probability $e^{-\Omega(n)}$, we colour the graph optimally using the planar separator theorem, and this takes $O(c^{\sqrt{n}})$ time. This amounts to a quadratic expected running time. One might wonder whether there is a faster colouring algorithm or a shorter proof of 4-colourability specific to random planar graphs; however, the appearances theorem suggests this to be unlikely. With the above in mind, we might ask what more can be said about colouring random planar graphs. We take several tacks.

One natural direction, towards a probabilistic analogue of the Ringel–Youngs theorem, is to consider graphs embeddable in a fixed surface S. Let S_n be a graph chosen uniformly at random from the set of graphs embeddable in S with vertex-set $[n]$. These graph classes were first considered in [69], where it was shown, for example, that they all possess a growth constant, and indeed all have the *same* growth constant γ (as we met for planar graphs). It then follows by an appearances statement (linearly many pendant copies of each connected planar graph) that $\chi(S_n) \geq 4$ a.a.s. due to K_4 subgraphs. With additional ingredients from topological graph theory and generating function analysis, Chapuy, Fusy, Giménez, Mohar and Noy [22] obtained more precise results on S_n. One consequence of their work was that a.a.s. S_n has arbitrarily large *face-width*, indicating local planarity. A result of Thomassen [85] then implies that a.a.s. S_n is properly 5-colourable. Thus the following holds.

Theorem 6.2 *A.a.s.* $\chi(S_n)$ *is* 4 *or* 5.

It was conjectured in [22] that 4 is a.a.s. the correct value. Recall from Chapter 6 that the *choosability* $\mathrm{ch}(G)$ of a graph G is the least k for which, for any assignment of lists of at least k colours at each vertex, there is always a proper colouring of G so that each vertex is given a colour from its list. (We discuss choosability further in Section 7.) The choosability of S_n has been determined exactly: a list-colouring extension [29] of the aforementioned result of Thomassen, together with the appearances statement for some non-4-choosable (connected) planar graph, imply that $\mathrm{ch}(S_n) = 5$ a.a.s.

From surfaces, one could continue further by studying graphs chosen uniformly from a minor-closed class, as was done in [70]. We comment on one particular aspect of this study which is relevant to chromatic graph theory. In the Introduction, we mentioned that $G_{n,1/2}$ generates a plethora of counter-examples to an over-strengthened form of Hadwiger's conjecture. Let us rather consider Hadwiger's conjecture itself from the viewpoint of random graphs. Recall that this conjecture asserts that, for any positive integer k, each k-chromatic graph has K_k as a minor. It has been proved for $k \leq 6$. Fix an integer $k \geq 7$, and let $\mathrm{Ex}(K_k)$ denote the class of graphs without K_k as a minor. Since the class $\mathrm{Ex}(K_k)$ is 'addable' (meaning that it satisfies some simple decomposition closure properties), it then has a growth constant and so admits an appearances theorem (linearly many pendant copies of each connected member of $\mathrm{Ex}(K_k)$). So, if Hadwiger's conjecture fails for k, then it should be 'easy' to find counter-examples: if G_n is chosen uniformly at random from the set of graphs in $\mathrm{Ex}(K_k)$ on the vertex-set $[n]$, then $\chi(G_n) \geq k$ with probability $1 - e^{-\Omega(n)}$.

Next, we consider random planar graphs of a given density. Let $P_{n,m}$ be a graph chosen uniformly at random from the set of all planar graphs on $[n]$ with m edges; any such graph satisfies $m \leq 3n - 6$. This model is directly inspired by $G_{n,m}$ or $G_{n,p}$, although the condition of planarity clearly imposes extra difficulties. As a by-product of their analytic work, Giménez and Noy exactly determined the growth constant for the asymptotic number of planar graphs on $[n]$ with cn edges, for $c \in (1, 3)$. Recently, M. Kang and Łuczak [50], investigating what happens around $c = 1$, proved (by a combination of counting and analytic methods) a phenomenon akin to, though qualitatively distinct from, the emergence of the 'giant component' in $G_{n,p}$. Around $c = 1$, something also occurs with respect to the chromatic number of $P_{n,m}$.

Theorem 6.3
If $\limsup_{n\to\infty} m/n < 1$*, then* $\chi(P_{n,m}) \leq 3$ *a.a.s.*
If $\liminf_{n\to\infty} m/n > 1$*, then* $\chi(P_{n,m}) = 4$ *a.a.s. and there is an expected quadratic-time algorithm for an optimal colouring of* $P_{n,m}$*.*

Kang and Łuczak showed the first part as a corollary to structural properties that they proved for $P_{n,m}$ in the sparse regime. The second part follows from work of Dowden; in particular, the argument is similar to that above for P_n, but with an appearances theorem for random planar graphs of a given edge density (up to and including 3) (see [31, Theorem 61] and [32]). This demonstrates a phase transition in the behaviour of $\chi(P_{n,m})$. It would be of interest to understand more precisely how $\chi(P_{n,m})$ behaves when $m/n \to 1$.

Our last tack is to consider the stability number of P_n. Obviously, $\alpha(P_n) \geq \frac{1}{4}n$, but is it a.a.s. much higher? We do not determine $\alpha(P_n)$, but we shall quickly show that the answer is *yes*. Recall that the *core* (or 2-*core*) $\mathrm{core}(G)$ of a graph G is the unique maximal subgraph with minimum degree at least 2. Call $\mathrm{fr}(G) = G \setminus \mathrm{core}(G)$ the *fringe* of G. Take G to be a 4-colourable graph with n vertices. Since $\mathrm{core}(G)$ is also

4-colourable, it has a bipartite induced subgraph B with $|B| \geq \frac{1}{2}(n - |\operatorname{fr}(G)|)$. The vertices of $\operatorname{fr}(G)$ can be added one by one to B so that it remains bipartite, yielding a bipartite graph B' with $|B'| \geq \frac{1}{2}(n - |\operatorname{fr}(G)|) + |\operatorname{fr}(G)| = \frac{1}{2}(n + |\operatorname{fr}(G)|)$. Thus $\alpha(G) \geq \frac{1}{4}(n + |\operatorname{fr}(G)|)$. Applying this to P_n, using the fact shown in [71] that $|\operatorname{fr}(P_n)|/n \to \delta \approx 0.038157$ a.a.s., we conclude that $\alpha(P_n) \geq (0.25 + 0.019)n = 0.269n$ a.a.s. The question remains, *how large is* $\alpha(P_n)$ *a.a.s.*?

7. Other colourings

In this section we hint at the variety of research, even within random graph theory, on other types of graph colouring. We have had to limit ourselves to brief accounts for a small selection of topics, showing the tip of the iceberg. Moreover, we mainly consider just the symmetric case $G_{n,1/2}$, the dense Erdős–Rényi random graph, as in Section 2. Some of the results mentioned below extend to other (sparser) choices of p, to varying extents.

Strengthened colourings

Equitable chromatic number

A proper colouring is *equitable* if the orders of the colour classes differ by at most 1. The celebrated Hajnal–Szemerédi theorem states that every graph G has an equitable proper k-colouring, for each $k \geq \Delta(G) + 1$. The least k for which G has an equitable proper k-colouring is the *equitable chromatic number* $\chi_=(G)$ of G. The property of equitable k-colourability is not monotone in k, but in the light of the above statement the following definition is natural. The least k for which G has an equitable proper k'-colouring for every $k' \geq k$ is the *equitable chromatic threshold* $\chi_=^*(G)$ of G. Note that $\chi(G) \leq \chi_=(G) \leq \chi_=^*(G)$ always.

Krivelevich and Patkós [56] re-purposed the greedy processes, both of Grimmett and McDiarmid and of Bollobás, to prove the following.

Theorem 7.1
A.a.s. $\chi_=(G_{n,1/2}) \leq (1 + o(1))\chi(G_{n,1/2})$
and $\chi_=^*(G_{n,1/2}) \leq (2 + o(1))\chi(G_{n,1/2})$.

They obtained partial results for other values of p and have furthermore conjectured that $\chi_=^*(G_{n,p}) \leq (1 + o(1))\chi(G_{n,p})$ a.a.s. if $np \to \infty$. Further results have been obtained by Rombach and Scott [80].

Choosability

We defined the choosability $\operatorname{ch}(G)$ of a graph G in Section 6 – see also Chapter 6. In a first paper on choosability in 1979, Erdős, Rubin and Taylor conjectured that

$\mathrm{ch}(G_{n,1/2}) = o(n)$ a.a.s. This conjecture was proved by Alon [4] in 1992. Shortly thereafter, Kahn proved the following theorem (see [5]).

Theorem 7.2 *A.a.s.* $\mathrm{ch}(G_{n,1/2}) = (1 + o(1))\chi(G_{n,1/2})$.

Kahn's proof is elegant. The idea is to extract repeatedly (using Theorem 2.4) a large colour class for any colour that appears on more than $n/\log^2 n$ vertices, and at the end to match the remaining vertices with the remaining colours using Hall's theorem. Later, Krivelevich [55] extended Theorem 7.2 to hold for p as low as $n^{-\alpha}$ with $\alpha < \frac{1}{4}$, which was soon after improved to $\alpha < \frac{1}{3}$ in [58]. Alon, Krivelevich and Sudakov [8] conjectured that Theorem 7.2 extends to the case $np \to \infty$. They obtained partial progress on this conjecture, getting the correct value up to a constant multiple; this was proved independently by Vu [86].

Edge-colourings and total colourings

Chromatic index

Recall from Chapter 5 that a *proper edge-colouring* of a graph G is an assignment of colours to the edges so that incident edges receive different colours; the *chromatic index* $\chi'(G)$ is the least number of colours needed for such a colouring. Vizing's theorem states that $\chi'(G)$ is $\Delta(G)$ or $\Delta(G) + 1$; those graphs with the smaller value are of *class 1* while those with the larger are of *class 2*. By noticing that $G_{n,1/2}$ a.a.s. has a unique vertex of maximum degree, Erdős and Wilson [37] in 1977 proved the following.

Theorem 7.3 *A.a.s.* $G_{n,1/2}$ *is of class 1.*

A decade later, Frieze, Jackson, McDiarmid and Reed [42] improved on this by showing that the probability of $G_{n,1/2}$ being of class 2 satisfies, for any fixed $\varepsilon > 0$ and large enough n,

$$n^{-(1/2+\varepsilon)n} \le \mathbb{P}(G_{n,1/2} \text{ is of class 2}) \le n^{-(1/8-\varepsilon)n}.$$

They proved the upper bound by analyzing a certain polynomial-time edge-colouring algorithm and bounding its failure probability. The lower bound relied on bounding the probability that $G_{n,1/2}$ is regular.

Total chromatic number

A *total colouring* of a graph G is an assignment of a colour to each vertex and each edge so that no pair of adjacent vertices or adjacent edges receive the same colour and every edge is coloured differently from its endpoints. The *total chromatic number* $\chi''(G)$ is the least number of colours needed for such a colouring. It is obvious that $\chi''(G) \ge \Delta(G) + 1$ for all G. The total colouring conjecture, due independently to Behzad and Vizing, asserts that $\chi''(G) \le \Delta(G) + 2$ for all G. Molloy and Reed have shown by probabilistic means that there is a (not incredibly large) constant C for which $\chi''(G) \leq \Delta(G) + C$ for all G (see [75, Chs. 17–18]). Before that, McDiarmid and Reed [73] in 1993 studied $\chi''(G_{n,1/2})$. Partly by adapting the above-mentioned algorithm for edge-colouring and partly by invoking the probabilistic method, they proved the following.

Theorem 7.4

$$\mathbb{P}(\chi''(G_{n,1/2}) > \Delta(G_{n,1/2}) + 1) \leq n^{-(1/8-\varepsilon)n}.$$

$$\mathbb{P}(\chi''(G_{n,1/2}) > \Delta(G_{n,1/2}) + 2) = e^{-\Omega(n^2)}.$$

The second bound in particular implies that $G_{n,1/2}$ satisfies the total colouring conjecture up to a minuscule failure probability.

Generalized colourings

Hereditary properties

A graph property is *hereditary* if it is closed under vertex deletion. (Note, for example, that every monotone graph property is hereditary.) Given some fixed hereditary property $\mathcal{P}$, the *(generalized)* $\mathcal{P}$*-chromatic number* $\chi_{\mathcal{P}}(G)$ of a graph G is the least number of parts in a vertex partition of G for which each part induces a subgraph satisfying $\mathcal{P}$. By taking $\mathcal{P}$ to be the set of all edgeless graphs, we recover the ordinary chromatic number. If $\mathcal{P}$ is the set of all graphs or is finite, then $\chi_{\mathcal{P}}(G)$ is uninteresting; so we assume otherwise, calling such properties *non-trivial*. In 1992 Scheinerman [81] discovered that a non-trivial general statement is possible. In particular, for non-trivial hereditary $\mathcal{P}$, $\chi_{\mathcal{P}}(G_{n,1/2}) = O(n/\log n)$ a.a.s. A few years later, Bollobás and Thomason [19] proved the following tighter result.

Theorem 7.5 *For any non-trivial hereditary property $\mathcal{P}$, there is a constant $c = c_{1/2}(\mathcal{P})$ for which $\chi_{\mathcal{P}}(G_{n,1/2}) = (c + o(1))n/(2\log n)$ a.a.s.*

To prove this, they used a well-known asymptotic characterization of hereditary properties according to a particular extremal parameter. Given non-negative integers a and b, we say that an (a,b)*-colouring* of a graph G is a partition of the vertex-set into a cliques and b stable sets. The *colouring number* $\tau(\mathcal{P})$ of a hereditary property $\mathcal{P}$ is the largest k for which there are a and b with $a + b = k$ for which every (a,b)-colourable graph has property $\mathcal{P}$. It was known that the parameter $\tau(\mathcal{P})$ governs the asymptotic behaviour of the cardinality $|\mathcal{P}_n|$ of graphs on $[n]$ satisfying $\mathcal{P}$. Bollobás and Thomason showed moreover that $\tau(\mathcal{P})$ determines the constant c in Theorem 7.5. In follow-up work [20], they extended this theorem to other fixed values of $p \in (0,1)$, with $c_p(\mathcal{P})$ in the place of $c_{1/2}(\mathcal{P})$. This problem gave exception to the rule that a result for $G_{n,1/2}$ is easily extendable to the entire dense regime. An application of Szemerédi's regularity lemma was needed to approximate the given hereditary property structurally. Marchant and Thomason [61] recently improved this structural approximation to shed more light on how to compute specific values of $c_p(\mathcal{P})$.

Improper colourings

Given a non-negative integer t, a *t-improper colouring* of G is a vertex partition in which each part induces a subgraph of maximum degree at most t; the *t-improper chromatic number* $\chi^t(G)$ is the least number of parts needed for such a partition. This type of chromatic parameter can be used to measure the effects of a controlled amount of error in our graph colourings, which can be helpful in certain applications, such as in channel assignment or in distributed computing. Clearly, $t = 0$ corresponds to the ordinary chromatic number. It is known (and fairly easy to see) that

$$\frac{\chi(G)}{t+1} \leq \chi^t(G) \leq \min\left\{\left\lceil \frac{\Delta(G)+1}{t+1} \right\rceil, \chi(G)\right\}.$$

The authors in [51] considered $\chi^t(G_{n,1/2})$ when t is allowed to increase as a function of n. In particular, the above theorem for hereditary properties does not apply in this situation. If $t = o(\log n)$, then $\chi^t(G_{n,1/2})$ has the same a.a.s. first-order behaviour as $\chi(G_{n,1/2})$, while $\chi^t(G_{n,1/2}) \sim np/t$ a.a.s. if $t = \omega(\log n)$. This implies that the above upper bound on $\chi^t(G)$ for general G is usually the right asymptotic answer for $G_{n,1/2}$. The following is a fuller and more detailed result.

Theorem 7.6 *There exists a function $\kappa = \kappa_{1/2}(\tau)$ that is continuous and strictly increasing for $\tau \in [0, \infty)$, with $\kappa_{1/2}(0) = 2/\ln 2$ and $\kappa_{1/2}(\tau) \sim 2\tau$ as $\tau \to \infty$, for which, if $t(n) = o(n)$, then a.a.s.*

$$\chi^t(G_{n,1/2}) \sim \frac{n}{\kappa_{1/2}(t/\ln n)\ln n}.$$

This result was proved by a careful study of the nearly maximum *t-stable sets* – vertex subsets that induce subgraphs of maximum degree at most t – using tools from large deviations analysis. Such tools were critical in computing both the first and second moments, especially at the threshold $t = \Theta(\log n)$. Other more precise results on both the t-stability and t-improper chromatic numbers of $G_{n,p}$ can be found in [38] and [39].

References

1. D. Achlioptas and E. Friedgut, A sharp threshold for k-colorability, *Random Structures Algorithms* **14** (1999), 63–70.
2. D. Achlioptas and C. Moore, The chromatic number of random regular graphs, *Proc. 7th International Workshop on Approximation Algorithms for Combinatorial Optimization Problems (APPROX) and 8th International Workshop on Randomization and Computation (RANDOM)*, Lecture Notes in Computer Science, Springer (2004), 219–228.
3. D. Achlioptas and A. Naor, The two possible values of the chromatic number of a random graph, *Ann. of Math. (2)* **162** (2005), 1335–1351.

4. N. Alon, Choice numbers of graphs: a probabilistic approach, *Combin. Probab. Comput.* **1** (1992), 107–114.
5. N. Alon, Restricted colorings of graphs, *Surveys in Combinatorics, 1993* (Keele) (ed. K. Walker), *London Math. Soc. Lecture Notes* **187**, Cambridge Univ. Press (1993), 1–33.
6. N. Alon, The chromatic number of random Cayley graphs, *Europ. J. Combin.* **34** (2013), 1232–1243.
7. N. Alon and M. Krivelevich, The concentration of the chromatic number of random graphs, *Combinatorica* **17** (1997), 303–313.
8. N. Alon, M. Krivelevich and B. Sudakov, List coloring of random and pseudo-random graphs, *Combinatorica* **19** (1999), 453–472.
9. N. Alon and J. Spencer, *The Probabilistic Method* (3rd edn.), with an appendix on the life and work of Paul Erdős, Wiley–Interscience Series in Discrete Mathematics and Optimization, 2008.
10. S. Ben-Shimon and M. Krivelevich, Random regular graphs of non-constant degree: concentration of the chromatic number, *Discrete Math.* **309** (2009), 4149–4161.
11. E. A. Bender and E. R. Canfield, The asymptotic number of labeled graphs with given degree sequences, *J. Combin. Theory (A)* **24** (1978), 296–307.
12. E. A. Bender, Z. Gao and N. C. Wormald, The number of labeled 2-connected planar graphs, *Electron. J. Combin.* **9**, Research Paper 43, 13 pp. (electronic), 2002.
13. B. Bollobás, A probabilistic proof of an asymptotic formula for the number of labelled regular graphs, *Europ. J. Combin.* **1** (1980), 311–316.
14. B. Bollobás, The chromatic number of random graphs, *Combinatorica* **8** (1988), 49–55.
15. B. Bollobás, *Random Graphs* (2nd edn.), Cambridge Studies in Advanced Math. **73**, 2001.
16. B. Bollobás, How sharp is the concentration of the chromatic number?, *Combin. Probab. Comput.* **13** (1) (2004), 115–117.
17. B. Bollobás and P. Erdős, Cliques in random graphs, *Math. Proc. Cambridge Philos. Soc.* **80** (1976), 419–427.
18. B. Bollobás and A. Thomason, Threshold functions, *Combinatorica* **7** (1987), 35–38.
19. B. Bollobás and A. Thomason, Generalized chromatic numbers of random graphs. *Random Structures Algorithms* **6** (1995), 353–356.
20. B. Bollobás and A. Thomason, The structure of hereditary properties and colourings of random graphs, *Combinatorica* **20** (2000), 173–202.
21. A. Braunstein, R. Mulet, A. Pagnani, M. Weigt and R. Zecchina, Polynomial iterative algorithms for coloring and analyzing random graphs, *Phys. Rev. E (3)* **68** (2003), 036702, 15.
22. G. Chapuy, É. Fusy, O. Giménez, B. Mohar and M. Noy., Asymptotic enumeration and limit laws for graphs of fixed genus, *J. Combin. Theory (A)* **118** (2011), 748–777.
23. V. Chvátal, Almost all graphs with 1.44n edges are 3-colorable, *Random Structures Algorithms* **2** (1991), 11–28.
24. A. Coja-Oghlan, Upper-bounding the k-colorability threshold by counting covers, *Electron. J. Combin.* **20** (2013), paper 32, 28.
25. A. Coja-Oghlan, C. Efthymiou and S. Hetterich, On the chromatic number of random regular graphs, ArXiv *e-prints*, Aug. 2013.
26. A. Coja-Oghlan, K. Panagiotou and A. Steger, On the chromatic number of random graphs, *J. Combin. Theory (B)* **98** (2008), 980–993.
27. A. Coja-Oghlan and D. Vilenchik, Chasing the k-colorability threshold, ArXiv *e-prints*, Apr. 2013.
28. C. Cooper, A. Frieze, B. Reed and O. Riordan, Random regular graphs of non-constant degree: independence and chromatic number, *Combin. Probab. Comput.* **11** (2002), 323–341.

29. M. DeVos, K. Kawarabayashi and B. Mohar, Locally planar graphs are 5-choosable, *J. Combin. Theory (B)* **98** (2008), 1215–1232.
30. J. Díaz, A. C. Kaporis, G. D. Kemkes, L. M. Kirousis, X. Pérez-Giménez and N. C. Wormald, On the chromatic number of a random 5-regular graph, *J. Graph Theory* **61** (2009), 157–191.
31. C. Dowden, *Uniform Random Planar Graphs with Degree Constraints*, Ph.D. thesis, University of Oxford, 2008, *http://ora.ox.ac.uk/objects/uuid%3Af8a9afe3-30ad-4672-9a6c-4fb9ac9af041*.
32. C. Dowden, The evolution of uniform random planar graphs, *Electron. J. Combin.* **17** (2010), R7, 20 pp.
33. M. Dyer, A. Frieze and C. Greenhill, On the chromatic number of a random hypergraph, ArXiv *e-prints*, Aug. 2012.
34. P. Erdős, Graph theory and probability, *Canad. J. Math.* **11** (1959), 34–38.
35. P. Erdős and S. Fajtlowicz, On the conjecture of Hajós, *Combinatorica* **1** (1981), 141–143.
36. P. Erdős and A. Rényi, On the evolution of random graphs, *Magyar Tud. Akad. Mat. Kutató Int. Közl.* **5** (1960), 17–61.
37. P. Erdős and R. J. Wilson, On the chromatic index of almost all graphs, *J. Combin. Theory (B)* **23** (1977), 255–257.
38. N. Fountoulakis, R. J. Kang and C. McDiarmid, The t-stability number of a random graph, *Electron. J. Combin.* **17** (2010), R59, 29 pp.
39. N. Fountoulakis, R. J. Kang and C. McDiarmid, Largest sparse subgraphs of random graphs, *Europ. J. Combin.* **35** (2014), 232–244.
40. E. Friedgut, Sharp thresholds of graph properties, and the k-SAT problem, with an appendix by Jean Bourgain, *J. Amer. Math. Soc.* **12** (1999), 1017–1054.
41. A. Frieze, On the independence number of random graphs, *Discrete Math.* **81** (1990), 171–175.
42. A. Frieze, B. Jackson, C. McDiarmid and B. Reed, Edge-colouring random graphs, *J. Combin. Theory (B)* **45** (1988), 135–149.
43. A. Frieze and T. Łuczak, On the independence and chromatic numbers of random regular graphs, *J. Combin. Theory (B)* **54** (1992), 123–132.
44. A. Frieze and E. Vigoda, A survey on the use of Markov chains to randomly sample colourings, *Combinatorics, Complexity, and Chance* (eds. G. Grimmett and C. McDiarmid), Oxford Lecture Ser. Math. Appl. **34**, Oxford Univ. Press (2007), 53–71.
45. O. Giménez and M. Noy, Asymptotic enumeration and limit laws of planar graphs, *J. Amer. Math. Soc.* **22** (2009), 309–329.
46. B. Green, On the chromatic number of random Cayley graphs, ArXiv *e-prints*, Aug. 2013.
47. B. Green and R. Morris, Counting sets with small sumset and applications, ArXiv *e-prints*, May 2013.
48. G. Grimmett and C. McDiarmid, On colouring random graphs, *Math. Proc. Cambridge Philos. Soc.* **77** (1975), 313–324.
49. S. Janson, T. Łuczak and A. Ruciński, *Random Graphs*, Wiley–Interscience Series in Discrete Math. and Optimization, 2000.
50. M. Kang and T. Łuczak, Two critical periods in the evolution of random planar graphs, *Trans. Amer. Math. Soc.* **364** (2012), 4239–4265.
51. R. J. Kang and C. McDiarmid, The t-improper chromatic number of random graphs, *Combin. Probab. Comput.* **19** (2010), 87–98.
52. R. M. Karp, The probabilistic analysis of some combinatorial search algorithms, *Algorithms and Complexity* (Carnegie-Mellon Univ., 1976) (ed. J. F. Traub), Academic Press (1976), 1–19.
53. G. Kemkes, X. Pérez-Giménez and N. C. Wormald, On the chromatic number of random d-regular graphs, *Adv. Math.* **223** (2010), 300–328.

54. J. H. Kim and V. H. Vu, Sandwiching random graphs: universality between random graph models, *Adv. Math.* **188** (2004), 444–469.
55. M. Krivelevich, The choice number of dense random graphs, *Combin. Probab. Comput.* **9** (2000), 19–26.
56. M. Krivelevich and B. Patkós, Equitable coloring of random graphs, *Random Structures Algorithms* **35** (2009), 83–99.
57. M. Krivelevich, B. Sudakov, V. H. Vu and N. C. Wormald, Random regular graphs of high degree, *Random Structures Algorithms* **18** (2001), 346–363.
58. M. Krivelevich, B. Sudakov, V. H. Vu and N. C. Wormald, On the probability of independent sets in random graphs, *Random Structures Algorithms* **22** (2003), 1–14.
59. T. Łuczak, The chromatic number of random graphs, *Combinatorica* **11** (1991), 45–54.
60. T. Łuczak, A note on the sharp concentration of the chromatic number of random graphs, *Combinatorica* **11** (1991), 295–297.
61. E. Marchant and A. Thomason, The structure of hereditary properties and 2-coloured multigraphs, *Combinatorica* **31** (2011), 85–93.
62. D. W. Matula, On the complete subgraphs of a random graph, *Proc. Second Chapel Hill Conf. on Combinatorial Mathematics and its Applications* (Univ. of North Carolina, 1970), Univ. of North Carolina, Chapel Hill (1970), 356–369.
63. D. W. Matula, The employee party problem, *Notices Amer. Math. Soc.* **19** (1972), A–382.
64. D. W. Matula, Expose-and-merge exploration and the chromatic number of a random graph, *Combinatorica* **7** (1987), 275–284.
65. D. W. Matula and L. Kučera, An expose-and-merge algorithm and the chromatic number of a random graph, *Random Graphs '87* (Poznań, 1987) (eds. M. Karoński, J. Jaworski and A. Ruciński), Wiley (1990), 175–187.
66. C. McDiarmid, Colouring random graphs badly, *Graph Theory and Combinatorics* (Open Univ., Milton Keynes, 1978) (ed. R. J. Wilson), Research Notes in Math. **34**, Pitman (1979), 76–86.
67. C. McDiarmid, On the chromatic number of random graphs, *Random Structures Algorithms* **1** (1990), 435–442.
68. C. McDiarmid, Random channel assignment in the plane, *Random Structures Algorithms* **22** (2003), 187–212.
69. C. McDiarmid, Random graphs on surfaces, *J. Combin. Theory (B)* **98** (2008), 778–797.
70. C. McDiarmid, Random graphs from a minor-closed class, *Combin. Probab. Comput.* **18** (2009), 583–599.
71. C. McDiarmid, Random graphs from a weighted minor-closed class, *Electron. J. Combin.* **20** (2013), R52, 39 pp.
72. C. McDiarmid and T. Müller, On the chromatic number of random geometric graphs, *Combinatorica* **31** (2011), 423–488.
73. C. McDiarmid and B. Reed, On total colourings of graphs, *J. Combin. Theory (B)* **57** (1993), 122–130.
74. C. McDiarmid, A. Steger and D. Welsh, Random planar graphs, *J. Combin. Theory (B)* **93** (2005), 187–205.
75. M. Molloy and B. Reed, *Graph Colouring and the Probabilistic Method, Algorithms and Combinatorics* **23**, Springer-Verlag, 2002.
76. R. Mulet, A. Pagnani, M. Weigt and R. Zecchina, Coloring random graphs, *Phys. Rev. Lett.* **89** (2002), 268701, 4.
77. T. Müller, Two-point concentration in random geometric graphs, *Combinatorica* **28** (2008), 529–545.

78. K. Panagiotou and A. Steger, A note on the chromatic number of a dense random graph, *Discrete Math.* **309** (2009), 3420–3423.
79. M. Penrose, *Random Geometric Graphs*, Oxford Studies in Probability **5**, Oxford University Press, 2003.
80. M. P. Rombach and A. Scott, Equitable colourings of random graphs, in preparation, 2014.
81. E. R. Scheinerman, Generalized chromatic numbers of random graphs, *SIAM J. Discrete Math.* **5** (1992), 74–80.
82. E. Shamir and J. Spencer, Sharp concentration of the chromatic number on random graphs $G_{n,p}$, *Combinatorica* **7** (1987), 121–129.
83. L. Shi and N. C. Wormald, Colouring random 4-regular graphs, *Combin. Probab. Comput.* **16** (2007), 309–344.
84. L. Shi and N. C. Wormald, Colouring random regular graphs, *Combin. Probab. Comput.* **16** (2007), 459–494.
85. C. Thomassen, Five-coloring maps on surfaces, *J. Combin. Theory (B)* **59** (1993), 89–105.
86. V. H. Vu, On some simple degree conditions that guarantee the upper bound on the chromatic (choice) number of random graphs, *J. Graph Theory* **31** (1999), 201–226.
87. N. C. Wormald, Differential equations for random processes and random graphs, *Ann. Appl. Probab.* **5** (1995), 1217–1235.
88. N. C. Wormald, Models of random regular graphs, *Surveys in Combinatorics, 1999* (Canterbury) (eds. J. D. Lamb and D. A. Preece), *London Math. Soc. Lecture Notes* **267**, Cambridge Univ. Press (1999), 239–298.

11

Hypergraph colouring

CSILLA BUJTÁS, ZSOLT TUZA and VITALY VOLOSHIN

We discuss the colouring theory of finite set systems. This is not merely an extension of results from collections of 2-element sets (graphs) to larger sets. The wider structure (hypergraphs) offers many interesting new kinds of problems, which either have no analogues in graph theory or become trivial when we restrict them to graphs.

1. Introduction

In this introductory section we give the most important definitions required to study hypergraph colouring, and briefly survey the half-century history of this topic. For more details on the material of Sections 1 and 2 we refer to Berge [8], Zykov [76] and Duchet [27].

Let $V = \{v_1, v_2, \ldots, v_n\}$ be a finite set of elements called *vertices*, and let $\mathcal{E} = \{E_1, E_2, \ldots, E_m\}$ be a family of subsets of V called *edges* or *hyperedges*. The pair $\mathcal{H} = (V, \mathcal{E})$ is called a *hypergraph* with vertex-set $V = V(\mathcal{H})$ and edge-set $\mathcal{E} = \mathcal{E}(\mathcal{H})$. The hypergraph $\mathcal{H} = (V, \mathcal{E})$ is sometimes called a *set system*. If each edge of a hypergraph contains precisely two vertices, then it is a graph. As in graph theory, the number $|V| = n$ is called the *order* of the hypergraph. Edges with fewer than two elements are usually allowed, but will be disregarded here. Thus, throughout this chapter we assume that each edge $E \in \mathcal{E}$ contains at least two vertices, unless stated explicitly otherwise. Edges that coincide are called *multiple edges*.

In a hypergraph, two vertices are said to be *adjacent* if there is an edge containing both of these vertices. The adjacent vertices are sometimes called *neighbours* of each other, and the set of neighbours of a given vertex v is called the (open) *neighbourhood* $N(v)$ of v. If $v \in E$, then the vertex v and the edge E are *incident* with each other. For an edge E, the number $|E|$ is called the *size* or *cardinality* of E.

If every edge of $\mathcal{H}$ is of size r, then $\mathcal{H}$ is called an *r-uniform hypergraph;* evidently, a simple graph is a 2-uniform hypergraph. For $2 \leq r \leq n$, we define the *complete r-uniform hypergraph* to be the hypergraph $\mathcal{K}_n^r = (V, \mathcal{E})$ for which $|V| = n$ and $\mathcal{E}$ is the family of all subsets of V of size r. Thus, the complete 2-uniform hypergraph $\mathcal{K}_n^2$ is the complete graph K_n. Traditionally, for a set X and an integer $k \geq 1$, the family of all k-element subsets of X is denoted by $\binom{X}{k}$. With this notation we have $\mathcal{K}_n^r = (V, \binom{V}{r})$.

A subset of vertices $S \subseteq V$ is called *stable* (or *independent*) if no edge of $\mathcal{H}$ is a subset of S. The maximum cardinality of a stable set is termed the *stability number* (or *independence number*) $\alpha(\mathcal{H})$ of $\mathcal{H}$.

A *proper λ-colouring* of a hypergraph $\mathcal{H} = (V, \mathcal{E})$ is a mapping $c : V \to \{1, 2, \ldots, \lambda\}$ for which every edge $E \in \mathcal{E}$ has at least two vertices of different colours. The number of proper λ-colourings of $\mathcal{H}$ is a polynomial in λ; it is denoted by $P(\mathcal{H}, \lambda)$ and is called the *chromatic polynomial.* The minimum value of λ for which there exists a proper λ-colouring of a hypergraph $\mathcal{H}$ is called the *chromatic number* of $\mathcal{H}$, denoted by $\chi(\mathcal{H})$. A hypergraph $\mathcal{H}$ is *k-colourable* if $\chi(\mathcal{H}) \leq k$, and is *k-chromatic* if $\chi(\mathcal{H}) = k$. When $\chi(\mathcal{H}) \leq 2$, the hypergraph $\mathcal{H}$ is called *bicolourable*; in parts of the literature the term 'bipartite' is also used.

As for graphs, proper colourings generate partitions of the vertex-set into a number of stable (monochromatic) non-empty subsets called *colour classes*, with as many classes as the number of colours actually used.

A graph $G = (V, E)$ is called a *host graph* of a hypergraph $\mathcal{H} = (V, \mathcal{E})$ if each edge of $\mathcal{H}$ induces a connected subgraph in G. Some structural subclasses of hypergraphs are identified referring to this notion. In particular, *interval hypergraphs, hyperstars, hypertrees* and *circular hypergraphs* are defined as hypergraphs with a host graph that is a path, star, tree or cycle. Since 1-element edges are excluded, a 'hyperstar' means that the intersection of all edges is non-empty.

Historical background

The above definitions represent natural generalizations of the respective concepts from graph theory. It was realized in the early 1960s that many graph-theoretical methods can be successfully extended to the more general structure of set systems. The term 'hypergraph' was first suggested at the seminar series of Claude Berge [9].

The colouring of hypergraphs started in 1966 when Erdős and Hajnal [31] introduced the notions of colouring and the chromatic number of a hypergraph, and obtained the first important results. In particular, they defined the *colouring number,* whose graph analogue is also called the 'Szekeres–Wilf number' and is equal to the 'degeneracy' + 1; this became an important tool in studying the chromatic number of a hypergraph. From the definition, in a proper colouring *no edge is allowed to be monochromatic*. In the literature these colourings are sometimes called *weak colourings*.

This generalization of graph colourings initiated a wide area of further research. First, some old problems in set systems were formulated as colouring problems and many results in graph colouring were extended to hypergraphs. For example, using a natural generalization of the degree of a vertex, the classic theorem of Brooks (see Chapter 2) was shown to hold for hypergraphs (see [8], [76]). Another generalization of the degree of a vertex allowed Tomescu [61] to extend to hypergraphs some advanced upper bounds on the chromatic number from graph results of Welsh and Powell [71]. Berge [8] introduced the β-degree of a vertex and showed that by using it many previous bounds follow, including those related to the chromatic number. In addition, it provided an algorithm for finding a bound on the chromatic number of a hypergraph.

It is well known that bipartite graphs play a significant role in graph theory. A simple but very important characterization of them is the one by König, that a graph is bipartite if and only if it contains no odd cycles. For hypergraphs, there is no such simple condition for bicolourability, although some criteria in terms of transversal hypergraphs were found by Berge [8]. In the next section we present some further important sufficient conditions on bicolourability. Also, the bicolourable hypergraphs were the first application of the widely used Lovász local lemma [32]. Many more classes of hypergraphs generalizing bipartite graphs have been investigated; see [8] for details.

It is generally accepted that graph colouring started in 1852 with the question of whether any map can be coloured using just four colours (see Chapter 1). It eventually led to the concept of a planar graph. In 1974 Zykov [76] introduced the notion of a planar hypergraph and obtained the first results on them. He also generalized the notions of connection–contraction and of the Hadwiger number of a graph (see Chapter 4). The latter makes it possible to consider hypergraph versions of problems related to Hadwiger's conjecture.

Sometimes a hypergraph approach permits problems to be solved more easily than a graph approach. A celebrated example is Berge's weak perfect graph conjecture, which was first proved by Lovász [46] using so-called normal hypergraphs.

Many generalizations of graph colourings to hypergraphs are more specific and more restrictive than the Erdős–Hajnal approach. We discuss a few of these in the following sections. However, there is one thing that they have in common: they all have *graph analogues*, because they all emerged from a single concept – graph colourings: this means that they all exclude monochromatic edges. As we see next, a new approach that is even more general than the Erdős–Hajnal generalization allows us to introduce colourings that have *no analogues for graph colouring*. In part, this is because, by assumption, the possibility of monochromatic edges is not excluded.

We may observe further that the Erdős–Hajnal classic concept of hypergraph colouring is asymmetric: with the chromatic number as its central notion, this theory focuses on the *minimum* number of colours, while the *maximum* number of colours has no mathematical interest since a totally multicoloured vertex-set is always feasible. This asymmetry of classical hypergraph colouring was inherited from graph colouring.

A different direction for research concerns the maximum number of colours, excluding polychromatic edges. In hypergraph terms this was proposed by Berge and was first investigated in detail by Sterboul [59] in the early 1970s. We may say that the so-called 'sub-Ramsey numbers' and 'rainbow numbers' also fit here, but those studies went along a different track and were not investigated as widely. We discuss the hypergraph version under the name '$\mathcal{C}$-colouring'. Although the first published paper on this is more than forty years old, it gained its importance only two decades ago when a more general model was born.

In 1993 Voloshin [66], [67] introduced the concept of a mixed hypergraph colouring, which eliminated the above asymmetry and opened up an entirely new direction of research. Instead of $\mathcal{H} = (V, \mathcal{E})$, the basic idea is to consider a structure $\mathcal{H} = (V, \mathcal{C}, \mathcal{D})$, termed a *mixed hypergraph*, with *two* families of subsets called $\mathcal{C}$-edges and $\mathcal{D}$-edges. By definition, a *proper* λ*-colouring* of a mixed hypergraph $\mathcal{H} = (V, \mathcal{C}, \mathcal{D})$ is a mapping $c : V \to \{1, 2, \ldots, \lambda\}$ for which two conditions hold:

- every $C \in \mathcal{C}$ has at least two vertices of a $\mathcal{C}$ommon colour
- every $D \in \mathcal{D}$ has at least two vertices of $\mathcal{D}$ifferent colours.

Formally, this is a combination of classical proper colourings and $\mathcal{C}$-colourings, but mixed hypergraphs were introduced independently.

The concept of a mixed hypergraph colouring has led to the discovery of new principal properties of colourings that do not exist in classical graph and hypergraph colourings: uncolourability in its most general setting, chromatic polynomials of degree less than n, phantom vertices, gaps in the chromatic spectrum, and hypergraph perfection, to name just a few. Moreover, it led to the introduction and study of new classes of hypergraphs, and further types of constraints that are imposed on the edges of a hypergraph when we colour the vertices. It also brings a new look at some classical graph and hypergraph colouring problems, such as colouring planar hypergraphs, chromatic polynomials, edge-colourings, colouring and probability, colouring algorithms, and so on. Finally, it has generated, and continues to generate, a wide variety of combinatorial problems and applications that formerly had no analogues. (The newest, and perhaps most unexpected, application of mixed hypergraphs is the modelling of problems arising in distributed computing and cybersecurity – see [37].) A significant number of subsequent new ideas, results and publications have led to a situation where colouring theory as a whole is taking a new shape.

In the next section we describe a collection of significant results in classical hypergraph colouring, where monochromatic edges are excluded. We then consider $\mathcal{C}$-colourings, which may be viewed as the counterpart on excluding multicoloured edges. In later sections we survey some of the new trends which, in their turn, give rise to new challenges in this fast-developing area. In particular, Section 4 is devoted to mixed hypergraphs, which opened up this new dimension of hypergraph colouring. For further information on these, we refer to the research monograph [68] and the regularly updated website [69]. Some results and many open problems are collected in the surveys [64] and [2]. Finally, we discuss an even more general colouring model of hypergraphs, introduced in 2007 by Bujtás and Tuza [15], [13]. In that setting, lower and upper bounds on the number of different colours, and those on the size of the largest monochromatic vertex-subset, can be prescribed for each edge independently. A hypergraph with these colouring restrictions is called a *stably bounded hypergraph*.

2. Proper vertex- and edge-colourings

In this section, we survey the most significant results in classical hypergraph colouring, their relationship with graph colouring, and some further general hypergraph concepts and parameters.

We mention first that, for each $k \geq 2$, it is NP-complete to decide in general whether a hypergraph is k-colourable. It remains NP-complete for $k = 2$ on 3-uniform hypergraphs (see Lovász [47]). For all $k \geq 2$ and $r \geq 2$, it is NP-complete also on r-uniform hypergraphs in which any two edges share at most one vertex, except for bipartite graphs when $k = r = 2$ (see Phelps and Rödl [55]).

In fact, the decision problem for bicolourability had already been proved to be NP-hard in 1972 by the following result of Woodall [73].

Theorem 2.1 *Let G be a graph, and construct a hypergraph $\mathcal{H}$ as follows: each vertex of $\mathcal{H}$ corresponds to an edge of G, and each odd cycle of G forms an edge of $\mathcal{H}$. Then G is 4-colourable if and only if $\mathcal{H}$ is 2-colourable.*

For *uniform* hypergraphs, a sufficient condition for bicolourability can be given in terms of the number of edges. Erdős [29] observed that every r-uniform hypergraph with at most 2^{r-1} edges is bicolourable: in fact, assigning one of two colours to each vertex randomly and independently, we obtain a proper 2-colouring with positive probability. This bound on the number of edges is not tight and has been improved a couple of times. The current record is held by Radhakrishnan and Srinivasan [56].

Theorem 2.2 *For sufficiently large values of r, every r-uniform hypergraph with at most $0.7 \times 2^r \sqrt{r/\ln r}$ edges is bicolourable, and can be properly 2-coloured by a polynomial-time algorithm.*

On the other hand, for every $r \geq 2$, non-bicolourable r-uniform hypergraphs exist with fewer than $r^2\, 2^{r+1}$ edges (see Erdős [30]).

Hypergraphs that are r-uniform and r-regular are interesting combinatorial objects – for instance, they describe the neighbourhood structures (both open and closed) of regular graphs. Concerning their bicolourability, Alon and Bregman [1] proved the sufficient condition $r \geq 8$, and it was recently proved by Henning and Yeo [36] that $r \geq 4$ is sufficient. (The Fano plane shows that the assertion is not valid for $r = 3$.)

A *cycle of length* s ($s \geq 2$) in a hypergraph $\mathcal{H} = (V, \mathcal{E})$ is a sequence $v_0E_0v_1E_1v_2 \ldots v_{s-1}E_{s-1}v_s$ with $v_0 = v_s$, consisting of s distinct vertices and s distinct edges, such that $v_i, v_{i+1} \in E_i$ for all $i = 0, 1, \ldots, s-1$. Depending on the parity of s, the cycle is called *odd* or *even*. Moreover, the odd cycle $v_0E_0v_1E_1v_2 \ldots v_{s-1}E_{s-1}v_s$ is an *anti-Sterboul cycle* if any two non-consecutive edges are disjoint and if $|E_i \cap E_{i+1}| = 1$ for every $i = 0, 1, \ldots, s-2$ (but not necessarily for $E_{s-1} \cap E_0$). A *Sterboul hypergraph* is a hypergraph containing no anti-Sterboul cycle.

The next two theorems give sufficient conditions for bicolourability in terms of conditions imposed on cycles. The first one, due to Fournier and Las Vergnas [33], was the deepest theorem of this kind for several decades. Its formulation with 3-critical hypergraphs is taken from Zykov [76]. (As for graphs, a k-chromatic hypergraph is *critical* if the removal of any edge decreases the chromatic number.)

Theorem 2.3 *If every cycle of odd length in a hypergraph $\mathcal{H}$ contains three edges with a vertex in common, then $\mathcal{H}$ is bicolourable. Equivalently, every 3-chromatic critical hypergraph contains an odd cycle in which no three edges have a vertex in common.*

This theorem was recently strengthened by Défossez [24], who proved an old conjecture of Sterboul.

Theorem 2.4 *Every Sterboul hypergraph is bicolourable.*

There are also lower bounds on the number of edges in critical hypergraphs. The first one was proved by Seymour [58] in 1974; the second is more recent, by Kostochka and Stiebitz [40].

Theorem 2.5 *If a hypergraph on n vertices is 3-chromatic critical, then it has at least n edges.*

Theorem 2.6 *If a hypergraph on n vertices is $(k+1)$-chromatic critical and contains no 2-element edges, then the number of its edges is at least $(k - 3/\sqrt[3]{k})\, n$.*

Let $\mathcal{H} = (V, \mathcal{E})$ be a hypergraph, and let T be a subset of V. Then the hypergraph $\mathcal{H}/T = (T, \mathcal{E}')$ is the *restriction of $\mathcal{H}$ to T*, where $\mathcal{E}'$ consists of all non-empty intersections of edges of $\mathcal{E}$ with T. A hypergraph is said to be *balanced* if every odd cycle has an edge that contains three vertices of the cycle. Berge [7] proved the following characterization.

Theorem 2.7 *A hypergraph is balanced if and only if every restriction is 2-colourable.*

Let $\mathcal{H} = (V, \mathcal{E})$ be a hypergraph. The *chromatic index* $\chi'(\mathcal{H})$ of $\mathcal{H}$ is the least number of colours necessary to colour the edges of $\mathcal{H}$ in such a way that any two intersecting edges have distinct colours.

For a vertex v, the number of all edges containing v is the *degree of* v, denoted by $d(v)$. The maximum vertex-degree in a hypergraph $\mathcal{H} = (V, \mathcal{E})$ is $\Delta(\mathcal{H}) = \max_{v \in V} d(v)$. For any hypergraph $\mathcal{H}$, it is clear that $\chi'(\mathcal{H}) \geq \Delta(\mathcal{H})$. A hypergraph $\mathcal{H}$ has the *edge-colouring property* if $\chi'(\mathcal{H}) = \Delta(\mathcal{H})$. In this connection, the following famous result was proved by Baranyai [6].

Theorem 2.8 *If n is a multiple of r, then $\chi'(\mathcal{K}_n^r) = \Delta(\mathcal{K}_n^r) = \binom{n-1}{r-1}$.*

In a hypergraph $\mathcal{H} = (V, \mathcal{E})$, for any subfamily $\mathcal{F} \subseteq \mathcal{E}$ we call the hypergraph $\mathcal{H}_{\mathcal{F}} = (V, \mathcal{F})$ the *partial subhypergraph* of $\mathcal{H}$. We say that a hypergraph $\mathcal{H}$ is *normal* if every partial subhypergraph of $\mathcal{H}$ satisfies the edge-colouring property. The following result was proved by Fournier and Las Vergnas [33].

Theorem 2.9 *Every normal hypergraph is bicolourable.*

A subset $T \subseteq V$ is called a *transversal* of a hypergraph $\mathcal{H} = (V, \mathcal{E})$ if $|T \cap E| \geq 1$ for each edge $E \in \mathcal{E}$. The cardinality of a minimum transversal is denoted by $\tau(\mathcal{H})$. In a hypergraph $\mathcal{H}$, a set of edges that pairwise have no vertices in common is called a *matching*. The maximum size of a matching (over all matchings) is denoted by $\nu(\mathcal{H})$.

Since any matching is a set of pairwise non-intersecting edges, any transversal must have at least one vertex from each edge of the matching. This implies that $\tau(\mathcal{H}) \geq \nu(\mathcal{H})$ for any hypergraph $\mathcal{H}$. We say that $\mathcal{H}$ satisfies the *König property* if $\tau(\mathcal{H}) = \nu(\mathcal{H})$. The next theorem, due to Lovász [46], is sometimes called the *perfect graph theorem* because it implies that the complement of a perfect graph is perfect. The latter statement was formerly known as the *weak perfect graph conjecture*, formulated by Claude Berge in the early 1960s (see Chapter 7).

Theorem 2.10 *A hypergraph $\mathcal{H}$ is normal if and only if every partial subhypergraph $\mathcal{H}'$ has the König property.*

Besides normal and balanced hypergraphs, there are many other classes of hypergraphs that are either bicolourable or generalize the bipartite graphs: hypergraphs without odd cycles, unimodular hypergraphs, hypertrees (also called arboreal hypergraphs), Mengerian hypergraphs, paranormal hypergraphs, and others. The hierarchy of these and further related classes is exhibited in [8, p.163]. Some of them have min-max properties arising from bipartite graphs and are used in polyhedral optimization (see [8], [27]).

It may appear that increasing both the size of edges and the length of the shortest cycle decreases the chromatic number of a hypergraph and eventually makes it bicolourable. This is not true: there exist high-chromatic hypergraphs for any values of these parameters. The following theorem was first proved by Erdős and Hajnal [31], and constructions were given later by Lovász [45] and by Nešetřil and V. Rödl [52].

Theorem 2.11 *For any integers $r, s, t \geq 2$, there exists an r-uniform hypergraph $\mathcal{H}$ with no cycles shorter than s, for which $\chi(\mathcal{H}) \geq t$.*

For a hypergraph $\mathcal{H} = (V, \mathcal{E})$, we define the *bipartite (König) representation* of $\mathcal{H}$ to be the bipartite graph $B(\mathcal{H})$ with partite sets V and $\mathcal{E}$, where a vertex $v \in V$ is adjacent to a vertex $E \in \mathcal{E}$ in $B(\mathcal{H})$ if and only if the vertex $v \in V$ is incident with the edge $E \in \mathcal{E}$ in $\mathcal{H}$.

A hypergraph $\mathcal{H} = (V, \mathcal{E})$ is said to be *planar* if $B(\mathcal{H})$ is a planar graph (see [68], [76]). This means that any planar hypergraph can be drawn in the plane in the following way: vertices are points; each edge is a closed curve whose interior region contains the points of the edge, and any two edges intersect only at small neighbourhoods of their common vertices. The remaining connected regions of the plane form the faces of such a plane embedding. Planar graphs are special cases of planar hypergraphs: we can replace the curves corresponding to graph edges with closed curves encircling the adjacent vertices.

We next cite two results, the first by Bulitko and the second by Burshtein and Kostochka [76, p. 138].

Theorem 2.12 *The four-colour theorems for planar graphs and for planar hypergraphs are equivalent.*

Theorem 2.13 *If a planar hypergraph contains at most one edge of size* 2, *then* $\chi(\mathcal{H}) \leq 2$.

There are several types of more restrictive hypergraph colourings that are regularly encountered in the literature (see [8], [76], [68]). Here we list some of them.

Strong colouring

A *strong* λ*-colouring* of $\mathcal{H}$ is a partition of V into λ stable sets S_i $(i = 1, 2, \ldots, \lambda)$ such that the inequality $|E_j \cap S_i| \leq 1$ holds for each edge E_j and for every i. The *strong chromatic number* $\gamma(\mathcal{H})$ is the smallest number λ for which there exists a strong λ-colouring of $\mathcal{H}$. It follows that $\gamma(\mathcal{H}) \geq \chi(\mathcal{H})$, because every strong colouring is also a weak colouring. Note that the strong and weak colourings coincide when $\mathcal{H}$ is a graph. Also, $\gamma(\mathcal{H})$ is the chromatic number of the 2*-section graph* of $\mathcal{H}$ – that is, the graph obtained from $\mathcal{H}$ on replacing each edge E_j by the complete graph with vertex-set E_j.

Equitable colouring

An *equitable* λ*-colouring* of $\mathcal{H}$ is a partition of V into λ stable sets S_i $(i = 1, 2, \ldots, \lambda)$ such that the following inequalities hold for each edge E_j and for each i:

$$\lfloor |E_j|/\lambda \rfloor \leq |E_j \cap S_i| \leq \lceil |E_j|/\lambda \rceil .$$

Good colouring

A *good* λ*-colouring* of $\mathcal{H}$ is a partition of V into λ stable sets S_i $(i = 1, 2, \ldots, \lambda)$ such that each edge E_j has $\min\{|E_j|, \lambda\}$ colours. If $\lambda \geq \max_{j \in J} |E_j|$, then a good

λ-colouring is a strong λ-colouring. For each λ, any equitable λ-colouring is also a good λ-colouring.

I-regular colouring

For each edge E_j in $\mathcal{H}$, let a_j and b_j be integers with $0 \le a_j \le b_j < |E_j|$. An *I-regular λ-colouring* of $\mathcal{H}$ is a partition of V into λ stable sets S_i $(i = 1, 2, \ldots, \lambda)$, in such a way that $a_j \le |E_j \cap S_i| \le b_j$ for each edge E_j and each $i = 1, 2, \ldots, \lambda$. Notice that each weak colouring is an I-regular colouring if $a_j = 0$ and $b_j = |E_j| - 1$, that every strong colouring is an I-regular colouring with $a_j = 0$ and $b_j = 1$, and that every equitable λ-colouring is an I-regular colouring with $a_j = \lfloor |E_j|/\lambda \rfloor$ and $b_j = \lceil |E_j|/\lambda \rceil$.

List-colouring

As for graphs, list-colouring and related notions (see Chapter 6) can also be defined for hypergraphs; for example, it is also true that the choice number of any hypergraph is equal to its fractional chromatic number (see [48]). Since a hypergraph can represent a set system of any generality, this theorem implies the corresponding equality for many kinds of colourings, including edge- and total-colourings, generalized colourings defined in terms of hereditary properties, and so on.

There are further connections between hypergraph colourings and extremal problems, Kneser's problem, Ramsey-type problems, etc., and many applications. These issues are not discussed here; for information we refer to [8] and [27].

3. $\mathcal{C}$-colourings

The classical proper colouring of hypergraphs discussed in the previous section requires that two vertices have different colours inside each edge. A *$\mathcal{C}$-colouring* of a hypergraph applies an opposite colouring constraint, prescribing the presence of two vertices with a $\mathcal{C}$ommon colour inside each edge. Assigning the same colour to all vertices always yields a $\mathcal{C}$-colouring. Thus, the essential parameter here is not the minimum possible number of colours, but the maximum. This is called the *upper chromatic number* of a hypergraph $\mathcal{H}$ and is denoted by $\bar{\chi}(\mathcal{H})$.

Historically, the upper chromatic number appears in works of Sterboul [59], Berge [8] (who called $\bar{\chi}(\mathcal{H})+1$ the 'cochromatic number'), Voloshin [67] and other authors. Recent results are surveyed in [20].

The following are some basic facts concerning $\mathcal{C}$-colourings.

- The inequality $\bar{\chi}(\mathcal{H}) \le \alpha(\mathcal{H})$ holds for every hypergraph: indeed, select one vertex from each colour class of a $\bar{\chi}$-colouring of $\mathcal{H}$. The $\bar{\chi}$-element vertex-set obtained in this way is independent, as otherwise we would have a polychromatic edge.

- Denote by $\tau_2(\mathcal{H})$ the minimum cardinality of a '2-*transversal*', a vertex-set $T \subseteq V$ for which $|T \cap E| \geq 2$ for each $E \in \mathcal{E}$. We obtain a $\mathcal{C}$-colouring of $\mathcal{H}$ by assigning colour 1 to each element of T and colouring the remaining vertices with colours $2, 3, \ldots, n - \tau_2 + 1$ pairwise differently. This proves that $\bar{\chi}(\mathcal{H}) \geq n - \tau_2(\mathcal{H}) + 1$.
- Let $\mathcal{S} = (V, \mathcal{E})$ be a hyperstar, so the intersection $\bigcap_{E \in \mathcal{E}} E$ is not empty. In addition, assume that $|E| \geq 3$ for all $E \in \mathcal{E}$, and consider a vertex z contained in all edges. Then on deleting z from each $E \in \mathcal{E}$ we obtain a hypergraph $\mathcal{S}^-$ for which $\bar{\chi}(\mathcal{S}) = \alpha(\mathcal{S}^-) + 1$. Moreover, if $\mathcal{S}$ is 3-uniform, then $\mathcal{S}^-$ is simply a graph. It is thus easy to see that the determination of the upper chromatic number is an $\mathcal{NP}$-hard problem, even for the class of 3-uniform hypertrees. Another consequence of this simple construction is that the difference $\alpha(\mathcal{S}) - \bar{\chi}(\mathcal{S})$ can be arbitrarily large, since $\alpha(\mathcal{S}) = |V| - 1$.

In connection with the upper chromatic number, Voloshin [67] introduced the notion of a *$\mathcal{C}$-perfect hypergraph*. Analogously to the well-known definition of a perfect graph, we start with an inequality that is valid for all objects and prescribe that it hereditarily holds with equality. More precisely, let $\mathcal{H} = (V, \mathcal{E})$ be a hypergraph, and let T be a subset of V. Then the hypergraph $\mathcal{H}' = (T, \mathcal{E}')$ is called the *induced subhypergraph* if $\mathcal{E}'$ consists of all edges of $\mathcal{E}$ that lie entirely in T. A hypergraph $\mathcal{H}$ is called *$\mathcal{C}$-perfect* if $\bar{\chi}(\mathcal{H}') = \alpha(\mathcal{H}')$ for every induced subhypergraph $\mathcal{H}' \subseteq \mathcal{H}$; otherwise, $\mathcal{H}$ is *$\mathcal{C}$-imperfect*. Finally, $\mathcal{C}$-imperfect hypergraphs, all of whose proper induced subhypergraphs are $\mathcal{C}$-perfect, are called *minimally $\mathcal{C}$-imperfect*.

The following are some basic facts concerning $\mathcal{C}$-perfect and $\mathcal{C}$-imperfect hypergraphs.

- A typical example of a $\mathcal{C}$-perfect hypergraph is a hyperstar $\mathcal{H}$ with $|\bigcap_{E \in \mathcal{E}} E| \geq 2$. Indeed, if $\mathcal{H}$ is of order n, then $\alpha(\mathcal{H}) = \bar{\chi}(\mathcal{H}) = n - 1$, and every induced subhypergraph of it is either edgeless or has at least two vertices contained in the common intersection of the edges.
- A *monostar* $\mathcal{H}$ is a hyperstar with $|\bigcap_{E \in \mathcal{E}} E| = 1$. It is always $\mathcal{C}$-imperfect, since $\bar{\chi}(\mathcal{H}) < \alpha(\mathcal{H}) = n - 1$. (For example, the hypergraphs in Fig. 1 are monostars.)

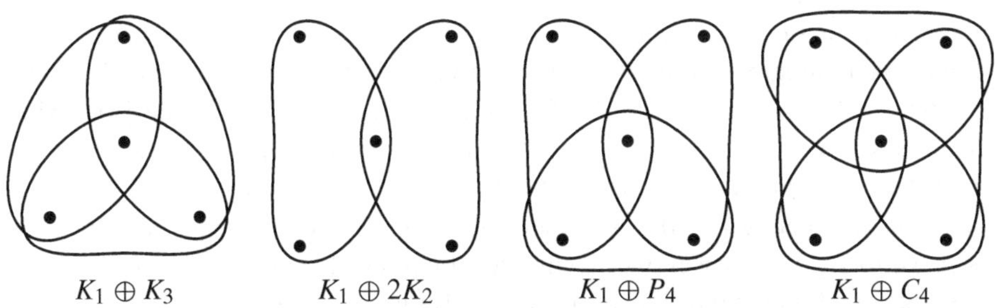

Fig. 1. The four minimal $\mathcal{C}$-imperfect 3-uniform monostars

- A *cycloid* C_n^r is an r-uniform circular hypergraph with n vertices and n edges (where $n > r \geq 3$) for which each edge contains r consecutive vertices of the host cycle. (See Fig. 2, which is the cycloid C_5^3.) Every cycloid of the form C_{2r-1}^r (with $r \geq 3$) is minimally $\mathcal{C}$-imperfect, since $\alpha(C_{2r-1}^r) = 2r-3$ and $\bar{\chi}(C_{2r-1}^r) = 2r-4$; moreover, each of its non-edgeless proper induced subhypergraphs is a $\mathcal{C}$-perfect hyperstar. Otherwise, C_n^r is $\mathcal{C}$-perfect if $n \leq 2r-2$, and $\mathcal{C}$-imperfect (but not minimally) if $n \geq 2r$.

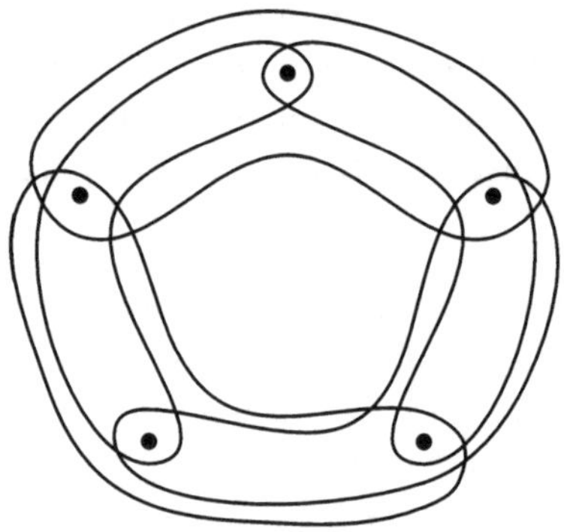

Fig. 2. The cycloid C_5^3

Clearly, a hypergraph is $\mathcal{C}$-perfect if and only if it has no induced subhypergraph that is minimally $\mathcal{C}$-imperfect. For some structural subclasses of hypergraphs, forbidden subhypergraph characterizations have been proved for $\mathcal{C}$-perfectness. The following statements appear in [23], [18] and [19], respectively.

Theorem 3.1

- *An interval hypergraph is $\mathcal{C}$-perfect if and only if it contains no induced monostar.*
- *Let $\mathcal{H}$ be a circular hypergraph in which no edge contains any other edge as a subset. Then $\mathcal{H}$ is $\mathcal{C}$-perfect if and only if it contains no induced monostar and is not isomorphic to the cycloid C_{2r-1}^r for any $r \geq 3$.*
- *A hypertree is $\mathcal{C}$-perfect if and only if it contains no induced monostar.*

It has been conjectured that the number of r-uniform minimally $\mathcal{C}$-imperfect hypergraphs is finite for each r, but in fact we do not even know whether there is a 3-uniform minimally $\mathcal{C}$-imperfect hypergraph different from the six examples shown in Figs. 1, 2 and 3. Here, $C_6(1,2,5) \cup C_6(1,3,5)$ is Král's example [41], which consists of six edges obtained by rotating clockwise the edge on the left plus two more edges on the right (they should be in the same vertex-set).

On the number of r-uniform minimally $\mathcal{C}$-imperfect hypergraphs, finiteness has been verified in the cases where also the transversal number is bounded. More precisely, the following result was proved by Bujtás and Tuza [18].

Theorem 3.2 *For each fixed $r \geq 3$ and $t \geq 1$, there are only finitely many r-uniform minimally $\mathcal{C}$-imperfect hypergraphs with transversal number t.*

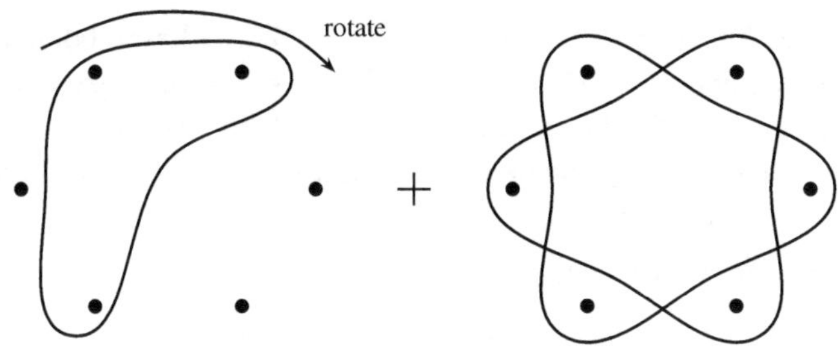

Fig. 3. A minimally $\mathcal{C}$-imperfect hypergraph $C_6(1,2,5) \cup C_6(1,3,5)$ that is neither a monostar nor a cycloid

In the same paper, Bujtás and Tuza characterized minimally $\mathcal{C}$-imperfect hypergraphs with transversal numbers $\tau = 2$ and $\tau = 3$ and constructed r-uniform examples with $\tau = 2$ and 3 for which the number of minimally $\mathcal{C}$-imperfect constructions increases without bound as $r \to \infty$. We do not know any minimally $\mathcal{C}$-imperfect hypergraphs with $\tau \geq 4$.

The $\mathcal{C}$-perfection of a hypergraph may lead to efficient algorithms for solving some problems that are NP-complete in general. For example, although the determination of the upper chromatic number is an NP-complete problem, even for just the class of 3-uniform monostars, the $\mathcal{C}$-perfect hypertrees (even without any restrictions on edge sizes) can still be $\bar{\chi}$-coloured in polynomial time (see [19]). On the other hand, the recognition problem for $\mathcal{C}$-perfect hypertrees is co-NP-complete, but it becomes polynomial-time solvable if edge-sizes are bounded from above. Further results on time complexity can be found in [43] and [19], which show that the situation is more complicated than in the case of graph perfectness.

The minimum number of r-edges, for $\bar{\chi} < r$

By the pigeonhole principle, for every r-uniform hypergraph $\mathcal{H}$ any vertex colouring with $r-1$ colours yields a $\mathcal{C}$-colouring; thus, $\bar{\chi}(\mathcal{H}) \geq r-1$ always holds. This observation leads to the extremal problem of determining the minimum number of r-element edges on n vertices for which $\bar{\chi} = r-1$. We denote this minimum number by $f(n,r)$, where $n \geq r \geq 2$. Alternatively, we can define $f(n,r)$ to be the minimum number of r-element subsets of an n-element underlying set V, selected so that, for each r-partition $(V_1, V_2, \ldots, V_r)$ of V, there exists a selected r-element set S that intersects each V_i in precisely one element.

For $r = 2$, $f(n,2)$ is the minimum number of edges in a connected graph of order n, so $f(n,2) = n-1$. Taking an n-element vertex-set V and fixing a vertex $z \in V$, we obtain a general upper bound for $f(n,r)$ from the hyperstar $\mathcal{H} = (V, \mathcal{E})$ with edge-set

$$\mathcal{E} = \left\{ E \in \binom{V}{r} : z \in E \right\}.$$

Observe that $\bar{\chi}(\mathcal{H}) = r-1$ and $|\mathcal{E}| = \binom{n-1}{r-1}$. For a lower bound, consider an r-uniform hypergraph $\mathcal{H} = (V, \mathcal{E})$ with $\bar{\chi}(\mathcal{H}) = r-1$, and take a set U of $r-2$ vertices. Since the graph with vertex-set $V \setminus U$ and edge-set $\{E \setminus U : U \subseteq E \in \mathcal{E}\}$ must be connected for every choice of U, a lower bound on $f(n, r)$ can be calculated. These estimates appeared first in a paper of Sterboul [60].

Theorem 3.3 *For every $n \geq r \geq 3$,*

$$\frac{2}{n-r+2}\binom{n}{r} \leq f(n, r) \leq \binom{n-1}{r-1}.$$

The lower bound in this theorem is tight for 3-uniform hypergraphs – that is, $f(n, 3) = \lceil \frac{1}{3}n(n-2) \rceil$, for each $n \geq 3$. The simplest proof appeared in [25], and further references to a series of papers proving the same theorem can be found in the survey [20]. For $r \geq 4$ and n large, however, the lower bound is no longer tight (see [17]).

Let us mention a further equality from [60]:

$$f(n, n-2) = \binom{n}{2} - \mathrm{ex}(n, \{C_3, C_4\}),$$

where the last term is the Turán number for graphs of order n and girth 5. Its determination (exact or asymptotic) has been an open problem for several decades. The ratio of known upper and lower bounds is still about $\sqrt{2}$ for any large n. This indicates that finding an exact general formula for $f(n, k)$ is a hopeless task.

Recently, Bujtás and Tuza [17] gave an asymptotic solution:

Theorem 3.4 *If $n > k > 2$, then*

- *for all fixed k and all n,*

$$f(n, k) \leq \frac{2}{n-1}\binom{n-1}{k} + \frac{n-1}{k-1}\left(\binom{n-2}{k-2} - \binom{n-k-1}{k-2}\right).$$

- *for all $k = o(n^{1/3})$ as $n \to \infty$,*

$$f(n, k) = (1 + o(1))\frac{2}{k}\binom{n-2}{k-1}.$$

$\mathcal{C}$-type colourings of graphs

For graphs (that is, 2-uniform hypergraphs), $\mathcal{C}$-colouring imposes the simple restriction that each connected component of the graph must be monochromatic, so there is little to explore in this direction. But the following defective version of $\mathcal{C}$-colouring introduced by Bujtás *et al.* [11] raises interesting questions, even for graphs.

For each integer $k \geq 1$, a *k-improper $\mathcal{C}$-colouring* is a colouring of vertices of a graph G such that, for each vertex v, at most k vertices in the neighbourhood $N(v)$ receive colours different from that of v; the *k-improper upper chromatic number*

$\bar{\chi}_{k\text{-imp}}(G)$ is the maximum number of colours that can be achieved in such a colouring. Equivalently, $\bar{\chi}_{k\text{-imp}}(G)$ is precisely the maximum number of components that can be obtained from G by deleting at most k edges at each vertex. As proved in [11], the Nordhaus–Gaddum-type inequality

$$\bar{\chi}_{k\text{-imp}}(G) + \bar{\chi}_{k\text{-imp}}(\overline{G}) \leq n + 1$$

holds for every $k > 0$ and for every graph G of order $n \geq 8k + 1$. Moreover, if both G and $\overline{G}$ have colourings with more than one colour, then the upper bound is $4k + 2$, independently of n.

Some further versions of graph colouring that can be described by $\mathcal{C}$-type colouring constraints are surveyed in [20].

4. Colourings of mixed hypergraphs

We recall from the end of Section 1 that a *mixed hypergraph* is a triple $\mathcal{H} = (V, \mathcal{C}, \mathcal{D})$, where both $\mathcal{C}$ and $\mathcal{D}$ are set-systems over the vertex-set V; their members $C \in \mathcal{C}$ and $D \in \mathcal{D}$ are termed *$\mathcal{C}$-edges* and *$\mathcal{D}$-edges*, respectively. A vertex λ-colouring $c : V \to \{1, 2, \ldots, \lambda\}$ is *proper* if

$$|c(C)| < |C| \text{ for all } C \in \mathcal{C}, \quad \text{and} \quad |c(D)| > 1 \text{ for all } D \in \mathcal{D},$$

where $c(Y)$ denotes, for any $Y \subseteq V$, the set of colours assigned by the mapping c to the vertices in Y. We say that $\mathcal{H}$ is a *bi-hypergraph* if $\mathcal{C} = \mathcal{D}$, and for any mixed hypergraph a set in $\mathcal{C} \cap \mathcal{D}$ is called a *bi-edge*. Note that the case $\mathcal{C} = \varnothing$ (termed a *$\mathcal{D}$-hypergraph*) corresponds to a proper vertex-colouring in the classical sense, whereas $\mathcal{D} = \varnothing$ (termed a *$\mathcal{C}$-hypergraph*) was discussed in the previous section with respect to $\mathcal{C}$-colouring.

In our historical introduction we have already indicated briefly that mixed hypergraphs opened up a new dimension in hypergraph colouring theory. In this section we give more details.

Example. Consider the complete r-uniform hypergraph $\mathcal{K}_n^r$ of order n, and consider all r-tuples as bi-edges. Then we may use no more than $r - 1$ colours (because $\mathcal{C} = \binom{V}{r}$) and no colour may occur on more than $r - 1$ vertices (because $\mathcal{D} = \binom{V}{r}$). Thus, for $n > (r - 1)^2$, the hypergraph does not admit any proper colouring. In particular, for $r = 2$ and $n = 2$, the vertex pair, which would simultaneously be a $\mathcal{C}$-edge and a $\mathcal{D}$-edge, leads to a contradiction of colouring requirements.

Arising from this example, we call a mixed hypergraph *colourable* if it admits at least one proper colouring, and call it *uncolourable* otherwise. Uncolourable mixed hypergraphs were studied in detail by Tuza and Voloshin [63], who presented the following construction for uncolourable hypergraphs, based on a principle substantially different from the one above.

Example. Let $G = (V,E)$ be a graph with chromatic number k, and define $\mathcal{D} = E$. Moreover, let $\mathcal{C} \subseteq \binom{V}{k}$ be the k-uniform set system whose members are the vertex-sets of P_k subpaths in G. Then $\mathcal{H} = (V,\mathcal{C},\mathcal{D})$ is uncolourable, by a theorem of Gallai and Roy (see [34], [57], [62]).

A further construction from the same paper can be thought of as 'list-colouring without lists'. For any instance of the list-colouring problem with a graph $G = (V,E)$ and lists L_v on its vertices, we can extend V with the set of colours to obtain a mixed hypergraph $\mathcal{H}$ on the vertex-set $V \cup \bigcup_{v \in V} L_v$. The $\mathcal{D}$-edges of $\mathcal{H}$ are the edges of G, and the $\mathcal{C}$-edges are the sets $L_v \cup \{v\}$, for all $v \in V$. Then $\mathcal{H}$ is colourable if and only if G admits a list-colouring.

Algorithmically it is an NP-complete problem to recognize colourable mixed hypergraphs, and so there is no hope of finding easily identifiable obstacles against colourability. Quantitatively, only the very strong requirement $\bar{\chi}(\mathcal{C}) - \chi(\mathcal{D}) \geq n-3$ is sufficient to imply colourability; even replacing $n-3$ by $n-4$ yields some uncolourable hypergraphs (see Tuza and Voloshin [63]).

The *chromatic inversion* of a mixed hypergraph $\mathcal{H} = (V,\mathcal{C},\mathcal{D})$ is the mixed hypergraph $\mathcal{H}^c = (V,\mathcal{C}^c,\mathcal{D}^c)$, with $\mathcal{C}^c = \mathcal{D}$ and $\mathcal{D}^c = \mathcal{C}$. Nearly two decades ago, Voloshin [67] asked whether there is a connection between the colourability of $\mathcal{H}$ and that of $\mathcal{H}^c$. A negative answer was recently given by Hegyháti and Tuza [35], who proved that it is NP-complete to test whether the chromatic inversion of a colourable 3-uniform mixed hypergraph is colourable, and co-NP-complete to test whether the chromatic inversion of an uncolourable 3-uniform mixed hypergraph is uncolourable. This negative result is derived from the positive theorem that if $\mathcal{D} = \binom{V}{3}$ and $\mathcal{H}$ is 3-uniform, then the colourability of $\mathcal{H}$ is decidable in polynomial time.

Properties of colourable mixed hypergraphs

For the rest of this section we restrict our attention to mixed hypergraphs $\mathcal{H} = (V,\mathcal{C},\mathcal{D})$ that are colourable. Under this assumption it is interesting to study the *feasible set* $\Phi(\mathcal{H})$ of $\mathcal{H}$, defined as the set of those integers k for which $\mathcal{H}$ admits a proper k-colouring with exactly k-colours.

Among the elements of $\Phi(\mathcal{H})$, the minimum $\chi(\mathcal{H})$ and the maximum $\bar{\chi}(\mathcal{H})$ are termed the *lower chromatic number* and *upper chromatic number*, respectively. In some classes of hypergraphs, where $\bar{\chi}$ is near to n, the *decrement*, defined as $\mathrm{dec}(\mathcal{H}) = n - \bar{\chi}(\mathcal{H})$, may be informative.

Further information, more detailed than $\Phi(\mathcal{H})$, about the colourability properties of a mixed hypergraph, is provided by the *chromatic spectrum*. It is defined as the sequence $(r_1, r_2, \ldots, r_n)$, where each r_k is the number of *feasible partitions* – that is, the partitions of the vertex-set into colour classes induced by proper colourings using precisely k colours. A *gap* in the chromatic spectrum and in the feasible set is an integer k for which $\chi(\mathcal{H}) < k < \bar{\chi}(\mathcal{H})$ and $r_k = 0$.

If $\mathcal{H}$ is a $\mathcal{D}$-hypergraph, then its feasible set is an interval of integers: $\Phi(\mathcal{H}) = \{\chi(\mathcal{H}), \chi(\mathcal{H})+1, \ldots, n\}$. Similarly, if $\mathcal{H}$ is a $\mathcal{C}$-hypergraph, then its feasible set is the interval $\Phi(\mathcal{H}) = \{1, 2, \ldots, \bar{\chi}(\mathcal{H})\}$. Somewhat unexpectedly, the feasible sets of non-1-colourable mixed hypergraphs are rather unrestricted – namely, for any finite set $S \subset \mathbb{N} \setminus \{1\}$ of integers, there exists a mixed hypergraph $\mathcal{H}$ for which $\Phi(\mathcal{H}) = S$ (see Jiang *et al.* [38]). Even more generally, for any finite sequence $(r_k)_{k=a}^{b}$ of non-negative integers with $2 \leq a \leq b$ and $r_a \neq 0 \neq r_b$, there exists a mixed hypergraph $\mathcal{H}$ for which $\chi(\mathcal{H}) = a$, $\bar{\chi}(\mathcal{H}) = b$, and for each $a \leq k \leq b$, the number of feasible partitions is exactly r_k (see Král' [42]). A similar statement is valid for feasible sets of r-uniform mixed hypergraphs with only one additional restriction: if an integer $k < r - 1$ belongs to Φ, then all integers between k and $r - 1$ also have to be contained in Φ (see Bujtás and Tuza [14]). The case of chromatic spectra with $r_1 = 0$ and $r_k = 0$ or 1 for all $k \geq 2$ was recently studied by Zhao, Diao and Wang [74], [75], especially concerning the minimum number of vertices and edges.

Hypertrees are more restrictive: colourability implies that $\chi = 2$ (unless $\mathcal{D} = \varnothing$, in which case $\chi = 1$) and that the feasible set is gap-free. More generally, no gap occurs if the hypergraph admits a host graph in which any two cycles are vertex-disjoint (see Král *et al.* [43]). A proper 2-colouring of a hypertree is also easy to find: as long as there exist $\mathcal{C}$-edges of size 2, contract them to single vertices, and then colour the contracted host tree properly in the usual sense. A hypertree is uncolourable if and only if this procedure yields a contracted $\mathcal{D}$-edge of size 1 (see Tuza and Voloshin [63]).

The upper chromatic number turns out to be a harder issue. Although Bulgaru and Voloshin [23] proved that it can be expressed with a nice formula on mixed *interval* hypergraphs, the determination of $\bar{\chi}$ on hypertrees is NP-hard (see [43]). Moreover, various results on its non-approximability are known (see [22]).

For the feasible set of a planar mixed hypergraph $\mathcal{H}$, Kobler and Kündgen [39] proved that either $\Phi(\mathcal{H})$ is a gap-free interval with minimum value 4 or less, or the lower chromatic number is $\chi = 2$ and the only gap occurs at 3.

On the boundary of colourability and uncolourability, those mixed hypergraphs have been located that admit just one feasible partition – that is, those with just one $r_k = 1$ in their chromatic spectrum, and zeros otherwise. The systematic study of these *uniquely colourable* mixed hypergraphs was initiated by Tuza, Voloshin and Zhou [65]. This class is NP-hard to recognize (it is co-NP-complete when the input hypergraph is given, together with a proper colouring) and it does not admit a characterization in terms of forbidden induced subhypergraphs; in fact, every mixed hypergraph can be embedded as an induced subhypergraph into a uniquely colourable one. Moreover, Bujtás and Tuza [12] proved that it is also NP-complete to decide whether $\mathcal{H}$ admits a vertex ordering $v_1, v_2, \ldots, v_n$ such that, for all $1 \leq i \leq n$, the subhypergraph induced by $\{v_1, v_2 \ldots, v_i\}$ is uniquely colourable.

Some uniquely colourable subclasses admit a good characterization and efficient recognition algorithm – for instance, those with $\chi = n - 1$ and $\chi = n - 2$ (see [54]), uniquely colourable mixed hypertrees (see [53]) and circular mixed hypergraphs (see [70]). Moreover, the possible size distributions of colour classes in uniquely colourable *r-uniform bi-hypergraphs* have also been characterized (see [5]).

An analogy with complete graphs is that uniquely colourable separators (vertex subsets that induce a uniquely colourable subhypergraph and whose removal makes $\mathcal{H}$ disconnected) can be applied to derive a recursive formula to compute the chromatic polynomial (see [65]). On the other hand, colourable *graphs* do not satisfy the requirement put on uniquely colourable mixed hypergraphs; instead, they form a subclass of *weakly uniquely colourable* mixed hypergraphs; this means that $r_\chi = r_{\bar{\chi}} = 1$. They have also been studied to some extent in [65].

$\mathcal{C}$-perfect mixed hypergraphs

$\mathcal{C}$-perfectness can be defined for mixed hypergraphs $\mathcal{H} = (V, \mathcal{C}, \mathcal{D})$ in the same way as for $\mathcal{C}$-colouring, by requiring that $\bar{\chi}(\mathcal{H}') = \alpha_{\mathcal{C}}(\mathcal{H}')$ for all *induced* subhypergraphs $\mathcal{H}' = (V', \mathcal{C}', \mathcal{D}')$ of $\mathcal{H}$, where $\alpha_{\mathcal{C}}(\mathcal{H}')$ is the independence number of the hypergraph $(V, \mathcal{C}')$. Generalizing the class of monostars, one can obtain $\mathcal{C}$-imperfect mixed hypergraphs by taking some $\mathcal{C}$-edges with non-empty intersection, and inserting each 2-element subset of their common intersection as a $\mathcal{D}$-edge. (These constructs are called *polystars*.) But the characterization problem for $\mathcal{C}$-perfect mixed hypergraphs looks even harder than that for $\mathcal{C}$-perfect hypergraphs. Some examples given in [19] indicate that the situation already becomes quite complicated for mixed hypertrees.

Steiner systems and finite projective planes

A *Steiner system* $\mathrm{S}(t, k, v)$ is a k-uniform hypergraph of order v, for which each t-tuple of vertices is contained in precisely one edge. To mention some celebrated examples, a system $\mathrm{S}(2, 3, v)$ is a *Steiner triple system* $\mathrm{STS}(v)$, an $\mathrm{S}(3, 4, v)$ is a *Steiner quadruple system* $\mathrm{SQS}(v)$, and an $\mathrm{S}(2, q+1, q^2+q+1)$ is a *finite projective plane* of order q.

The edges are traditionally called *blocks*. We may view each block as a $\mathcal{C}$-edge (where an $\mathrm{STS}(v)$ is denoted by $\mathrm{CSTS}(v)$) or as a bi-edge – that is, a $\mathcal{C}$-edge and a $\mathcal{D}$-edge at the same time (where an $\mathrm{STS}(v)$ is denoted by $\mathrm{BSTS}(v)$). The notations $\mathrm{CS}(t, k, v)$, $\mathrm{BS}(t, k, v)$, $\mathrm{CSQS}(v)$ and $\mathrm{BSQS}(v)$ are derived for the respective systems in a similar way.

The study of the upper chromatic number in Steiner triple systems started with a paper by Milazzo and Tuza [49], where the authors proved that

$$\bar{\chi}(BSTS(v)) \leq \bar{\chi}(CSTS(v)) \leq s,$$

for all $v \leq 2^s - 1$. This upper bound on $\bar{\chi}$ is tight for all $s \geq 2$, and the systems attaining equality were also characterized. More generally,

$$\bar{\chi}(BS(t, t+1, v)) \leq \bar{\chi}(CS(t, t+1, v)) = O(\ln v)$$

holds for any fixed $t \geq 2$ as $v \to \infty$ (see [50]). Interestingly, from below it is not even known whether $\bar{\chi}(BSQS(v))$ can tend to infinity with v. Moreover, no uncolourable *BSQS* has so far been found.

Concerning the decrements of projective planes $\Pi(q)$ of order q, Bacsó and Tuza [4] proved that $\mathrm{dec}(\Pi(q)) \geq 2q + \sqrt{q}/2 - o(\sqrt{q})$. For an infinite sequence of values q this is provably optimal, even in the order $\Theta(\sqrt{q})$ of its second term: if q is a square, then the union of two disjoint Baer subplanes in the Galois plane $PG(2,q)$ meets each line in more than one point, and this allows us to construct a colouring; we deduce that $\mathrm{dec}(PG(2,q)) \leq 2q + 2\sqrt{q} + 1$. Recently, Bacsó, Héger and Szőnyi [3] gave sufficient conditions to ensure that $\mathrm{dec}(PG(2,q)) = \tau_2((PG(2,q))) - 1$. (The right-hand side is always an upper bound if there exists a 2-transversal of minimum cardinality that is independent – see the section on $\mathcal{C}$-colouring.)

For projective planes $\Pi(q)$ without an underlying algebraic structure, a general upper bound for $\mathrm{dec}(\Pi(q))$ is $3q - 2$. Indeed, one can pick three lines L, L', L'' in general position (that is, with empty intersection), and assign colour 1 to $L' \cap L''$ and to all points of $L \setminus (L' \cup L'')$, and colour 2 to all points of $(L' \cup L'') \setminus (L' \cap L'')$. Then every line contains two points of the same colour, and so each remaining point can have its distinct colour.

More details about colourings of block designs can be found in the survey [51].

5. Colour-bounded and stably bounded hypergraphs

In this section we consider generalizations of mixed hypergraphs; we shall view them as hypergraphs $\mathcal{H} = (V, \mathcal{E})$ with additional constraints prescribed for the edges. For convenience, we assume that $\mathcal{E} = \{E_1, E_2 \ldots, E_m\}$. These structures include colour-bounded hypergraphs, stably bounded hypergraphs and pattern hypergraphs.

The most general class is the third one, introduced by Dvořák *et al.* [28]. For each edge E_i, a family $\mathcal{P}_i$ of partitions is specified, and a vertex-colouring is considered to be *proper* if its colour classes induce a partition of each E_i which is a member of $\mathcal{P}_i$. Even in this very general model, the authors managed to develop a theory concerning feasible sets with or without gaps.

The other two classes are closer to the flavour of mixed hypergraphs; their conditions on the edges are given quantitatively. For the more general one, a *stably bounded hypergraph* required four functions $s, t, a, b : \mathcal{E} \to \mathbb{N}$ to be given: we write $s_i = s(E_i)$, $t_i = t(E_i)$, $a_i = a(E_i)$, $b_i = b(E_i)$, for all $1 \leq i \leq m$. Given these parameters, a colouring is *proper* if, for each $E_i \in \mathcal{E}$:

- the number of different colours assigned to the vertices of E_i is at least s_i
- the number of different colours assigned to the vertices of E_i is at most t_i
- there exists a colour assigned to at least a_i vertices of E_i
- no colour occurs on more than b_i vertices of E_i.

So the functions s and t impose bounds on the largest polychromatic subsets of the edges, while the functions a and b impose bounds on their largest monochromatic subsets.

These four colour-bounding functions can capture many kinds of partitioning problems, allowing a concise model description. A practical example of this is resource allocation where, loosely speaking, the lower bounds (ensuring several kinds of resources or multiple copies of one resource) increase the security and stability of a system, while the upper bounds correspond to constructing the system in an economical way (since more pieces of resource increase the cost).

Functional subclasses

Colour-bounding functions of types $s_i = 1$, $t_i = |E_i|$, $a_i = 1$, and $b_i = |E_i|$ are called *non-restrictive* on the edge E_i, because they express no real restrictions on proper colourings. A colour-bounding function is *non-restrictive* in a hypergraph $\mathcal{H}$ if it is so on all edges of $\mathcal{H}$; otherwise, it is called *restrictive*. In notation, we write capital letters to indicate the functions that are allowed to be restrictive in a hypergraph. For instance, an (S, T)-hypergraph means that only s and t are given – that is, only the numbers of different colours occurring in the edges are bounded; an (S, T)-hypergraph is also called a *colour-bounded hypergraph*. Similarly, in S-hypergraphs only the minimum number of required colours is given for each edge. Note that every S-hypergraph is an (S, T)-hypergraph at the same time, because t is not required to be restrictive for the latter.

Chromatic polynomials

As in the classical theory of graph colourings, for a hypergraph $\mathcal{H} = (V, \mathcal{E})$ and an integer $\lambda \in \mathbb{N}$, we denote by $P(\mathcal{H}, \lambda)$ the number of proper colourings $c : V \to \{1, 2, \ldots, \lambda\}$. This $P(\mathcal{H}, \lambda)$ is a polynomial in λ for every $\mathcal{H}$ (in the most general class, pattern hypergraphs, too), and is thus termed the *chromatic polynomial* of $\mathcal{H}$. Voloshin [66] extended the concept of $P(\mathcal{H}, \lambda)$ from the classical one to mixed hypergraphs, and discovered some new properties of $P(\mathcal{H}, \lambda)$. For example, he showed that the degree of a chromatic polynomial is not determined by the number of vertices in general, as it is equal to the upper chromatic number $\bar{\chi}(\mathcal{H})$; only in classical graph and hypergraph colourings do the two invariants coincide. Another unusual property is that some mixed hypergraphs contain vertices whose removal from $\mathcal{H}$ changes nothing with respect to colourability: all colourings and $P(\mathcal{H}, \lambda)$ remain the same. Such vertices are called *phantom vertices*; they are invisible for $P(\mathcal{H}, \lambda)$, and $\mathcal{H}$ may contain any number of them.

The famous connection–contraction algorithm for finding the chromatic polynomial, as originated by Birkhoff [10] in 1912, formalized by Whitney [72] in 1932, and generalized to hypergraphs by Zykov [76] in 1974, was extended to the splitting–contraction algorithm for mixed hypergraphs by Voloshin [66] in 1993. Further, some properties of $P(\mathcal{H}, \lambda)$ in S-hypergraphs have been described in [26].

It is not known which polynomials occur as chromatic polynomials of mixed hypergraphs. For non-1-colourable structures, however, Bujtás and Tuza [15] gave the following necessary and sufficient conditions. Here $S(n,k)$ is the Stirling number of the second kind, the number of partitions of n elements into precisely k non-empty sets.

Theorem 5.1 *Let $P(\lambda) = \sum_{k=0}^{\ell} a_k \lambda^k$ be a non-zero polynomial for which $P(1) = 0$ (so $\sum_{k=0}^{\ell} a_k = 0$). Then the following properties are equivalent.*

- *$P(\lambda)$ is the chromatic polynomial of a colour-bounded hypergraph*
- *$P(\lambda)$ is the chromatic polynomial of a mixed hypergraph*
- *$P(\lambda)$ satisfies the following conditions:*
  - *all coefficients a_k of $P(\lambda)$ are integers*
  - *the leading coefficient a_ℓ is positive*
  - *the constant term a_0 is 0*
  - *for each positive integer $j \leq \ell$, $\sum_{k=j}^{\ell} a_k S(k,j) \geq 0$.*

These characterizations remain the same if we replace the pair (S,T) (colour-bounded hypergraphs) by any of $(S,A), (T,B)$ and (A,B) (those that admit uncolourable hypergraphs), or by (S,T,A,B) itself, or by pattern hypergraphs.

One way to describe relations between functional classes (defined in terms of the subsets of $\{S,T,A,B\}$) is to consider the classes of chromatic polynomials that occur in them. In Fig. 4 we exhibit their Hasse diagram; we indicate the positions of the classes of C-, D-, and mixed hypergraphs by $\mathcal{P}_\mathcal{C}$, $\mathcal{P}_\mathcal{D}$ and $\mathcal{P}_M$, respectively. The sets of chromatic polynomials belonging to non-1-colourable (S,T,A,B)-hypergraphs and to non-1-colourable (S,T)-hypergraphs are the same; the difference shown in Fig. 4 arises from the fact that there are 1-colourable (S,T,A,B)-hypergraphs whose chromatic spectra do not occur for any (S,T)-hypergraphs. Note that there are examples of problems (such as algorithmic problems), and of classes, where the problem is provably harder on the class located lower in the hierarchy of chromatic polynomials.

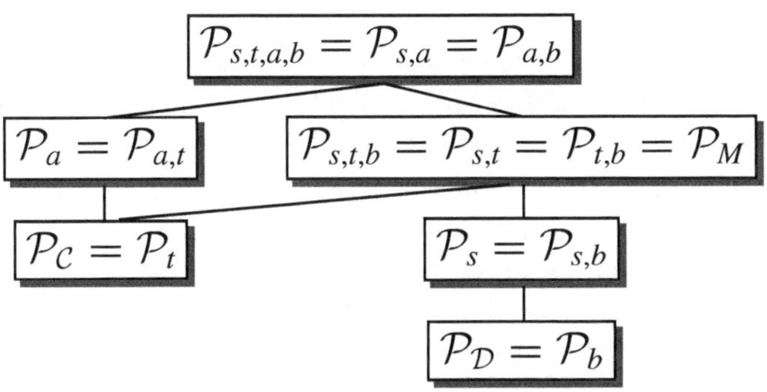

Fig. 4. The Hasse diagram for classes of chromatic polynomials

Interval hypergraphs and hypertrees

The colouring properties of some well-structured hypergraph classes have been analysed in detail. Many results are known about the computational complexity of the problems of deciding colourability and of determining χ and $\bar{\chi}$ for hypertrees, interval hypergraphs and some of their subclasses. For a summary of what is known and what is still open, we refer to the three handy tables in Bujtás and Tuza [21]. Below we mention a few of these results.

The cases of just one restrictive function are quite well understood, except in interval A-hypergraphs, where the algorithmic complexity of determining $\bar{\chi}$ is still an open problem. The interval S- and B-hypergraphs admit optimal *periodic* colourings, with $\chi = \max s_i$ and $\chi = \max\lceil |E_i|/b_i \rceil$ colours, respectively. We observe that $\bar{\chi}$ for interval T-hypergraphs can be computed efficiently.

Concerning two restrictive functions, the simplest case is (S, B), for which $\chi = \max(s_i, \lceil |E_i|/b_i \rceil)$. For each of the pairs (T, B), (A, B) and (S, A), however, it is $\mathcal{NP}$-complete to decide whether a given interval hypergraph is colourable. (This implies intractability for any three restrictive functions, while colourability and the values of χ and $\bar{\chi}$ can be determined efficiently, even for (S, T, A, B) on classes of hypergraphs with bounded edge size.) The important unsolved case is to decide colourability and to determine $\bar{\chi}$ for an interval (S, T)-hypergraph. But interestingly enough, as proved in [16], once we know that it is colourable, we immediately obtain $\chi = \max s_i$, and if a proper colouring is given also, then it can be transformed efficiently to an optimal one; moreover, the feasible set is gap-free. For other pairs of restrictive functions, only a few upper bounds for χ have been found for interval hypergraphs (for example, $\chi \leq \max |E_i|$, even for (S, T, A, B)), and it is not known whether some interval (S, A)-hypergraphs, (T, B)-hypergraphs or (A, B)-hypergraphs can have any gaps between $\chi + 1$ and $\max |E_i| - 1$.

Colour-bounded and stably bounded *hypertrees* turn out to be much more complex than mixed hypertrees. The decision problem for colourability is $\mathcal{NP}$-complete for any non-trivial pair of restrictive functions (that is, other than (S, B) and (T, A)), even in the 3-uniform case. Recalling that every mixed hypertree has a gap-free feasible set, we report that a further dramatic change is illustrated by the following theorem from [16].

Theorem 5.2 *Let F be a finite set of positive integers. Then there exists an (S, T)-hypertree with feasible set F if and only if* $\min(F) = 1$ *or* $\min(F) = 2$ *and F contains all integers between* $\min(F)$ *and* $\max(F)$, *or* $\min(F) \geq 3$.

Finally, we mention that the results on interval (S, T)-hypergraphs can be partially extended to *circular* (S, T)-hypergraphs (see [16]); namely, if a hypergraph from this class is colourable, then it satisfies the inequality $\chi \leq 2(\max s_i) - 1$ and no gaps can occur between $2(\max s_i) - 1$ and $\bar{\chi}$. For $\max s_i \geq 3$ we do not know, however, whether there can exist any gaps between χ and $2(\max s_i) - 1$.

6. Conclusion

In this chapter we described the current state of hypergraph colourings as a separate subject. With so many relations to graph colouring and new fundamental ideas, results and applications, this area is developing fast. The impact of every result may be to give rise to a set of new problems, often without any analogues from the past. While the number of publications continues to grow, we observe that some possible subdirections remain untouched. We predict that hypergraph generalizations of such areas as colouring graphs on surfaces, edge-colourings, colouring and probability, Hadwiger's conjecture, colouring games, orientations and flows (to name just a few) have great research potential in the observable future.

References

1. N. Alon and Z. Bregman, Every 8-uniform 8-regular hypergraph is 2-colorable, *Graphs Combin.* **4** (1988), 303–306.
2. G. Bacsó, Cs. Bujtás, Zs. Tuza and V. Voloshin, New challenges in the theory of hypergraph coloring, *Advances in Discrete Mathematics and Applications*, RMS Lecture Notes **13** (2010), 45–57.
3. G. Bacsó, T. Héger and T. Szőnyi, The 2-blocking number and the upper chromatic number of $PG(2, q)$, *J. Combin. Designs* **21** (2013), 585–602.
4. G. Bacsó and Zs. Tuza, Upper chromatic number of finite projective planes, *J. Combin. Designs* **16** (2008), 221–230.
5. G. Bacsó, Zs. Tuza and V. Voloshin, Unique colorings of bi-hypergraphs, *Austral. J. Combin.* **27** (2003), 33–45.
6. Zs. Baranyai, On the factorization of the complete uniform hypergraph, *Infinite and Finite Sets* (Keszthely, 1973) (eds. A. Hajnal, R. Rado and V. T. Sós), Colloq. Math. Soc. J. Bolyai **10**, North-Holland (1975), 91–107.
7. C. Berge, Sur certains hypergraphes généralisant les graphes bipartites, *Combinatorial Theory and its Applications, I* (Balatonfüred, 1969), North-Holland (1970), 119–133.
8. C. Berge, *Hypergraphs: Combinatorics of Finite Sets*, North-Holland, 1989.
9. C. Berge, Motivations and history of some of my conjectures, *Discrete Math.* **165/166** (1997), 61–70.
10. G. D. Birkhoff, A determinant formula for the number of ways of colouring a map, *Ann. of Math.* **14** (1912), 42–46.
11. Cs. Bujtás, E. Sampathkumar, Zs. Tuza, L. Pushpalatha and R. C. Vasundhara, Improper C-colorings of graphs, *Discrete Applied Math.* **159** (2011), 174–186.
12. Cs. Bujtás and Zs. Tuza, Orderings of uniquely colorable mixed hypergraphs, *Discrete Applied Math.* **155** (2007), 1395–1407.
13. Cs. Bujtás and Zs. Tuza, Color-bounded hypergraphs, III: Model comparison, *Applic. Anal. and Discrete Math.* **1** (2007), 36–55.
14. Cs. Bujtás and Zs. Tuza, Uniform mixed hypergraphs: The possible numbers of colors, *Graphs Combin.* **24** (2008), 1–12.
15. Cs. Bujtás and Zs. Tuza, Color-bounded hypergraphs, I: General results, *Discrete Math.* **309** (2009), 4890–4902.

16. Cs. Bujtás and Zs. Tuza, Color-bounded hypergraphs, II: Interval hypergraphs and hypertrees, *Discrete Math.* **309** (2009), 6391–6401.
17. Cs. Bujtás and Zs. Tuza, Smallest set-transversals of k-partitions, *Graphs Combin.* **25** (2009), 807–816.
18. Cs. Bujtás and Zs. Tuza, C-perfect hypergraphs, *J. Graph Theory* **64** (2010), 132–149.
19. Cs. Bujtás and Zs. Tuza, Voloshin's conjecture for C-perfect hypertrees, *Australas. J. Combin.* **48** (2010), 253–267.
20. Cs. Bujtás and Zs. Tuza, Maximum number of colors: C-coloring and related problems, *J. Geometry* **101** (2011), 83–97.
21. Cs. Bujtás and Zs. Tuza, Color-bounded hypergraphs, VI: Structural and functional jumps in complexity, *Discrete Math.* **313** (2013), 1965–1977.
22. Cs. Bujtás and Zs. Tuza, Approximability of the upper chromatic number of hypergraphs, manuscript, 2013.
23. E. Bulgaru and V. Voloshin, Mixed interval hypergraphs, *Discrete Applied Math.* **77** (1997), 24–41.
24. D. Défossez, A sufficient condition for the bicolorability of a hypergraph, *Discrete Math.* **308** (2008), 2265–2268.
25. K. Diao, G. Liu, D. Rautenbach and P. Zhao, A note on the least number of edges of 3-uniform hypergraphs with upper chromatic number 2, *Discrete Math.* **306** (2006), 670–672.
26. E. Drgas-Burchardt and E. Łazuka, Chromatic polynomials of hypergraphs, *Applied Math. Letters* **20** (2007), 1250–1254.
27. P. Duchet, Hypergraphs, *Handbook of Combinatorics*, Vol. 1 (eds. R. L. Graham, M. Grötschel and L. Lovász), Elsevier (1995), 381–432.
28. Z. Dvořák, J. Kára, D. Král' and O. Pangrác, Pattern hypergraphs, *Electron. J. Combin.* **17** (2010), #R15.
29. P. Erdős, On a combinatorial problem, *Nordisk Mat. Tidskr.* **11** (1963), 5–10.
30. P. Erdős, On a combinatorial problem II, *Acta Math. Acad. Sci. Hungar.* **15** (1964), 445–447.
31. P. Erdős and A. Hajnal, On chromatic number of graphs and set-systems, *Acta Math. Acad. Sci. Hungar.* **17** (1966), 61–99.
32. P. Erdős and L. Lovász, Problems and results on 3-chromatic hypergraphs and some related questions, *Infinite and Finite Sets* (Keszthely, 1973) (eds. A. Hajnal, R. Rado and V. T. Sós), Colloq. Math. Soc. J. Bolyai **10**, North-Holland (1975), 609–627.
33. J.-C. Fournier and M. Las Vergnas, Une class d'hypergraphes bichromatiques, *Discrete Math.* **2** (1972), 407–410.
34. T. Gallai, On directed paths and circuits, *Theory of Graphs* (Tihany, 1966) (eds. P. Erdős and G. O. H. Katona), Academic Press, San Diego (1968), 115–118.
35. M. Hegyháti and Zs. Tuza, Colorability of mixed hypergraphs and their chromatic inversions, *J. Combin. Optim.* **25** (2013), 737–751.
36. M. A. Henning and A. Yeo, 2-colorings in k-regular k-uniform hypergraphs, *Europ. J. Combin.* **34** (2013), 1192–1202.
37. A. Jaffe, T. Moscibroda and S. Sen, On the price of equivocation in Byzantine agreement, *Proc.* 2012 *ACM Symp. on Principles of Distributed Computing*, ACM, New York (2012), 309–318.
38. T. Jiang, D. Mubayi, V. Voloshin, Zs. Tuza and D. West, The chromatic spectrum of mixed hypergraphs, *Graphs Combin.* **18** (2002), 309–318.
39. D. Kobler and A. Kündgen, Gaps in the chromatic spectrum of face-constrained plane graphs, *Electron. J. Combin.* **8** (2001), #N3.
40. A. V. Kostochka and M. Stiebitz, On the number of edges in colour-critical graphs and hypergraphs, *Combinatorica* **20** (2000), 521–530.

41. D. Král', A counter-example to Voloshin's hypergraph co-perfectness conjecture, *Australas. J. Combin.* **27** (2003), 25–41.
42. D. Král', On feasible sets of mixed hypergraphs, *Electron. J. Combin.* **11** (2004), #R19.
43. D. Král', J. Kratochvíl, A. Proskurowski and H.-J. Voss, Coloring mixed hypertrees, *Discrete Applied Math.* **154** (2006), 660–672.
44. D. Král', J. Kratochvíl and H.-J. Voss, Mixed hypercacti, *Discrete Math.* **286** (2004), 99–113.
45. L. Lovász, On chromatic number of finite set-systems, *Acta Math. Acad. Sci. Hungar.* **19** (1968), 59–67.
46. L. Lovász, Normal hypergraphs and the perfect graph conjecture, *Discrete Math.* **2** (1972), 253–267.
47. L. Lovász, Coverings and colourings of hypergraphs, *Congr. Numer.* **8** (1973), 3–12.
48. P. Mihók, Zs. Tuza and M. Voigt, Fractional $\mathcal{P}$-colourings and $\mathcal{P}$-choice-ratio, *Tatra Mountains Math. Publ.* **18** (1999), 69–77.
49. L. Milazzo and Zs. Tuza, Upper chromatic number of Steiner triple and quadruple systems, *Discrete Math.* **174** (1997), 247–259.
50. L. Milazzo and Zs. Tuza, Logarithmic upper bound for the upper chromatic number of $S(t, t+1, v)$ systems, *Ars Combin.* **92** (2009), 213–223.
51. L. Milazzo, Zs. Tuza and V. I. Voloshin, Strict colorings of Steiner triple and quadruple systems: a survey, *Discrete Math.* **261** (2003), 399–411.
52. J. Nešetřil and V. Rödl, A short proof of the existence of highly chromatic hypergraphs without short cycles, *J. Combin. Theory (B)* **27** (1979), 225–227.
53. A. Niculitsa and V. Voloshin, About uniquely colorable mixed hypertrees, *Discuss. Math. Graph Theory* **20** (2000), 81–91.
54. A. Niculitsa and H.-J. Voss, A characterization of uniquely colorable mixed hypergraphs of order n with upper chromatic numbers $n-1$ and $n-2$, *Australas. J. Combin.* **21** (2000), 167–177.
55. K. T. Phelps and V. Rödl, On the algorithmic complexity of coloring simple hypergraphs and Steiner triple systems, *Combinatorica* **4** (1984), 79–88.
56. J. Radhakrishnan and A. Srinivasan, Improved bounds and algorithms for hypergraph 2-coloring, *Random Struct. Alg.* **16** (2000), 4–32.
57. B. Roy, Nombre chromatique et plus longs chemins d'un graphe, *Rev. AFIRO* **1** (1967), 127–132.
58. P. D. Seymour, On the two-colouring of hypergraphs, *Quart. J. Math. Oxford (3)* **25** (1974), 303–312.
59. F. Sterboul, A new combinatorial parameter, *Infinite and Finite Sets* (Keszthely, 1973) (eds. A. Hajnal, R. Rado and V. T. Sós), Colloq. Math. Soc. J. Bolyai **10**, North-Holland (1975), 1387–1404.
60. F. Sterboul, Un problème extrémal pour les graphes et les hypergraphes, *Discrete Math.* **11** (1975), 71–78.
61. I. Tomescu, Sure le probléme du coloriage des graphes généralisés, *C. R. Acad. Sci. (Paris)* **267**:6 (1968), 250–252.
62. Zs. Tuza, Graph coloring in linear time, *J. Combin. Theory (B)* **55** (1992), 236–243.
63. Zs. Tuza and V. Voloshin: Uncolorable mixed hypergraphs, *Discrete Applied Math.* **99** (2000), 209–227.
64. Zs. Tuza and V. Voloshin, Problems and results on colorings of mixed hypergraphs, *Horizons of Combinatorics* (eds. E. Győri, G. Katona and L. Lovász), Bolyai Society Math. Studies **17**, Springer-Verlag (2008), 235–255.
65. Zs. Tuza, V. I. Voloshin and H. Zhou, Uniquely colorable mixed hypergraphs, *Discrete Math.* **248** (2002), 221–236.

66. V. I. Voloshin, The mixed hypergraphs, *Computer Sci. J. Moldova* **1** (1993), 45–52.
67. V. I. Voloshin, On the upper chromatic number of a hypergraph, *Australas. J. Combin.* **11** (1995), 25–45.
68. V. I. Voloshin, *Coloring Mixed Hypergraphs: Theory, Algorithms and Applications*, Fields Institute Monographs **17**, Amer. Math. Soc., 2002.
69. V. Voloshin, Mixed Hypergraph Coloring Web Site, *http://spectrum.troy.edu/voloshin/mh.html.*
70. V. Voloshin and H.-J. Voss, Circular mixed hypergraphs I: colorability and unique colorability, *Congr. Numer.* **144** (2000), 207–219.
71. D. J. A. Welsh and M. B. Powell, An upper bound for the chromatic number of a graph and its application to timetabling problems, *Comp. J.* **10** (1967), 85–86.
72. H. Whitney, The colouring of graphs, *Ann. of Math.* **33** (1932), 688–718.
73. D. R. Woodall, Property B and the four-colour problem, *Combinatorics*, Proc. 1972 Oxford Combinatorial Conference (eds. D. J. A. Welsh and D. R. Woodall), Inst. Math. Appl. (1972), 322–340.
74. P. Zhao, K. Diao and K. Wang, The smallest one-realization of a given set, *Electron. J. Combin.* **19** (2012), #P19.
75. P. Zhao, K. Diao and K. Wang, The smallest one-realization of a given set, II, *Discrete Math.* **312** (2012), 2946–2951.
76. A. A. Zykov, Hypergraphs, *Uspekhi Mat. Nauk* **29** (1974), 89–154 (in Russian).

12

Chromatic scheduling

DOMINIQUE DE WERRA and ALAIN HERTZ

Variations and extensions of the basic vertex-colouring and edge-colouring models have been developed to deal with increasingly complex scheduling problems. We present and illustrate them in specific situations where additional requirements are imposed. We include list-colouring, mixed graph colouring, co-colouring, colouring with preferences and bandwidth colouring, and we present applications of edge-colourings to open shop, school timetabling and sports scheduling problems. We also discuss balancing and compactness constraints which often appear in practical situations.

1. Introduction

We show here how graph colouring models may provide a natural tool for dealing with a variety of scheduling problems. Starting from the basic vertex-colouring model, we will introduce some variations and extensions that are motivated by their applications to some scheduling issues. In each case we give references for further results and for extensions of the various models presented. For algorithms, see Chapter 13.

In chromatic scheduling problems we have a collection V of items, such as operations of jobs to be performed. In V there are some pairs v, w that are subject to an incompatibility condition and we call E the set of such incompatibility pairs. These data are represented by the graph $G = (V, E)$ in which the items are associated with the vertices and the incompatible pairs v, w with the edges vw between the corresponding vertices.

We also have a set $C = \{1, 2, \ldots, k\}$ of time periods (of unit duration). Assuming that each item (considered as an operation) has unit completion time, we may ask whether we can find a schedule taking the incompatibilities into account and using at most k periods of time. This is precisely the vertex k-colouring problem: there exists a feasible schedule if and only if the set V of vertices can be partitioned into subsets $S_1, S_2, \ldots, S_k$, where each S_i contains no two incompatible items.

In some instances, we may try to find the smallest set C of periods (that is, the smallest k) for which a schedule in time $k = |C|$ exists. This is the usual chromatic number $\chi(G)$ of the associated graph G. A classical application of this model is the basic school timetabling problem: the items are one-hour classes, and two classes are incompatible if they cannot be simultaneously given because they involve the same students or the same teacher.

We recall that for arbitrary graphs, the vertex k-colouring problem is NP-complete (see Garey and Johnson [21]).

2. Colouring with weights on the vertices

Previously we have assumed that all items v have the same unit completion time, which we denote by $\omega(v)$. More generally, we may have arbitrary completion times for the items; we identify these completion times with weights $\omega(v)$ which we assume for simplicity to be non-negative integers.

As before we have a set C of unit periods with $k = |C|$, and in order to complete item v we have to assign to it a subset $c(v)$ of $\omega(v)$ periods from C. In order to respect the incompatibility requirements, the assignments of $\omega(v)$ periods to the items v must be such that $c(v) \cap c(w) = \varnothing$ for each incompatible pair vw in E. This is precisely the *multicolouring problem* which has been studied, for instance, by Halldórsson and Kortsarz [28].

In the multicolouring model we observe that if $\omega(v) = r > 1$ for every v in a graph G, then the minimum number of colours needed for a multicolouring of G may be strictly smaller than $r\,\chi(G)$, as can be seen for the pentagon with $r = 2$ in Fig. 1, where the weights are indicated inside each vertex: here only $5 (< 2\,\chi(G) = 6)$ colours are needed.

We observe immediately that the multicolouring problem in a graph G is equivalent to the basic vertex-colouring problem in the graph G' obtained from

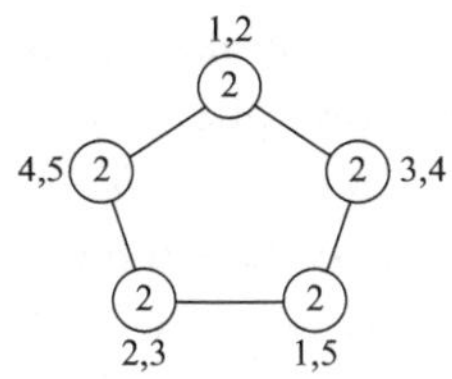

Fig. 1.

G by replacing each vertex v by vertices $v_1, v_2, \ldots, v_{\omega(v)}$, each one linked to all neighbours of v in G. We also introduce all possible edges between $v_1, v_2, \ldots, v_{\omega(v)}$.

The above multicolouring model may produce schedules in which the $\omega(v)$ periods during which a particular item v is being processed are not consecutive. We say that the schedule has some *preemptions* – that is, interruptions during the execution of some of the items. This may be acceptable in some types of application such as school timetabling: if v represents the $\omega(v)$ lectures of a same topic given by a teacher to a specific set of students, a schedule will be improved if these $\omega(v)$ lectures are spread throughout the week. We see later how one can impose such a requirement for these $\omega(v)$ lectures.

There are also contexts in which interruptions during the completion of an item v are not allowed: this means that v has to be assigned a set $c(v)$ of $\omega(v)$ consecutive time periods in C. If this is required for every item v, we have an *interval colouring problem*. This model has been discussed by various authors (see Čangalović and Schreuder [41] and Bouchard *et al.* [3]). It occurs in particular when there are prohibitive set-up times needed to start or restart the processing of an item.

As an illustration, the multicolouring with $k = 4$ colours in Fig. 2(a) (where weights appear inside each vertex) is not an interval colouring. An interval colouring with $k = 5$ colours is shown in Fig. 2(b). Notice that no such interval colouring can be found with fewer than five colours.

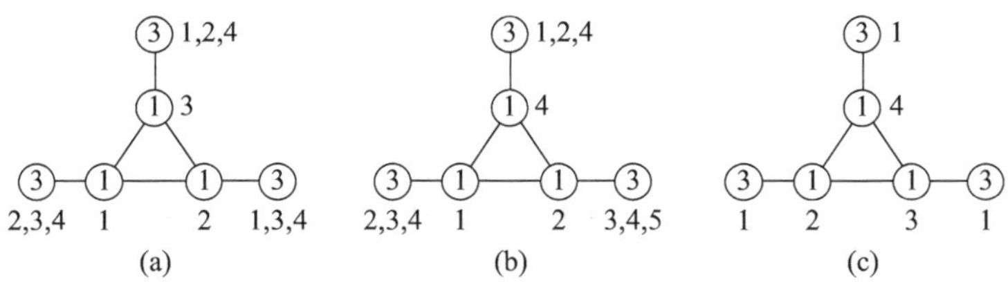

Fig. 2.

In some circumstances, the schedules constructed are designed to be repeated, day after day (say). In such cases it may be convenient to consider that the set $C = \{1, 2, \ldots, k\}$ is cyclically ordered, this means in particular that period 1 follows period k.

In the interval colouring context, we also allow intervals that contain periods k and 1. The possible intervals are now called *cyclic intervals* and the corresponding problem is the *cyclic interval colouring problem*: assign to each vertex v of the graph G a cyclic interval $c(v)$ in C of length $\omega(v)$, in such a way that for any two incompatible items v, w we have $c(v) \cap c(w) = \varnothing$. Cyclic interval colourings have been considered by de Werra and Solot [14]. The multicolouring in Fig. 2(a) is a cyclic interval colouring with four colours.

A related situation occurs in the batch scheduling problem: as before we have a set V of items with a family E of pairwise incompatibilities; each item v again has an integral weight $\omega(v)$ that corresponds to its processing time. We will not allow the items to start their processing individually, but we rather partition the set V into a number k (to be determined) of batches $S_1, S_2, \cdots, S_k$, where each S_i is a subset of items that have no incompatibilities, and therefore whose items can be processed simultaneously. The completion time of a batch S_i is then defined as $p(S_i) = \max_{v \in S_i} \omega(v)$. It is then required to find a value k and a partition into batches $S_1, S_2, \ldots, S_k$ for which $p(S_1) + \ldots + p(S_k)$ is minimum.

Such a problem arises when we have items v to sterilize by placing them into an oven for a duration of at least $\omega(v)$ consecutive time units. Incompatibility requirements do not allow us to place certain specified pairs of items simultaneously in the oven. Finding such a schedule is called the *batch colouring problem*: it has been studied (under the name of 'weighted colouring') in Demange *et al.* [15]. An optimum batch colouring (minimizing $\sum_i p(S_i)$) is given in Fig. 2(c); here $\sum_i p(S_i) = 3 + 1 + 1 + 1 = 6$. It uses $4 (> \chi(G) = 3)$ colours and no colouring with three colours can be as good.

As a final observation we mention that if in all the above types of colouring each weight $\omega(v)$ is equal to 1, then we get back to the basic vertex-colouring problem. So all of these models are generalizations of this problem and they are motivated by various types of application.

We now consider more situations in which extensions of the basic vertex-colouring problem arise.

3. List-colouring

We return to the basic vertex-colouring model in which each weight $\omega(v) = 1$. Here we have a set $C = \{1, 2, \ldots, k\}$ of periods and each item v has a list $L(v) \subseteq C$ of periods to which it could be assigned. In the course scheduling context this means that each class v can be offered only at one of the periods in $L(v)$. Finding a basic vertex-colouring in G such that each vertex v gets a colour $c(v) \in L(v)$ is the well-known *list-colouring problem* which has been extensively studied, for reasons related to its many applications (see Chapter 6 and Voigt [42] for some basic results).

Worthy of interest is the special case in which $|L(v)| = 1$ for a particular subset V^* of items and $L(v) = C$ for all items $v \in V - V^*$. It means that the vertices $v \in V^*$

have already been coloured, and it is required to determine whether the colouring can be extended to the whole graph. This is the *precolouring extension problem*; it is often met in applications related to class scheduling. In such situations some classes are scheduled initially (in order to satisfy some requirements that may be intractable in a simple graph colouring model) and we wish to construct the rest of the schedule while keeping the precolouring fixed.

This problem is also a generalization of the basic vertex-colouring model (it is the case $V^* = \varnothing$) and is NP-hard for general graphs. Surprisingly, for a bipartite graph G with $V^* = \{v_1, v_2, v_3\}$ and $L(v_1) = \{1\}$, $L(v_2) = \{2\}$, $L(v_3) = \{3\}$, $C = \{1, 2, 3\}$, the precolouring extension problem is difficult (see Even *et al.* [20]).

We may observe that if the set $C = \{1, 2, \ldots, k\}$ of colours to be used is specified, then a list-colouring problem may be transformed into a basic vertex-colouring problem. Starting from $G = (V, E)$ and a collection of lists $L(v) \subseteq C = \{1, 2, \ldots, k\}$ for each $v \in V$, we add a clique K on new vertices $w_1, w_2, \ldots, w_k$ and join each vertex $v \in V$ to all vertices $w_i \in K$ for which $i \notin L(v)$. Then any colouring with k colours in the resulting graph gives a list-colouring of G if we call i the colour $c(w_i)$ for $i = 1, 2, \ldots, k$.

It is also appropriate to recall that in most school timetabling contexts, we usually also have a list $\bar{L}(v)$ of unavailable periods for each teacher v. But since the set $C = \{1, 2, \ldots, k\}$ of teaching periods in the week is generally specified, we easily get $L(v) = C - \bar{L}(v)$ as a list of possible periods (colours) for the classes of teacher v.

In connection with list-colouring, there is another concept that is of interest in the context of timetabling and scheduling: we say that a graph $G = (V, E)$ is *p-choosable* if there is a list-colouring of G for all assignments of lists $L(v)$ with $|L(v)| \geq p$ for each $v \in V$. Choosability has been extensively studied from a theoretical point of view (see Chapter 6, Erdős *et al.* [19] and Gutner [26]). The *choice number* $ch(G)$ is the minimum number p for which G is p-choosable. Graphs G with $ch(G) = 2$ have been characterized (see Erdős *et al.* [19]), but for $ch(G) > 2$ a complete characterization is still elusive.

In terms of timetabling, choosability is a natural concept: consider a school timetabling problem, where the dean in charge of the timetable has assigned courses to the teachers, and on this basis the graph G of conflicts between courses can be constructed. A colouring of G with k colours gives an assignment of classes to periods in the set $C = \{1, 2, \ldots, k\}$ of teaching periods. But teachers are not always available and may be invited to give a list of periods in which their own classes can be scheduled. In order to find a feasible timetable, one has to ask teachers to give lists of sufficient size. This is a problem for the dean who has to find a value p for which an acceptable timetable can be constructed, whatever set of at least p periods is given by each teacher. The dean naturally has to determine a value p that is as small as possible to avoid a rebellion of teachers. This is why it is so useful to be able to determine $ch(G)$.

4. Mixed graph colouring

Suppose now that in a basic class scheduling problem we have not only incompatibility requirements preventing some pairs of classes from being assigned to the same period, but also *precedence constraints*: these require that, for some pairs v, w of classes, v must be scheduled for some period $c(v)$ occurring earlier than the period $c(w)$ assigned to w. Such a requirement is represented in the graph G by an arc oriented from v to w. A graph $G = (V, E)$ with a set A of oriented arcs will be denoted by $G = (V, E, A)$, and is called a *mixed graph*. It follows that, given $G = (V, E, A)$ and $C = \{1, 2, \ldots, k\}$, a mixed graph colouring is an assignment of a period $c(v) \in C$ to each item v in such a way that $c(v) \neq c(w)$ for each pair $vw \in E$ and $c(v) < c(w)$ for each (ordered) pair $vw \in A$.

Mixed graph colourings have been extensively studied. We can see easily that, for a bipartite mixed graph $G = (V, E, A)$ with $C = \{1, 2, 3\}$, the problem of finding whether there is a mixed graph colouring is difficult: the precolouring extension problem at the end of Section 3 is equivalent to a mixed graph colouring problem obtained from G by introducing a set $A = \{v_1u_1, u_1u_2, u_3v_2, v_2u_4, u_5u_6, u_6v_3\}$ of arcs, where the u_i are new vertices (see Fig. 3). Properties of mixed graph colourings can be found in Ries [35], Ries and de Werra [36], Hansen *et al.* [29] and Sotskov *et al.* [40].

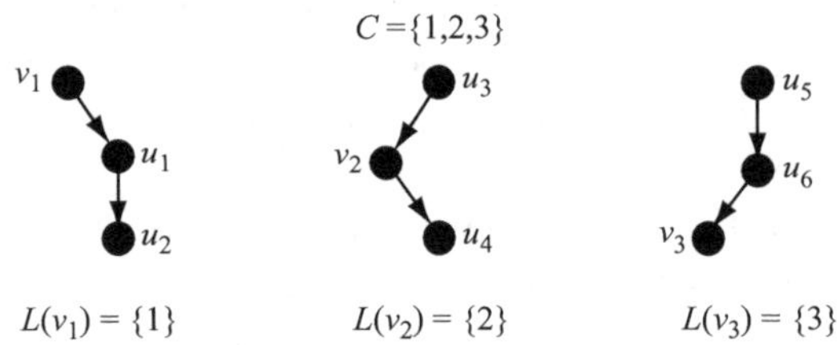

Fig. 3.

Notice that what makes the mixed graph colouring problem difficult is the presence of edges, not of arcs: if we have a graph $G = (V, E, A)$ with $E = \varnothing$, then there is a vertex-colouring satisfying $c(v) < c(w)$ for each arc vw if and only if G contains no oriented cycle. The colour of a vertex v is simply the length of a longest path ending in v.

There are strong connections between colourings and orientations of graphs: it is not difficult to see that finding a colouring with a minimum number of colours is equivalent to finding an acyclic orientation of all its edges for which the longest oriented path is as short as possible.

In this spirit we recall the well-known *Gallai–Roy theorem* which states that, for any oriented graph $\vec{G}$, the chromatic number of its unoriented copy G does not exceed the maximum number of vertices in an elementary oriented path of $\vec{G}$ (see Roy [38]).

The graph colouring problem can also be considered as formulated in the oriented graph obtained by replacing each edge vw in the initial graph G by a pair of opposite arcs *vw* and *wv* forming a *disjunctive pair*. Let $\vec{G}$ be the resulting graph: then finding a vertex-colouring of G with k colours amounts to choosing exactly one arc from each disjunctive pair in such a way that the resulting graph G^* has no oriented cycles and no oriented path has more than k vertices.

Now for a mixed vertex-colouring, we can construct from G a graph $\vec{G}$ in the same way, while keeping the original arcs of G as they are. The mixed colouring problem is then similar to a problem with some preassignments: the choice has already been made for the arcs appearing in the original graph, and the question is whether one can extend the partial solution to the whole of $\vec{G}$.

5. Co-colouring

Another type of vertex-colouring has been introduced under the name of *co-colouring* as a natural generalization of the basic model (see Gimbel *et al.* [24]). Up to now the colour classes of colourings (the sets of vertices of the same colour) have been independent (or stable) sets. We now allow each colour class to be either an independent set or a clique, so a partition of the vertex-set V of a graph $G = (V, E)$ into p cliques and q independent sets is a co-colouring with $p+q$ colours. The smallest value of $p+q$ for which such a partition exists is the *co-chromatic number* of G. This type of colouring is motivated in particular by an application in robotics (see Demange *et al.* [17]).

Suppose that we wish to schedule the moves of a robot which is required to pick up a collection of items of different sizes along a storage line. The robot may pick up several items during a trip and it has to pile them up: this implies that, to ensure stability, larger items must be placed below smaller ones; in other words, during a trip along the storage line the robot must pick up items in decreasing order of size.

Different assumptions can be made on the possible moves of the robot along the storage line. We can number the stored items from 1 to n in order of decreasing size, so the alignment of the items along the line corresponds to a permutation Π of $\{1, 2, \dots, n\}$. We construct the permutation graph $G(\Pi)$ corresponding to Π by assigning a vertex to each item and by linking vertices i and j if $i > j$ and i comes before j when we move from left to right along the line. So, in a move from left to right, the robot can pick up items in increasing order of their numbers. The vertices of $G(\Pi)$ corresponding to the items picked up during one trip form a independent set, so minimizing the number of left-to-right trips to pick up all the items is equivalent to partitioning the vertex-set V of $G(\Pi)$ into the minimum number of independent sets. It is the basic vertex-colouring model.

If we now assume that the robot can also move from right to left, then the items picked during any trip from right to left correspond to a clique in $G(\Pi)$. So minimizing the number of trips (with both directions allowed) amounts to finding the co-chromatic number of $G(\Pi)$.

As an illustration, suppose that there are nine items stored in line in the order 926415738. A robot moving from left to right needs four trips to pick up all the items:

$$138 \qquad 2457 \qquad 9 \qquad 6.$$

Since the problem corresponds to a permutation graph, there is a polynomial algorithm to find a minimum collection of trips. If we also allow trips from right to left, we have a solution with three trips:

$$24578 \qquad 369 \qquad 1.$$

This corresponds to a minimum co-colouring. The problem is NP-hard, even in the class of comparability graphs; there is, however, a polynomial 2-approximation algorithm which guarantees that we use at most twice as many trips as necessary (see Demange *et al.* [16]).

In Demange *et al.* [17] additional variations on the possible moves of the robot are studied extensively and complexity results are given for these problems. They consider in particular the situation in which each robot (starting from the left or from the right) is allowed to make one move in one direction followed by a move in the other direction; the items picked during such a double move form a clique and an independent set in $G(\Pi)$; each colour class thus defines a *split graph* – that is, a graph whose vertex-set can be partitioned into a clique and an independent set. Partitioning the vertex-set of a graph G into split graphs is the *split colouring problem* which has been defined and studied in Ekim and de Werra [18].

6. Colouring with preferences

In the models examined so far we had essentially rigid constraints that could not be violated. But in almost all real situations there are some 'soft' constraints that may be satisfied or not. When they are violated, a certain penalty is incurred and the problem consists in finding a solution that minimizes the total penalty, or at least keeps it within a given bound. This is precisely what can be done in the vertex-colouring problem with preferences, which is another extension of the basic model.

We start with a graph $G = (V, E)$ and we introduce a subset P of pairs vw (not in E) on which some requirements are added. So we now have a graph $G = (V, E; P)$ in which P is a set of pairs with *preferences*, which we represent by dotted lines. Each edge in E is called a *strong* edge. For each pair $vw \in P$ we have a positive penalty $\omega(vw)$ which is counted whenever the preference is not satisfied.

In this first model (called M_1), the preference on vw could be that v and w should have the same colour, otherwise the penalty $\omega(vw)$ is incurred. We want to find a vertex-colouring with the smallest possible total penalty or with a total penalty that is below a given value W, and a minimum number of colours. As we can see, the number k of colours need not be specified in M_1. Figure 4 shows a colouring of a graph with $k = 3$ colours and a total penalty of 2. The colours are indicated besides the vertices and penalties are shown on the preference edges. No solution can be

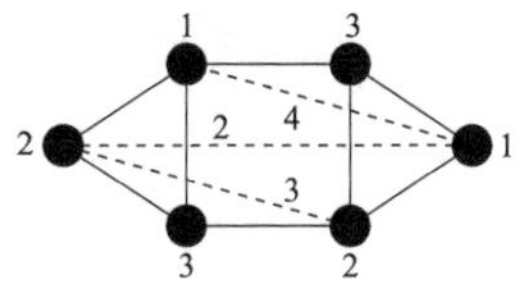

Fig. 4.

found with a smaller penalty, even if we use more colours, since there are cycles with exactly one strong edge.

When does a solution without penalty exist? This question has been studied for a long time, and we have the classical result of Cartwright and Harary [7] that in a graph $G = (V,E;P)$ there exists a vertex-colouring satisfying all preferences if and only if G contains no cycle with exactly one strong edge.

In M_1, minimizing the number of colours remains a difficult problem as can be expected. We can however recognize the situations in which there is a colouring with two colours that satisfies all preferences. As shown in Cartwright and Harary [7], a graph $G = (V,E;P)$ has a 2-colouring satisfying all preferences if and only if every cycle has an even number of strong edges.

The model M_1 is used in compilation, where one has to minimize the number of registers used. The model with $G = (V,E;P)$ has been exploited in Robillard [37]: each variable of the computer programme to be compiled is associated with a vertex of an associated graph. Whenever two variables have to be alive simultaneously (because they appear in the same instruction of the programme), they are joined by an edge. E is the set of these edges, and two variables whose vertices are joined by an edge cannot be assigned to the same register. So finding a vertex-colouring of G minimizing the number of colours gives us an assignment of variables to registers with a minimum number of registers.

Specialists in computation use an operation called *spilling*. It amounts to introducing, for some variable x, another variable x' which may be assigned to another register; practically, one copies the content of variable x at some stage of the execution of the programme into another variable x', and uses x' instead of x in the next instructions. This operation is represented in G by splitting x into two vertices x and x', and linking x and x' by a preference edge with penalty $\omega(xx')$ equal to the cost of copying the content of x into x'. In the resulting graph $G = (V,E;P)$ we may find, as before, whether there is a vertex-colouring without penalty. If so, it means that we can find a register allocation without any spilling. Such an allocation may however require a larger number of registers (colours) than are available, so we may impose a bound on k (the number of available registers) and find a vertex-colouring with at most k colours and a minimum penalty. Such a solution will require some spillings. For example, if three variables x,y,z have to be simultaneously alive, and if no spilling is allowed, then three registers are needed. By introducing an additional variable x' (by splitting x into x and x') we may use only two registers, but with a penalty for the spilling since x and x' get different colours. This is illustrated in Fig. 5.

Fig. 5.

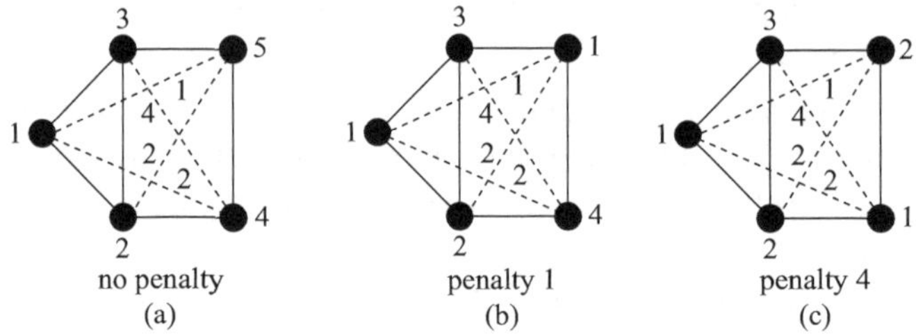

Fig. 6.

A second model (M_2) can also be considered. Here, in $G = (V, E; P)$, the preference edges vw in P are such that we would prefer to have $c(v) \neq c(w)$ in a vertex-colouring, but if we happen to have $c(v) = c(w)$ then a penalty $\omega(vw)$ is incurred. In this model M_2 the number k of colours is fixed, since by using a sufficient number of colours one may always satisfy all preferences.

Such a situation occurs rather naturally in scheduling: two operations v and w which are normally not simultaneous could be assigned to the same operator, but one may get into trouble in case of any delay in v or w; so we prefer to assign v and w to different operators, but if this is not possible then we pay a penalty $\omega(vw)$. The smallest number k of colours for which a vertex-colouring without penalty exists is the chromatic number of the graph obtained by introducing all pairs of P into E.

Finding a colouring with minimum penalty, for a fixed number k, is the *robust colouring problem* which is NP-hard in general (see Yánez and Ramírez [43]). A 5-colouring without penalty is shown in Fig. 6(a). In Figs. 6(b) and 6(c) an optimal robust 4-colouring and an optimal robust 3-colouring are shown, with penalties 1 and 4. This problem is studied by Archetti *et al.* [2].

7. Bandwidth colouring

The central model of vertex-colouring is an ideal instrument for capturing incompatibilities between pairs of items to be scheduled, but it may not be sufficient because of the scheduling applications considered. The notion of incompatibility may be refined by saying more than just 'a pair v, w of items must not be scheduled at the same time period – that is, the corresponding vertices must not be assigned the same colour'.

We may also consider more quantified requirements by saying, for instance, that two items v, w must be scheduled at periods that are separated from each other by at least t_{vw} periods. This is precisely the idea of *bandwidth colouring* (see Prestwich [33]). One may imagine many applications of scheduling in which operations requiring some set-up times must be separated by sufficiently many time units.

A graph $G = (V, E)$ is given, together with a collection of non-negative integer values t_{vw} assigned to the edges vw of E. We wish to find a number k (generally as small as possible) of colours to be used, and a partition of V into independent sets $S_1, S_2, \ldots, S_k$ such that $|c(v) - c(w)| > t_{vw}$ for each edge $vw \in E$, where $c(u)$ is the colour of u (that is, $c(u) = j$ if $u \in S_j$). If $t_{vw} = 0$ for each edge vw, the requirement simply means that $S_1, S_2, \ldots, S_k$ is a vertex-colouring of G.

Such a model may be extended in several ways. First, we could assign to each vertex v a number $\omega(v)$ of consecutive colours; this is the *interval colouring model*, which is able to handle items or operations that have different processing times. In such a case we would have requirements of the form $|f(v) - f(w)| \geq t_{vw}$, where $f(u)$ is the first colour assigned to the vertex u.

In a more general formulation we could consider that each edge vw is an oriented arc from v to w, and we introduce for each such arc a set T_{vw} of integer values that are forbidden for the difference $f(w) - f(v)$ – that is, for each arc vw we require that $f(w) - f(v) \notin T_{vw}$. For instance, if $T_{vw} = \{0\}$ for all arcs vw, and $\omega(v) = 1$ for all $v \in V$, we get the basic vertex-colouring model. If we have an interval vertex-colouring, then to avoid overlap of items v and w in the schedule, we must impose the condition:

$$f(v) + \omega(v) - 1 < f(w) \quad \text{or} \quad f(w) + \omega(w) - 1 < f(v).$$

This means that $T_{vw} = \{-\omega(w) + 1, \ldots, -1, 0, 1, \ldots, \omega(v) - 1\}$ ensures that items v and w are not simultaneously in process. To separate the two processing intervals for v and w, we may just choose a larger set than before.

This general model has been considered initially for frequency assignment problems with possible interferences. The requirements are that on neighbouring emitters the frequencies that have to be assigned to each one have forbidden values for their differences.

Observe that the orientation given to each initial edge vw is arbitrary. If we choose to have an arc wv instead of the arc vw, then we set:

$$T_{wv} = -T_{vw} = \{-a \;:\; a \in T_{vw}\}.$$

For such types of colourings, one can easily extend many classical upper bounds for the chromatic number (see de Werra and Gay [12]).

In these frequency assignment problems it is interesting to distinguish the *span* of the colouring constructed (the difference between the largest and the smallest frequencies) and the *order* (the number of different frequencies). Consider, for

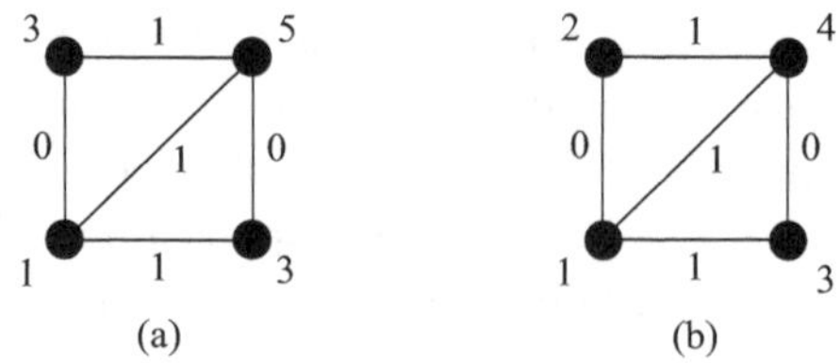

Fig. 7.

example, the graphs in Fig. 7, where the numbers on the edges vw correspond to the values t_{vw} (where $|c(v) - c(w)| > t_{vw}$): Fig. 7(a) shows that a bandwidth colouring c with minimum order 3 has span(c) $= 5 - 1 = 4$, while Fig. 7(b) shows that a bandwidth colouring c with minimum span 3 has order 4.

8. Edge-colouring

A well-known special case of the basic vertex-colouring problem is the edge-colouring model. We are given a multigraph $G = (V, E)$ and we want to partition the edge-set into matchings $M_1, M_2, \ldots, M_p$, so that no two edges in the same subset M_i have a vertex in common. The smallest number p for which such a partition exists is the chromatic index $\chi'(G)$ of G.

Let us consider the line graph $L(G)$ of G, the graph obtained by assigning a vertex v_e to each edge e of G and joining two vertices v_e and v_f in $L(G)$ by an edge if the corresponding edges e,f in G are adjacent. Then there is a one-to-one correspondence between the edge-colourings of G and the vertex-colourings of $L(G)$.

Many applications of edge-colourings have been studied. Among these, the class–teacher timetabling problem and the preemptive open shop scheduling models are the most famous.

Although it would be conceivable to study edge-colourings as special vertex-colourings, it seems to be extremely convenient to consider the edge-colourings of G directly. One reason is that the situation is more naturally visualized in some applications where G has a specific structure (such as being bipartite). Another reason is that there are techniques (such as alternating chain methods) that provide useful construction tools for edge-colourings, and these procedures are easily visualized without going to $L(G)$.

General properties of edge-colourings are presented in Chapter 5. Here we restrict our attention to specific applications related to scheduling and timetabling. The class–teacher timetabling problem is undoubtedly the most basic model, to which a variety of additional constraints may be added, according to the needs of concrete applications. We have a set T of teachers $t_1, t_2, \ldots, t_n$, a set CL of classes $c_1, c_2, \ldots, c_m$ (here a class is a group of students who follow exactly the same programme), and a collection of one-period lectures; each lecture is given by a specific teacher to a specific class.

The data of this problem may be represented by a bipartite multigraph. We associate each teacher t_j and each class c_i to a vertex of the graph, and each one-period lecture given by teacher t_j to class c_i is represented by an edge between c_i and t_j. We thus obtain a bipartite multigraph G with CL as the left-hand set of vertices, and T as the right-hand set. Let $C = \{1,2,\ldots,k\}$ be the set of periods available. A timetable is an assignment of one period in C to each lecture in such a way that no teacher and no class is involved in more than one lecture at any period. This is an edge-colouring of G.

If no other requirements are imposed, it is known that a timetable exists if and only if the number of available periods is no less than the maximum load of all teachers and all classes; here the load of a teacher or of a class is the number of lectures in which the teacher or class is involved, and is the degree of the corresponding vertex of G. This statement follows from König's theorem which asserts that, for a bipartite graph $G = (V,E)$, $\chi'(G) = \Delta(G) = \max_{v \in V} d(v)$. Many variations and extensions of this basic timetabling model have been studied (see Burke *et al.* [6]).

Another classical application of edge-colouring in bipartite multigraphs is the preemptive open shop scheduling problem: we are given a collection $\mathcal{P} = \{P_1, P_2,\ldots,P_m\}$ of processors together with a set $\mathcal{J} = \{J_1,J_2,\ldots,J_n\}$ of jobs. Each job J_j consists of operations $O_{1j}, O_{2j},\ldots,O_{mj}$ to be processed (with interruptions allowed) on the processors in any order. We are given the processing time p_{ij} of O_{ij} for each operation on each processor. We usually assume that p_{ij} is a non-negative integer; if $p_{ij} = 0$, the operation O_{ij} does not exist. A *schedule* is an assignment of p_{ij} time units on P_i (not necessarily consecutive) for each operation O_{ij}, in such a way that at any moment no processor is involved in more than one operation and no two operations of the same job are in process.

We usually want to find a schedule S with a minimum total completion time $C_{\max}(S)$. Clearly, $C_{\max}(S) \geq \max\left\{\max_i \sum_j p_{ij}, \max_j \sum_i p_{ij}\right\}$. We may now associate a vertex to each processor P_i and a vertex to each job J_j. Each operation O_{ij} is represented by p_{ij} parallel edges between the vertices associated to P_i and to J_j. We recognize that if we identify the processors P_i of the open shop scheduling model with the classes c_i of the class–teacher timetabling model, and if we similarly identify the jobs J_j with the teachers t_j, then by considering O_{ij} as the set of lectures of t_j to c_i, we can use the edge-colouring model to solve the open shop scheduling problem. From König's theorem we see that $C_{\max}(S) = \max\left\{\max_i \sum_j p_{ij}, \max_j \sum_i p_{ij}\right\}$, and that without any loss of generality we may restrict the occurrence of preemptions to integer times.

Notice that there is also a non-preemptive open shop scheduling problem, where no interruption is allowed during the execution of an operation O_{ij}. The model to use would then be an interval edge-colouring analogous to the interval vertex-colouring. Minimizing $C_{\max}(S)$ is then an NP-hard problem as soon as the number of processors is at least 3 (see Gonzales and Sahni [25]). For two processors the problem remains easy (see de Werra [11]).

9. Sports scheduling

Another classical application of edge-colouring is the tournament scheduling problem. In its most elementary version we are given a league of $2n$ teams numbered from 1 to $2n$, and each team has to play exactly one game against every other team. This situation is represented by a complete graph K_{2n} on $2n$ vertices associated with the teams, while each edge corresponds to a game involving the two teams associated to its endpoints.

No team can play more than one game on each day, so the games that can be played on the same day form a matching in K_{2n}. Thus, finding a schedule for the $n(2n-1)$ games of the league corresponds to constructing an edge-colouring of K_{2n}. At least $2n-1$ days are needed, and it is known that $\chi'(K_{2n}) = 2n-1$, so a schedule in $2n-1$ days can be found; it corresponds to an edge-colouring of K_{2n} with matchings $M_1, M_2, \ldots, M_{2n-1}$ and $|M_i| = n$ for $i = 1, 2, \ldots, 2n-1$.

A simple way of constructing such a schedule is to define each set M_i (the games played on day i) as follows: team $2n$ plays againt team i, and team $i+k$ plays against team $i-k$ for $k = 1, 2, \ldots, n-1$, where the numbers $i+k$ and $i-k$ are taken (modulo $2n-1$) as one of the numbers $1, 2, \ldots, 2n-1$. Figure 8 shows a five-day schedule for a league of six teams.

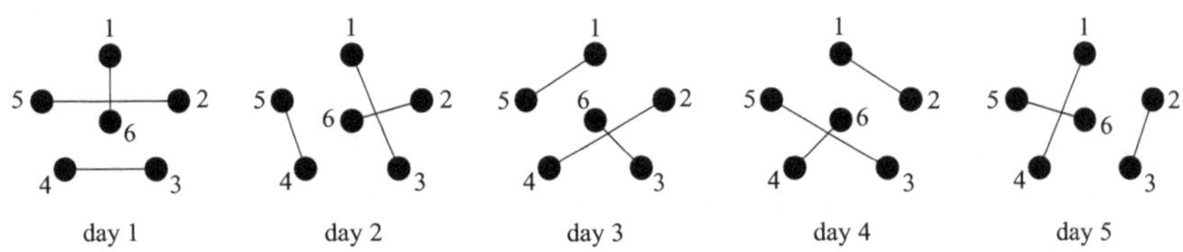

Fig. 8.

We now introduce the idea of an oriented edge-colouring to formulate and solve some related problems that arise in many situations. We assume that each of the $2n$ teams has its own stadium. When a team plays in its home stadium, it is a *Home game* (H) for this team, and when a team plays elsewhere it is an *Away game* (A). When a schedule is constructed, we have an edge-colouring $M_1, M_2, \ldots, M_{2n-1}$, but we still have to decide in which stadium each game should be played. We represent a game played between teams i and j in the home stadium of team j by an arc ij oriented from i to j. So constructing a schedule for such a sports league consists of determining an edge-colouring of K_{2n} and an orientation for each edge.

In general, one desires to have for each team a sequence of games that alternate as regularly as possible between Home games and Away games. Perfect alternation for all teams is not possible; as can be seen easily, we cannot have more than two teams with perfect alternation. When two games for a team i scheduled on consecutive days d and $d+1$ are both Home or both Away, we say that team i has a break on day $d+1$. In other words, the minimum number of breaks in a schedule for $2n$ teams is at least $2n-2$, and in fact, one can always construct a schedule where $2n-2$ teams have

M_1	$\overleftarrow{61}$	$\overrightarrow{25}$	$\overleftarrow{34}$
M_2	$\overrightarrow{62}$	$\overrightarrow{31}$	$\overleftarrow{45}$
M_3	$\overleftarrow{63}$	$\overrightarrow{42}$	$\overleftarrow{51}$
M_4	$\overrightarrow{64}$	$\overrightarrow{53}$	$\overleftarrow{12}$
M_5	$\overleftarrow{65}$	$\overrightarrow{14}$	$\overleftarrow{23}$

oriented colouring

		teams					
		1	2	3	4	5	6
days	1	A	A	H	A	H	H
	2	H	H	A	H	A	A
	3	A	$\underline{\text{H}}$	$\underline{\text{A}}$	A	H	H
	4	H	A	H	H	A	A
	5	A	H	A	$\underline{\text{H}}$	$\underline{\text{A}}$	H

Home–Away pattern

Fig. 9.

exactly one break each and the two remaining teams have no break; this is clearly optimal (see de Werra [10]). The construction is tabulated in Fig. 9 for a league of $2n = 6$ teams; the arrow above each game indicates where the game is played (where $\vec{ij}$ stands for the arc ij). We have underlined the Hs and As corresponding to breaks: teams 1 and 6 have no break, and all others have one break each.

In practice there are many additional requirements to be taken into account. One of these is the fact that some stadia are not always available: for each day with n games, n stadia are used and n stadia are unused. If the stadium of team i is not available on day d, it means that the game played by team i must be an (A) for i, and the edge adjacent to vertex i that receives colour d should be an arc ij for some j. In addition, it may be necessary to schedule some games between teams i and j on one of certain specific days; this gives another kind of list-colouring. There may be additional constraints related to the geography of the league: the sequence of opponents of a given team i may have to be arranged in such a way that team i does not have too many long distances to travel. And there are many more. So while the problem is basically still an edge-colouring problem with orientations to determine, its difficulty may be greatly increased, and other techniques such as integer programming or constraint programming have to be used (see Rassmussen and Trick [34] and Kendall *et al.* [31]).

10. Balancing requirements

We have not considered the use of resources in the scheduling applications discussed so far. However, one usually has some resources available in limited amounts. For this reason, in our chromatic scheduling models consisting of a sequence of k one-period schedules, it may be wise to balance the consumption of resources among the k periods so as to reduce as much as possible the maximum consumption in the different periods.

In the basic class–teacher timetabling problem, for instance, the number of classrooms is limited. We consider them as the resources consumed by the lectures. Assuming that the classrooms are all available during the k teaching periods of the week, we should construct a timetable (here it is an edge-colouring $M_1, M_2, \ldots, M_k$) in such a way that $\max_i |M_i|$ is as small as possible, where the size $|M_i|$ of the matching M_i is precisely the number of classrooms needed during period i.

It is well known that whenever an edge-colouring $M_1, M_2, \ldots, M_k$ exists, there is a *balanced edge-colouring* $M'_1, M'_2, \ldots, M'_k$ with $-1 \leq |M'_i| - |M'_j| \leq 1$, for all $i, j \leq k$. The construction of such a colouring can be easily achieved by trying to balance the cardinalities of two matchings M_i and M_j with $|M_j| - |M_i| \geq 2$. If $G = (V, E)$ is the bipartite graph associated to the class−teacher timetabling problem, and if $k \geq \Delta(G)$ is the number of teaching periods, we need $\lceil |E|/k \rceil$ classrooms. If the number c of classrooms is given, then we can find a timetable in $\max\{\Delta(G), \lceil |E|/c \rceil\}$ periods.

We notice at this stage that the above reasoning assumes that lectures can be moved arbitrarily inside the set C of teaching periods. As soon as we have lists $L(e)$ of available colours for some edges e, this may not work any more.

The above discussion shows that perfect balancing is always possible for the basic edge-colouring problem. The same holds for vertex-colouring in claw-free graphs (graphs that contain no $K_{1,3}$ as an induced subgraph). The reason is that the union of any two colour classes S_i, S_j in a vertex-colouring of a claw-free graph induces a subgraph that consists of even cycles and elementary paths, so that balancing between S_i and S_j is possible. It follows that perfect balancing is possible for any vertex-colouring $S_1, S_2, \ldots, S_k$.

However, for arbitrary graphs perfect balancing is not possible in general: in $K_{1,3}$ any vertex-colouring with $k = 2$ colours has $\big||S_1| - |S_2|\big| = 2$. But using more colours may help the balancing procedure; it has been shown by Hajnal and Szemerédi [27] that, for any graph G, there exists a balanced vertex-colouring with k colours when $k \geq \Delta(G) + 1$; a shorter proof is given by Kierstead and Kostochka [32]. Notice that the set of integers k for which a graph G admits a balanced vertex k-colouring is not necessarily an interval; for example, the graph $K_{3,3}$ in Fig. 10 has a balanced 2-colouring with $S_1 = \{a, b, c\}$, $S_2 = \{d, e, f\}$, and a balanced 4-colouring with $S_1 = \{a, b\}$, $S_2 = \{c\}$, $S_3 = \{d, e\}$, $S_4 = \{f\}$, but no balanced 3-colouring.

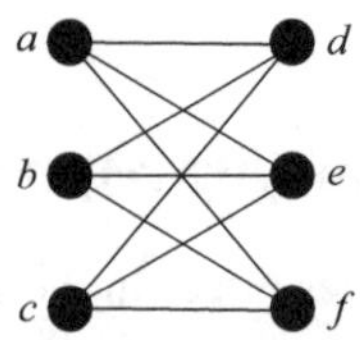

Fig. 10.

For the practice of timetabling, we should also be able to solve the following problem: for a fixed number k of periods (colours), find a schedule (a vertex-colouring) $S_1, S_2, \ldots, S_k$ such that $\max_{1 \leq i \leq k} |S_i| - \min_{1 \leq j \leq k} |S_j|$ is minimum among all possible vertex-colourings with k colours. Finding this optimal value is an NP-hard problem in general, but polynomial algorithms exist for some types of graphs such as trees (see Hertz and Ries [30]). Partial results on balancing in vertex-colouring can be found in de Werra [8]. It has been shown in particular that if G is a graph in which no vertex belongs to more than q maximal cliques, then any vertex-colouring $S_1, S_2, \ldots, S_k$ can be transformed to satisfy:

$$|S_i| - |S_j| \leq (q-2)\min\left\{|S_i|, |S_j|\right\} + 1, \text{ for all } i, j \leq k.$$

Another way of formulating this is to say that, for any $k \geq \chi(G)$, a graph G has a k-colouring $S_1, S_2, \cdots, S_k$ with

$$|S_1| \leq |S_2| \leq \ldots \leq |S_k| \leq (q-1)|S_1| + 1.$$

More generally, we may wonder what can be done in the case where the amount of resource available is not the same for each of the k periods of the schedule; for instance, if h_i is the amount of resource available at period i, does there always exist a schedule represented by a vertex-colouring $S_1, S_2, \ldots, S_k$ with $|S_i| \leq h_i$, for $i = 1, 2, \ldots, k$? This is generally a difficult problem. It has been shown that, even if G is the line graph of a bipartite graph G with $\Delta(G) = 3$, the problem of deciding whether a colouring S_1, S_2, S_3 exists with $|S_i| \leq h_i$ for $i = 1, 2, 3$ is NP-complete (see Even *et al.* [20]). However, if G is the line graph of a tree of bounded degree, then a polynomial algorithm exists for finding such a colouring if it exists (see de Werra *et al.* [13]).

Let us now return to the basic class–teacher timetabling problem for which the model is an edge-colouring $M_1, M_2, \ldots, M_k$ in a bipartite multigraph $G = (V, E)$. We recall that the left-hand set of vertices corresponds to the m classes c_i, and that the right-hand set corresponds to the n teachers t_j. For each pair c_i, t_j, let $m_G(c_i, t_j)$ be the number of parallel edges between c_i and t_j in G: these represent the one-period lectures of teacher t_j to class c_i. We assume now that we have to construct a timetable for a week of k days and that s is the number of teaching periods in any day of the week.

If G is the bipartite multigraph representing the data of our basic class–teacher problem, then for a solution to exist we must have:

$$ks \geq \Delta(G). \tag{1}$$

A timetable corresponds to an edge-colouring $M_1, M_2, \ldots, M_{ks}$. For $r = 1, 2, \ldots, k$, let $N_r = \bigcup_{p=(r-1)s+1}^{rs} M_p \subseteq E$. Then N_r is the set of lectures (edges) assigned to the s periods of day r.

For a subset N_r of edges, we denote by $d_{N_r}(x)$ the degree of the vertex x, and by $m_{N_r}(x, y)$ the number of edges between the vertices x and y in the subgraph of $G = (V, E)$ generated by $N_r \subseteq E$.

In a basic class–teacher timetabling problem, represented by a bipartite multigraph $G = (V, E)$ with a number k of days and s of teaching periods in a day satisfying (2), we say that an edge-colouring $M_1, M_2, \ldots, M_{ks}$ is *perfectly balanced* (on k days) if it satisfies:

$$\lfloor |E|/ks \rfloor \leq |M_t| \leq \lceil |E|/ks \rceil \quad (t = 1, 2 \ldots, ks) \tag{2}$$

(the numbers of lectures scheduled at any period are all within 1 of each other) and, for each $r = 1, 2, \dots, k$,

$$\lfloor |E|/k \rfloor \le |N_r| \le \lceil |E|/k \rceil \tag{3}$$

(the total numbers of lectures scheduled on any day are all within 1 of each other).

Also,

$$\lfloor d_G(c_i)/k \rfloor \le d_{N_r}(c_i) \le \lceil d_G(c_i)/k \rceil \quad (i = 1, 2, \dots, m) \tag{4}$$

(the numbers of lectures involving class c_i on any day are all within 1 of each other),

$$\left\lfloor d_G(t_j)/k \right\rfloor \le d_{N_r}(t_j) \le \left\lceil d_G(t_j)/k \right\rceil \quad (j = 1, 2, \dots, n) \tag{5}$$

(the numbers of lectures involving teacher t_j on any day are all within 1 of each other), and

$$\left\lfloor m_G(c_i, t_j)/k \right\rfloor \le m_{N_r}(c_i, t_j) \le \left\lceil m_G(c_i, t_j)/k \right\rceil \; (i = 1, 2, \dots, m; j = 1, 2, \dots, n) \tag{6}$$

(the lectures of teacher t_j to clas c_i are spread uniformly among the k days, for all teachers t_j and all classes c_i).

By analogy we say that the timetable associated with such an edge-colouring is also perfectly balanced.

One can show (see de Werra [9]) that for the basic class–teacher model there exists a perfectly balanced timetable on k days with s teaching periods per day, for any choice of k, s satisfying $ks \ge \Delta(G)$. As an example, consider the problem represented in Fig. 11(a) by the array giving $m_G(c_i, t_j)$ in entry j, i; we have $k = 3$ and $s = 2$.

(a)

	c_1	c_2	c_3	c_4
t_1	4		1	
t_1		1		4
t_1		4	1	

(b)

N_1

2			
	1		1
	1		

N_2

1		1	
			1
	2		

N_3

1			
			2
	1	1	

(c)

M_1

1			
	1		

M_2

1			
			1
	1		

M_3

1			
	1		

M_4

		1	
			1
	1		

M_5

			1
	1		

M_6

1			
			1
		1	

Fig. 11.

Fig. 11(b) shows the sets N_r of lectures assigned to each day $r \leq k$ $(= 3)$, and Fig. 11(c) shows the edge-colouring with $ks = 6$ colours representing the lectures assigned for each period of the week; it is obtained by constructing a balanced edge-colouring of each of the subgraphs generated by N_r. We see that the resulting edge-colouring $M_1, M_2, \ldots, M_{ks}$ of G satisfies (2).

Notice also that the above example admits a balanced timetable $M'_1, M'_2, \ldots, M'_5$ with $\Delta(G) = 5$ periods $(|M'_t| = 3, t = 1, 2, \ldots, 5)$, but then the constraints (3) do not hold with $k = 3$.

11. Compactness

There are many quality criteria that should be considered when we estimate the value of a schedule. One of these is the *compactness* (see Brélaz *et al.* [5]). Going back to the basic (preemptive or non-preemptive) open shop problem, we may require that the working periods of each processor be consecutive – that is, we may have idle periods at the beginning and/or end of the set of k periods used. Similarly, we may require that the periods during which some operation of a fixed job J_j is in process be consecutive (this does not imply that the periods of processing a single operation are consecutive). If such a schedule can be found, we say that the schedule is *compact*, because each job and each processor has a compact schedule; such schedules are sometimes called 'no wait' schedules. But such a schedule (represented by an edge-colouring of a graph G) may not always exist. For example, consider the case when we have two processors and three jobs consisting each of one operation, with unit processing time on each processor: in any edge-colouring with three colours, there must be one job with a non-compact schedule (see Fig. 12(a) with an idle period for J_3). Observe, however, that there is a compact schedule that uses four colours, as illustrated in Fig. 12(b).

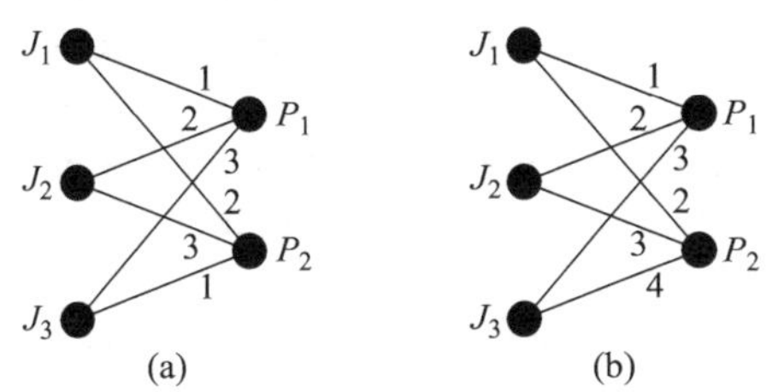

Fig. 12.

In class–teacher timetabling, it is also desirable to have schedules that are compact for each class (so idle times for the pupils between lectures are to be avoided), and hopefully also for the teachers. So, if compact schedules cannot always be found, we have to introduce a measure of distance to compactness that can be minimized, or at least bounded.

For an edge-colouring c we denote by $C(v, c)$ the set of colours assigned to the edges adjacent to vertex v. The *deficiency* $D(v, c)$ of a colouring c at a vertex v is the minimum number of integers to be added to $C(v, c)$ to form an interval. The *deficiency* $D(c)$ of an edge-colouring c of a graph $G = (V, E)$ is the sum $\sum_{v \in V} D(v, c)$. It has been shown by Giaro [22] that finding an edge-colouring with minimum deficiency is NP-hard, even if G is bipartite.

An edge-colouring of G with $D(c) = 0$ (so that $D(v, c) = 0$ for each $v \in V$) is sometimes called a *consecutive colouring*. We define the *deficiency* def(G) of G as the minimum value of $D(c)$ taken over all edge-colourings of G. An edge-colouring c with $D(c) =$ def(G) is called *optimal*. Figure 12 illustrates the fact that def($K_{2,3}$) $= 0$, even if the deficiency of each edge-colouring of $K_{2,3}$ with three colours is strictly positive. Beside partial results obtained for bipartite graphs, k-regular graphs, odd cycles, wheels and complete graphs (see Giaro *et al.* [23], Schwartz [39] and Bouchard *et al.* [4]), very little is known about the deficiency of an arbitrary graph.

Analogous to what was done for interval colourings, we may introduce the idea of *cyclic compactness*: assuming that the schedule is to be repeated, we may require that the colours assigned to edges adjacent to any vertex form a cyclic interval in $C = \{1, 2, \ldots, k\}$; this problem has been formulated and studied in de Werra and Solot [14] and Altinakar *et al.* [1]. Some classes of graphs have been characterized for which cyclic compact interval edge-colourings do exist, whatever the numbers of parallel edges between pairs of vertices. This provides some families of non-preemptive open shop problems with cyclic compactness requirements which can be solved by means of edge-colouring techniques.

12. Conclusion

We have shown through the motivation of applications how numerous variations and extensions of the basic vertex-colouring and edge-colouring models have been introduced and studied. Our purpose was to give an idea of the many directions that have been explored. The development of technologies will undoubtedly create a need for unsuspected colouring models in the future, thus opening original avenues of research.

In this chapter we have not concentrated on solution techniques; this was not our aim. But we should not forget that for all the models presented here there is an urgent need to develop efficient algorithmic procedures that yield either 'optimal' solutions (provided that one can reasonably define a concept of optimality), or at least good approximations to such solutions. The necessity of providing such procedures is enhanced by the fact that in all these applications we are dealing with graphs of large size.

We hope that the above colouring models, which may be viewed as the core of chromatic scheduling, will stimulate the interest of future researchers and users.

References

1. S. Altinakar, G. Caporossi and A. Hertz, On compact k-edge colorings: a polynomial time reduction from linear to cyclic, *Discrete Optimization* **8** (2011), 502–512.
2. C. Archetti, N. Bianchessi and A. Hertz, A branch-and-price algorithm for the robust graph coloring problem, *Les Cahiers du Gerad,* G-2011-75, Montréal, 2011.
3. M. Bouchard, M. Čangalović and A. Hertz, About equivalent interval colorings of weighted graphs, *Discrete Appl. Math.* **157** (2009), 3615–3624.
4. M. Bouchard, A. Hertz and G. Desaulniers, Lower bounds and a tabu search algorithm for the minimum deficiency problem, *J. Combin. Optimization* **17** (2009), 168–191.
5. D. Brélaz, Y. Nicolier and D. de Werra, Compactness and balancing in scheduling, *Zeitschrift für Operations Research* **21** (1977), 65–73.
6. E. K. Burke, D. de Werra and J. Kingston, Applications to timetabling, *Handbook of Graph Theory* (eds. J. L. Gross and J. Yellen), CRC Press (2004), 445–474.
7. D. Cartwright and F. Harary, Structural balance: a generalization of Heider's theory, *Psychological Review* **63** (1956), 277–293.
8. D. de Werra, A note on graph coloring, *RAIRO, Revue Française d'Automatique, d'Informatique et de Recherche Opérationnelle* **R-1** (1974), 49–53.
9. D. de Werra, A few remarks on chromatic scheduling, *Combinatorial programming, Methods and Applications* (ed. B. Roy), D. Reidel (1975), 337–342.
10. D. de Werra, Some models of graphs for scheduling sports competitions, *Discrete Appl. Math.* **21** (1988), 47–65.
11. D. de Werra, Graph theoretical models for preemptive scheduling, *Advances in Project Scheduling* (eds. R. Slowinski and J. Weglarz), Elsevier (1989), 171–185.
12. D. de Werra and Y. Gay, Chromatic scheduling and frequency assignment, *Discrete Appl. Math.* **49** (1994), 165–174.
13. D. de Werra, A. Hertz, D. Kobler and N. V. R. Mahadev, Feasible edge colorings of trees with cardinality constraints, *Discrete Math.* **222** (2000), 61–72.
14. D. de Werra and Ph. Solot, Compact cylindrical chromatic scheduling, *SIAM J. Discrete Math.* **4** (1991), 528–534.
15. M. Demange, D. de Werra, J. Monnot and V. Paschos, Time slot scheduling of compatible jobs, *J. Scheduling* **10** (2007), 111–127.
16. M. Demange, T. Ekim and D. de Werra, On the approximation of min split coloring and min cocoloring, *J. Graph Algorithms and Appl.* **10** (2006), 297–315.
17. M. Demange, T. Ekim and D. de Werra, A tutorial on the use of graph coloring for some problems in robotics, *Europ. J. Oper. Res.* **192** (2009), 41–55.
18. T. Ekim and D. de Werra, On split-coloring problems, *J. Combin. Optimization* **10** (2005), 211–225.
19. P. Erdős, A. L. Rubin and H. Taylor, Choosability in graphs, *Congr. Numer.* **26** (1979), 125–157.
20. S. Even, A. Itai and A. Shamir, On the complexity of timetable and multicommodity flow problems, *SIAM J. Comput.* **5** (1976), 691–703.
21. M. R. Garey and D. S. Johnson, *Computers and Intractability: A Guide to the Theory of NP-completeness*, Freeman, 1979.
22. K. Giaro, The complexity of consecutive Δ-coloring of bipartite graphs: 4 is easy, 5 is hard, *Ars Combin.* **47** (1997), 287–298.
23. K. Giaro, M. Kubale and M. Malafiejski, Consecutive colorings of the edges of general graphs, *Discrete Math.* **236** (2001), 131–143.
24. J. Gimbel, D. Kratsch and L. Stewart, On cocolourings and cochromatic numbers of graphs, *Discrete Appl. Math.* **48** (1994), 111–127.

25. T. Gonzales and S. Sahni, Open shop scheduling to minimize finish time, *J. ACM* **23** (1976), 665–679.
26. S. Gutner, The complexity of planar graph choosability, *Discrete Math.* **159** (1996), 119–130.
27. A. Hajnal and E. Szemerédi, Proof of a conjecture of P. Erdős, *Combinatorial Theory and its Applications*, Balatonfüred (eds. P. Erdős, A. Rényi and V. T. Sós), North-Holland (1970), 601–623.
28. M. M. Halldórsson and G. Kortsarz, Multicoloring: problems and techniques, *Math. Foundations of Computer Science*, LNCS **3153** (2004), 25–41.
29. P. Hansen, J. Kuplinski and D. de Werra, Mixed graph coloring, *Math. Methods of Operations Research* **45** (1997), 145–160.
30. A. Hertz and B. Ries, On *r*-equitable colorings of trees and forests, *Les Cahiers du Gerad,* G-2011-40, Montréal, 2011.
31. G. Kendall, S. Knust, C. C. Ribeiro and S. Urrutia, Scheduling in sports: an annotated bibliography, *Computers Oper. Res.* **37** (2010), 1–19.
32. H. A. Kierstead and A. V. Kostochka, A short proof of the Hajnal–Szemerédi theorem on equitable coloring, *Combin. Probab. Comput.* **17** (2008), 265–270.
33. S. Prestwich, Constrained bandwidth multicoloration neighborhoods, *Proc. Computational Symp. on Graph Coloring and its Generalizations*, Ithaca (eds. D. S. Johnson, A. Mehrotra and M. Trick), (2002), 126–133.
34. R. V. Rassmussen and M. A. Trick, The timetable constrained distance minimization problem, *Annals of Operations Research* **171** (2006), 167–181.
35. B. Ries, Coloring some classes of mixed graph, *Discrete Appl. Math.* **155** (2007), 1–6.
36. B. Ries and D. de Werra, On two coloring problems in mixed graphs, *Europ. J. Combin.* **29** (2008), 712–725.
37. B. Robillard, *Vérification Formelle et Optimisation de l'Allocation de Registres*, Ph.D. Thesis, CNAM, Paris, 2010.
38. B. Roy, Nombre chromatique et plus longs chemins d'un graphe, *RAIRO, Revue Française d'Automatique, d'Informatique et de Recherche Opérationnelle* **1** (1967), 129–132.
39. A. Schwartz, The deficiency of a regular graph, *Discrete Math.* **306** (2006), 1947–1954.
40. Y. N. Sotskov, A. Dolgui and F. Werner, Mixed graph coloring for unit-time job-shop scheduling, *Internat. J. Math. Algorithms* **4** (2001), 289–323.
41. M. Čangalović and J. A. M. Schreuder, Exact colouring algorithm for weighted graphs applied to timetabling problems with lectures of different lengths, *Europ. J. Operational Research* **51** (1991), 248–258.
42. M. Voigt, List colourings of planar graphs, *Discrete Math.* **120** (1993), 215–219.
43. J. Yánez and J. Ramírez, The robust coloring problem, *Europ. J. Operational Research* **148** (2003), 546–558.

13

Graph colouring algorithms

THORE HUSFELDT

This chapter presents an introduction to graph colouring algorithms. The focus is on vertex-colouring algorithms that work for general classes of graphs with worst-case performance guarantees in a sequential model of computation. The presentation aims to demonstrate the breadth of available techniques and is organized by algorithmic paradigm.

1. Introduction

A straightforward algorithm for finding a vertex-colouring of a graph is to search systematically among all mappings from the set of vertices to the set of colours, a technique often called *exhaustive* or *brute force*:

Algorithm X (*Exhaustive search*) *Given an integer $q \geq 1$ and a graph G with vertex-set V, this algorithm finds a vertex-colouring using q colours if one exists.*

X1 [Main loop] For each mapping $f\colon V \to \{1, 2, \ldots, q\}$, do Step X2.

X2 [Check f] If every edge vw satisfies $f(v) \neq f(w)$, terminate with f as the result. ■

This algorithm has few redeeming qualities, other than its being correct. We consider it here because it serves as an opportunity to make explicit the framework in which we present more interesting algorithms.

Model of computation

If G has n vertices and m edges, then the number of operations used by Algorithm X can be asymptotically bounded by $O(q^n(n+m))$, which we call the *running time* of the algorithm.

To make such a claim, we tacitly assume a computational model that includes primitive operations, such as iterating over all mappings from one finite set A to another finite set B in time $O(|B|^{|A|})$ (Step X1), or iterating over all edges in time $O(n+m)$ (Step X2). For instance, we assume that the input graph is represented by an array of sequences indexed by vertices; the sequence stored at vertex v contains the neighouring vertices $N(v)$, see Fig. 1. This representation allows us to iterate over the neighbours of a vertex in time $O(\deg v)$. (An alternative representation, such as an incidence or adjacency matrix, would not allow this.) Note that detecting whether two graphs are isomorphic is *not* a primitive operation. The convention of expressing computational resources using asymptotic notation is consistent with our somewhat cavalier attitude towards the details of our computational model. Our assumptions are consistent with the behaviour of a modern computer in a high-level programming language. Nevertheless, we will explain our algorithms in plain English.

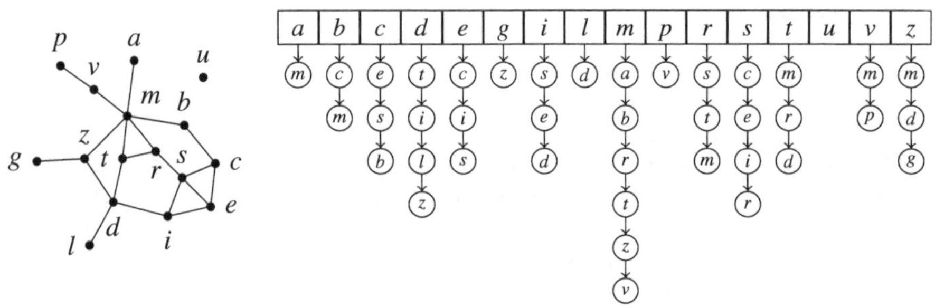

Fig. 1. A graph and its representation as an array of sequences

Worst-case asymptotic analysis

Note that we could have fixed the colouring of a specific vertex v as $f(v) = 0$, reducing Algorithm X's running time to $O(q^{n-1}(n+m))$. A moment's thought shows that this reasoning can then be extended to cliques of size $r \geq 1$: search through all $\binom{n}{r}$ induced subgraphs until a clique of size r is found, arbitrarily map these vertices to $\{1, 2, \ldots, r\}$ and then let Algorithm X colour the remaining vertices. This reduces the running time to $O(q^{n-\omega(G)}n^{\omega(G)}(n+m))$, where $\omega(G)$ is the clique size. This may be quite useful for some graphs. Another observation is that in the best case the running time is $O(n+m)$. However, we will normally not pursue this kind of argument. Instead, we are maximally pessimistic about the input and the algorithm's underspecified choices. In other words, we understand running times as worst-case performance guarantees, rather than 'typical' running times or average running times over some distribution.

Sometimes we may even say that Algorithm X requires time $q^n \operatorname{poly}(n)$, where we leave the polynomial factor unspecified in order to signal the perfunctory attention we extend to these issues.

Overview and notation

Straightforward variants of Algorithm X can be used to solve some other graph colouring problems. For instance, to find a list-colouring, we restrict the range of values for each $f(v)$ to a given list; to find an edge-colouring, we iterate over all mappings $f: E \to \{1, 2, \ldots, q\}$.

Another modification is to count the number of colourings instead of finding just one. These extensions provide baseline algorithms for list-colouring, edge-colouring, the chromatic polynomial, the chromatic index, and so forth. However, for purposes of exposition, we present algorithms in their *least* general form, emphasizing the algorithmic idea rather than its (sometimes quite pedestrian) generalizations. The algorithms are organized by algorithmic technique rather than problem type, graph class, optimality criterion, or computational complexity. These sections are largely independent and can be read in any order, except perhaps for Algorithm G in Section 2. The penultimate section takes a step back and relates the various colouring problems to each other.

2. Greedy colouring

The following algorithm, sometimes called the *greedy* or *sequential* algorithm, considers the vertices one by one and uses the first available colour.

Algorithm G (*Greedy vertex-colouring*) *Given a graph G with maximum degree* Δ *and an ordering* $v_1, v_2, \ldots, v_n$ *of its vertices, this algorithm finds a vertex-colouring with* $\max_i |\{j < i : v_j v_i \in E\}| + 1 \le \Delta + 1$ *colours.*

G1 [Initialize] Set $i = 0$.

G2 [Next vertex] Increment i. If $i = n + 1$, terminate with f as the result.

G3 [Find the colours $N(v_i)$] Compute the set $C = \bigcup_{j<i} f(v_j)$ of colours already assigned to the neighbours of v_i.

G4 [Assign the smallest available colour to v_i] For increasing $c = 1, 2, \ldots$, check whether $c \in C$. If not, set $f(v_i) = c$ and return to Step G2. ■

For the number of colours, it is clear that in Step G4, the value of c is at most $|C|$, which is bounded by the number of neighbours of v_i among $v_1, v_2, \ldots, v_{i-1}$. In particular, Algorithm G establishes that $\chi(G) \le \Delta(G) + 1$.

For the running time, note that both Steps G3 and G4 take at most $O(1 + \deg v_i)$ operations. Summing over all i, the total time spent in Steps G3 and G4 is asymptotically bounded by $n + (\deg v_1 + \deg v_2 + \cdots + \deg v_n) = n + 2m$. Thus, Algorithm G takes time $O(n + m)$.

Optimal ordering

The size of the colouring computed by Algorithm G depends heavily on the vertex ordering. Its worst-case behaviour is poor. For instance, it spends $\frac{1}{2}n$ colours on the 2-colourable *crown graph* shown in Fig. 2.

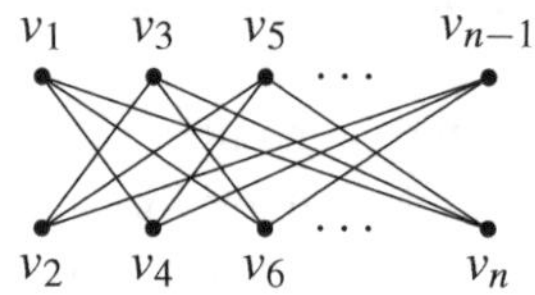

Fig. 2. The crown graph

On the other hand, for every graph there exists an ordering for which Algorithm G uses an optimal number of colours; indeed, any ordering that satisfies $f(v_i) \le f(v_{i+1})$ for an optimal colouring f has this property. Since there are $n!$ different orderings, this observation is algorithmically quite useless. An ordering is *perfect* for a graph if, for every induced subgraph, Algorithm G results in an optimal colouring; triangulated graphs and comparability graphs always admit such an ordering, as shown by Chvátal [11].

Randomness

Algorithm G performs quite well on random graphs, whatever the vertex ordering. For almost all n-vertex graphs, it uses $n/(\log n - 3 \log\log n)$ colours, which is roughly twice the optimum value (see [15]).

This suggests the following randomized algorithm. For a graph G, choose a vertex ordering at random and then execute Algorithm G. For many problems, it is a sound algorithmic design strategy to trade good average-case behaviour for good (expected) worst-case behaviour in this way. However, for Algorithm G the result is quite poor: for every $\varepsilon > 0$ there exist graphs with chromatic number n^ε for which the randomized algorithm uses $\Omega(n/\log n)$ colours with high probability, as shown by Kučera [26].

Other orderings

In the *largest-first* vertex-degree ordering introduced by Welsh and Powell [38], the vertices are ordered such that $\deg v_1 \ge \deg v_2 \ge \cdots \ge \deg v_n$. This establishes the bound $\chi(G) \le 1 + \max_i \min\{\deg v_i, i - 1\}$, which is sometimes better than $1 + \Delta$, such as in Fig. 3.

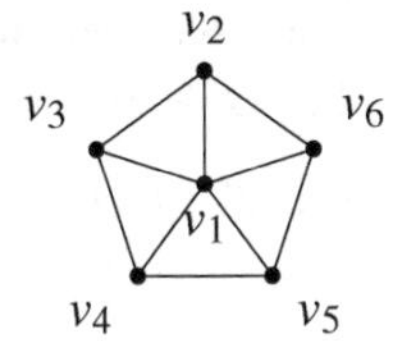

Fig. 3.

Closely related in spirit is Matula's *smallest-last* ordering [32], given as follows: choose as the last vertex v_n a vertex of minimum degree in G, and proceed recursively with $G - v_n$, see Fig. 4. With this ordering, the size of the resulting colouring is bounded by the Szekeres–Wilf bound [36],

$$\chi(G) \leq \text{dgn}(G) + 1\,,$$

where the *degeneracy* dgn(G) is the maximum over all subgraphs H of G of the minimum degree $\delta(H)$. This ordering optimally colours crown graphs and many other classes of graphs, and uses six colours on any planar graph.

Fig. 4.

Other orderings are dynamic, in the sense that the ordering is determined during the execution of the algorithm, rather than in advance. For example, Brélaz [6] suggests choosing the next vertex from among those adjacent to the largest number of different colours. Many other orderings have been investigated (see the surveys of Kosowski and Manuszewski [25] and Maffray [31]). Many of them perform quite well on instances that one may encounter 'in practice', but attempts at formalizing what this means are quixotic.

2-colourable graphs

Of particular interest are those vertex orderings in which every vertex v_i is adjacent to some vertex v_j with $j < i$. Such orderings can be computed in time $O(m+n)$ using basic graph-traversal algorithms. This algorithm is sufficiently important to be made explicit.

Algorithm B (*Bipartition*) *Given a connected graph G, this algorithm finds a 2-colouring if one exists. Otherwise, it outputs an odd cycle.*

B1 [Initialize] Let $f(v_1) = 1$ and let Q (the 'queue') be an empty sequence. For each neighbour w of v_1, set $p(w) = v_1$ (the 'parent' of w) and add w to Q.

B2 [Next vertex] If Q is empty, go to Step B3. Otherwise, remove the first vertex v from Q and set $f(v)$ to the colour not already assigned to $p(v)$. For each neighbour w of v, if w is not yet coloured and does not belong to Q, then set $p(w) = v$ and add w to the end of Q. Repeat Step B2.

B3 [Verify 2-colouring] Iterate over all edges to verify that $f(v) \neq f(w)$ for every edge vw. If so, terminate with f as the result.

B4 [Construct odd cycle] Let vw be an edge with $f(v) = f(w)$ and let u be the nearest common ancestor of v and w in the tree defined by p. Output the path $w, p(w), p(p(w)), \ldots, u$, followed by the reversal of the path $v, p(v), p(p(v)), \ldots,$ u, followed by the edge vw. ■

Figure 5 shows an execution of Algorithm B finding a 2-colouring.

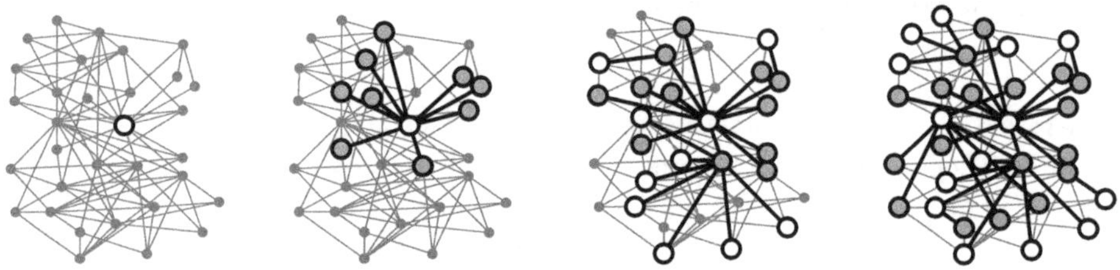

Fig. 5. Execution of Algorithm B

Algorithm B is an example of a 'certifying' algorithm: an algorithm that produces a witness to certify its correctness, in this case an odd cycle if the graph is not 2-colourable. To see that the cycle constructed in Step B4 has odd length, note that on the two paths $w, p(w), p(p(w)), \ldots, u$ and $v, p(v), p(p(v)), \ldots, u$, each vertex has a different colour from its predecessor. Since the respective endpoints of both paths have the same colour, they must contain the same number of edges modulo 2. In particular, their total length is even. With the additional edge vw, the length of the resulting cycle is odd.

The order in which the vertices are considered by Algorithm B depends on the first-in first-out behaviour of the queue Q. The resulting ordering is called *breadth-first*. An important variant uses a last-in first-out 'stack' instead of a queue; the resulting ordering is called *depth-first*. Figure 6 shows the resulting behaviour on the graph from Fig. 5.

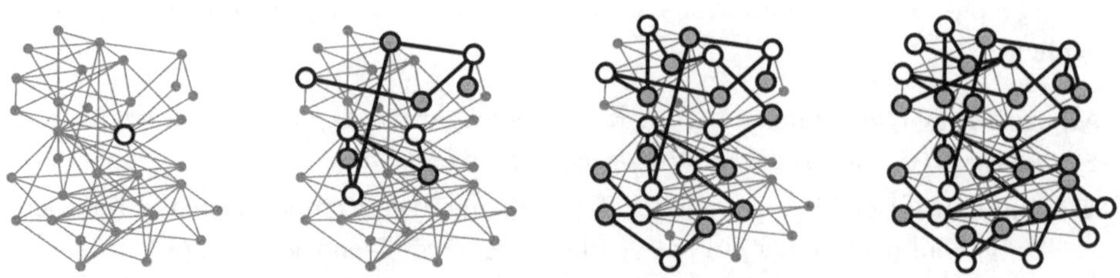

Fig. 6. Execution of Algorithm B using depth-first search

Algorithm B works also for the list-colouring problem, provided that for each vertex v the available list of colours $L(v)$ has size at most 2. This observation leads to a simple randomized exponential-time algorithm for 3-colouring, due to Beigel and Eppstein [1].

Algorithm P (*Palette restriction*) *Given a graph, this algorithm finds a 3-colouring if one exists.*

P1 [Forbid one colour at each vertex] For each vertex v, select a list $L(v)$ of colours available at v uniformly and independently at random from the three lists $\{1, 2\}$, $\{2, 3\}$ and $\{1, 3\}$.

P2 [Attempt a 2-colouring] Try to solve the list-colouring instance given by L using Algorithm B, setting $f(v_1) = \min L(v_1)$ in Step B1. If successful, terminate with the resulting colouring. Otherwise, return to Step P1. ■

To analyze the running time, consider a 3-colouring f. For each vertex v, the colour $f(v)$ belongs to $L(v)$ with probablity $\frac{2}{3}$. Thus, with probability at least $(\frac{2}{3})^n$, the list colouring instance constructed in step P1 has a solution. It follows that the expected number of repetitions is $(\frac{3}{2})^n$, each of which takes polynomial time.

Wigderson's algorithm

Algorithms B and G appear together in Wigderson's algorithm [40].

Algorithm W (*Wigderson's algorithm*) *Given a 3-chromatic graph G, this algorithm finds a vertex-colouring with $O(\sqrt{n})$ colours.*

W1 [Initialize] Let $c = 1$.

W2 [$\Delta(G) \geq \lceil\sqrt{n}\rceil$] Consider a vertex v in G with $\deg v \geq \lceil\sqrt{n}\rceil$; if no such vertex exists, go to Step W3. Use Algorithm B to 2-colour the neighbourhood $G[N(v)]$ with colours c and $c + 1$. Remove $N(v)$ from G and increase c by $\chi(G[N(v)])$. Repeat Step W2.

W3 [$\Delta(G) < \lceil\sqrt{n}\rceil$] Use Algorithm G to colour the remaining vertices with the colours $c, c + 1, \ldots, c + \lceil\sqrt{n}\rceil$. ■

Fig. 7 shows an execution of Algorithm W finding a 5-colouring of the 16-vertex instance from Fig. 1.

The running time is clearly bounded by $O(n + m)$. To analyze the number of colours, we first need to verify Step W2. Since G is 3-colourable, so is the subgraph induced by $N(v) \cup \{v\}$. Now, if $G[N(v)]$ requires three colours, then $G[N(v) \cup \{v\}]$ requires four, so $G[N(v)]$ is 2-colourable and therefore Step W2 is correct. Note that Step W2 can be run at most $O(\sqrt{n})$ times, each using at most two colours. Step W3 expends another $\lceil\sqrt{n}\rceil$ colours, according to Algorithm G.

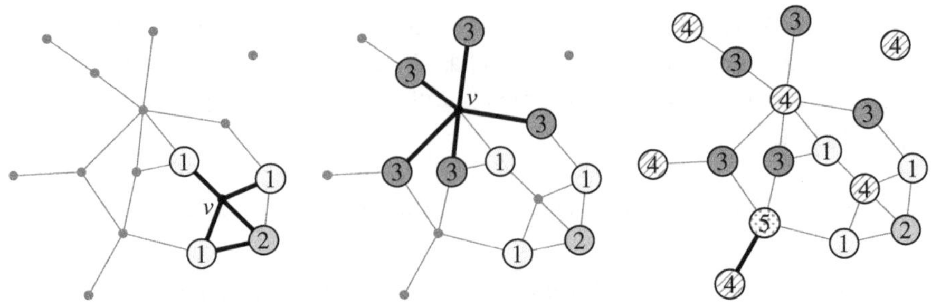

Fig. 7. Execution of Algorithm W

Algorithm W extends naturally to graphs with $\chi(G) > 3$. In this case, Step W2 calls Algorithm W recursively to colour $(\chi(G)-1)$-colourable neighbourhoods. The resulting algorithm uses $O(n^{1-1/(1-\chi(G))})$ colours.

3. Recursion

Recursion is a fundamental algorithmic design technique. The idea is to reduce a problem to one or more simpler instances of the same problem.

Contraction

The oldest recursive construction for graph colouring expresses the chromatic polynomial $P(G,q)$ and the chromatic number $\chi(G)$ in terms of edge-contractions: for non-adjacent vertices v, w and integer $q = 0, 1, \ldots, n$,

$$P(G,q) = P(G \cup vw, q) + P(G/vw, q)\,,$$
$$\chi(G) = \min\{\chi(G \cup vw), \chi(G/vw)\}$$

(see Chapter 3, Section 2.1). These 'addition–contraction' recurrences immediately imply a recursive algorithm. For instance,

$P($$, q) = P($$, q) + P($$, q)$

$$= P(K_4, q) + P(K_3, q) = q(q-1)(q-2)\big((q-3)(q-4)+1\big)\,.$$

Note that the graphs at the end of the recursion are complete.

For sparse graphs, it is more useful to express the same idea as a 'deletion–contraction' recurrence, which deletes and contracts edges until the graph is empty:

$$P(G,q) = P(G/e,q) - P(G-e,q) \qquad (e \in E)\,.$$

Many other graph problems beside colouring can be expressed by a deletion–contraction recurrence. The most general graph invariant that can be defined in this fashion is the Tutte polynomial (see [5] and [18] for its algorithmic aspects).

The algorithm implied by these recursions is sometimes called *Zykov's algorithm* [42]. Here is the deletion–contraction version.

Algorithm C (*Contraction*) *Given a graph G, this algorithm returns the sequence of coefficients* $(a_0, a_1, \ldots, a_n)$ *of the chromatic polynomial* $P(G,q) = \sum_{i=0}^{n} a_i q^i$.

C1 [Base] If G has no edges then return the coefficients $(0,0,\ldots,0,1)$, corresponding to the polynomial $P(G,q) = q^n$.

C2 [Recursion] Pick an edge e and construct the graphs $G' = G/e$ and $G'' = G - e$. Call Algorithm C recursively to compute $P(G',q)$ and $P(G'',q)$ as sequences of coefficients $(a'_0, a'_1, \ldots, a'_n)$ and $(a''_0, a''_1, \ldots, a''_n)$. Return $(a'_0 - a''_0, a'_1 - a''_1, \ldots, a'_n - a''_n)$, corresponding to the polynomial $P(G/e,q) - P(G-e,q)$. ■

To analyze the running time, let $T(r)$ be the number of executions of Step C1 for graphs with n vertices and m edges, where $r = n + m$. The two graphs constructed in Step C2 have size $n - 1 + m - 1 = r - 2$ and $n + m - 1 = r - 1$, respectively, so T satisfies $T(r) = T(r-1) + T(r-2)$. This is a well-known recurrence with solution $T(r) = O(\varphi^r)$, where $\varphi = \frac{1}{2}(1 + \sqrt{5})$ is the golden ratio. Thus, Algorithm C requires $\varphi^{n+m}\,\mathrm{poly}(n) = O(1.619^{n+m})$ time. A similar analysis for the algorithm implied by the deletion–addition recursion gives $\varphi^{n+\overline{m}}\,\mathrm{poly}(n)$, where $\overline{m} = \binom{n}{2} - m$ is the number of edges in the complement of G.

These worst-case bounds are often very pessimistic. They do not take into account that recurrences can be stopped as soon as the graph is a tree (or some other easily recognized graph whose chromatic polynomial is known as a closed formula), or that P factorizes over connected components. Moreover, we can use graph isomorphism heuristics and tabulation to avoid some unnecessary recomputation of isomorphic subproblems (see [18]). Thus, Algorithm C is a more useful algorithm than its exponential running time may indicate.

Vertex partitions and dynamic programming

We turn to a different recurrence, which expresses $\chi(G)$ in terms of induced subgraphs of G. By taking S to be a colour class of an optimal colouring of G, we observe that every graph has an independent set of vertices S for which $\chi(G) = 1 + \chi(G - S)$. Thus, we have

$$\chi(G) = 1 + \min \chi(G - S)\,, \tag{1}$$

where the minimum is taken over all non-empty independent sets S in G.

The recursive algorithm implied by (1) is too slow to be of interest. We expedite it by using the fundamental algorithmic idea of *dynamic programming*. The central observation is that the subproblems $\chi(G - S)$ for various vertex-subsets S appearing in (1) are computed over and over again. It thus makes sense to store these 2^n values in a table when they are first computed. Subsequent evaluations can then be handled by consulting the table.

We express the resulting algorithm in a bottom-up fashion:

Algorithm D (*Dynamic programming*) *Given a graph G, this algorithm computes a table T with $T(W) = \chi(G[W])$, for each $W \subseteq V$.*

D1 [Initialize] Construct a table with (initially undefined) entries $T(W)$ for each $W \subseteq V$. Set $T(\varnothing) = 0$.

D2 [Main loop] List all vertex-subsets $W_1, W_2, \ldots, W_{2^n} \subseteq V$ in non-decreasing order of their size. Do Step D3 for $W = W_2, W_3, \ldots, W_{2^n}$, then terminate.

D3 [Determine $T(W)$] Set $T(W) = 1 + \min T(W \setminus S)$, where the minimum is taken over all non-empty independent sets S in $G[W]$. ■

The ordering of subsets in the main loop D2 ensures that each set is handled before any of its supersets. In particular, all values $T(W \setminus S)$ needed in Step D3 will have been previously computed, so the algorithm is well defined. The minimization in Step D3 is implemented by iterating over all $2^{|W|}$ subsets of W. Thus, the total running time of Algorithm D is within a polynomial factor of

$$\sum_{W \subseteq V} 2^{|W|} = \sum_{k=0}^{n} \binom{n}{k} 2^k = 3^n . \tag{2}$$

This rather straightforward application of dynamic programming already provides the non-trivial insight that the chromatic number can be computed in time exponential in the number of vertices, rather than depending exponentially on m, $\chi(G)$, or a superlinear function of n.

Maximal independent sets

To pursue this idea a little further we notice that S in (1) can be assumed to be a *maximal* independent (stable) set – that is, it is not a proper subset of another independent set. To see this, let f be an optimal colouring and consider the colour class $S = f^{-1}(1)$. If S is not maximal, then repeatedly pick a vertex v that is not adjacent to S, and set $f(v) = 1$.

By considering the disjoint union of $\frac{1}{3}k$ triangles, we see that there exist k-vertex graphs with $3^{k/3}$ maximal independent sets. It is known that this is also an upper bound, and that the maximal independent sets can be enumerated within a polynomial factor of that bound (see [7], [34] and [37]). We therefore have the following result:

Theorem 3.1 *The maximal independent sets of a graph on k vertices can be listed in time $O(3^{k/3})$ and polynomial space.*

We can apply this idea to Algorithm D. The minimization in Step D3 now takes the following form:

D3′ [Determine $T(W)$] Set $T(W) = 1 + \min T(W \setminus S)$, where the minimum is taken over all maximal independent sets S in $G[W]$.

Using Theorem 3.1 with $k = |W|$ for the minimization in Step D3′, the total running time of Algorithm D comes within a polynomial factor of

$$\sum_{k=0}^{n} \binom{n}{k} 3^{k/3} = (1 + 3^{1/3})^n = O(2.443^n).$$

For many years, this was the fastest-known algorithm for the chromatic number.

3-colouring

Of particular interest is the 3-colouring case. Here, it makes more sense to let the outer loop iterate over all maximal independent sets and check whether the complement is bipartite.

Algorithm L (*Lawler's algorithm*) *Given a graph G, this algorithm finds a 3-colouring if one exists.*

L1 [Main loop] For each maximal independent set S of G, do Step L2.

L2 [Try $f(S) = 3$] Use Algorithm B to find a colouring $f\colon V \setminus S \to \{1, 2\}$ of $G - S$ if one exists. In that case, extend f to all of V by setting $f(v) = 3$ for each $v \in S$, and terminate with f as the result. ■

The running time of Algorithm L is dominated by the number of executions of L2, which is $3^{n/3}$, by Theorem 3.1. Thus, Algorithm L decides 3-colourability in time $3^{n/3}\,\mathrm{poly}(n) = O(1.442^n)$ and polynomial space.

The use of maximal independent sets goes back to Christofides [10], while Algorithms D and L are due to Lawler [28]. A series of improvements to these ideas have further reduced these running times. At the time of writing, the best-known time bound for 3-colouring is $O(1.329^n)$ by Beigel and Eppstein [1].

4. Subgraph expansion

The *Whitney expansion* [39] of the chromatic polynomial is

$$P(G,q) = \sum_{A \subseteq E} (-1)^{|A|} q^{k(A)};$$

see Chapter 3, Section 2 for a proof. It expresses the chromatic polynomial as an alternating sum of terms, each of which depends on the number of connected components $k(A)$ of the edge-subset $A \subseteq E$. Determining $k(A)$ is a well-studied algorithmic graph problem, which can be solved in time $O(n+m)$ (for example, by depth-first search). Thus, the Whitney expansion can be evaluated in time $O(2^m(n+m))$.

A more recent expression (see [2]) provides an expansion over *induced* subgraphs:

Theorem 4.1 *For $W \subseteq V$, let $g(W)$ be the number of non-empty independent sets in $G[W]$. Then G can be q-coloured if and only if*

$$\sum_{W \subseteq V} (-1)^{|V \setminus W|} \big(g(W)\big)^q > 0 \,. \tag{3}$$

Proof For each $W \subseteq V$, the term $\big(g(W)\big)^q$ counts the number of ways of selecting q non-empty independent sets $S_1, S_2, \ldots, S_q$, where $S_i \subseteq W$. For $U \subseteq V$, let $h(U)$ be the number of ways of selecting q non-empty independent sets whose union is U. Then $(g(W))^q = \sum_{U \subseteq W} h(U)$, so

$$\begin{aligned} \sum_{W \subseteq V} (-1)^{|V \setminus W|} \big(g(W)\big)^q &= \sum_{W \subseteq V} (-1)^{|V \setminus W|} \sum_{U \subseteq W} h(U) \\ &= \sum_{U \subseteq V} h(U) \sum_{W \supseteq U} (-1)^{|V \setminus W|} = h(V) \,. \end{aligned}$$

For the last step, note that the inner sum (over W, with $U \subseteq W \subseteq V$) vanishes except when $U = V$, because there are as many odd-sized as even-sized sets sandwiched between different sets, by the principle of inclusion–exclusion.

If $h(V)$ is non-zero, then there exist independent sets $S_1, S_2, \ldots, S_q$ whose union is V. These sets correspond to a colouring: associate a colour with the vertices in each set, breaking ties arbitrarily. ■

For each $W \subseteq V$, we can compute the value $g(W)$ in time $O(2^{|W|}m)$ by constructing each non-empty subset of W and testing it for independence. Thus, the total running time for evaluating (3) is within a polynomial factor of 3^n, just as in the analysis (2) for Algorithm D; however, the space requirement here is only polynomial. We can further reduce the running time to $O(2.247^n)$ by using dedicated algorithms for evaluating $g(W)$ from the literature (see [3]).

If exponential space is available, we can do even better. To that end, we first introduce a recurrence for g.

Theorem 4.2 *Let $W \subseteq V$. We have $g(\varnothing) = 0$, and, for every $v \in W$,*

$$g(W) = g(W \setminus \{v\}) + g(W \setminus N(v)) + 1 \,, \tag{4}$$

where $N(v)$ is the neighbourhood of V.

Proof Fix $v \in W$. The non-empty independent sets $S \subseteq W$ can be partitioned into two classes with $v \notin S$ and $v \in S$. In the first case, S is a non-empty independent set with $S \subseteq W \setminus \{v\}$ and is thus accounted for by the first term of (4). Consider the second case. Since S contains v and is independent, it contains no vertex from $N(v)$. Thus, S

is a non-empty independent set with $\{v\} \subseteq S \subseteq W \setminus N(v)$. The number of such sets is the same as the number of (not necessarily non-empty) independent sets S' with $S' \subseteq W \setminus N(v)$, because of the bijective mapping $S \mapsto S'$ where $S' = S \setminus \{v\}$. By induction, the number of such sets is $g(W \setminus N(v)) + 1$, where the '+1' term accounts for the empty set. ■

This leads to the following algorithm, due to Björklund *et al.* [3]:

Algorithm I (*Inclusion–exclusion*) *Given a graph G and an integer $q \geq 1$, this algorithm determines whether G can be q-coloured.*

I1 [Tabulate g] Set $g(\varnothing) = 0$. For each non-empty subset $W \subseteq V$ in inclusion order, pick $v \in W$ and set $g(W) = g(W \setminus \{v\}) + g(W \setminus N[v]) + 1$.

I2 [Evaluate (3)] If $\sum_{W \subseteq V} (-1)^{|V \setminus W|} (g(W))^q > 0$ output 'yes', otherwise 'no'. ■

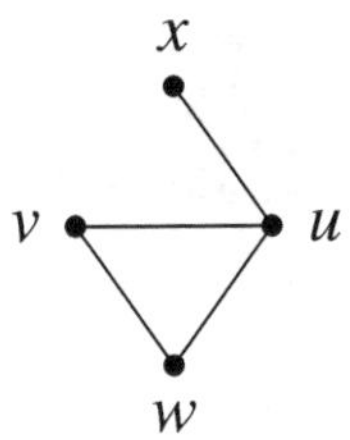

W	g	a_2	a_3
$\varnothing$	0	0	0
$\{u\}$	1	-1	-1
$\{v\}$	1	-1	-1
$\{w\}$	1	-1	-1
$\{x\}$	1	-1	-1
$\{u,v\}$	2	4	8
$\{u,w\}$	2	4	8
$\{u,x\}$	2	4	8

W	g	a_2	a_3
$\{v,w\}$	2	4	8
$\{v,x\}$	3	9	27
$\{w,x\}$	3	9	27
$\{u,v,w\}$	3	-9	-27
$\{u,v,x\}$	4	-16	-64
$\{u,w,x\}$	4	-16	-64
$\{v,w,x\}$	5	-25	-125
V	6	36	216

Fig. 8. Execution of Algorithm I

Both Steps I1 and I2 take time $2^n \operatorname{poly}(n)$, and the algorithm requires a table with 2^n entries. Fig. 8 shows the computations of Algorithm I on a small graph for $q = 2$ and $q = 3$, with $a_q(W) = (-1)^{|V \setminus W|}(g(W))^q$. The sum of the entries in column a_2 is 0, so there is no 2-colouring. The sum of the entries in column a_3 is 18, so a 3-colouring exists.

With slight modifications, Algorithm I can be made to work for other colouring problems such as the chromatic polynomial and list-colouring, also in time and space $2^n \operatorname{poly}(n)$ (see [3]); currently, this is the fastest-known algorithm for these problems. For the chromatic polynomial, the space requirement can be reduced to $O(1.292^n)$, while maintaining the $2^n \operatorname{poly}(n)$ running time (see [4]).

5. Local augmentation

Sometimes a non-optimal colouring can be improved by a local change that recolours some vertices. This general idea is the basis of many local search heuristics and also several central theorems.

Kempe changes

An important example for edge-colouring establishes Vizing's theorem, $\Delta(G) \le \chi'(G) \le \Delta(G) + 1$. Chapter 5 gives a modern and more general presentation of the underlying idea, and our focus in the present chapter is to make the algorithm explicit.

A colour is *free* at v if it does not appear on an edge at v. (We consider an edge-colouring with $\Delta(G) + 1$ colours, so every vertex has at least one free colour.) A (Vizing) *fan* around v is a maximal set of edges $vw_0, vw_1, \ldots, vw_r$, where vw_0 is not yet coloured and the other edges are coloured as follows. For $j = 0, 1, \ldots, r$, no colour is free at both v and w_j. For $j = 1, 2, \ldots, r$, the jth fan edge vw_j has colour j and the colours appearing around w_j include $1, 2, \ldots, j$ but not $j + 1$ (see Fig. 9(a)). Such a fan allows a recolouring by moving colours as follows: remove the colour from vw_j and set $f(vw_0) = 1, f(vw_1) = 2, \ldots, f(vw_{j-1}) = j$. This is called *downshifting from j* (see Fig. 9(b)).

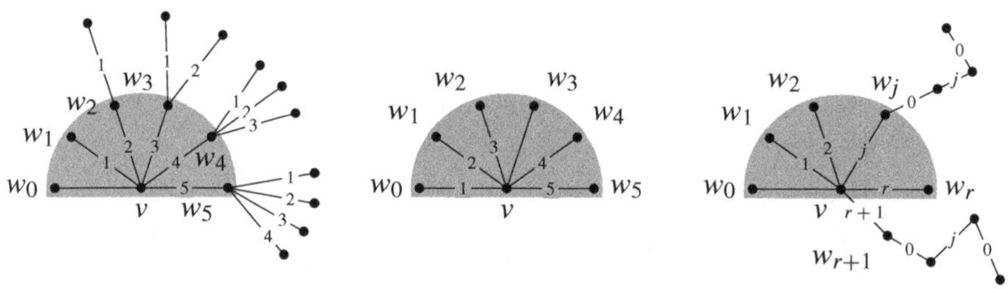

Fig. 9. (a) A fan (b) Downshifting from 3 (c) Step V7: colour j is free at w_{r+1}

Algorithm V (*Vizing's algorithm*) *Given a graph G, this algorithm finds an edge colouring with at most $\Delta(G) + 1$ colours in time $O(nm)$.*

V1 [Initialize] Order the edges arbitrarily $e_1, e_2, \ldots, e_m$. Let $i = 0$.

V2 [Extend colouring to next edge] Increment i. If $i = m + 1$, then terminate. Otherwise, let $vw = e_i$.

V3 [Easy case] If a colour c is free at both v and w, then set $f(vw) = c$ and return to Step V2.

V4 [Find w_0 and w_1] Let $w_0 = w$. Pick a free colour at w_0 and call it 1. Let vw_1 be the edge incident with v coloured 1. (Such an edge exists because 1 is not also free at v.)

V5 [Find w_2] Pick a free colour at w_1 and call it 2. If 2 is also free at v, then set $f(vw_0) = 1, f(vw_1) = 2$, and return to Step V2. Otherwise, let vw_2 be the edge incident with v coloured 2. Set $r = 2$.

V6 [Extend fan to w_{r+1}] Pick a free colour at w_r and call it $r+1$. If $r+1$ is also free at v, then downshift from r, recolour $f(vw_r) = c_{r+1}$ and return to Step V2. Otherwise, let vw_{r+1} be the edge incident with v coloured $r+1$. If each colour $1, 2, \ldots, r$ appears around w_{r+1}, then increment r and repeat Step V6.

V7 [Build a $\{0, j\}$-path from w_j or from w_{r+1}] Let $j \in \{1, 2, \ldots, r\}$ be a free colour at w_{r+1} and let 0 be a colour free at v and different from j. Construct two maximal $\{0, j\}$-coloured paths P_j and P_{r+1} from w_j and w_{r+1}, respectively, by following edges of alternating colours $0, j, 0, j, \ldots$ (see Fig. 9(c)). (The paths cannot both end in v.) Let $k = j$ or $r+1$ so that P_k does not end in v.

V8 [Flip colours on P_k] Recolour the edges on P_k by exchanging 0 and j. Downshift from k, recolour $f(vw_k) = 0$, and return to Step V2. ■

To see that this algorithm is correct, one needs to check that the recolourings in Steps V6 and V8 are legal. A careful analysis was given by Misra and Gries [33].

For the running time, first note that Step V6 is repeated at most $\deg v$ times, so the algorithm eventually has to leave that step. The most time-consuming step is Step V7: a $\{0, j\}$-path can be constructed in time $O(n)$ if for each vertex we maintain a table of incident edges indexed by colour. Thus the total running time of Algorithm V is $O(mn)$.

Another example from this class of algorithms appears in the proof of Brooks's theorem (see Chapter 2 and [8]), which relies on an algorithm that follows Algorithm G but attempts to re-colour the vertices of bichromatic components whenever a fresh colour is about to be introduced.

Random changes

There are many other graph colouring algorithms that fall under the umbrella of local transformations. Of particular interest are local search algorithms that recolour individual vertices at random. This idea defines a random process on the set of colourings called the *Glauber* or *Metropolis* dynamics, or the natural Markov chain Monte Carlo method. The aim here is not merely to find a colouring (since $q > 4\Delta$, this would be easily done by Algorithm G), but to find a colouring that is uniformly distributed among all q-colourings.

Algorithm M (*Metropolis*) *Given a graph G with maximum degree Δ and a q-colouring f_0 for $q > 4\Delta$, this algorithm finds a uniform random q-colouring f_T in polynomial time.*

M1 [Outer loop] Set $T = \lceil qn \ln 2n/(q - 4\Delta) \rceil$. Do Step M2 for $t = 1, 2, \ldots, T$, then terminate.

M2 [Recolour a random vertex] Pick a vertex $v \in V$ and a colour $c \in \{1, 2, \ldots, q\}$ uniformly at random. Set $f_t = f_{t-1}$. If c does not appear among v's neighbours, then set $f_t(v) = c$. ■

An initial colouring f_0 can be provided in polynomial time because $q > \Delta + 1$ – for example, by Algorithm G. To see that the choice of initial colouring f_0 has no influence on the result f_T, we consider two different initial colourings f_0 and f'_0 and execute Algorithm M on both, using the same random choices for v and c in each step.

Let $d_t = |\{ v\colon f_t(v) \neq f'_t(v) \}|$ be the number of *disagreeing* vertices after t executions of Step M2. Each step can change only a single vertex, so $|d_t - d_{t-1}| = 1$, 0, or -1. We have $d_t = d_{t-1} + 1$ only if $f_{t-1}(v) = f'_{t-1}(v)$ but $f_t(v) \neq f'_t(v)$, so exactly one of the two processes rejects the colour change. In particular, v must have a (disagreeing) neighbour w with $c = f_{t-1}(w) \neq f'_{t-1}(w)$ or $f_{t-1}(w) \neq f'_{t-1}(w) = c$. There are d_{t-1} choices for w and therefore $2\Delta d_{t-1}$ choices for c and v. Similarly, we have $d_t = d_{t-1} - 1$ only if $f_{t-1}(v) \neq f_{t-1}(v)$ and c does not appear in v's neighbourhood in either f_{t-1} or f'_{t-1}. There are at least $(q - 2\Delta)d_{t-1}$ such choices for c and v.

Thus, the expected value of d_t can be bounded as follows:

$$\mathbf{E}[d_t] \leq \mathbf{E}[d_{t-1}] + \frac{(q-2\Delta)\mathbf{E}[d_{t-1}]}{qn} - \frac{2\Delta\mathbf{E}[d_{t-1}]}{qn} = \mathbf{E}[d_{t-1}]\left(1 - \frac{q-4\Delta}{qn}\right).$$

Iterating this argument and using $d_0 \leq n$, we have

$$\mathbf{E}[d_T] \leq n\left(1 - \frac{q-4\Delta}{qn}\right)^T \leq n\exp\left(-\frac{T(q-4\Delta)}{qn}\right) \leq n\exp(-\ln 2n) = \tfrac{1}{2}\,.$$

By Markov's inequality, and because d_T is a non-negative integer, we conclude that

$$\Pr(f_T = f'_T) = \Pr(d_T = 0) \geq 1 - \Pr(d_T \geq 1) \geq 1 - \mathbf{E}[d_T] \geq \tfrac{1}{2}\,.$$

We content ourselves with this argument, which shows that the process is 'sufficiently random' in the sense of being memoryless. Informally, we can convince ourselves that f_T is uniformly distributed because we can assume that f'_0 in the above argument was sampled according to such a distribution. This intuition can be formalized using standard coupling arguments for Markov chains; our calculations above show that the 'mixing time' of Algorithm M is $O(n \log n)$.

Algorithm M and its variants have been well studied, and the analysis can be much improved (see the survey of Frieze and Vigoda [13]). Randomized local search has wide appeal across disciplines, including simulations in statistical physics and heuristic methods in combinatorial optimization.

6. Vector colouring

We now turn to a variant of vertex-colouring that is particularly interesting from an algorithmic point of view.

Vector chromatic number

Let $S^{d-1} = \{\mathbf{x} \in \mathbb{R}^d : \|\mathbf{x}\| = 1\}$. A *vector q-colouring* in $d \le n$ dimensions is a mapping $x\colon V \to S^{d-1}$ from the vertex-set to the set of d-dimensional unit vectors for which neighbouring vectors are 'far apart', in the sense that their scalar product satisfies

$$\langle x(v), x(w)\rangle \le -\frac{1}{q-1}, \qquad \text{for } vw \in E.$$

The smallest such number q is called the *vector chromatic number* $\vec{\chi}(G)$, which need not be an integer. For instance, the vertices of the 3-chromatic cycle graph C_5 can be laid out on the unit circle in the form of a pentagram . Then the angle between vectors corresponding to neighbouring vertices is $\frac{4}{5}\pi$, corresponding to the scalar product $-1/(\sqrt{5}-1)$, so $\vec{\chi}(C_5) \le \sqrt{5} < 3$.

Theorem 6.1 *If G has clique number $\omega(G)$, then $\omega(G) \le \vec{\chi}(G) \le \chi(G)$.*

Proof For the first inequality, let W be a clique in G of size $r = \omega(G)$ and consider a vector q-colouring x of G. Let $\mathbf{y} = \sum_{v \in W} x(v)$. Then

$$0 \le \langle \mathbf{y}, \mathbf{y}\rangle \le r \cdot 1 + r(r-1)\cdot\left(-\frac{1}{q-1}\right),$$

which implies that $r \le q$.

For the second inequality, place the vertices belonging to each colour class at the corners of a $(q-1)$-dimensional simplex. To be specific, let $f\colon V \to \{1, 2, \ldots, q\}$ be an optimal q-colouring and define $x(v) = (x_1, x_2, \ldots, x_n)$ by

$$x_i = \begin{cases} \big((q-1)/q\big)^{1/2} & \text{if } i = f(v), \\ -\big(q(q-1)\big)^{-1/2} & \text{if } i \ne f(v) \text{ and } i \le q, \\ 0 & \text{if } i > q. \end{cases}$$

Then we have

$$\langle x(v), x(v)\rangle = \frac{q-1}{q} + \frac{q-1}{q(q-1)} = 1,$$

and for v and w with $f(v) \ne f(w)$ we have

$$\langle x(v), x(w)\rangle = 2\left(\frac{q-1}{q}\right)^{1/2}\left(-\left(\frac{q}{q-1}\right)^{1/2}\right) + \frac{q-2}{q(q-1)} = -\frac{1}{q-1}.$$

Thus, x is a vector q-colouring, so $\vec{\chi}(G)$ is at most q. ■

What makes vector colourings interesting from the algorithmic point of view is that they can be found in polynomial time, at least approximately, using algorithms based on semidefinite programming. The details behind those constructions lie far outside the scope of this chapter (see Gärtner and Matoušek [14]).

Theorem 6.2 *Given a graph G with $\vec{\chi}(G) = q$, a vector $(q+\varepsilon)$-colouring of G can be found in time polynomial in n and* $\log(1/\varepsilon)$.

For a graph with $\omega(G) = \chi(G)$, 6.1 shows that the vector chromatic number equals the chromatic number. In particular, it is an integer, and can be determined in polynomial time using 6.2 with $\varepsilon < \frac{1}{2}$. This shows that the chromatic numbers of perfect graphs can be determined in polynomial time. The theory behind this result counts as one of the highlights of combinatorial optimization (see Grötschel, Lovász and Schrijver [16]).

How does the vector chromatic number behave for general graphs? For $q = 2$, the vectors have to point in exactly opposite directions. In particular, there can be only two vectors for each connected component, so vector 2-colouring is equivalent to 2-colouring.

But already for $q=3$ the situation becomes more interesting, since there exist vector 3-colourable graphs that are not 3-colourable. For instance, the Grötzsch graph, the smallest triangle-free graph with chromatic number 4, admits the vector 3-colouring shown in Fig. 10 as an embedding on the unit sphere. More complicated constructions (that we cannot visualize) show that there exist vector 3-colourable graphs with chromatic number at least $n^{0.157}$ (see [12] and [22]).

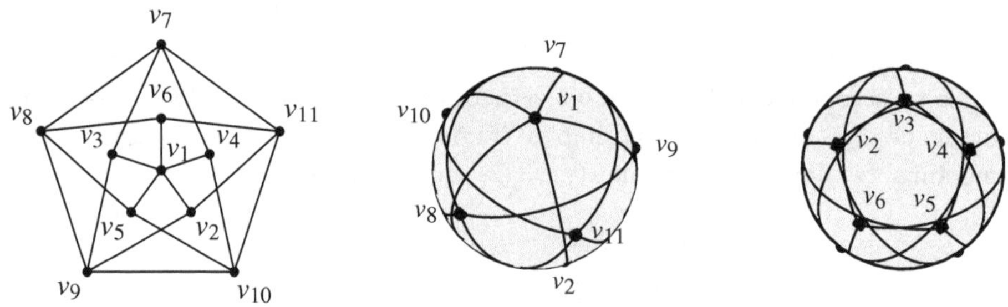

Fig. 10. Left: the Grötzsch graph Middle and right: a vector 3-colouring

Randomized rounding

Even though the gap between $\vec{\chi}$ and χ can be large for graphs in general, vector colouring turns out to be a useful starting point for (standard) colouring. The next algorithm, due to Karger, Motwani and Sudan [22], translates a vector colouring into a (standard) vertex-colouring using random hyperplanes.

Algorithm R (*Randomized rounding of vector colouring*) *Given a 3-chromatic graph G with maximum degree Δ, this algorithm finds a q-colouring in polynomial time, where the expected size of q is* $\mathbf{E}[q] = O(\Delta^{0.681} \log n)$.

R1 [Vector colour] Set $\varepsilon = 2 \cdot 10^{-5}$ and compute a vector $(3+\varepsilon)$-colouring x of G using semidefinite programming. Let $\alpha \geq \arccos(-1/(2+\varepsilon))$ be the minimum angle in radians between neighbouring vertices.

R2 [Round] Set

$$r = \lceil \log_{\pi/(\pi-\alpha)}(2\Delta) \rceil$$

and construct r random hyperplanes $H_1, H_2, \ldots, H_r$ in $\mathbb{R}^n$. For each vertex v, let $f(v)$ be the binary number $b_r b_{r-1} \cdots b_1$, where $b_i = 1$ if and only if $x(v)$ is on the positive side of the ith hyperplane H_i.

R3 [Handle monochromatic edges recursively] Iterate over all edges to find the set of monochromatic edges $M = \{vw \in E : f(v) = f(w)\}$. Recolour these vertices by running Algorithm R recursively on $G[M]$, with fresh colours. ■

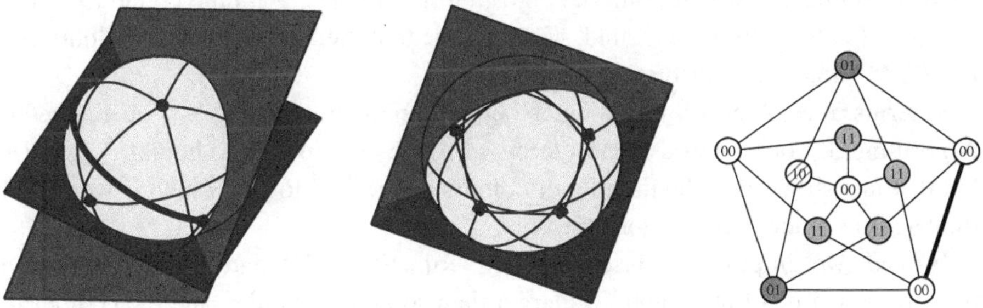

Fig. 11. Left and middle: two hyperplanes Right: the corresponding colouring

Figure 11 illustrates the behaviour of Algorithm R on the vector 3-colouring of the Grötzsch graph from Fig. 10. Two hyperplanes separate the vertices into four parts. The resulting vertex-colouring with colours from $\{0, 1\}^2$ is shown to the right. In this example, the set M of monochromatic edges determined in Step M3 contains only the single edge $v_{10}v_{11}$, drawn bold in the figure.

Algorithm R algorithm runs in polynomial time, because 6.2 ensures that Step R1 can be performed in polynomial time.

We proceed to analyze the size of the final colouring. Step R2 uses the colours $\{0, 1, \ldots, 2^{r-1}\}$, so the number of colours used in each Step R2 is

$$2^r \le (2\Delta)^{-1/\log(\pi/(\pi-\alpha))} < (2\Delta)^{0.631}. \tag{5}$$

What is more difficult is to bound the total number of recursive invocations. To this end, we need to understand how fast the instance size, determined by the size of M in Step R3, shrinks.

Let e be an edge whose endpoints received the vector colours $\mathbf{x}$ and $\mathbf{y}$. Elementary geometrical considerations establish the following result.

Theorem 6.3 *Let $\mathbf{x}, \mathbf{y} \in \mathbb{R}^d$ with angle φ (in radians). A random hyperplane in $\mathbb{R}^d$ fails to separate $\mathbf{x}$ and $\mathbf{y}$ with probability $1 - \varphi/\pi$.*

The angle between the vectors $\mathbf{x}$ and $\mathbf{y}$ is at most α. (To gain some intuition of this, if we ignore the error term ε, 6.3 shows that $\mathbf{x}$ and $\mathbf{y}$ end up on the same side of a random hyperplane with probability $1 - \alpha/\pi \le 1 - \arccos(-\frac{1}{2})/\pi = 1 - 2\pi/3\pi$

$= \frac{1}{3}$.) The edge e is monochromatic if all r independent random hyperplanes fail to separate $\mathbf{x}$ and $\mathbf{y}$ in Step R2. Thus,

$$\Pr(e \in M) \leq (1 - \alpha/\pi)^r \leq (\pi/(\pi - \alpha))^{-r} \leq 1/2\Delta\,.$$

By linearity of expectation, the expected size of M is

$$\mathbf{E}[|M|] = \sum_{e \in E} \Pr(e \in M) \leq m/2\Delta \leq \tfrac{1}{4}n\,.$$

Since each edge has two vertices, the expected number of vertices in the recursive instance $G[M]$ is at most $\frac{1}{2}n$, and in general, for $i > 2$, the expected number of vertices n_i in the ith instance satisfies $n_i \leq \frac{1}{2}n_{i-1}$. In particular, $n_t \leq 1$ after $t = O(\log n)$ rounds, at which point the algorithm terminates. With the bound (5) on the number of colours used per round, we conclude that the total number of colours used is $O(\Delta^{0.631} \log n)$ in expectation.

In terms of Δ, Algorithm R is much better than the bound of $\Delta + 1$ guaranteed by Algorithm G. For an expression in terms of n, we are tempted to bound Δ by $O(n)$, but that just shows that the number of colours is $O(n^{0.631} \log n)$, which is worse than the $O(\sqrt{n})$ colours from Algorithm W.

Instead, we employ a hybrid approach. Run Steps W1 and W2 as long as the maximum degree of the graph G is larger than some threshold d, and then colour the remaining graph using Algorithm R. The number of colours used by the combined algorithm is of the order of $(2n/d) + (2d)^{0.631} \log n$, which is minimized around $d = n^{1/1.631}$ with value $O(n^{0.387})$.

Variants of Algorithm R for general q-colouring and with intricate rounding schemes have been investigated further (see Langberg's survey [27]). The current best polynomial-time algorithm for colouring a 3-chromatic graph based on vector colouring uses $O(n^{0.208})$ colours, due to Chlamtac [9].

7. Reductions

The algorithms in this chapter are summarized in Table 1.

Not only do these algorithms achieve different running times and quality guarantees, they also differ in which specific problem they consider. Let us now be more precise about the variants of the graph colouring problem:

Decision Given a graph G and an integer q, decide whether G can be q-coloured.

Chromatic number Given a graph G, compute the chromatic number $\chi(G)$.

Construction Given a graph G and an integer q, construct a q-colouring of G.

Counting Given a graph G and an integer q, compute the number $P(G, q)$ of q-colourings of G.

Sampling Given a graph G and an integer q, construct a random q-colouring of G.

Algorithm	Time	Problem
B Bipartition	$O(n+m)$	2-colouring
C Contraction	$O(1.619^{n+m})$	$P(G,q)$
D Dynamic programming	$3^n \operatorname{poly}(n)$	$\chi(G)$
G Greedy	$O(n+m)$	$(\Delta(G)+1)$-colouring
I Inclusion–exclusion	$2^n \operatorname{poly}(n)$	$\chi(G)$
L Lawler's algorithm	$O(1.443^n)$	3-colouring
M Metropolis dynamics	$\operatorname{poly}(n)$	random q-colouring ($q > 4\Delta$)
P Palette restriction	$1.5^n \operatorname{poly}(n)$	3-colouring
R Rounded vector colouring	$\operatorname{poly}(n)$	$O(\Delta^{0.681} \log n)$-colouring for $\chi(G)=3$
V Vizing's algorithm	$O(mn)$	edge-$(\Delta(G)+1)$-colouring
W Wigderson's algorithm	$O(n+m)$	$O(\sqrt{n})$-colouring for $\chi(G)=3$
X Exhaustive search	$q^n \operatorname{poly}(n)$	$P(G,q)$

Table 1. Algorithms discussed in this chapter

Chromatic polynomial Given a graph G, compute the chromatic polynomial – that is, the coefficients of the integer polynomial $q \mapsto P(G,q)$.

Some of these problems are related by using fairly straightforward reductions. For example, the decision problem is easily solved using the chromatic number by comparing q with $\chi(G)$; conversely, $\chi(G)$ can be determined by solving the decision problem for $q = 1, 2, \ldots, n$. It is also clear that if we can construct a q-colouring, then we can decide that one exists. What is perhaps less clear is the other direction. This is seen by a self-reduction that follows the contraction algorithm, Algorithm C.

Reduction C (*Constructing a colouring using a decision algorithm*) Suppose that we have an algorithm that decides whether a given graph G can be q-coloured. If $G = K_n$ and $n \le q$, give each vertex its own colour and terminate. Otherwise, select two non-adjacent vertices v and w in G. If $G \cup vw$ cannot be q-coloured, then every q-colouring f of G must have $f(v) = f(w)$. Thus we can identify v and w and recursively find a q-colouring for G/vw. Otherwise, there exists a q-colouring of G with $f(v) \ne f(w)$, so we recursively find a colouring for $G \cup vw$. ■

Some of our algorithms work only for a specific fixed q, such as Algorithm B for 2-colourability or Algorithm L for 3-colourability. Clearly, they both reduce to the decision problem where q is part of the input. But what about the other direction? The answer turns out to depend strongly on q: the decision problem reduces to 3-colorability, but not to 2-colorability.

Reduction L (*q-colouring using* 3*-colouring*) Given a graph $G = (V,E)$ and an integer q, this reduction constructs a graph H that is 3-colourable with colours $\{0,1,2\}$ if and only if G is q-colourable with colours $\{1,2,\ldots,q\}$.

First, to fix some colour names, the graph H contains a triangle with the vertices $0, 1, 2$. We assume that vertex i has colour i, for $i = 0, 1, 2$.

For each vertex $v \in V$, the graph H contains $2q$ vertices $v_1, v_2, \dots, v_q$ and $v'_1, v'_2, \dots, v'_q$. Our intuition is that the v_is act as indicators for a colour in G in the following sense: if v_i has colour 1 in H then v has colour i in G. The vertices are arranged as in Fig. 12(a); the right-most vertex is 1 or 2, depending on the parity of q. The vertices $v_1, v_2, \dots, v_q$ are all adjacent to 2, and so must be coloured 0 or 1. Moreover, at least one of them must be coloured 1, since otherwise, the colours for $v'_1, v'_2, \dots, v'_q$ are forced to alternate as $1, 2, 1, \dots$, conflicting with the colour of the right-most vertex.

Now consider an edge vw in G. Let $v_1, v_2, \dots, v_q$ and $w_1, w_2, \dots, w_q$ be the corresponding 'indicator' vertices in H. For each colour $i = 1, 2, \dots, q$, the vertices v_i and w_i are connected by a 'fresh' triangle as shown in Fig. 12(b). This ensures that v_i and w_i cannot both be 1. In other words, v and w cannot have received the same colour. ■

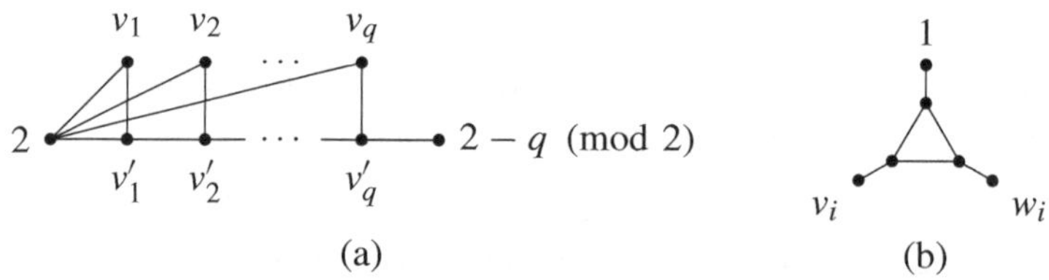

Fig. 12.

The above reduction, essentially due to Lovász [30], can easily be extended to a larger fixed $q > 3$, because G is q-colourable if and only if G with an added 'apex' vertex adjacent to all other vertices is $(q+1)$-colourable. For instance, 4-colourability is not easier than 3-colourability for general graphs.

Thus, all q-colouring problems for $q \geq 3$ are (in some sense) equally difficult. This is consistent with the fact that the case $q = 2$ admits a very fast algorithm (Algorithm B), whereas none of the others does.

Many constructions have been published that show the computational difficulty of colouring for restricted classes of graphs. We will sketch an interesting example due to Stockmeyer [35]: the restriction of the case $q = 3$ to planar graphs. Consider the subgraph in Fig. 13(a), called a *planarity gadget*. One can check that this subgraph has the property that every 3-colouring f satisfies $f(\mathrm{E}) = f(\mathrm{W})$ and $f(\mathrm{N}) = f(\mathrm{S})$. Moreover, every partial assignment f to $\{\mathrm{N}, \mathrm{S}, \mathrm{E}, \mathrm{W}\}$ that satisfies $f(\mathrm{E}) = f(\mathrm{W})$ and $f(\mathrm{N}) = f(\mathrm{S})$ can be extended to a 3-colouring of the entire subgraph.

The gadget is used to transform a given (non-planar) graph G as follows. Draw G in the plane and for each edge vw replace each edge intersection by the planarity gadget. The outer vertices of neighbouring gadgets are identified, and v is identified with W in its neighbouring gadget (see Fig. 13(b)). The resulting graph is planar, and it can be checked that it is 3-chromatic if and only if G is 3-chromatic. Thus, the restriction to planar instances does not make 3-colourability computationally easier. Unlike the case for non-planar graphs, this construction cannot be generalized to larger $q > 3$, since the decision problem for planar graphs and every $q \geq 4$ has answer 'yes' because of the four-colour theorem.

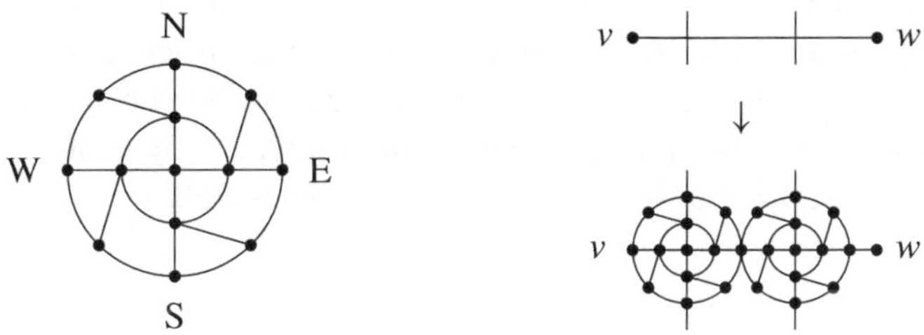

(a) (b)

Fig. 13. A planarity gadget

Computational complexity

The field of computational complexity relates algorithmic problems from various domains to one another in order to establish a notion of computational difficulty. The chromatic number problem was one of the first to be analyzed in this fashion. The following reduction, essentially from the seminal paper of Karp [23], shows that computing the chromatic number is 'hard for the complexity class NP' by reducing from the NP-hard satisfiability problem for Boolean formulas on conjunctive normal form (CNF). This implies that all other problem in the class NP reduce to the chromatic number.

The input to *CNF-Satisfiability* is a Boolean formula consisting of s clauses $C_1, C_2, \ldots, C_s$. Each clause C_j consists of a disjunction $C_j = (l_{j1} \vee l_{j2} \vee \cdots \vee l_{jk})$ of literals. Every literal is a variable $x_1, x_2, \ldots, x_r$ or its negation $\overline{x}_1, \overline{x}_2, \ldots, \overline{x}_r$. The problem is to find an assignment of the variables to 'true' and 'false' that makes all clauses true.

Reduction K (*Satisfiability using chromatic number*) Given an instance $C_1, C_2, \ldots,$ C_s of CNF-Satisfiability over the variables $x_1, x_2, \ldots, x_r$, this reduction constructs a graph G on $3r + s + 1$ vertices such that G can be coloured with $r + 1$ colours if and only the instance is satisfiable.

The graph G contains a complete subgraph on $r + 1$ vertices $\{0, 1, \ldots, r\}$. In any colouring these vertices receive different colours, say $f(i) = i$. The intuition is that the colour 0 represents 'false', while the other colours represent 'true'. For each variable x_i ($1 \leq i \leq r$) the graph contains two adjacent 'literal' vertices v_i and $\overline{v}_i$, both adjacent to all 'true colour' vertices $\{1, 2, \ldots, r\}$ except i. Thus, one of the two vertices $v_i, \overline{v}_i$ must be assigned the 'true' colour i, and the other must be coloured 0. The construction is completed with 'clause' vertices w_j, one for each clause C_j ($1 \leq j \leq s$). Let $x_{i_1}, x_{i_2}, \ldots, x_{i_k}$ be the variables appearing (positively or negatively) in C_j. Then w_j is adjacent to $\{0, 1, \ldots, r\} \setminus \{i_1, i_2, \ldots, i_k\}$. This ensures that only the 'true' colours $\{i_1, i_2, \ldots, i_k\}$ are available at w_j. Furthermore, if x_i appears positive in C_j, then w_j is adjacent to $\overline{v}_i$; if x_i appears negative in C_j, then w_j is adjacent to v_i. Figure 14 shows the reduction for a small instance consisting of just the clause $C_1 = (x_1 \vee \overline{x}_2 \vee \overline{x}_3)$ and a valid colouring corresponding to the assignment $x_1 = x_3 = $ true, $x_2 = $ false; the edges of the clique on $\{0, 1, 2, 3\}$ are not shown. Thus, the only colours available to w_j are those chosen by its literals. ■

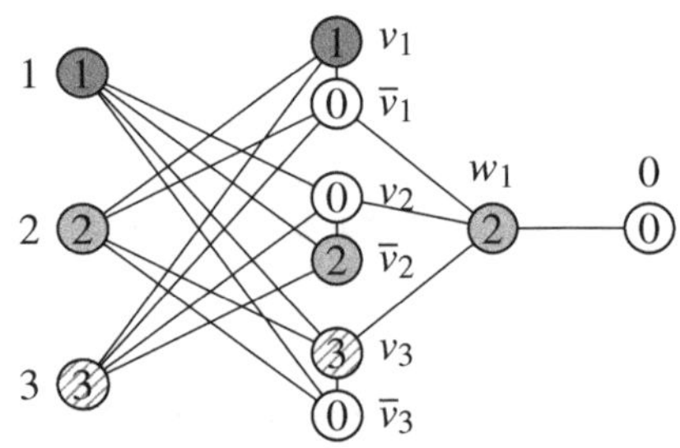

Fig. 14. A 4-colouring instance corresponding to $C_1 = (x_1 \vee \overline{x}_2 \vee \overline{x}_3)$

Edge-colouring

A mapping $f \colon E \to \{1, 2, \ldots, q\}$ is an edge-colouring of G if and only if it is a vertex-colouring of the line graph $L(G)$ of G. In particular, every vertex-colouring algorithm can be used as an edge-colouring algorithm by running it on $L(G)$. For instance, Algorithm I computes the chromatic index in time $2^m \operatorname{poly}(n)$, which is the fastest currently known algorithm. Similarly, Algorithm G finds an edge-colouring with $2\Delta - 1$ colours, but this is worse than Algorithm V. In fact, since $\Delta \leq \chi'(G) \leq \Delta + 1$, Algorithm V determines the chromatic index within an additive error of 1. However, deciding which of the two candidate values for $\chi'(G)$ is correct is an NP-hard problem, as shown by Holyer [19] for $\chi'(G) = 3$, and by Leven and Galil [29] for $\chi'(G) > 3$.

Approximating the chromatic number

Algorithm V shows that the chromatic index can be very well approximated. In contrast, approximating the chromatic *number* is much harder. In particular, it is NP-hard to 4-colour a 3-chromatic graph (see [17]). This rules out an approximate vertex-colouring algorithm with a performance guarantee as good as Algorithm V, but is far from explaining why the considerable machinery behind (say) Algorithm R results only in a colouring of size n^c for 3-chromatic graphs. The best currently known exponent is $c = 0.204$ (see [24]).

For sufficiently large fixed q, it is NP-hard to find an $\exp(\Omega(q^{1/3}))$-colouring for a q-colourable graph. If q is not fixed, even stronger hardness results are known. We saw in Section 6 that the polynomial-time computable function $\vec{\chi}(G)$ is a lower bound on $\chi(G)$, even though the gap can sometimes be large (say, $\chi(G) \geq n^{0.157}\vec{\chi}(G)$) for some graphs. Can we guarantee a corresponding upper bound for $\vec{\chi}$? If not, maybe there is some other polynomial-time computable function g so that we can guarantee, for example, $g(G) \leq \chi(G) \leq n^{0.999}g(G)$? The answer turns out to be 'no' under standard complexity-theoretic assumptions: For every $\varepsilon > 0$, it is NP-hard to approximate $\chi(G)$ within a factor $n^{1-\varepsilon}$, as shown by Zuckerman [41].

Counting

The problem of counting the q-colourings is solved by evaluating $P(G, q)$. Conversely, because the chromatic polynomial has degree n, it can be interpolated using Lagrangian interpolation from the values of the counting problem at $q = 0, 1, \ldots, n$. Moreover, note that $\chi(G) \geq q$ if and only if $P(G, q) > 0$, so it is NP-hard to count the number of q-colourings simply because the decision problem is known to be hard. In fact, the counting problem is hard for Valiant's counting class #P.

On the other hand, an important result in counting complexity [21] relates the estimation of the size of a finite set to the problem of uniformly sampling from it. In particular, a uniform sampler such as Algorithm M serves as a 'fully polynomial randomized approximation scheme' (FPRAS) for the number of colours. Thus, provided that $q > 4\Delta$, Algorithm M can be used to compute a value $g(G)$ for which

$$(1 - \varepsilon)g(G) \leq P(G, q) \leq (1 + \varepsilon)g(G)$$

with high probability in time polynomial in n and $1/\varepsilon$, for any $\varepsilon > 0$. Much better bounds on q are known (see the survey of Frieze and Vigoda [13]). Without some bound on q, such an FPRAS is unlikely to exist because, with $\varepsilon = \frac{1}{2}$, it would constitute a randomized algorithm for the decision problem and would therefore imply that all of NP can be solved in randomized polynomial time.

8. Conclusion

Together, the algorithms and reductions presented in this survey give a picture of the computational aspects of graph colouring. For instance, 2-colouring admits a polynomial time algorithm, while 3-colouring does not. In the planar case, 4-colouring is trivial, but 3-colouring is not. An almost optimal edge-colouring can be found in polynomial time, but vertex-colouring is very difficult to approximate. If q

is sufficiently large compared to $\Delta(G)$ then the set of colourings can be sampled and approximately counted, but not counted exactly. Finally, even the computationally hard colouring problems admit techniques that are much better than our initial Algorithm X.

None of these insights is obvious from the definition of graph colouring, so the algorithmic perspective on chromatic graph theory has proved to be a fertile source of questions with interesting answers.

References

1. R. Beigel and D. Eppstein, 3-coloring in time $O(1.3289^n)$, *J. Algorithms* **111** (2005), 168–204.
2. A. Björklund and T. Husfeldt, Exact algorithms for exact satisfiability and number of perfect matchings. *Algorithmica* **52** (2008), 226–249.
3. A. Björklund, T. Husfeldt and M. Koivisto, Set partitioning via inclusion–exclusion, *SIAM J. Comput.* **39** (2009), 546–563.
4. A. Björklund, T. Husfeldt, P. Kaski and M. Koivisto, Covering and packing in linear space, *Inform. Process. Lett.* **111** (2011), 1033–1036.
5. A. Björklund, T. Husfeldt, P. Kaski and M. Koivisto, Computing the Tutte polynomial in vertex-exponential time, *Proc. 49th Annual IEEE Symposium on Foundations of Computer Science (FOCS 2008)*, IEEE Computer Society (2008), 677–686.
6. D. Brélaz, New methods to color the vertices of a graph, *Comm. Assoc. Comput. Mach.* **22** (1979), 251–256.
7. C. Bron and J. Kerbosch, Algorithm 457: finding all cliques of an undirected graph, *Comm. Assoc. Comput. Mach.* **16** (1973), 575–577.
8. R. L. Brooks, On colouring the nodes of a network, *Proc. Cambridge Philos. Soc.* **37** (1941), 194–197.
9. E. Chlamtac, Approximation algorithms using hierarchies of semidefinite programming relaxations, *Proc. 48th Annual IEEE Symposium on Foundations of Computer Science (FOCS 2007)*, IEEE Computer Society (2007), 691–701.
10. N. Christofides, An algorithm for the chromatic number of a graph. *Comput. J.* **14** (1971), 38–39.
11. V. Chvátal, Perfectly ordered graphs, *Topics on Perfect Graphs* (eds. C. Berge and V. Chvátal), *Ann. Discrete Math.* **21** (1984), 63–68.
12. U. Feige, M. Langberg and G. Schechtman, Graphs with tiny vector chromatic numbers and huge chromatic numbers, *SIAM J. Comput.* **33** (2004), 1338–1368.
13. A. Frieze and E. Vigoda, A survey on the use of Markov chains to randomly sample colorings, *Combinatorics, Complexity and Chance*, Oxford University Press, 2007.
14. B. Gärtner and J. Matoušek, *Approximation Algorithms and Semidefinite Programming*, Springer, 2012.
15. G. Grimmett and C. McDiarmid, On colouring random graphs, *Math. Proc. Cambridge Philos. Soc.* **77** (1975), 313–324.
16. M. Grötschel, L. Lovász and A. Schrijver, *Geometric Algorithms and Combinatorial Optimization*, Springer, 1988.
17. V. Guruswami and S. Khanna, On the hardness of 4-coloring a 3-colorable graph, *SIAM J. Discrete Math.* **18** (2004), 30–40.
18. G. Haggard, D. J. Pearce and G. Royle, Computing Tutte polynomials, *Assoc. Comput. Mach. Math. Software* **37** (2010), Article 24.
19. I. Holyer, The NP-completeness of edge-coloring, *SIAM J. Comput.* **10** (1981), 718–720.

20. S. Huang, Improved hardness of approximating chromatic number, *Approximation, Randomization, and Combinatorial Optimization. Algorithms and Techniques*, Springer Lecture Notes Comput. Sci. **8096** (2013), 233–243.
21. M. Jerrum, L. G. Valiant and V. V. Vazirani, Random generation of combinatorial structures from a uniform distribution, *Theor. Comput. Sci.* **43** (1986), 169–188.
22. D. R. Karger, R. Motwani and M. Sudan, Approximate graph coloring by semidefinite programming, *J. Assoc. Comput. Mach.* **45** (1998), 246–265.
23. R. M. Karp, Reducibility among combinatorial problems, *Complexity of Computer Computations* (eds. R. E. Miller and J. W. Thatcher), Plenum (1972), 85–103.
24. K. Kawarabayashi and M. Thorup, Combinatorial coloring of 3-colorable graphs, *Proc. 53rd Annual IEEE Symposium on Foundations of Computer Science (FOCS 2012)*, IEEE Computer Society (2012), 68–75.
25. A. Kosowski and K. Manuszewski, Classical coloring of graphs, *Graph Colorings* (ed. M. Kubale), *Amer. Math. Soc. Contemp. Math.* **352** (2004), 1–20.
26. L. Kučera, The greedy coloring is a bad probabilistic algorithm, *J. Algorithms* **12** (1991), 674–684.
27. M. Langberg, Graph coloring, *Encyclopedia of Algorithms* (ed. M. Kao), Springer (2008), 368–371.
28. E. L. Lawler, A note on the complexity of the chromatic number problem, *Inform. Process. Lett.* **5** (1976), 66–67.
29. D. Leven and Z. Galil, NP completeness of finding the chromatic index of regular graphs, *J. Algorithms* **4** (1983), 35–44.
30. L. Lovász, Coverings and coloring of hypergraphs, *Proc. Fourth Southeastern Conference on Combinatorics, Graph Theory, and Computing*, Boca Raton, *Congr. Numer.* **8** (1973), 3–12.
31. F. Maffray, On the coloration of perfect graphs, *Recent Advances in Algorithms and Combinatorics*, CMS Books Math., Springer (2003), 65–84.
32. D. W. Matula, A min–max theorem for graphs with applications to graph coloring, *SIAM Rev.* **10** (1968), 481–482.
33. J. Misra and D. Gries, A constructive proof of Vizing's Theorem, *Inform. Proc. Lett.* **41** (1992), 131–133.
34. J. W. Moon and L. Moser, On cliques in graphs, *Israel J. Math.* **3** (1965), 23–28.
35. L. Stockmeyer, Planar 3-colorability is polynomial complete, *Assoc. Comput. Mach. SIGACT News* **5** (1973), 19–25.
36. G. Szekeres and H. S. Wilf, An inequality for the chromatic number of a graph, *J. Combin. Theory* **4** (1968), 1–3.
37. E. Tomita, A. Tanaka and H. Takahashi, The worst-case time complexity for generating all maximal cliques and computational experiments, *Theor. Comput. Sci.* **363** (2006), 28–42.
38. D. J. A. Welsh and M. B. Powell, An upper bound for the chromatic number of a graph and its application to timetabling problems, *Comput. J.* **10** (1967), 85–86.
39. H. Whitney, A logical expansion in mathematics, *Bull. Amer. Math. Soc.* **38** (1932), 572–579.
40. A. Wigderson, Improving the performance guarantee for approximate graph coloring, *J. Assoc. Comput. Mach.* **30** (1983), 729–735.
41. D. Zuckerman, Linear degree extractors and the inapproximability of Max Clique and Chromatic Number, *Theory of Computing* **3** (2007), 103–128.
42. A. A. Zykov, On some properties of linear complexes (in Russian), *Math. Sbornik* **24** (1949), 163–188.

14

Colouring games

ZSOLT TUZA and XUDING ZHU

There are various kinds of two-person games in which the players construct a colouring of a graph (or of a subgraph), and some additional games are closely related to graph colourings. Here we describe several variants, most of which deal with 'maker–breaker games'. We also consider the game version of list-colourings.

1. Introduction

The connection between graph colourings and game theory has its roots in the works of P. M. Grundy in the latter field as early as the 1930s. His research later motivated the introduction of some graph invariants, one of which is discussed in detail in Berge's monograph [9]. For undirected graphs the analogous parameter is the *Grundy number*, also called the *online chromatic number*, which we will consider in Section 9. But although Grundy colouring may be formulated in terms of games, it remains a solitaire game in which the player aims at finding a solution that is as bad as possible.

In this chapter we mainly consider non-cooperative two-person games that either construct colourings or are otherwise closely related to some kind of colouring concept, even when they do not colour the graph.

A good starting point is the following *maker–breaker game*, which was introduced by Brams [24] for planar graphs in 1981, and independently by Bodlaender [12] for general graphs in 1991. Given a graph G, two players alternately colour vertices of G from a set of colours, say $\{1, 2, \ldots, k\}$. The two players are Alice (the maker) and Bob (the breaker); each of them colours one uncoloured vertex at a time, under the condition that monochromatic edges must not occur, Alice moving first. Alice wins if a proper k-colouring of the entire graph G is eventually obtained, whereas Bob wins if a situation is reached where the neighbourhood of some uncoloured vertex contains all the k colours of the palette. Certainly, if the number of colours is sufficiently large (for example, if $k = |V(G)|$), then Alice surely wins the game.

This leads to the following definition. The *game-chromatic number* $\chi_g(G)$ of G is the smallest number k for which Alice has a winning strategy.

For every graph the inequalities

$$\chi(G) \leq \chi_g(G) \leq \Delta(G) + 1 \tag{1}$$

hold for the chromatic number χ, the game-chromatic number χ_g and the maximum vertex-degree Δ: a proper colouring with $\chi_g(G)$ colours is constructed by the end of the game whenever Alice wins, and a free colour is available for any uncoloured vertex whenever the number of colours exceeds the number of neighbours. The path P_4 of length 3 has $\chi = 2$, but $\chi_g = 3$, because if Alice colours a vertex with colour 1 in her first move, then Bob can assign colour 2 to the vertex at distance 2 from the vertex coloured first. It follows that Brooks's theorem cannot be extended directly to the game-chromatic number, and that χ_g may exceed χ, even when G is a path.

For some trees T, $\chi_g(T)$ can be as large as $\chi(T) + 2$ – that is, $\chi_g(T) = 4$ may occur. For example, consider the colouring game played with colours α, β, γ on the caterpillar of diameter 5 on 14 vertices (see Fig. 1(a)). No matter which vertex v Alice colours first, Bob can assign the same colour α to some vertex w at distance 3 from v, giving the subtree in Fig. 1(b).

If Alice colours a vertex on the central path, its colour β must differ from α, and then Bob can assign the third colour γ to a leaf adjacent to the last (uncoloured) vertex of the central path, thereby winning the game. If Alice assigns colour β to a leaf, Bob wins by assigning γ to the other leaf adjacent to the uncoloured middle vertex. If Alice assigns colour α to a leaf, Bob assigns colour β to a leaf adjacent to

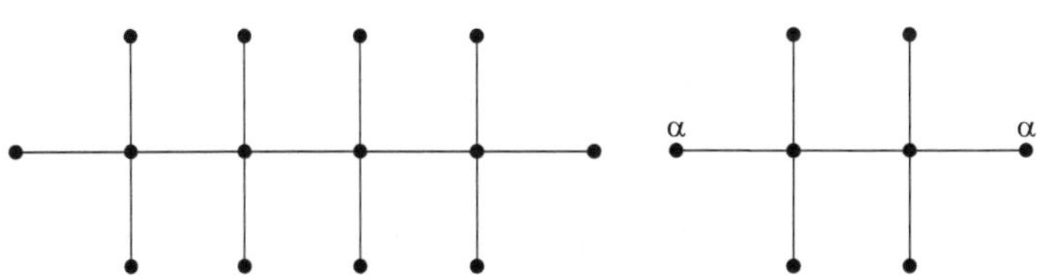

(a) (b)

Fig. 1. A tree T with $\chi_g(T) = 4$

the other vertex of the central path. This creates a double threat on the neighbour of colour β, because Bob has a chance to assign colour γ to any of its two uncoloured neighbours. The only way for Alice to eliminate this threat is to assign colour γ to the neighbour of colour β. Now the other vertex on the central path has colours α, γ in its neighbourhood, and still has an uncoloured leaf neighbour to which Bob can assign colour β on his next move, hence winning the game. Actually, the upper bound of 4 is tight on trees. We show this in a stronger form in the next section.

We note that a completely different game would be obtained if the first move were made by Bob. For example, the 'cocktail-party graph' $K_{k,k} - kK_2$, obtained by omitting a perfect matching from $K_{k,k}$, has $\chi_g = k$, since Bob can win for any fewer than k colours by assigning to the matching pair of the vertex just coloured the same colour that Alice used in her latest move. On the other hand, if Bob makes the first move, then Alice can assign a different colour to the matching pair of the coloured vertex, and hence win with just two colours.

Most studies on the game-chromatic number concentrate on upper bounds for classes of graphs. For a class $\mathcal{H}$ of graphs, let

$$\chi_g(\mathcal{H}) = \max\{\chi_g(G) : G \in \mathcal{H}\}.$$

Zhu [61] proved that if $\mathcal{H}$ is closed under the operations of taking disjoint unions and adding K_1 as a component, then $\chi_g(\mathcal{H})$ does not depend on who has the first move. Nevertheless, as the cocktail-party graph shows, for individual graphs it can make a difference who starts the game. In this chapter we concentrate on games started by Alice.

We observe further that $\chi_g(K_{k,k}) = 3$. Indeed, after Alice's first move Bob's strongest answer is to assign a second colour to a vertex that is not adjacent to the first one; but then Alice can assign a third colour in the other vertex-class and thus ensure that the colouring can be completed with three colours, no matter how Bob plays. In fact, if we re-insert just one edge of the omitted 1-factor into $K_{k,k} - kK_2$, we already obtain a graph with $\chi_g = 3$.

As a consequence of these examples, we deduce that the game-chromatic numbers of subgraphs of G are not bounded above by $\chi_g(G)$.

A general upper bound

The *acyclic chromatic number* $\chi_a(G)$ of a graph G is the smallest integer k for which G has a proper vertex k-colouring such that no cycle is 2-coloured. Dinski and Zhu [19] proved the following result.

Theorem 1.1 *For any graph G,*

$$\chi_g(G) \leq \chi_a(G)(\chi_a(G) + 1). \tag{2}$$

This upper bound is usually not sharp, but still the result offers a general approach to proving that the members of some graph classes have bounded game-chromatic number; for example, it trivially implies that $\chi_g \leq 6$ holds for all trees. More interestingly, applying Borodin's result [13] that $\chi_a(G) \leq 5$ holds for every planar graph G, we deduce that the game-chromatic number of every planar graph is at most 30.

Algorithmic complexity

Although we do not devote a separate section to this issue, we should say something about it, as this was the main topic of Bodlaender's seminal paper [12]. However, Bodlaender concentrated on a somewhat different game on a graph, where a linear ordering $v_1, v_2, \ldots, v_n$ of the vertices is also given, together with the number of colours, and the vertices have to be coloured in this fixed order. He used the term 'sequential colouring construction game'.

Bodlaender proved that, if the number of colours is at least 3, then it is PSPACE-complete to decide which of the two players has a winning strategy in the sequential colouring construction game, even when the input is restricted to graphs with maximum degree at most 5. On the other hand, if just two colours are available, the winner can be determined in time $O(n + m)$.

We will return to complexity issues in Section 9.

2. Marking games

The *colouring number* $\mathrm{col}(G)$ of a graph G is defined as

$$\mathrm{col}(G) = \min \max_{1 \leq i \leq n} (d^-(v_i) + 1),$$

where the minimum is taken over all vertex orderings $v_1, v_2, \ldots, v_n$, and $d^-(v_i)$ denotes the number of neighbours preceding v_i in the vertex order under consideration. The colouring number is an upper bound for $\chi(G)$ – a rather bad one in some cases. The point is that, when we colour the graph in the order of increasing subscripts according to an order that attains $\mathrm{col}(G)$, each vertex has fewer than $\mathrm{col}(G)$ coloured neighbours at the time that it has to be coloured, and so there is always a free colour available.

For the game version, consider the game in which Alice and Bob alternately select vertices of G, constructing a vertex order in which the vertices with odd index are chosen by Alice and those with even index are chosen by Bob. Alice's goal is to minimize $\max_{1 \leq i \leq n} d^-(v_i) + 1$, while Bob's goal is to maximize it. This leads to define the *game-colouring number* of a graph G, denoted by $\mathrm{col}_g(G)$, to be the value of $\max_{1 \leq i \leq n} d^-(v_i) + 1$ when both players play optimally.

This is an interesting parameter in its own right. However, the main purpose of introducing it here is because it is an upper bound for the game-chromatic number.

Theorem 2.1 *For any graph G, $\chi_g(G) \leq \mathrm{col}_g(G)$.*

In the play of a marking game we say that a vertex is *marked* if it has been chosen by a player. If $\mathrm{col}_g(G) = k$, Alice has a strategy for playing so that, at any moment of the game, any unmarked vertex has at most $k-1$ marked neighbours. In the maker–breaker game Alice uses her strategy for the marking game to choose vertices to be coloured, and she can then ensure that, at any moment of the game, any uncoloured vertex has at most $k-1$ coloured neighbours. Thus, with k colours available, all the vertices are eventually coloured. This argument proves Theorem 2.1.

As a significant difference between χ_g and col_g, it was proved by Wu and Zhu [56] that, if H is a subgraph of G, then $\mathrm{col}_g(H) \le \mathrm{col}_g(G)$. Perhaps this more natural behaviour – which is not satisfied by χ_g – is one reason why many upper bounds on χ_g are proved by showing that they are valid for col_g.

The game-colouring numbers for many classes of graphs were determined or estimated by Faigle, Kern, Kierstead and Trotter [23]. The first result of this kind was that $\mathrm{col}_g(F) \le 4$ for every forest F. We sketch Alice's winning strategy.

Alice must ensure that, after each of her moves, each unmarked subtree of F (that is, each component of the subgraph of F induced by the currently unmarked vertices) is joined by edges to at most two marked vertices; this property evidently holds after the first move, since just one vertex is marked. At any later stage Bob marks a vertex inside some unmarked component T of F. After Bob's move, the uncoloured vertices of T induce one or more components. In any case, at most one of the components has three coloured neighbours, while all the other components have at most two. If Bob leaves a component with three marked neighbours, then Alice marks the unique vertex separating them. This splits the component into at least three unmarked components, and three of the components have two marked neighbours each. If Bob leaves each component with at most two marked neighbours, then in some component Alice marks a vertex adjacent to a marked vertex, or marks an arbitrary vertex if the component has no marked neighbours. This strategy works because only one tree component of the unmarked subforest requires special attention – namely, the one for which Bob has created three marked neighbours.

Sidorowicz [53] showed that, if G contains cycles no two of which have a common edge (a cactus), then $\mathrm{col}_g(G) \le 5$ and this upper bound is tight also for $\chi_g(G)$. More generally, Junosza-Szaniawski and Rożej [32] showed that if each edge is contained in at most c cycles of G, then $\mathrm{col}_g(G) \le c+4$.

The activation strategy

A frequently used strategy in establishing upper bounds for col_g is the so-called 'activation strategy'. To use it for a graph G, we first need to fix a vertex order $v_1, v_2, \ldots, v_n$ of G. For a vertex v, let $N^<(v)$ be the set of neighbours preceding v in the vertex order and $N^>(v)$ be the set of neighbours occurring after v. During the play of the game Alice constructs a directed graph D with $V(D) = V(G)$; this digraph helps her to choose vertices in later moves. Initially, D has no arcs, and during play arcs are added to D.

By 'processing' a vertex v, we mean the following self-calling procedure: If there are no unmarked vertices, then stop (without doing anything). Assume that there are unmarked vertices.

Step 1. If $N^<(v)$ contains no unmarked vertices and v is unmarked, then mark v and stop.

Step 2. If $N^<(v)$ contains no unmarked vertices and v is marked, then mark the unmarked vertex of smallest index and stop.

Step 3. Otherwise, let x be the unmarked vertex of smallest index in $N^<(v)$. If x has an in-neighbour in D, then add an arc from v to x, mark x and stop. If x has no in-neighbour in D, then add an arc from v to x and process x (that is, go to Step 1, with v replaced by x).

The activation strategy is as follows:
On her first move, Alice marks v_1. Later, when Bob marks a vertex v, Alice processes v.

Note that, when processing a vertex v, the procedure may repeatedly call itself to process a sequence of vertices. In this sequence the vertices have decreasing indices, so the procedure eventually ends. The procedure always ends by marking an unmarked vertex, provided that there are still such vertices.

Suppose that Alice has just completed a move. We say that a vertex v is *active* if v is incident to an arc of D, or if v and all vertices in $N^<(v)$ are marked.

The following observations follow easily from the processing procedure and the activation strategy.

(a) Each marked vertex is active.
(b) If v is active and unmarked, then v has out-degree 1 in D.
(c) If v is active and $w \in N^<(v)$ is unmarked, then v has out-degree 1 in D, and the out-neighbour of v precedes or equals w.
(d) If a vertex w has in-degree 2 in D, then w is marked. Moreover, once a vertex w is marked, no more arcs incident to w can be added to D. Hence each vertex of D has in-degree at most 2.

To derive an upper bound on $\mathrm{col}_g(G)$ by using the activation strategy, we need the following result:
Assume that Alice has just completed a move and that v is unmarked, and suppose that there are two sets $A(v), B(v)$ of vertices for which the following properties hold:

(i) $N^<(v) \subseteq A(v)$; and
(ii) if u precedes v and there is a vertex $w \notin B(v)$ for which $u, v \in N^<(w)$, then $u \in A(v)$.

Then v has at most $3|A(v)| + |B(v)|$ marked neighbours.

To prove this, we let $P(v)$ be the set of marked vertices in $N^<(v)$ and $Q(v)$ be the set of marked vertices in $N^>(v)$. By (i), $P(v) \subseteq A(v)$. By (ii) and the observation above, if $w \in Q(v) - B(v)$, then there is an arc wu in D such that either $u = v$ or $u \in A(v)$. If there is an arc wv in D, then v is active, and so there is an arc from v to $A(v)$; so there are at least $|Q(v)| - |B(v)|$ arcs from $(Q(v) - B(v)) \cup \{v\}$ to $A(v)$. Since each vertex $u \in A(v)$ has in-degree at most 2, we conclude that $|Q(v)| - |B(v)| \leq 2|A(v)|$. Thus the total number of marked neighbours of v is

$$|P(v)| + |Q(v)| \leq 3|A(v)| + |B(v)|.$$

By this result, if v is unmarked at any moment of the game, and if the sets $A(v)$ and $B(v)$ satisfy the conditions listed above, then v has at most $3|A(v)| + |B(v)| + 1$ marked neighbours (since if Bob has just completed a move he could have marked a further neighbour of v). Therefore, if $A(v), B(v)$ satisfy the conditions above for each vertex v of G, then

$$\mathrm{col}_g(G) \leq \max_{v \in V(G)} 3|A(v)| + |B(v)| + 2.$$

A good vertex order of $V(G)$ is crucial for good performance of the activation strategy. If G is a chordal graph, then it has a vertex order in which $N^{<}(v)$ induces a complete subgraph for each v. Under this vertex order, taking $A(v) = N^{<}(v)$ and $B(v) = \varnothing$, we can easily verify that (i) and (ii) are satisfied. Hence

$$\mathrm{col}_g(G) \leq \max_{v \in V(G)} 3|A(v)| + 2 = 3\omega(G) - 1.$$

In particular, if G is a k-tree, then $\mathrm{col}_g(G) \leq 3k + 2$. By the monotonicity of $\mathrm{col}_g(G)$ this implies that the game-colouring number of any partial k-tree is bounded above by $3k + 2$. This upper bound is sharp for all $k \geq 2$ (see [56]).

Planar and outerplanar graphs

When Brams invented the colouring game he was interested only in planar graphs. One can easily see that four colours are not enough for the octahedron graph and five colours are not enough for the icosahedron graph. Robert High then found a planar graph for which six colours are not enough. Whether there is a universal finite upper bound on the game-chromatic number of all planar graphs remained an open problem for some time. The existence of such a bound was conjectured by Bodlaender and proved by Kierstead and Trotter [35] with bound 33. This was subsequently improved in a sequence of papers, and the current record of 17 is due to Zhu [60].

A further tool handling the game-colouring number is based on edge decomposition [58].

Theorem 2.2 *If G_1 and G_2 are subgraphs of G with $E(G) = E(G_1) \cup E(G_2)$, then*

$$\mathrm{col}_g(G) \leq \mathrm{col}_g(G_1) + \Delta(G_2). \tag{3}$$

For the more restricted class of C_4-free planar graphs, Borodin *et al.* [14] proved that the edge-set can be decomposed into a forest and a subgraph with maximum degree at most 6; $\mathrm{col}_g \le 10$ then follows by inequality (3). For the class of planar graphs with girth at least 8, it was proved by Montassier *et al.* [43], and independently by Wang [55], that the edge-set can be decomposed into a forest and a matching, so $\mathrm{col}_g \le 5$.

From below, it is known that there exist planar graphs with $\mathrm{col}_g \ge 11$ (see Wu and Zhu [56]) and with $\chi_g \ge 8$ (see Kierstead and Trotter [35]).

As outerplanar graphs are all partial 2-trees, their game-colouring number is bounded by 8. By observing that the edge-set of an outerplanar graph can be decomposed into a forest and a graph of maximum degree 3, Guan and Zhu [26] improved this to $\mathrm{col}_g \le 7$ (using inequality (3)), and this is sharp (see [38]). Since $\chi_g(G) \le \mathrm{col}_g(G)$, the game-chromatic number of outerplanar graphs is at most 7. Kierstead and Trotter [35] showed that there are outerplanar graphs with game-chromatic number 6. It remains an open question as to whether there exists an outerplanar graph with game-chromatic number 7.

Cartesian product graphs

Recall that the Cartesian product $G \,\Box\, H$ of two graphs G and H has vertex-set $V(G) \times V(H)$ and edge-set $\{(vw, v'w') : v = v' \in V(G) \text{ and } ww' \in E(H)\} \cup \{(vw, v'w') : vv' \in E(G) \text{ and } w = w' \in V(H)\}$. Various graph invariants have the property that their value on $G \,\Box\, H$ is bounded above by some function of their values on G and H.

Bartnicki *et al.* [8] proved that this is not the case for colouring games: χ_g is unbounded on the Cartesian products of complete bipartite graphs $K_{q,q}$ (recall that $\chi_g(K_{q,q}) = 3$ for all $q \ge 2$) and col_g is unbounded on the Cartesian products of stars $K_{1,q}$. On the other hand, Zhu [61] has proved that if H has acyclic chromatic number k, and if joining to each vertex of G a set of $|V(H)|$ neighbours of degree 1 gives a graph with game-colouring number m, then $\chi_g(G \,\Box\, H) \le k(k + m - 1)$. As a consequence, the Cartesian product of two forests has game-chromatic number at most 10, and the Cartesian product of two planar graphs has game-chromatic number at most 105.

Game-perfect graphs

Recall from Chapter 7 that a perfect graph is defined in terms of the equality $\chi = \omega$, which is required to hold for every induced subgraph. In this vein, Andres [7] investigated the structure of graphs G for which $\chi_g(G') = \omega(G')$ or $\mathrm{col}_g(G') = \omega(G')$ is valid for every induced subgraph G' of G. We call these graphs (ω, χ_g)-*perfect* and (ω, col_g)-*perfect*, respectively. (The terminology of [7] is different, because other types of game-perfectness are also studied there.) It turns out that these graphs admit a concise characterization in terms of forbidden induced subgraphs.

Theorem 2.3

A graph is (ω, χ_g)-perfect if and only if it contains none of the graphs P_4, C_4, $K_3 + 3K_1$, $K_2 + 2P_3$, $(K_1 + 2P_3) \cup (K_2 + 3K_1)$, $2(K_1 + 2P_3)$ or $2(K_2 + 3K_1)$ as an induced subgraph.

A graph is (ω, col_g)-perfect if and only if it contains none of the graphs P_4, C_4, $2(K_4 - e)$ or $K_1 + (K_4 - e)$ as an induced subgraph.

3. Greedy colouring games

In each round of the maker–breaker game each player chooses an uncoloured vertex and assigns a colour to it, whereas in the marking game each player just chooses an unmarked vertex and no colour is involved. The *greedy colouring game* is, in some sense, a game lying between these two. Here, each player chooses any uncoloured vertex, but the chosen vertex has to be coloured with the smallest positive integer not assigned to any of its neighbours. So the colours are positive integers and the chosen vertices are coloured greedily. Alice's goal is to minimize the largest used colour, while Bob's goal is to maximize it. The *game Grundy number* $\Gamma_g(G)$ of a graph G is the maximum colour used in playing the game when both players use optimal strategies.

When playing the greedy colouring game, the player does not choose the colour. However, the colour is a critical concern. It follows from the definition that $\Gamma_g(G) \leq \mathrm{col}_g(G)$, for every graph G. Therefore, $\Gamma_g(F) \leq 4$ for any forest F, and $\Gamma_g(G) \leq 8$ for any partial 2-tree G. Havet and Zhu [28] proved that $\Gamma_g(F) \leq 3$ for any forest F, and this bound is tight. For any partial 2-tree G, $\Gamma_g(G) \leq 7$. Since there are trees T for which $\chi_g(T) = 4$, there are graphs G for which $\Gamma_g(G)$ is strictly less than $\chi_g(G)$: indeed, Havet and Zhu also proved that, for any positive integer $k \geq 2$, there is a graph G for which $\Gamma_g(G) = k$ and $\chi_g(G) \geq 2^k - 1$. However, it is unknown whether $\Gamma_g(G)$ is always bounded by $\chi_g(G)$. It is also an open question as to whether $\chi_g(G)$ is bounded by a function of $\Gamma_g(G)$.

4. Playing on the edge-set

The edge-colouring version – equivalent to playing the original vertex version on line graphs – was investigated first by Cai and Zhu [15]. We adopt the commonly used 'prime' notation for the edge versions of parameters and denote by $\chi'_g(G)$ the smallest number of colours sufficient for Alice to win.

The particular case of (1) when applied to line graphs means that

$$\Delta \leq \chi'(G) \leq \chi'_g(G) \leq 2\Delta - 1,$$

since each edge can share a vertex with at most $2\Delta - 2$ other edges.

Vizing's classic theorem tells us that the upper bound for $\chi'(G)$ can be improved to $\Delta + 1$ for all graphs (see Chapter 5). Estimates of this strength are not known (and are not even valid) for the colouring game, although bounds of the form $\Delta + c$, for a constant c, have been established for several graph classes. One such result by Cai and Zhu [15] states that, if G is a d-degenerate graph, then

$$\text{col}'_g(G) \leq \Delta(G) + 3d - 1.$$

In particular, this inequality yields the upper bounds $\Delta + 6k - 4$ for graphs of arboricity k, $\Delta + 3k - 1$ for partial k-trees, $\Delta + 14$ for planar graphs, $\Delta + 5$ for outerplanar graphs and $\Delta + 2$ for forests. It may also be the case that $\chi'_g(F) \leq \Delta(F) + 1$ holds for all forests F, and this has been proved except for $\Delta = 4$ (see Erdős *et al.* [22] and Andres [5]).

It is not true, however, that $\chi'_g \leq \Delta + c$ must hold for some suitably chosen constant c. Beveridge *et al.* [10] proved the existence of an infinite family of graphs G with any large maximum degree, for which $\chi'_g \geq 1.008\Delta(G)$. An important consequence of this result is that it is substantially harder to win the colouring game for some graphs than it is to list-colour them, since lists of size $\Delta + o(\Delta)$ suffice for the latter, as $\Delta \to \infty$. On the positive side, one can prove under some minimum-degree conditions that $\chi'_g \leq (2 - \varepsilon)\Delta$, for some $\varepsilon > 0$ (see [10] for details). It remains a conjecture that, for a suitably chosen $\varepsilon > 0$, the upper bound $\chi'_g(G) \leq (2 - \varepsilon)\Delta(G)$ is valid for all graphs G.

The *game Grundy index* $\Gamma'_g(G)$ of a graph G is the game Grundy number of its line graph. The *Grundy index* $\Gamma'(G)$ of G is the Grundy number of its line graph. Just as for the game-chromatic index, we have

$$\Delta \leq \Gamma'_g(G) \leq \Gamma'(G) \leq 2\Delta - 1.$$

Zhang and Zhu [57] proved that if T is a tree with $\Delta \geq 5$, then $\Gamma'_g(T) \leq \Delta + 1$;
if G is a partial 2-tree with $\Delta \geq 11$, then $\Gamma'_g(G) \leq \Delta + 4$;
and if G is an outerplanar graph with $\Delta \geq 14$, then $\Gamma'_g(G) \leq \Delta + 3$.

5. Oriented and directed graphs

An *oriented graph* is obtained from an undirected graph by assigning an orientation to each of its edges; there is thus at most one arc between any two vertices, whereas a *directed graph* may contain directed cycles of length 2. Below we consider two oriented kinds of colouring games. For an oriented or directed graph we use the notation $D = (V, A)$, where V is the vertex-set and A is the arc-set.

Oriented graphs

A colour assignment $\varphi : V \to \{1, 2, \ldots, k\}$ of an oriented graph $D = (V, A)$ is an *oriented colouring* if it is a proper vertex-colouring and all arcs between any two colour classes are in the same direction; that is, if $vw \in A$, then there does not exist $v'w' \in A$ with $\varphi(v') = \varphi(w)$ and $\varphi(w') = \varphi(v)$. Note that digraphs containing cycles of length 2 do not admit any such colouring.

In accordance with this definition, in the oriented colouring game Alice and Bob alternately assign colours $\varphi(v)$ to vertices v of D under the following restrictions:

(i) If $vw \in A$, then $\varphi(w) \neq \varphi(v)$;
(ii) If $uw \in A$ and $wv \in A$, then $\varphi(u) \neq \varphi(v)$;
(iii) If $vw \in A$ and $v'w' \in A$, and if $\varphi(v) = \varphi(w')$, then $\varphi(v') \neq \varphi(w)$.

Clearly, (i) and (iii) together require that the partial colouring obtained at each stage should be an oriented colouring. The role of (ii) is to ensure that Bob cannot win the game in a trivial way.

The *oriented game-chromatic number* $\vec{\chi}_g(D)$ of D is the smallest number k for which Alice has a winning strategy when the game is played with k colours; this parameter was introduced by Nešetřil and Sopena [44]. Unlike χ_g for undirected graphs, the maximum value of $\vec{\chi}_g(T)$ for oriented trees is not known.

Motivated by the marking game we may consider, for any uncoloured vertex v, the set of coloured neighbours and the set of coloured vertices w at (directed) distance 2 from v (in either direction) for which the middle vertices of every length-2 path between w and v is uncoloured. Using the strategy that proves $\chi_g(T) \leq 4$ for trees, Alice can ensure that both of these sets have at most three elements. Combining this idea with the construction of tournaments satisfying a certain adjacency/orientation property, we can prove that $\vec{\chi}_g$ is bounded above (by at most 19) on the class of trees. Also, by using marking games, Kierstead and Tuza [37] proved the following theorem.

Theorem 5.1 *For each positive integer k, there is a constant $C(k)$ for which the oriented game-chromatic number of oriented partial k-trees is at most $C(k)$.*

Moreover, Kierstead and Trotter [36] proved the following theorem.

Theorem 5.2 *For any positive integer k, orientations of graphs containing no subdivision of K_k have bounded oriented game-chromatic number. In particular, the oriented game-chromatic number of oriented planar graphs is bounded by a constant.*

The problem of determining the exact value of $\vec{\chi}_g$ appears to be much more difficult than that of χ_g. To indicate the complications, we quote the following result of Nešetřil and Sopena [44]: every oriented path and cycle has $\vec{\chi}_g \leq 7$, and this bound is sharp since there exist oriented paths with $\vec{\chi}_g = 7$.

Directed graphs

The oriented colouring game cannot be played on digraphs that contain 2-cycles. For such digraphs, a natural game is given by the following rule: each vertex must be assigned a colour different from those of its coloured in-neighbours. This rule eliminates the danger originating from a monochromatic pair at distance 2, which arose in the oriented version above.

This kind of colouring game was introduced by Andres [6]. It includes the undirected version as a particular case, by replacing each edge vw in an undirected graph by the two arcs vw and wv (a cycle of length 2), so we need not introduce any new terminology: on a digraph D the minimum number of colours sufficient for a winning strategy for Alice is called the *game-chromatic number* $\chi_g(D)$. Analogously, we introduce the *game-colouring number* $\mathrm{col}_g(D)$, based upon the directed version of the marking game. It is clear that, for all digraphs D and their underlying undirected graphs G_D, we have the inequalities

$$\chi_g(D) \leq \chi_g(G_D) \qquad \text{and} \qquad \mathrm{col}_g(D) \leq \mathrm{col}_g(G_D).$$

Here, strict inequality may hold when some arcs of D are not part of a symmetric pair. For example, as proved by Andres [6], Alice can win the marking game with three colours on any forest without 2-cycles – that is, $\chi_g(D) \leq \mathrm{col}_g(D) \leq 3$, whenever D is an oriented forest. Andres also observed that inequality (3) has its analogue for the directed game: if (D_1, D_2) is an arc decomposition of D, then

$$\mathrm{col}_g(D) \leq \mathrm{col}_g(D_1) + \Delta^+(D_2),$$

where Δ^+ denotes the maximum out-degree.

6. Asymmetric games

In this variant, one or both players may colour (or mark) more than one vertex at a time – that is, in the game parametrized by a pair (a, b) of natural numbers, at each step Alice colours or marks a vertices and then Bob colours or marks b vertices.

The first systematic study of asymmetric colouring games was carried out by Kierstead [34], but its roots date back to the earlier paper of Faigle, Kern, Kierstead and Trotter [23]. The authors considered the case $(a, b) = (1, 3)$ on trees, and proved that Alice has a strategy in the marking game to keep the score below $c \log n$, for some constant c; this logarithmic upper bound is sharp, apart from the value of c. It was also shown in [23] how this upper bound can be extended for any fixed $b \geq 3$; these results were applied there to prove that $\mathrm{col}_g = O(\log n)$ in every class of graphs whose edge-sets decompose into a bounded number of forests.

Kierstead [34] considered (a, b)-game-colourings and marking of forests. Let $\chi_g(\mathcal{F}; a, b)$ and $\mathrm{col}_g(\mathcal{F}; a, b)$ be the supremum taken over all forests, needed for Alice to win in the colouring game and in the marking game. The characteristic ranges for all positive integers a and b can be summarized as follows; they are the same if one considers trees instead of forests.

Theorem 6.1 *Assume that a, b are positive integers.*
If $a < b$, then $\chi_g(\mathcal{F}; a, b) = \mathrm{col}_g(\mathcal{F}; a, b) = \infty$.
If $b \leq a < 2b$, or $(a, b) = (2, 1)$, then $\chi_g(\mathcal{F}; a, b) = \mathrm{col}_g(\mathcal{F}; a, b) = b + 3$.
If $2b \leq a < 3b$ and $b > 1$, then $\chi_g(\mathcal{F}; a, b) = b + 2$ and $\mathrm{col}_g(\mathcal{F}; a, b) = b + 3$.
If $3b \leq a$, then $\chi_g(\mathcal{F}; a, b) = \mathrm{col}_g(\mathcal{F}; a, b) = b + 2$.

Note that the range $2b \leq a < 3b$ exhibits cases where $\chi_g < \mathrm{col}_g$.

For graphs in general, Kierstead and Yang [38] proved the following result.

Theorem 6.2 *If G has an orientation with maximum out-degree k, then* $\mathrm{col}_g(G; k, 1) \leq 2k + 2$.

By letting $k = 1$, we see that the game-colouring number of any forest (as well as that of any graph with unicyclic components) is at most 4.

7. Relaxed games

For a positive integer d, Chou, Wang and Zhu [18] introduced the *d-relaxed game*. It is played by Alice and Bob with the same rules as the original game, except that the subgraphs induced by the monochromatic sets need not be independent but are allowed to have maximum degree at most d. Let $\chi_g^d(G)$ denote the minimum integer k for which Alice has a winning strategy on G when k colours are available.

Not surprisingly, the extended flexibility provided by the parameter d makes it easier for Alice to win. A typical example is the class of trees and forests for which $\chi_g^2 \leq 2$, by a theorem of He, Wu and Zhu [29]. In fact, there is a smooth transition from the maximum game-chromatic number 4 for trees, which can be expressed in the following uniform way: in the range $0 \leq d \leq 2$ all forests have $\chi_g^d \leq 4 - d$, and this upper bound is sharp for each d.

For partial k-trees G, Dunn and Kierstead [20] proved that $\chi_g^d(G) \leq 6k + 1$, under the assumption $d \geq 4k - 1$; the particular case $k = 2$ implies that $\chi_g^7(G) \leq 3$ for every outerplanar graph G. On the other hand, if $d \geq 2$ and G is outerplanar, then $\chi_g^d(G) \leq 5$, by [29]. It is not known in general, however, whether the assumption $d \geq 7 - k$ implies that $\chi_g^d(G) \leq k$ for all $2 \leq k \leq 7$ and all outerplanar graphs G. The sharpness of this upper bound on $\chi_g^d(G)$ (if valid) is not known either. It is known that the assumption $d = 4$ is not strong enough to ensure that $\chi_g^d(G) \leq 2$.

8. Paintability

We next consider the list-colouring version of the colouring game; list-colouring and the choice number of a graph are discussed in Chapter 6. Let L be a list assignment for a graph G: without loss of generality, we may assume that the set of colours is $\bigcup_{v \in V(G)} L(v) = \{1, 2, \ldots, q\}$. For $i = 1, 2, \ldots, q$, let $L_i = \{v : i \in L(v)\}$. Then the sequence $(L_1, L_2, \ldots, L_q)$ is another way of specifying the list assignment. An L-colouring of G is equivalent to a sequence $(S_1, S_2, \ldots, S_q)$ of independent sets of vertices that form a partition of $V(G)$, and such that $S_i \subseteq L_i$ for all $i = 1, 2, \ldots, q$. This approach to list-colouring motivates the definition of the following game on a graph G, which was introduced by Schauz [50], [51].

Let G be a graph and let $f : V(G) \to \mathbb{N}$ be a mapping, where $f(v)$ is the number of permissible colours for v. The *f-painting game on G* is played by two players: Lister and Painter.

1. In the ith round, Lister releases the set L'_i of uncoloured vertices that permit colour i (that is, the yet-uncoloured subset of L_i). Painter decides which of these vertices are coloured with colour i, and hence specifies S_i.

2. If, at the end of some round an uncoloured vertex runs out of permissible colours – that is, a vertex v appeared in $f(v)$ sets L'_i and remains uncoloured – then the game ends with Lister as the winner.

3. Otherwise, at the end of some round all vertices are coloured, the game ends, and Painter is the winner.

Note that once coloured, a vertex does not participate further in the game and can be considered as having been removed from the graph.

We say that G is *f-paintable* if Painter has a winning strategy in this game, and that G is *k-paintable* if G is f-paintable for the constant function with $f(v) = k$ for all v. The *paint number* $\chi_p(G)$ of G is the minimum k for which G is k-paintable.

The following result can be easily derived from the definition, and is convenient for inductive proofs.

A graph G is f-paintable if and only if $f(v) \geq 1$ for each $v \in V(G)$ and, for any subset U of $V(G)$, there is an independent set X of G contained in U for which $G - X$ is $(f - \delta_U)$-paintable, where $\delta_U(v) = 1$ if $v \in U$ and $\delta_U(v) = 0$ otherwise.

It follows from the definition that $\chi_p(G) \geq \chi_l(G)$ for any graph G, where χ_l is the list-chromatic number. There exist graphs G with $\chi_p(G) > \chi_l(G)$. For example, for any integer $k \geq 1$, the theta graph $\theta_{2,2,2k}$, which consists of three paths of lengths $2, 2, 2k$ joining two vertices, is 2-choosable but not 2-paintable (see [62]). It is not known whether, for a graph G, the difference $\chi_p(G) - \chi_l(G)$ can be arbitrarily large.

Upper bounds

Many currently known upper bounds for the choice numbers of classes of graphs are also upper bounds for their paint numbers. For example, the paint number of any planar graph is at most 5 (see Schauz [50]), and the paint number of planar graphs with girth at least 5 is at most 3 (see [50], [17]). Indeed, if an upper bound for $\chi_l(G)$ is proved by induction, then it is likely that the proof can be adapted to a strategy for Painter, establishing the same upper bound for $\chi_p(G)$.

One method to prove an upper bound for $\chi_l(G)$ is the *kernel method*:

if G has an orientation D for which each induced subdigraph has a kernel, and if $f(v) = d^+_D(v) + 1$ for every vertex v, then G is f-choosable (see Chapter 6).

The proof of this result actually shows that, under the condition above, G is f-paintable: Painter simply selects a kernel S_i in the subdigraph induced by L'_i in each step. In particular, this implies that the line graph $L(G)$ of a bipartite graph G is $\Delta(G)$-paintable (see [50]).

Another method for proving an upper bound for $\chi_l(G)$ is to use the following result of Alon and Tarsi [2]:

if G has an orientation D for which the numbers of Eulerian subdigraphs with odd and even numbers of arcs are not equal, and if $f(v) = d_D^+(v) + 1$ for every vertex v, then G is f-choosable.

Schauz [52] proved that, under the same assumption, G is f-paintable, so any upper bound for $\chi_l(G)$ proved by using the Alon–Tarsi theorem is also an upper bound for $\chi_p(G)$. To mention one, Hladký, Král' and Schauz [30] gave a new proof of Brooks's theorem, based on the Alon–Tarsi theorem. This new proof yields an upper bound for the paint number of graphs: G is Δ-paintable if G is connected and is not a complete graph or an odd cycle.

Some upper bounds for $\chi_l(G)$ are proved by the probabilistic method. If only expectations are used in the proof, then the proof maybe de-randomized to give a winning strategy for Painter.

For complete bipartite graphs, Erdős, Rubin and Taylor [21] proved that $\chi_l(K_{n,n}) \le \log_2 n + 2$. The same upper bound also holds for the paint number. Let G be any bipartite graph of order n. The following simple strategy for Painter to win the $(\log_2 n + 1)$-painting game on G was given by Zhu [62]. Let A and B be the two partite sets of G.

1. Initially, assign weight 1 to each vertex of G.

2. Each time Lister presents a set U of uncoloured vertices, Painter chooses the independent set $S = A \cap U$ or $S = B \cap U$, according to whichever has the larger total weight, and then doubles the weight of each vertex in $U \setminus S$. The vertices in S are then coloured.

Since the weight of $U \setminus S$ does not exceed that of S, the total weight of uncoloured vertices is never increased; in particular, each uncoloured vertex has weight at most $2n$. On the other hand, each time that a vertex is given a permissible colour but is not coloured, its weight doubles. If an uncoloured vertex has been given k permissible colours, then its weight is $2^k \le n$, implying that $k \le \log_2 n$. So if a vertex is given $\log_2 n + 1$ colours, then it is coloured. The above strategy is thus a winning strategy for Painter in the $(\log n + 1)$-painting game on G. Thus $\chi_p(G) \le \log n + 1$, and in particular, $\chi_p(K_{n,n}) \le \log n + 2$.

A similar argument shows that if G is a k-chromatic n-vertex graph, then $\chi_p(G) \le k \log n + 1$.

The online version of Ohba's conjecture

A graph G is *chromatic-choosable* if $\chi(G) = \chi_l(G)$, and is *chromatic-paintable* if $\chi_p(G) = \chi(G)$. Since $\chi_p(G) \geq \chi_l(G) \geq \chi(G)$, all chromatic-paintable graphs are chromatic-choosable, but the converse is not true. It was conjectured by Ohba [46], and proved by Noel, Reed and Wu [45], that graphs G with $n \leq 2\chi(G) + 1$ are chromatic-choosable.

Kim *et al.* [39] proved that the complete multipartite graph G, with k parts of size 2 and one part of size 3, has $\chi_p(G) = k+2$. These are graphs G with $n = 2\chi(G)+1$ and are not chromatic-paintable. Nevertheless, Huang *et al.* [31] proposed the online version of Ohba's conjecture, as follows.

Online version of Ohba's conjecture. Every graph G with $\chi(G) \geq \frac{1}{2}n$ is chromatic-paintable.

This conjecture is largely open. Caraher *et al.* [16] proved the following result.

Theorem 8.1 *If G is k-paintable and if $n \leq kt/(t-1)$, then the graph G' obtained from G by adding an independent set T of t vertices is $(k+1)$-paintable.*

We may assume that each vertex of T is adjacent to every vertex of G. Denote by R_i the set of uncoloured vertices after the ith move. For each vertex v, let $f_i(v)$ be the number of remaining permissible colours of v after the ith move, and let

$$\rho_i(v) = \max\{0, |R_i \cap N_G(v)| - f_i(v) + 1\}.$$

So $f_0(v) = k + 1$ for each vertex v, and $\rho_0(v) = \max\{0, d_G(v) - k\}$. Note that if $\rho_i(v) = 0$, then after the ith move the number of permissible colours of v exceeds the number of its uncoloured neighbours. In this case, the vertex v can be ignored throughout the rest of the game, since it can never run out of permissible colours.

We may assume that, for any i, the set of uncoloured vertices with i as a permissible colour is not an independent set, since otherwise Painter can simply colour all these vertices with colour i. The strategy for Painter to win the online $(G', k+1)$-list-colouring game is simple:

In the ith move, let L_i be the set of vertices with i as a permissible colour presented by Lister. Painter ignores the vertices of T and uses his winning strategy for the k-painting game on G to colour an independent set S_i of G (contained in L_i), with one exception: if $\rho_{i-1}(v) = 0$ for each vertex $v \in T \setminus L_i$, then Painter colours the vertices in $T \cap L_i$ with colour i.

Since the exceptional move is performed only once, and since each vertex of G has one more permissible colour than is required in the k-painting game, the vertices of G can never run out of permissible colours. Also, once the exceptional move is made, each vertex in T is either coloured or can be ignored. So to show that this is indeed a winning strategy for Painter, it suffices to show that before the exceptional move is made, no vertex in T can run out of permissible colours.

At the ith round, if the exceptional move is not carried out then (by the description of the strategy), some vertex $v \in T$ is not contained in L_i and hence has $0 \leq \rho_i(v) \leq \rho_{i-1}(v) - 1$. Now suppose that a vertex $v \in T$ runs out of permissible colours before the exceptional move. Then there are moves $j_1, j_2, \ldots, j_{k+1}$ such that $v \in L_{j_s}$ for $s = 1, 2, \ldots, k+1$. For each such s, there is a vertex $w \in T \setminus \{v\}$ for which $0 \leq \rho_{j_s}(w) \leq \rho_{j_s-1}(w) - 1$. Therefore,

$$k + 1 \leq \sum_{w \in T - \{v\}} \rho_0(w) = (t-1)(n-k).$$

This contradicts our assumption that $n \leq kt/(t-1)$.

A consequence of Theorem 8.1 is that the online version of Ohba's conjecture holds for graphs with independence number 2. Kozik *et al.* [40] proved that the conjecture also holds for graphs with independence number 3.

The fractional paint number

A q-tuple colouring φ of G assigns to each vertex v a set $\varphi(v)$ of q colours so that adjacent vertices receive disjoint colour sets. The *fractional chromatic number* $\chi^*(G)$ of a graph G is the infimum of the ratios p/q for which there exists a q-tuple colouring that uses p colours in total.

A graph is called (a, b)*-choosable* if, for any list assignment L that assigns to each vertex a set of a permissible colours, there is a b-tuple colouring φ with $\varphi(v) \subseteq L(v)$ for each vertex v; in particular, '$(a, 1)$-choosable' means a-choosable. The *fractional choice number* $\chi_l{}^*(G)$ of G is the infimum of the ratios a/b for which G is (a, b)-choosable.

Given a graph G and two positive integers a, b, the *b-tuple a-painting game on G* is defined analogously to the f-painting game. In the ith round Lister reveals the information as to which vertices can be coloured with colour i, and Painter then decides which of these vertices should be so coloured. The difference is that each vertex needs to be assigned b colours. If, at any moment of the game, a vertex v has not been assigned b colours but runs out of permissible colours, then Lister wins the game. Otherwise, at some round, each vertex is assigned b colours and Painter wins the game.

The *fractional paint number* $\chi_p{}^*(G)$ of G is the infimum of the ratios a/b for which Painter has a winning strategy in the online $(G, (a, b))$-list-colouring game. It follows from the definition that, for any graph G,

$$\chi^*(G) \leq \chi_l{}^*(G) \leq \chi_p{}^*(G).$$

Alon, Tuza and Voigt [4] proved that the fractional choice number of a graph is always equal to its fractional chromatic number. Gutowski [25] extended this by proving the following result.

Theorem 8.2 *For any graph G, $\chi_p{}^*(G) = \chi_l{}^*(G)$.*

Suppose that $\chi^*(G) = a/b$ and that φ is a b-tuple a-colouring of G. We shall prove that, for any $\varepsilon > 0$, there exist sufficiently large integers a', b' for which $a'/b' < a/b + \varepsilon$ and Painter has a winning strategy for the b'-tuple a-painting game on G. The strategy for Painter is simple. Suppose that in the ith round, Lister presents a set L_i, and that there are p sets among $L_1, L_2, \ldots, L_{i-1}$ that are identical to L_i, and let $p' \equiv p \pmod{a}$ with $1 \le p' \le a$. Then Painter's choice is $S_i = L_i \cap \varphi^{-1}(p')$. We may consider the permissible colour i as identified with the colour p' in the b-tuple a-colouring of G. A key observation is the following:

for any vertex v and for any two different colours p' and p'', the number of permissible colours of v that are identified with p' and the number of permissible colours of v that are identified with p'' may differ by at most 2^n.

Since the graph G is fixed, the number 2^n is constant. When a' and b' are sufficiently large, this difference may be ignored – that is, counted in the ε factor. So, roughly speaking, about b/a of the permissible colours of v are eventually assigned to v, and Painter wins the game.

This argument extends the identity $\chi^* = \chi_l{}^*$ to $\chi^* = \chi_p{}^*$. Nevertheless, $\chi_l{}^*$ and $\chi_p{}^*$ behave in a somewhat different way – namely, as proved in [4], for each graph G there exists a pair (a, b) of integers for which G is (a, b)-choosable with $b = a/\chi^*(G)$. Consequently, in the definition of $\chi_l{}^*$ we can take 'minimum' instead of 'infimum'. This is not valid for $\chi_p{}^*$, as we really need the infimum for some graphs, despite the corresponding proof for $\chi_l{}^*$ being constructive.

9. Achievement and avoidance games

Unlike the games already discussed, in achievement and avoidance games the goal of the two players is symmetrical. The players construct partial colourings together, playing on a given graph G within a given colour bound k: at most k colours may be used and monochromatic edges are forbidden. The players alternately colour one uncoloured vertex at a time, and the game is over when no further legal move is possible – either because the entire vertex-set is coloured, or because the neighbourhood of each uncoloured vertex contains all the k colours.

The *achievement game* is *won* by the player who makes the last legal move, and the *avoidance game* is *lost* by the player who makes the last legal move. These colouring games were introduced by Harary and Tuza [27], motivated by the algorithmic decision problem called Precolouring Extension (see [11] for the latter). An instance of the algorithmic problem PrExt is a graph $G = (V, E)$, a colour bound k, and a precoloured set $W \subset V$ – that is, a partial colouring $\varphi_W : W \to \mathbb{N}$ is given, which is supposed to be a proper vertex-colouring of the subgraph induced by W. The question to decide is whether G admits a proper k-colouring in which each precoloured vertex $v \in W$ has its pre-assigned colour $\varphi_W(v)$.

The most interesting particular cases of these games correspond to small values of k. If $k = 1$, then the game terminates with a maximal independent set, and the parity of its cardinality determines who is the winner. Schaefer [49] developed hardness results and studied various games concerning Boolean satisfiability, and also gave three examples of natural games, one of which (Node Kayles) is exactly our achievement game with $k = 1$. He proved that it is PSPACE-complete to decide whether the first player has a winning strategy. The same complexity was proved by Bodlaender [12] for $k = 2$ – that is, where the game terminates with a non-extendable bipartite subgraph of G.

Unlike the games discussed in previous sections, the case $k = 1$ already raises enormous difficulties, even for some very simple types of graph. Recall that the players in this game sequentially construct a maximal independent set. On a path of given length the game is equivalent to a game known as *Dawson's chess*, created by T. R. Dawson in 1935. Dawson's chess is played on a $3 \times n$ chessboard, with n white pawns in the first row and n black pawns in the third row. Pawns move in the standard way, except that capture is mandatory whenever possible.

The game on paths, with $k = 1$, is also known as Node Kayles. In the terminology of game theory, achievement corresponds to 'normal play' (last player wins) whereas avoidance corresponds to 'misère play' (last player loses). (Dawson proposed the latter.) The former game was solved after two decades by Guy and Smith (1956), but the latter game has remained open, even after three-quarters of a century, in spite of considerable efforts in game theory. The game with normal play can be analyzed by computing the Grundy function; it turns out that the second player has a winning strategy if and only if $n \equiv 4, 8, 20, 24$ or $28 \pmod{34}$ or $n = 14$ or $n = 34$ – that is, the score is ultimately periodic modulo 34.

As a rather vague intuition concerning reasons why the achievement game is easier than the avoidance one, we mention a simple strategy for the former, based on symmetry, that can be applied under certain conditions. Suppose that G has a cut-vertex v with the property that $G - v$ is the union of two disjoint graphs G_1 and G_2 for which $G - G_1 \cong G - G_2$ and some isomorphism between these two subgraphs maps v onto itself. Then the first player can win the achievement game with any given colour bound k as follows: colour v first, fix an automorphism $f : G \to G$ with $f(v) = v$, $f(G_1) = G_2$ and $f(G_2) = G_1$, and then copy the moves of the second player: thus, if the second player assigns colour c to a vertex $w \in V(G_i)$, answer by assigning c to $f(w) \in V(G_{3-i})$. An analogous winning strategy works for the second player if G is the disjoint union of two isomorphic graphs, but there seems to be no similarly trivial strategy for the avoidance game.

For a deeper discussion about why misère play is much harder than normal play, and for substantial partial results concerning the former, see [48] and some of the papers cited there.

10. The acyclic orientation game

Finally, we discuss a game that involves no colours. Its original motivation was to enable us to check efficiently whether a 'hidden' colouring of a graph is proper or not. It is related to the fact that every proper colouring $\varphi : V \to \mathbb{N}$ defines an orientation without directed cycles in a natural way, from smaller colours to larger ones. Should a monochromatic edge occur, its orientation remains undetermined.

The first version of this game was introduced by Manber and Tompa [42]; here we consider the following variation, presented by Aigner, Triesch and Tuza [1]. The game is played on a simple undirected graph G by two players, Algy and Strategist. At each step, Algy selects an edge $vw \in E$ and Strategist orients it either from w to v or from v to w, as he chooses, but in such a way that no directed cycle occurs. The goal is to identify an acyclic orientation of G unambiguously, and the game is over when the actual partial orientation has a unique extension to an acyclic orientation of G. Algy's goal is to minimize the number of steps, while Strategist aims at maximizing it. We denote by $c(G)$ the number of steps when both players play optimally.

When played on a complete graph, the game ends with a linear ordering of the vertices, and is thus equivalent to comparison sorting, where $c(K_n) = (1 + o(1))n \log_2 n$. On the other hand, when played on a bipartite graph, Strategist may even promise at the beginning to orient all but one edge from the first vertex-class to the second; this leaves the possibility of orienting the last edge in either way, yielding $c(G) = |E|$. More generally, the same holds if G is the cover graph of a partially ordered set. (A graph G is a *cover graph* if it admits an orientation $\vec{G}$ for which there exists a partially ordered set $P = (V, <)$ over the vertex-set V of G such that vw is an arc in $\vec{G}$ if and only if v covers w in P – that is, $v > w$, and there is no $u \in V$ with $v > u > w$.)

As proved in [1], the following general bounds are valid for all graphs G of order n and independence number α :

$$n \log_2 n\alpha - n \log_2 e + 1 + \tfrac{1}{2} \log_2 n < c(G) < \alpha n\,(\log_2(n/\alpha) + 1). \qquad (4)$$

It follows in particular that, for any graph with n vertices and $\binom{n}{2} - o(n^2)$ edges, $c(G) = o(n^2)$ as $n \to \infty$. On the other hand, for each $\varepsilon < \frac{1}{2}$ there exists $\delta > 0$ such that, for every sufficiently large n, there exists a graph G with at least $(\frac{1}{2} - \varepsilon)n^2$ edges and with $c(G) \geq \delta n^2$.

A less explicit inequality is the information-theoretic lower bound $c(G) \geq \log_2 a(G)$, where $a(G)$ denotes the number of acyclic orientations of G. Interestingly, $a(G)$ can be expressed in terms of the chromatic polynomial $P(G, \lambda)$ of G, as $a(G) = |P(G, -1)|$, by a theorem of Stanley [54].

The above observation on bipartite graphs yields $c(K_{\lceil n/2 \rceil, \lfloor n/2 \rfloor}) = \lfloor \frac{1}{4}n^2 \rfloor$. Pikhurko [47] proved that no graph of order n can require substantially more steps – that is, $c(G) \leq \frac{1}{4}n^2 + o(n^2)$. But from below, it is not known whether or not $c(G) - \frac{1}{4}n^2$ can be arbitrarily large; so far, no example has been found where the difference exceeds 1.

The typical value of $c(G)$, however, grows much more slowly as a function of n: Alon and Tuza [3] proved that, for any constant $p \in (0, 1)$, the random graph $G_{n,p}$ of order n with edge probability p satisfies $c(G_{n,p}) = \Theta(n \log n)$, with probability $1 - o(1)$ as $n \to \infty$. A similar slowly growing (and slightly larger) upper bound holds for an arbitrary edge probability function $p = p(n)$ – namely, $c(G_{n,p}) = O(n \log^3 n)$

holds almost surely. The exponent 3 can be improved; it remains open as to whether $O(n \log n)$ is a universal upper bound for all probability functions $p(n)$.

We say that $G = (V, E)$ is an *exhaustive graph* if $c(G) = |E|$ – that is, if Strategist can force Algy to probe all edges of G. Unlike the value of $c(G)$, it is known (see [1]) that for $n \geq 6$, the number of edges in an exhaustive graph of order n is at most $\lfloor \frac{1}{4}n^2 \rfloor$, and if $n \geq 7$ then $K_{\lceil n/2 \rceil, \lfloor n/2 \rfloor}$ is the unique extremal graph.

The cover graphs mentioned earlier, which are necessarily triangle-free, show that exhaustiveness is not a local property. However, it is proved in [1] that the exclusion of induced cycles of length larger than 3 has a different effect: a chordal graph is exhaustive if and only if it contains no K_4 or $P_4 + K_1$ as an induced subgraph.

As regards time complexity, Algy can design in polynomial time a sequence of sufficient questions fewer than the upper bound in the inequality 4 (see [1]). On the other hand, Pikhurko [47] proved that $c(G)$ is hard to compute, and is even hard to approximate within arbitrary precision: $c(G)$ admits no polynomial-time approximation scheme unless P = NP. As an upper bound for the approximation ratio achievable in polynomial time, only a $O(n/\log n)$-approximation is known (see [47]), and hence a wide gap remains to be narrowed.

As with several other games, it may turn out to be PSPACE-complete to determine the winner for a generic input graph G and value k which is part of the input, but this seems not to have been proved so far; on the other hand, it is known that $a(G)$ is #P-hard to compute (see [41]). A further open problem is to determine the computational complexity of recognizing exhaustive graphs.

References

1. M. Aigner, E. Triesch and Zs. Tuza, Searching for acyclic orientations of graphs, *Discrete Math.* **144** (1995), 3–10.
2. N. Alon and M. Tarsi, Colorings and orientations of graphs, *Combinatorica* **12** (1992), 125–134.
3. N. Alon and Zs. Tuza, The acyclic orientation game on random graphs, *Random Structures Algorithms* **6** (1995), 261–268.
4. N. Alon, Zs. Tuza and M. Voigt, Choosability and fractional chromatic numbers, *Discrete Math.* **165/166** (1997), 31–38.
5. S. D. Andres, The game chromatic index of forests of maximum degree $\Delta \geq 5$, *Discrete Appl. Math.* **154** (2006), 1317–1323.
6. S. D. Andres, Lightness of digraphs in surfaces and directed game chromatic number, *Discrete Math.* **309** (2009), 3564–3579.
7. S. D. Andres, On characterizing game-perfect graphs by forbidden induced subgraphs, *Contrib. Discrete Math.* **7** (2012), 21–34.
8. T. Bartnicki, B. Brešar, J. Grytczuk, M. Kovše, Z. Miechowicz and I. Peterin, Game chromatic number of Cartesian product graphs, *Electron. J. Combin.* **15** (2008), #R72.
9. C. Berge, *Graphs*, North-Holland, 1985.
10. A. Beveridge, T. Bohman, A. Frieze and O. Pikhurko, Game chromatic index of graphs with given restrictions on degrees, *Theoret. Comput. Sci.* **407** (2008), 242–249.

11. M. Biró, M. Hujter and Zs. Tuza, Precoloring extension. I, Interval graphs, *Discrete Math.* **100** (1992), 267–279.
12. H. L. Bodlaender, On the complexity of some coloring games. *Intern. J. Foundat. Comput. Sci.* **2** (1991), 133–147.
13. O. V. Borodin, On acyclic colorings of planar graphs, *Discrete Math.* **25** (1979), 211–236.
14. O. V. Borodin, A. V. Kostochka, N. N. Sheikh and G. Yu, M-degrees of quadrangle-free planar graphs, *J. Graph Theory* **60** (2009), 80–85.
15. L. Cai and X. Zhu, Game chromatic index of k-degenerate graphs, *J. Graph Theory* **36** (2001), 144–155.
16. J. Carraher, S. Loeb, T. Mahoney, G. Puleo, M. Tsai and D. West, Extending graph choosability results to paintability, manuscript, 2012.
17. T. Chang and X. Zhu, On-line 3-choosable planar graphs, *Taiwanese J. Math.* **16** (2012), 511–519.
18. C.-Y. Chou, W.-F. Wang and X. Zhu, Relaxed game chromatic number of graphs, *Discrete Math.* **262** (2003), 89–98.
19. T. Dinski and X. Zhu, A bound for the game chromatic number of graphs, *Discrete Math.* **196** (1999), 109–115.
20. C. Dunn and H. A. Kierstead, A simple competitive graph coloring algorithm II, *J. Combin. Theory (B)* **90** (2004), 93–106.
21. P. Erdős, L. Rubin and H. Taylor, Choosability in graphs, *Congr. Numer.* **XXVI** (1979), 125–157.
22. P. L. Erdős, U. Faigle, W. Hochstättler and W. Kern, Note on the game chromatic index of trees, *Theoret. Comput. Sci.* **313** (2004), 371–376.
23. U. Faigle, U. Kern, H. A. Kierstead and W. T. Trotter, On the game chromatic number of some classes of graphs, *Ars Combin.* **35** (1993), 143–150.
24. M. Gardner, Mathematical games, *Sci. Amer.* (April 1981), 23.
25. G. Gutowski, Mr. Paint and Mrs. Corrector go fractional, *Electron. J. Combin.* **18** (2011), #R140.
26. D. J. Guan and X. Zhu, Game chromatic number of outerplanar graphs, *J. Graph Theory* **30** (1999), 67–70.
27. F. Harary and Zs. Tuza, Two graph-colouring games, *Bull. Austral. Math. Soc.* **48** (1993), 141–149.
28. F. Havet and X. Zhu, The game Grundy number of graphs, *J. Combin. Optim.* **25** (2013), 752–765.
29. W. He, J. Wu and X. Zhu, Relaxed game chromatic number of trees and outerplanar graphs, *Discrete Math.* **281** (2004), 209–219.
30. J. Hladký, D. Král and U. Schauz, Brooks' theorem via the Alon–Tarsi theorem, *Discrete Math.* **310** (2010), 3426–3428.
31. P. Huang, T. Wong and X. Zhu, Application of polynomial method to on-line colouring of graphs, *Europ. J. Combin.* **33** (2012), 872–883.
32. K. Junosza-Szaniawski and Ł. Rożej, Game chromatic number of graphs with locally bounded number of cycles, *Inform. Process. Letters* **110** (2010), 757–760.
33. H. A. Kierstead, A simple competitive graph colouring algorithm, *J. Combin. Theory (B)* **78** (2000), 57–68.
34. H. A. Kierstead, Asymmetric graph coloring games, *J. Graph Theory* **48** (2005), 169–185.
35. H. A. Kierstead and W. T. Trotter, Planar graph coloring with an uncooperative partner, *J. Graph Theory* **18** (1994), 569–584.
36. H. A. Kierstead and T. Trotter, Competitive colorings of oriented graphs, *Electron. J. Combin.* **8** (2001), #R1.
37. H. A. Kierstead and Zs. Tuza, Marking games and the oriented game chromatic number of partial k-trees, *Graphs Combin.* **19** (2003), 121–129.

38. H. A. Kierstead and D. Yang, Very asymmetric marking games, *Order* **22** (2005), 93–107.
39. S. Kim, Y. Kwon, D. Liu and X. Zhu, On-line list colouring of complete multipartite graphs, *Electron. J. Combin.* **19** (2012) #R41.
40. J. Kozik, P. Micek and X. Zhu, Towards on-line Ohba's Conjecture, *Europ. J. Combin.* **36** (2014), 110–121.
41. N. Linial, Hard enumeration problems in geometry and combinatorics, *SIAM J. Alg. Discrete Methods* **7** (1986), 331–335.
42. U. Manber and M. Tompa, The effect of the number of Hamiltonian paths on the complexity of a vertex coloring problem, *Proc. 22nd Symp. on Foundations of Computer Science (FOCS)* (Nashville, TN) (1981), 220–227.
43. M. Montassier, P. Ossona de Mendez, A. Raspaud and X. Zhu, Decomposing a graph into forests, *J. Combin. Theory (B)* **102** (2012), 38–52.
44. J. Nešetřil and E. Sopena, On the oriented game chromatic number, *Electron. J. Combin.* **8** (2001), #R14.
45. J. Noel, B. Reed and H. Wu, A proof of a conjecture of Ohba, *J. Graph Theory* (2013), submitted.
46. K. Ohba, On chromatic-choosable graphs, *J. Graph Theory* **40** (2002), 130–135.
47. O. Pikhurko, Finding an unknown acyclic orientation of a given graph, *Combin. Probab. Comput.* **19** (2010), 121–131.
48. T. E. Plambeck and A. N. Siegel, Misère quotients for impartial games, *J. Combin. Theory (A)* **115** (2008), 593–622.
49. T. J. Schaefer, On the complexity of some two-person perfect-information games, *J. Comput. Syst. Sci.* **16** (1978), 185–225.
50. U. Schauz, Mr. Paint and Mrs. Correct, *Electron. J. Combin.* **16/1** (2009), #R77.
51. U. Schauz, A paintability version of the combinatorial Nullstellensatz, and list colorings of k-partite k-uniform hypergraphs, *Electron. J. Combin.* **17/1** (2010), #R176.
52. U. Schauz, Flexible color lists in Alon and Tarsi's theorem, and time scheduling with unreliable participants, *Electron. J. Combin.* **17/1** (2010), #R13.
53. E. Sidorowicz, The game chromatic number and the game colouring number of cactuses, *Inform. Process. Letters* **102** (2007), 147–151.
54. R. P. Stanley, Acyclic orientations of graphs, *Discrete Math.* **5** (1973), 171–178.
55. Y. Wang and Q. Zhang, Decomposing a planar graph with girth at least 8 into a forest and a matching, *Discrete Math.* **311** (2011), 844–849.
56. J. Wu and X. Zhu, Lower bounds for the game colouring number of partial k-trees and planar graphs, *Discrete Math.* **308** (2008), 2637–2642.
57. W. Zhang and X. Zhu, The game Grundy indices of graphs, *J. Combin. Optim.*, to appear.
58. X. Zhu, The game coloring number of planar graphs, *J. Combin. Theory (B)* **75** (1999), 245–258.
59. X. Zhu, Game coloring number of pseudo partial k-trees, *Discrete Math.* **215** (2000), 245–262.
60. X. Zhu, Refined activation strategy for the marking game, *J. Combin. Theory (B)* **98** (2008), 1–18.
61. X. Zhu, Game colouring Cartesian product of graphs, *J. Graph Theory* **59** (2008), 261–278.
62. X. Zhu, On-line list colouring of graphs, *Electron. J. Combin.* **16** (2009), #R127.

15

Unsolved graph colouring problems

TOMMY R. JENSEN and BJARNE TOFT

Our book Graph Coloring Problems [85] *appeared in 1995. It contains descriptions of unsolved problems, organized into sixteen chapters. A large number of publications on graph colouring have appeared since then, and in particular around thirty of the* 211 *problems in that book have been solved. In this chapter we review some of our favourite problems that remain unsolved.*

1. Introduction

A main reason for the continued interest in the area of graph colouring is its wealth of interesting unsolved problems. Many of these are easy to state, but seemingly difficult to solve. However they are not impossible, as the literature in the field will testify.

The seven most striking results of the past twenty years are:

- the 5-list-colourability of planar graphs (dating back to V. G. Vizing in 1975 and to P. Erdős, A. L. Rubin and H. Taylor in 1979) by Thomassen [159]
- the confirmation by Robertson, Sanders, Seymour and Thomas [137] of the truth of the four-colour theorem (F. Guthrie and A. De Morgan (1852))
- the asymptotic solution by Reed [134] of the problem as to whether for $k \geq 9$ there are k-chromatic graphs without complete k-graphs and of maximum degree k (V. G. Vizing (1968) and O. V. Borodin and A. V. Kostochka (1977))

- the proof by Chudnovsky, Robertson, Seymour and Thomas [39] of the strong perfect graph conjecture of C. Berge around 1960
- the proof by Thomassen [161] of the weak 3-flow conjecture of W. T. Tutte (1954) and F. Jaeger (1988)
- the solution by Kostochka and Yancey [111] to the problem of critical graphs with few edges (due to T. Gallai (1963) and O. Ore (1967))
- the description found by Borodin, Dvořák, Kostochka, Lidický and Yancey [24] of all 4-critical planar graphs with exactly four triangles (B. Grünbaum (1963), V. A. Aksenov (1974) and P. Erdős (1990)).

In addition to these major achievements there are many other important results – in fact, thirty-one of the original 211 problems from the lists in Jensen and Toft [85] were solved by 2013. The area continues to surprise, and if we had been asked in 1995 which problems would be solved first our guesses would have been totally wrong. Although graph colouring has its origins in the four-colour problem, the latter's solution does not diminish the interest. To cite W. T. Tutte from his 1978 paper [166], where he describes several difficult colouring problems:

The Four Colour Theorem is the tip of the iceberg, the thin edge of the wedge and the first cuckoo of spring.

The subsequent advances bear witness to this.

2. Complete graphs and chromatic numbers

Hadwiger's conjecture

As we saw in Chapter 4, the following conjecture was proposed by Hadwiger [74] in 1943.

Every k-chromatic graph can be transformed into a complete k-graph by successive contractions of edges and deletions of edges and vertices.

If a graph H can be obtained from a graph G by successive contractions and deletions, then H is a *minor* of G. Hadwiger's conjecture thus suggests that any graph with no complete k-graph as a minor is $(k-1)$-colourable. Hadwiger's idea was to create a new combinatorial classification of graphs, different from the classical topological classification via embeddability in surfaces.

A more striking formulation of Hadwiger's conjecture is this, where a class of graphs is *proper* if it is not the class of all graphs:

For any proper minor-closed class of graphs the maximum chromatic number of graphs in the class equals the number of vertices in the largest complete graph belonging to the class.

In particular, this suggests that all planar graphs are 4-colourable, because the complete graph K_4 is the largest complete planar graph. Thus Hadwiger's conjecture may be seen as a far-reaching generalization of the four-colour theorem. In 1937 Wagner [170] proved that the 4-colourability of all planar graphs can be extended to the 4-colourability of all graphs without K_5 as a minor, thereby proving that the case $k = 5$ of Hadwiger's conjecture is equivalent to the four-colour theorem.

For $k = 4$, Hadwiger's conjecture was proved by Hadwiger in his 1943 paper, and independently by Dirac [43]. By a deep theorem of Robertson, Seymour and Thomas [138], the case $k = 6$ is also equivalent to the four-colour theorem. Kawarabayashi and Toft [96] proved that a 7-chromatic graph has K_7 or $K_{4,4}$ as a minor, and Kawarabayashi [90] subsequently proved the corresponding statement with $K_{4,4}$ replaced by $K_{3,5}$.

Perhaps the most interesting special unsolved case of Hadwiger's conjecture arises by assuming that G is a graph with independence number α at most 2, so G contains no three mutually non-adjacent vertices. In 1982 Duchet and Meyniel [46] observed in general that, if G is a graph with n vertices, then G contains as a minor a complete $\lceil n/(2\alpha - 1)\rceil$-graph. In particular, if $\alpha \leq 2$, then G contains a complete $\lceil \frac{1}{3}n\rceil$-graph as a minor.

The importance of this special case was first pointed out by W. Mader who noticed that, whereas the truth of Hadwiger's conjecture would imply that if $\alpha \leq 2$, then G contains a complete $\lceil \frac{1}{2}n\rceil$-graph as a minor, it seems very difficult to improve on the lower bound of $\lceil \frac{1}{3}n\rceil$ for the order of a largest complete minor. P. D. Seymour has asked for a proof of the existence of a number $\varepsilon > 0$ such that, if $\alpha \leq 2$, then G contains a complete $\lceil (\frac{1}{3} + \varepsilon)n\rceil$-graph as a minor. As far as we know, this problem is still open. Partial results have been obtained by Plummer, Stiebitz and Toft [132] and Pedersen and Toft [130]. Duchet and Meyniel's original bound of $\lceil n/(2\alpha - 1)\rceil$ was improved for $\alpha > 2$ by Kawarabayashi, Plummer and Toft [92] to $\lceil n/(2\alpha - \frac{3}{2})\rceil$, and further by Kawarabayashi and Song [94] to $\lceil n/(2\alpha - 2)\rceil$.

In an interesting paper, Kawarabayashi and Reed [93] showed that, for any fixed number $k \geq 2$, there exist a constant N and an $O(n^2)$ algorithm which, for any input graph G with at most n vertices, produces one of the following:

- a colouring of G with $k - 1$ colours
- a K_k-minor of G
- a minor H of G with at most N vertices, for which $\chi(H) \geq k$ and H has no K_k-minor.

In particular it follows that, for any fixed k, the corresponding case of Hadwiger's conjecture is a finite problem, in the sense that it is sufficient to check the finite set of graphs with order at most N to find or rule out a counter-example. However, for the case $k = 5$ this does not imply a positive answer to the finiteness part of the four-colour problem (see Section 3), because Kawarabayashi and Reed [93] used the four-colour theorem to prove the existence of the constant N above.

Borowiecki [29] asked about the extent to which Hadwiger's conjecture may be generalized to list-colourings. For $k \leq 4$, every graph with list-chromatic number k has K_k as a minor. Since there are planar graphs that are not list-4-colourable, as shown by Voigt [169], the same correspondence does not hold for $k > 4$. In general, Barát, Joret and Wood [13] proved that, for each $t \geq 1$, there exists a list-$(4t + 1)$-chromatic graph with no complete minor larger than K_{3t+1}.

The Erdős–Faber–Lovász problem

In 1972 Erdős, Faber and Lovász posed the following problem and conjectured that the answer is affirmative.

Let G be the union of k complete graphs, each with at most k vertices, and suppose that the intersection of any two of them is at most a single vertex. Can G be coloured with k colours?

It is surprising that this innocent sounding problem is still open. Some advances have been described by Romero and Sánchez-Arroyo [141].

A reformulation of the conjecture was studied by Klein and Margraf [103], who observed that any graph G is the line graph of a linear simple hypergraph H, where *linear* means that any two edges have at most one vertex in common, and *simple* means that no hyperedge is a multiple edge. The smallest value of $|V(H)|$ for such an H is called the *linear intersection number* $li(G)$ of G. They conjectured that $\chi(G) \leq li(G)$ holds for all graphs G, and proved that this is equivalent to the Erdős–Faber–Lovász conjecture. Moreover, Klein and Margraf proved for any graph G that their conjecture holds for at least one of G and $\overline{G}$, where $\overline{G}$ is the complement of G.

Another more direct and well-known equivalent version of the Erdős–Faber–Lovász conjecture is the statement that, for any simple linear hypergraph H, the chromatic index $\chi'(H)$ is bounded by the number of vertices in H.

Gyárfás's forbidden subgraph conjecture

Gyárfás [71] and Sumner [158] made the following conjecture. In its statement, the parameter $\omega(G)$ denotes the clique-number of G, the number of vertices in a largest complete subgraph of G.

For any forest F, there exists a function $f_F : \mathbb{N} \to \mathbb{N}$, *such that* $\chi(G) \leq f_F(\omega(G))$ *if G does not contain F as an induced subgraph.*

A family $\mathcal{G}$ of graphs is called *χ-bound* if there exists a function f for which

$$\chi(G) \leq f(\omega(G)), \quad \text{for all } G \in \mathcal{G}.$$

If $\mathcal{G}(H)$ denotes the family of all graphs that do not contain the graph H as an induced subgraph, then the above conjecture suggests that $\mathcal{G}(H)$ is χ-bound when H is a forest.

Erdős [48] proved that, for all $g > 0$ and all $k > 0$, there exist graphs of chromatic number k without cycles of length less than g. So, for a family $\mathcal{G}(H)$ to be χ-bound, the graph H has to be a forest.

For a forest F, Kierstead and Penrice [99] and Sauer [144] proved that the family $\mathcal{G}(F)$ is χ-bound if and only if $\mathcal{G}(T)$ is χ-bound for each connected component T of F. The conjecture was proved for stars, paths and brooms by Gyárfás [72], where a *broom* (also called a *comet*) is obtained by identifying an end-vertex of a path with a vertex of a star, and for trees of radius 2 by Kierstead and Penrice [100].

We may reformulate the Gyárfás–Sumner conjecture as follows.

The graphs G in the family of graphs $\mathcal{G}(T,K)$ that do not contain a fixed tree T or a fixed complete graph K as induced subgraphs have bounded chromatic number.

A set of forbidden subgraphs $\{T,K\}$ with this property is called *heroic*. Chudnovsky and Seymour [40] characterized all finite heroic sets of graphs, provided that the Gyárfás–Sumner conjecture is true. Their characterization states that a set of graphs is heroic if and only if it contains a forest, the complement of a forest (such as a complete graph), a complete multipartite graph or the complement of a complete multipartite graph (also called a *clique partition graph*). They also considered a corresponding problem for tournaments, where a heroic set may consist of just one graph, and where a colouring of a tournament has colour classes consisting of transitive subtournaments.

The classes of graphs $\mathcal{G}(P_n)$, where P_n is a path on n vertices, have been investigated with respect to their colouring complexity by Broersma, Golovach, Paulusma and Song [33]. The existence of a polynomial algorithm for k-colouring graphs in $\mathcal{G}(P_n)$ seems unknown for $k = 3$ and $n \geq 7$, for $k = 4$ and $n = 6$ and 7, for $k = 5$ and $n = 6$ and 7, and for $k \geq 6$ and $n = 6$.

Double-critical graphs

The following question arises as a special case of the Erdős–Lovász Tihany conjecture from 1968 (see [85]).

Let G be a connected graph in which the removal of any two adjacent vertices together with all their incident edges results in a $(\chi(G)-2)$-colourable graph. Must G be a complete graph?

A connected graph G for which $\chi(G-v-w) \leq \chi(G)-2$ for any edge $vw \in E(G)$ is called *double-critical*.

The Erdős–Lovász Tihany conjecture is the following statement.

Let $\omega(G) < \chi(G) = s+t-1$ $(s,t \geq 2)$. Then G contains two disjoint subgraphs G_1 and G_2 for which $\chi(G_1) \geq s$ and $\chi(G_2) \geq t$.

The case $s = 2$ is the question for double-critical graphs.

The Erdős–Lovász Tihany conjecture has been proved for quasi-line graphs (where the neighbourhood of each vertex is the complement of a bipartite graph), and also for graphs of independence number 2, by Balogh, Kostochka, Prince and Stiebitz [12]. Another relaxation of the conjecture has been obtained by Stiebitz [155], replacing the chromatic number by the so-called colouring number. The precise statement of the result is: let $\omega(G) < \chi(G) = s + t - 1$ $(s, t \geq 2)$. Then G contains two disjoint subgraphs G_1 and G_2 for which either $\chi(G_1) \geq s$ and $\mathrm{col}(G_2) \geq t$, or $\mathrm{col}(G_1) \geq s$ and $\chi(G_2) \geq t$.

The double-critical graph problem is open for $k \geq 6$. Instead of requiring that G be complete, one might relax the condition to G having a complete minor of the desired size. The cases $k = 6, 7$ and 8 have been obtained for this relaxation (see [91] and [4]).

An interesting equivalent version of the double-critical graph question was formulated by Stehlik [152] in terms of hypergraphs with the *Helly property* – that is, whenever some hyperedges intersect pairwise, then they all have a non-empty intersection. The *covering number* of a hypergraph is the smallest size of a set of vertices that intersects every hyperedge.

Let H be any hypergraph with the Helly property. If the covering number decreases by 2 whenever two disjoint hyperedges are removed from H, is it then true that H consists entirely of disjoint hyperedges?

Adaptive colouring

Let f be any integer-valued function defined on the edge-set of a graph G. An (improper) colouring of the vertices of G is called *adaptive to f* if, for every edge vw of G, the colour assigned to v or w differs from $f(vw)$. The *adaptive chromatic number* $\chi_{\mathrm{ad}}(G)$ is the least number k for which G allows, for each such function f, a colouring adaptive to f and using colours $1, 2, \ldots, k$. The following questions were posed by Hell and Zhu [78].

What is the value of the adaptive chromatic number $\chi_{\mathrm{ad}}(K_n)$ of the complete graph K_n with n vertices?

Is it true that K_n has the smallest adaptive chromatic number among all graphs with chromatic number n?

Is there a function F such that $\chi(G) \leq F(\chi_{\mathrm{ad}}(G))$ holds for all graphs G?

It is clear that $\chi_{\mathrm{ad}}(G) \leq \chi(G)$ is true for every graph, since a usual k-colouring is adaptive to all functions f as above. Hell and Zhu [78] proved a few other additional close ties and similarities between the two parameters:

- If G is connected, then $\chi_{\mathrm{ad}}(G) = 2$ if and only if $G - e$ is bipartite for some edge e of G.
- For any numbers $k, g \geq 3$, there exists a graph G without cycles of length less than g for which $\chi_{\mathrm{ad}}(G) = \chi(G) = k$.
- For each $k \geq 3$, it is an NP-hard problem to decide, with a graph G as the input, whether $\chi_{\mathrm{ad}}(G) \leq k$.

However, in contrast to Brooks's theorem [34] (see Chapter 2), the maximum adaptive chromatic number grows no faster than the square-root of the maximum degree. Hell and Zhu proved that $\chi_{ad}(G) \le \lceil\sqrt{e(2\Delta - 1)}\rceil$, where Δ is the maximum degree of G. This bound was improved by Molloy and Thron [125] to the best-possible $\chi_{ad}(G) \le (1+\varepsilon)\sqrt{\Delta}$, for every $\varepsilon > 0$. In particular, this implies an upper bound $\chi_{ad}(K_n) \le \sqrt{n}$ for the complete graph K_n. A similar upper bound for $\chi_{ad}(K_n)$ also follows from earlier results of Gyárfás and Erdős [73]. Robert Šámal and Alexander Kostochka independently proved the lower bound $\chi_{ad}(K_n) \ge \sqrt{n/\ln n}$ for large enough n.

In an informative paper, Hell, Pan, Wong and Zhu [77] proved that the four-fold direct (or 'Cartesian') product K_n^4 of the complete graph K_n has adaptive chromatic number n. This again is in contrast to the behaviour of the chromatic number, which never increases when we form direct products (see also Problem 11.1 of [85]). They conjectured that in general, for any graph G, there exists a number d for which $\chi_{ad}(G^d) = \chi(G)$.

If G is planar, then $\chi_{ad}(G) \le 4$; this follows from the four-colour theorem. It was also proved in a list-colouring version, without using the four-colour theorem, by Esperet, Montassier and Zhu [51], thereby answering a further question raised by Hell and Zhu [78].

The list-colouring version of the third question above was answered affirmatively by Molloy and Thron [124]. Adaptive list-colourings have also been studied by Kostochka and Zhu [112] and by Montassier, Raspaud and Zhu [126].

3. Graphs on surfaces

The four-colour problem

Interesting accounts have continued to appear with new views on different aspects of the four-colour problem, such as the books by Fritsch and Fritsch [58] and Wilson [172]. The following questions remain.

Is there a short proof of the four-colour theorem (that every planar graph is 4-colourable) which can be checked in detail by hand by a competent mathematician in at most two weeks?

Is there a short argument, independent of known proofs of the four-colour theorem, to demonstrate that the four-colour problem is a finite problem, in the sense that there exists a number N for which a smallest counter-example must have at most N vertices?

Is there a proof of the four-colour theorem, perhaps neither short nor easily checkable, which is not based on the method of finding an unavoidable set of reducible configurations?

Francis Guthrie's famous four-colour problem has a letter from Augustus De Morgan to W. R. Hamilton in 1852 as its earliest written source. It became the four-colour theorem 124 years later, when a computer-aided proof was obtained by Appel and Haken [10] and Appel, Haken and Koch [11]. One might have thought that the story would end here. But new and sometimes surprising facts are continually being discovered surrounding the early history, and also concerning the possibilities of finding proofs that are simpler according to specific criteria.

Recently McKay [120] discovered a printed publication of the four-colour problem which was earlier than the 1860 anonymous book review by De Morgan. A letter signed only by the initials 'F.G.' appeared in the *Athenæum* already in 1854 (see [60]), and read as follows:

Tinting maps. – In tinting maps, it is desirable for the sake of distinctness to use as few colours as possible, and at the same time no two conterminous divisions ought to be tinted the same. Now, I have found by experience that four colours are necessary and sufficient for this purpose, – but I cannot prove that this is the case, unless the whole number of divisions does not exceed five. I should like to see (or know where I can find) a general proof of this apparently simple proposition, which I am surprised never to have met with in any mathematical work. F.G.

The identity of the author is not known with certainty. Authors whose interests touched on the map-colour problem included Francis Guthrie's brother Frederick, and an illustrious explorer and mapmaker Francis Galton, both of whom shared the same initials. It appears likely that Francis or Frederick Guthrie was actually the author of the note.

In September 2004 Gonthier [67] and B. Werner completed a proof of the four-colour theorem, which they adapted from the proof of Robertson, Seymour, Sanders and Thomas [137]. It was written in a sufficiently basic formalization to make it checkable by the general mathematical theorem-proving software *Coq*, which had been developed during the 1980s and 1990s at several institutions in France. They reported that *Coq* was able to verify the correctness of this proof. (See also the article by Knight [105].)

The proof given by Robertson, Seymour, Sanders and Thomas is simpler than the original one by Appel and Haken [10], not only by exhibiting a much smaller unavoidable set of reducible configurations, but also by avoiding the use of 'ring-reduction' arguments first introduced by Bernhart [18]. A qualitatively even simpler proof was described by Steinberger [154] in 2010. Although Steinberger's proof involves a much larger unavoidable set of 2822 configurations, it avoids the use of more complicated C-reducibility techniques and applies only the simpler D-reducibility method; the existence of such a four-colour proof relying entirely on D-reducibility was first conjectured in 1975 by Stromquist [157] in his Ph.D. thesis.

The finiteness question for the four-colour problem may have been considered first by H. Heesch, who wrote a letter to G. Ringel in 1953 explaining the basic idea of step-wise extending an unavoidable set of configurations into a large enough finite such set, all members of which were reducible. According to Bigalke [19], Heesch told Ringel that it had occurred to him in 1935-36 that a proof of the finiteness of the four-colour problem might be obtained in this way. The idea, known as the *finitization of the four-colour problem* (see Wilson [172]), led Heesch in 1970 to propose a particular unavoidable set of 8900 configurations, so that a computational confirmation of the conjectured reducibility of all of them would produce a proof of the four-colour problem, after a large (but finite) amount of effort. The question about an upper bound N on the order of a smallest 5-chromatic counter-example differs from Heesch's question in the respect that, even if the existence of N were known and hence the proof of the four-colour theorem were reduced to a finite task, this does not immediately imply the truth of the four-colour theorem. (Compare this with the discussion in the previous section of the finiteness of Hadwiger's conjecture for graphs with a fixed chromatic number.)

Each proof of the four-colour theorem described above is based on an idea, implicitly used back in 1879 by A. B. Kempe. This is to establish by Euler's formula that at least one element of a collection of possible constituents is present in any given graph or map, and then to apply a 'reduction' to show that no such configuration appears in a smallest example that is not 4-colourable. It appears likely that such collections must be fairly large and that the actual calculations are infeasible without the aid of computers. Several such proofs are available by now, and it is a controversial issue among mathematicians (albeit to a much lesser extent than in the 1970s) as to whether computer-assisted proofs can be considered valid. Whether or not every correct proof of the four-colour theorem must be too long to be checked in every detail by a human, it would be interesting to see proofs based on different approaches, such as by algebraic means, as suggested by work of G. D. Birkhoff and (later) W. T. Tutte, or by using other powerful methods, such as advanced probabilistic methods.

Steinberg's 3-colour problem

As reported by Aksionov and Melnikov [3], the following question was asked by R. Steinberg in 1975.

Assume that no cycle in a planar graph has length 4 *or* 5*. Does it follow that the graph can be* 3*-coloured?*

Following a suggestion by P. Erdős in 1991, several authors have attempted to find the smallest possible value of k for which a planar graph must be 3-colourable if it contains no cycles of any length $4, 5, \ldots, k$. Abbott and Zhou [1] in 1991 were the first to prove the existence of such a number k, and they gave $k \leq 11$ as a bound. In 1995 the bound was improved to $k \leq 9$ by Sanders and Zhao [143], and then to $k \leq 8$ by Salavatipour [142] in 2002, and to $k \leq 7$ by Borodin, Glebov, Raspaud and Salavatipour [27] in 2005. In 2009 Borodin, Glebov, Montassier and Raspaud [26] proved the currently best result, that if a planar graph contains neither 5-cycles nor 7-cycles, nor any pair of 3-cycles that share an edge, then it can be 3-coloured.

A stronger conjecture by Borodin, Glebov, Jensen and Raspaud [25] suggests that if a planar graph contains no 3-cycle that shares an edge with a 4-cycle or a 5-cycle, then it can be 3-coloured; this suggests a common strengthening of

Steinberg's problem and the theorem of Grötzsch [69] that every planar graph with no 3-cycle is 3-colourable. They proved that if a planar graph contains no 3-cycle which shares an edge with a cycle of length at most 9, then it can be 3-coloured.

Several related theorems and conjectures are found in the informative survey by Borodin [23] on the general theme of planar graph colouring.

Zhao [175] extended Erdős's suggestion to higher surfaces, proving that for any fixed surface S there is a constant k such that, if a graph without any cycles of length $4, 5, \ldots, k$ is embeddable in S, then it can be 3-coloured.

Ringel's earth–moon problem

What is the largest chromatic number f_2 that any union of two planar graphs can have?

The equivalent dual problem, asked in 1959 by Ringel [135], deals with a map of 'countries', each consisting of one connected region on the earth and one connected region on the moon. The aim is to colour the regions, using as few colours as possible, so that the two parts of each country receive the same colour, and so that any two regions sharing a common border receive different colours.

Ringel [135] noted the bounds $8 \leq f_2 \leq 12$, where the lower bound was improved to 9 by T. Sulanke in 1974, as first reported by Gardner [64]. An infinite class of critical 9-chromatic graphs of thickness 2 was constructed by Boutin, Gethner and Sulanke [30]. In general, the *thickness* of a graph G is the smallest number t for which there are t planar graphs whose union is G. For the maximum chromatic number f_t of graphs of thickness t, the known bounds are $6t - 2 \leq f_t \leq 6t$, for $t \geq 3$.

Motivated by the difficulty of computing both the thickness (proved NP-hard by Mansfield [117]) and the chromatic number of a given graph, Albertson, Boutin and Gethner [6], [7] compiled catalogues of known bounds for both parameters, as well as the fractional chromatic number, when restricted to special classes of graphs.

Several extensions of Ringel's earth–moon problem to higher surfaces were discussed by Jackson and Ringel [82].

Barnette's conjecture

Every simple planar 3-colourable graph allows a partitioning of its vertex-set into the vertex-sets of two induced forests.

The original conjecture, posed by Barnette [14], was stated dually in terms of the existence of a Hamiltonian cycle in any 3-connected bipartite cubic planar graph.

An early partial result in 1975 by Goodey [68] implies the truth of the conjecture for maximal planar graphs in which each vertex has degree 4 or 6. This was extended by Feder and Subi [54], who proved that if a maximal planar graph G allows a 3-colouring in which all vertices of degree larger than 6 receive the same colour, then the conjecture is true for G.

Holton, Manvel and McKay [81] proved the conjecture in its original formulation for every 3-connected planar bipartite cubic graph with at most 64 vertices. Among other results, Aldred, Bau, Holton and McKay [8] showed that every 3-connected planar cubic graph with at most 176 vertices and with all faces of size at most 6 is Hamiltonian, independently of the bipartiteness condition. The bound 176 was improved to 250 by Brinkmann, McKay and von Nathusius [31].

Kelmans [97] proved that Barnette's conjecture is equivalent to the following interesting stronger statement: for any two edges e_1 and e_2 that belong to the boundary of a common face in a 3-connected bipartite cubic planar graph G, there exists a Hamiltonian cycle in G containing e_1 but not e_2.

The square of planar graphs

If a graph is planar with maximum degree 3, *can its vertices be coloured with seven colours, so that all vertices at distance* 1 *or* 2 *receive different colours?*

This question was asked in 1977 by Wegner [171], who observed that seven colours is best possible, by subdividing a non-triangular edge of a triangular prism, and who proved that eight colours would be sufficient.

Various authors have mentioned a positive solution to Wegner's problem by C. Thomassen, obtained in 2001. However, according to the survey paper published in 2008 by Kramer and Kramer [113], the problem is still considered open. It seems open as to whether six colours are always sufficient when we restrict attention to the special case of 3-connected graphs.

An interesting variation applies to planar graphs of arbitrary maximum degree. It requires, for some natural number d, that vertices are coloured differently when their distance is precisely equal to d. There is no upper bound on the number of colours needed when d is even, but the situation appears much more complicated when d is odd. For $d = 1$, we of course retrieve the four-colour problem. For $d = 3$ Nešetřil and Ossona de Mendez [127] have proved the first known (and very large) upper bound on the number of colours that are sufficient in a colouring, and they have presented an example for which six colours are needed for $d = 3$; this may be the best lower bound at present. In fact Nešetřil and Ossona de Mendez proved upper bounds for all odd $d \geq 3$; their bounds depend on d and are extremely large, even for $d = 3$.

The game-chromatic number and Brams's problem

In April 1981 Martin Gardner publicised the following problem as one of several short problems in his column in *Scientific American*.

Let G be a graph and let k be a positive integer, and suppose that the following two-person game is played on G. Alice and Bob alternate turns, with Alice having the first move. Both players know G and k from the beginning. A move consists in choosing a vertex v from G that has not yet been assigned a colour and assigning to v a colour from $1,2,\ldots,k$ *distinct from the colours assigned previously (by either player) to the vertices adjacent to v. The game stops with Alice the winner if G has been coloured after* $|V(G)|$ *moves. Bob is the winner if, at some point of the game, there is a vertex that cannot be assigned a colour – that is, the vertex is already adjacent to vertices coloured with all k colours.*

The game-chromatic number $\chi_g(G)$ *is the least number k for which Alice has a winning strategy. If Alice has a winning strategy for G using k colours, does she also then have a winning strategy with* $k+1$ *colours? What is the least number k for which Alice has a winning strategy for all planar graphs?*

The part of this problem that involves planar graphs is due to Steven Brams, a professor of politics at New York University. Brams is a game theory expert who invented the game for planar maps, thinking about possible alternative proofs of the four-colour theorem. The problem came up in correspondence with Martin Gardner [63] in late 1980. In a later letter to Gardner in October 1980 Lloyd Shapley showed that $k \geq 6$ (with five colours for the map of the dodecahedron: Bob wins if he always colours the opposite face of Alice's last choice with the same colour), and Robert High improved the bound to 7. Brams and Gardner contemplated that there might exist no least k for planar graphs.

As far as we know, a proof that there is a least number k for which Alice has a winning strategy for all planar graphs was first given by Kierstead and Trotter [101] in 1993. In a more recent paper, Bartnicki, Grytczuk, Kierstead and Zhu [15] gave 8 and 17 as the best-known lower and upper bounds for the least k for which Alice has a winning strategy for planar graphs, with a proof that $k = 18$ suffices.

For graphs in general, the delightful paper [15] contains the following intriguing remark:

It is not clear what influence increasing the number of colours has on the colouring game. Suppose that Alice wins with k colours on a graph G. Then it would seem trivial that she should also win on G with $k+1$ *colours. But can you prove it? As yet, nobody has been able to do so!*

For more on graphs and games, see Chapter 14.

Albertson's 4-colour problem

Let S be a surface. Is there a constant M_S *such that, if G is embeddable on S, then* $G - X$ *is 4-colourable for some subset* $X \subseteq V(G)$ *with at most* M_S *vertices? In particular, does this hold for the torus T with* $M_T = 3$*?*

These questions were asked by Albertson [5] in 1981. Not much recent progress has been achieved on either of them.

DeVos, Kawarabayashi and Mohar [42] proved the similar general statement for 5-list-colourability: there is a constant k such that, for every graph G embedded in S, there exists a vertex-set U with at most k vertices for which $G-U$ is 5-list-colourable.

The similar statement with 4-list-colourability instead of 5-list-colourability is false, since there exist non-4-list-colourable planar graphs that may be drawn in any number of copies on any surface, as proved by Voigt [169] and Mirzakhani [121].

The similar question for 3-colourability of triangle-free graphs on S, which by Grötzsch's theorem [69] has a positive answer when S is the plane, was answered affirmatively by Kawarabayashi and Thomassen [95].

We are not aware of similar results concerning the 4-list-colourability of triangle-free graphs, or the 3-list-colourability of graphs with girth at least 5. Both of these results hold for the plane; the former follows easily from Euler's formula and the latter was proved by Thomassen [160].

4. Degrees and colourings

Edge-disjoint placements

Let G_1 and G_2 be graphs with the same number n of vertices, and suppose that their maximum degrees $\Delta(G_1)$ and $\Delta(G_2)$ satisfy

$$(\Delta(G_1) + 1)(\Delta(G_2) + 1) \leq n + 1.$$

Does it follow that there exists a graph on n vertices which is an edge-disjoint union of a copy of G_1 and a copy of G_2?

This question was asked independently by Bollobás and Eldridge [20], [21] and by Catlin [35], [36]. A positive answer would extend considerably the classical colouring theorem of Hajnal and Szemerédi [75], which states that if G has maximum degree Δ and exactly $k(\Delta+1)$ vertices, then there exists a colouring of G with $\Delta+1$ colours in which each colour class has size k. A short proof of this theorem was given by Kierstead and Kostochka [98].

The case when one of the maximum degrees is at most 2 was solved affirmatively by Aigner and Brandt [2]. So if a graph on n vertices has minimum degree at least $\frac{1}{3}(2n - 1)$, then it contains as a subgraph every graph on n vertices of maximum degree at most 2. For maximum degree 3 and sufficiently large n a positive solution was given by Csaba, Shokoufandeh and Szemerédi [41].

Kaul, Kostochka and Yu [89] proved that the conclusion holds under the extra assumption that $\min\{\Delta(G_1), \Delta(G_2)\} \geq 300$, and the stronger condition

$$(\Delta(G_1) + 1)(\Delta(G_2) + 1) \leq \tfrac{3}{5}n + 1.$$

For graphs with large maximum degree their result improves a theorem of Sauer and Spencer [145] stating that the conclusion follows for all G_1 and G_2 satisfying

$$\Delta(G_1)\Delta(G_2) < \tfrac{1}{2}n.$$

A generalization of the problem to hypergraph packing was studied by Rödl, Ruciński and Taraz [140].

Reed's ω-Δ-χ conjecture

If G has maximum degree Δ and contains no complete subgraph with more than ω vertices, then

$$\chi(G) \leq \left\lceil \tfrac{1}{2}(\Delta + 1) + \tfrac{1}{2}\omega \right\rceil.$$

This conjecture was posed by Reed [133] in 1998 and quickly became popular. The idea behind it is to combine the obvious lower and upper bounds

$$\omega \leq \chi(G) \leq \Delta + 1$$

into an improved upper bound obtained as a convex combination of them. For $\omega \geq \Delta$ the conjectured bound is trivial, and for $\omega = \Delta - 1$ the truth follows from Brooks's theorem [34].

In the special case $\omega = \Delta = 2$ we must round up the average of ω and $\Delta + 1$, due to odd cycles. Classical constructions by Gallai [61] of 4-critical graphs with $\Delta = 4$ and $\omega = 2$, and graphs obtained as joins of these with complete graphs, show in general that either rounding up is necessary, or the weight of the $\Delta + 1$ term must be increased in the convex combination.

If ω is as small as possible – that is, in the non-trivial case when G is triangle-free and $\omega = 2$ – the best possible bound by A. Johansson (unpublished, see [123]) is

$$\chi(G) \leq \frac{c \cdot \Delta}{\log \Delta},$$

where $c > 0$ is a constant, and shows that the conjecture is true when Δ is large enough.

Reed [133] also made a second conjecture for $\Delta \geq 3$:

$$\chi(G) \leq \tfrac{2}{3}(\Delta + 1) + \tfrac{1}{3}\omega.$$

By giving more weight to the trivial upper bound $\Delta + 1$ in the convex combination, rounding is avoided. As one of the first positive results, Reed proved that there exists a constant a, with $0 < a < 1$, for which

$$\chi(G) \leq (1 - a)(\Delta + 1) + a\omega, \quad \text{for } \Delta \geq 3.$$

The following stronger 'local version' of Reed's conjecture was conjectured by King [102] in his Ph.D. thesis in 2009. For a vertex v of G, let $\omega(v)$ denote the size of a largest complete subgraph of G that contains v. King's conjecture is

$$\chi(G) \leq \max_{v \in V(G)} \left\{ \left\lceil \tfrac{1}{2}(d_G(v) + 1 + \omega(v)) \right\rceil \right\}.$$

Chudnovsky, King, Plumettaz and Seymour [38] proved this conjecture for *quasi-line graphs* – graphs with the property that the neighbourhood of each vertex induces the complement of a bipartite graph. It is easy to see that all line graphs are quasi-line graphs. Edwards and King [47] have suggested an even stronger 'superlocal version' of Reed's conjecture.

The total colouring conjecture

Let G be a simple graph with maximum degree Δ. *Then it is possible to colour the edges and vertices of G with* $\Delta + 2$ *colours in such a way that no two adjacent or incident elements in* $V(G) \cup E(G)$ *receive the same colour.*

There has been some controversy about the priority to this conjecture, as described by Soifer [150]. In 1964 the conjecture was proposed by V. G. Vizing in graph theory seminars at the Academy of Sciences in Akademgorodok (outside Novosibirsk), organized by A. A. Zykov. It was independently proposed by M. Behzad at the same time, and its first written appearence is in Behzad's Ph.D. thesis [17] at Michigan State University in 1965. The controversy arising from this has been addressed by Stiebitz, Scheide, Toft and Favrholdt [156].

In 1968 Vizing [167] published a paper containing a wealth of problems, including the *total graph colouring conjecture* for multigraphs, which suggests that $\Delta(G) + \mu(G) + 1$ colours suffice in a simultaneous colouring of the vertices and edges of a multigraph G with maximum degree $\Delta(G)$ and maximum multiplicity $\mu(G)$. A natural stronger conjecture is that $\chi'(G) + 1$ colours suffice for all multigraphs. Unfortunately, Vizing [167] used the term 'chromatic index' to denote the least number of colours needed in a simultaneous colouring of vertices and edges! This number is now commonly called the *total chromatic number*, denoted by χ'', and a simultaneous colouring of the vertices and edges of a graph is called a *total colouring*, following Behzad.

A further strengthening when $\chi' \geq \Delta + 3$ was proposed by Goldberg [66]:

A graph G with $\chi'(G) \geq \Delta(G) + 3$ *satisfies* $\chi''(G) = \chi'(G)$.

We observe that Goldberg's conjecture says that any graph G with $\chi'(G) \geq \Delta(G) + 2$ satisfies $\chi'(G) = \mathcal{W}(G)$, where $\mathcal{W}(G)$ is the *density* of G (see Section 5). Goldberg [66] noticed that any critical graph G with $\chi'(G) = \mathcal{W}(G)$ satisfies $\chi''(G) = \chi'(G)$: we first take a $(k-1)$-edge-colouring of $G - e$, where e is an edge of G and $k = \chi'(G)$. He then proved (see Stiebitz, Scheide, Toft and Favrholdt [156]) that none of the $k - 1$ colours is missing at two distinct vertices of G. So we can colour each vertex v of G, using a colour that is missing at v, and we can colour e with colour k. This results in a total colouring of G with k colours.

We may also ask, as an analogue of Goldberg's conjecture, whether every graph satisfies $\chi''(G) \leq \max\{\Delta(G) + 2, \mathcal{W}(G)\}$?

It is folklore that any graph G satisfies $\chi''(G) \leq \chi'^{\ell}(G) + 2$, where $\chi'^{\ell}(G)$ is the list-chromatic index of G (see Section 5). In 1997 Borodin, Kostochka and Woodall [28] used list-colouring arguments to show that $\chi''(G) \leq \frac{3}{2}\Delta + 1$ for graphs G with maximum degree $\Delta \geq 2$. One year later, Molloy and Reed [122] used probabilistic arguments to show that $\chi'' \leq \Delta + 10^{26}$ for simple graphs.

Clearly, there is a list variant for the total chromatic number, and Borodin, Kostochka and Woodall [28] conjectured that the total list-chromatic number equals the total chromatic number.

Critical graphs with many edges or high degrees

Let $k \geq 4$ be an integer. A graph is *k-critical* if it has chromatic number k and every proper subgraph is $(k-1)$-colourable.

If $F_k(n)$ is the largest number of edges of a k-critical graph on n vertices, does

$$\lim_{n\to\infty} F_k(n)/n^2$$

exist? Is $F_6(n) = \frac{1}{4}n^2 + n$, for $n \equiv 2 \,(\text{mod } 4)$?

If $\delta_k(n)$ is the largest minimum degree of a k-critical graph on n vertices, does there exist a constant $c > 0$ for which $\delta_4(n) \geq cn$? What are the orders of magnitude of $\delta_4(n)$ and $\delta_5(n)$?

Critical graphs were first defined and used by G. A. Dirac, when he studied for his Ph.D. degree at the University of London around 1950. In 1949 he met Paul Erdős, whose first reaction was to ask about the maximum number of edges in these minimal graphs. Dirac observed that the join of two odd cycles of equal length is a 6-critical graph with many edges: no better examples have since been found. For $|V(G)| = n$, the number of edges is $\frac{1}{4}n^2 + n$. Jensen and Toft [85] conjectured, however, that this is not the best possible value.

For $k = 4$, it is known that

$$\liminf_{n\to\infty} F_4(n)/n^2 \geq \tfrac{1}{16},$$

and for $k = 5$ the corresponding value is $\frac{4}{31}$, as proved by Toft [162]. Jensen [84] observed the general upper bound

$$F_k(n) \leq \frac{k-2}{2(k-1)}n^2, \quad \text{for all } n \text{ and } k \text{ with } n > k \geq 4.$$

For the largest minimum degree in critical graphs the above examples of Dirac show that $\delta_6(n) \geq \frac{1}{2}n$ for $n \equiv 2 \,(\text{mod } 4)$. Simonovits [149] and Toft [163] proved that $\delta_4(n) \geq cn^{1/3}$. This may be the best order of magnitude, but it seems difficult to prove this. As for the order of magnitude of $\delta_5(n)$, there is no reasonable guess.

Erdős [50] asked specifically about the existence of regular 4-critical graphs. The only 3-regular example is K_4, by Brooks's theorem [34]. On the other hand, Gallai [61] constructed an infinite family of 4-regular examples. An infinite family of 5-regular 4-critical graphs was constructed by Jensen [84]. Examples of regular 4-critical graphs of even degrees 6 to 16 were found by Dobrynin, Melnikov and Pyatkin [44], [45]. We are not aware of any examples of 4-critical graphs that are regular of odd degree 7 or more.

Probability colouring

Assume that $(p_1, p_2, \ldots, p_k)$ is a probability distribution on the set $\{1, 2, \ldots, k\}$ – that is, $p_i \geq 0$ for $1 \leq i \leq k$, and $\sum_{i=1}^{k} p_i = 1$. Let the vertices of G be coloured randomly and independently so that each vertex of G receives colour i with probability p_i, for $i = 1, 2, \ldots, k$. For $m \geq 0$, let P_G^m be the probability of at most m monochromatic edges. As usual, Δ denotes the maximum degree.

Is P_G^m maximized by the uniform distribution $p_i = 1/k$ when $k = k(G, m)$ is large enough – say, whenever $k \geq f(\Delta, m)$ for some function f?

Is there a structural characterization of those graphs G for which, for each $k > 0$, the probability P_G^0 of a proper colouring is maximized by the uniform distribution?

If, for all k and m, the probability P_G^m is maximized by the uniform distribution, then G is called *P-uniform*.

Is there a a structural characterization of P-uniform graphs?

We note that

$$P_G^m = \sum_f \prod_{v \in V(G)} p_{f(v)},$$

where the sum is over all mappings $f : V(G) \to \{1, 2, \ldots, k\}$ for which there are at most m edges xy with $f(x) = f(y)$. Permutations of the colours do not change this probability, and so P_G^m may be expressed as a symmetric polynomial in $p_1, p_2, \ldots, p_k$.

We also note in particular that

$$P_G^0 = \chi(G, k) k^{-|V(G)|} \quad \text{when} \quad (p_1, p_2, \ldots, p_k) = (1/k, 1/k, \ldots, 1/k),$$

where $\chi(G, k)$ is the chromatic polynomial of G.

Fadnavis [52] and her Ph.D. advisor P. Diaconis asked the first question above, and she proved that the answer is affirmative for $m = 0$ if the number of colours satisfies $k > 6.3 \times 10^5 \Delta^4$. For $G = K_{1,n}$ and $m = 0$ it suffices that $k \geq n + 1$. Diaconis had conjectured that the answer to the second question above is affirmative when G is claw-free – that is, when $K_{1,3}$ is not an induced subgraph. This conjecture was also proved by Fadnavis.

The graph $K_{1,4}$ is not P-uniform, as witnessed by $m = 0$ and $k = 2$. Fadnavis [52] exhibited a claw-free graph on 19 vertices which is not P-uniform (for values $m = 30$ and $k = 2$), but proved that all complete graphs and cycles are P-uniform.

5. Edge-colourings

In this and the following section we no longer restrict our attention to simple graphs. Accordingly the term 'graph' is used more generally to denote a multigraph – that is, two vertices may be joined by more than one edge, but loops do not occur.

Goldberg's conjecture

The *density* $\mathcal{W}(G)$ of a graph G is defined by

$$\mathcal{W}(G) = \max\{\left\lceil \frac{|E(H)|}{\lfloor\frac{1}{2}|V(H)|\rfloor} \right\rceil \ : \ H \subseteq G, |V(H)| \geq 2\}.$$

The following conjecture was proposed independently by Goldberg [65] and Seymour [148] in the 1970s. A detailed account of its history is given in Stiebitz, Scheide, Toft and Favrholdt [156].

Every graph G satisfies $\chi'(G) \leq \max\{\Delta(G) + 1, \mathcal{W}(G)\}$.

This conjecture can be stated in different equivalent ways, as follows.

Every graph G satisfies $\chi'(G) \in \{\Delta(G), \Delta(G) + 1, \mathcal{W}(G)\}$.

Every graph G satisfies, for each $r \in \mathbb{N}$,

$$\chi'(G) \leq \max\{\frac{(2r+3)\Delta(G) + 2r}{2r+2}, \mathcal{W}_1(G), \mathcal{W}_2(G), \ldots, \mathcal{W}_r(G)\},$$

where $\mathcal{W}_r(G) = \max\{\lceil |E(H)|/r \rceil \ : \ H \subseteq G, |V(H)| = 2r + 1\}$ (Gupta [70]).

Let G be a χ'-critical graph (that is, $\chi'(H) < \chi'(G)$ for every proper subgraph H of G), let $\chi'(G) = k + 1$ for an integer $k \geq \Delta(G) + 1$, let e be an edge of G, and let φ be an edge-colouring of $G - e$ with k colours. Then no colour is missing at two distinct vertices of G with respect to φ. (Andersen [9]).

Every χ'-critical graph G with $\chi'(G) \geq \Delta(G) + 2$ has odd order and satisfies

$$2|E(G)| = (\chi'(G) - 1)(|V(G)| - 1) + 2$$

(Stiebitz, Scheide, Toft and Favrholdt [156]).

The conjecture of Goldberg and Seymour is one of the most important conjectures in edge-colouring theory. It generalizes Vizing's theorem in a strong way, by showing that there are only three possible values for the chromatic index of an arbitrary graph. While it appears unknown whether the density $\mathcal{W}(G)$ can be computed in polynomial time, the parameter $\kappa(G) = \max\{\Delta(G), \mathcal{W}(G)\}$ can be determined in polynomial time: this follows from Edmonds's matching polytope theory and the theory of anti-blocking polyhedra (see Schrijver [146]). So Goldberg's conjecture supports the following conjecture of Hochbaum, Nishizeki and Shmoys [79]:

There is a polynomial-time algorithm that colours the edges of any graph G with at most $\chi'(G) + 1$ colours.

Results related to Goldberg's conjecture were discussed by Stiebitz, Scheide, Toft and Favrholdt [156]. In particular, Goldberg's conjecture was confirmed for graphs with at most 15 vertices, and also for graphs with maximum degree at most 15. Gupta's version of Goldberg's conjecture is true for $1 \le r \le 6$. Furthermore, it is known that every graph G satisfies

$$\chi'(G) \le \max\{\Delta(G) + \sqrt{(\Delta(G) - 1)/2}, \mathcal{W}(G)\}.$$

Whether Goldberg's conjecture holds for all planar graphs is unknown.

The Berge–Fulkerson conjecture

This conjecture is attributed to C. Berge (see Seymour [147]), but it first appeared in print in Fulkerson [59]:

Every bridgeless cubic graph G contains a family of six perfect matchings such that each edge of G is contained in precisely two of them.

An equivalent form of it is that $\chi'(2G) = 6$ for every bridgeless cubic graph. The conjecture was extended by Seymour [148] to the generalized Fulkerson conjecture:

Every r-graph G satisfies $\chi'(2G) = 2r$.

Here, an *r-graph* is an r-regular graph G such that, for each $X \subseteq V(G)$ with an odd number of vertices, there are at least r edges in G with precisely one endpoint in X. Since the fractional chromatic index of every r-graph G is $\chi'^*(G) = r$, there are integers t for which $\chi'(tG) = tr$ (see Stiebitz, Scheide, Toft and Favrholdt [156]). Seymour's conjecture suggests that in general $t = 2$ is one of these integers.

The following seemingly weaker version of the Berge–Fulkerson conjecture was proposed by Berge (personal communication from F. Jaeger in 1994). It is easy to see that this conjecture is implied by the Berge–Fulkerson conjecture, and Mazzuoccolo [119] proved that the two conjectures are actually equivalent.

Every bridgeless cubic graph G contains a family of five perfect matchings such that each edge of G is contained in at least one of them.

We note that Frink's proof [57] of Petersen's theorem [131] tells us that each edge of a cubic bridgeless graph is contained in a perfect matching (see also König's book [110]). On the other hand, it is not known whether Berge's conjecture holds, even with 'five' replaced by any other fixed number. For a positive integer k, define

$$m_k(G) = \max\left\{|\bigcup_{i=1}^{k} M_i|/|E(G)| \ : \ M_1, M_2, \ldots, M_k \text{ are perfect matchings of } G\right\}.$$

Clearly, $m_1(G) = \frac{1}{3}$ and Berge's conjecture suggests that $m_5(G) = 1$. For the Petersen graph P, we have

$$m_2(P) = \tfrac{3}{5},\ m_3(P) = \tfrac{4}{5},\ m_4(P) = \tfrac{14}{15},\ m_5(P) = 1,$$

and Patel [129] made the following conjecture:

Every bridgeless cubic graph G satisfies $m_k(G) \geq m_k(P)$, *for* $1 \leq k \leq 5$.

Patel proved that this conjecture is implied by the Berge–Fulkerson conjecture, and thus, by the result of Mazzuoccolo, the two conjectures are equivalent. Lower bounds on $m_k(G)$ for $2 \leq k \leq 5$ were established by Kaiser, Král' and Norine [87]: in particular, they verified the case $k = 2$ of the conjecture. The paper by Fouquet and Vanherpe [56] provides some partial results related to the perfect matching index of bridgeless cubic graphs – that is, the minimum number of perfect matchings that cover all edges.

For the Petersen graph P, only twelve edges can be covered with three perfect matchings, but thirteen edges can be covered with three matchings. Rizzi [136] proved that if G is a simple graph with $\Delta(G) \leq 3$, then there are three matchings M_1, M_2 and M_3 for which $|M_1 \cup M_2 \cup M_3| \geq \frac{6}{7}|E(G)|$. The *max edge t-colouring problem* is to colour the largest possible number of edges of a graph G in a proper way using t colours; the complexity of this problem was studied by Feige, Ofek and Wieder [55].

There is another attractive formulation of the Berge–Fulkerson conjecture in terms of even graphs – that is, graphs in which all vertices have even degree. That the following conjecture is equivalent to the Berge–Fulkerson conjecture was proved by Jaeger [83]:

Every bridgeless graph G contains a family of six even subgraphs such that each edge of G is contained in exactly four of them.

Hao, Niu, Wang, Zhang and Zhang [76] proved that a cubic graph G satisfies $\chi'(2G) = 6$ if and only if there are two edge disjoint matchings M_1 and M_2 such that $\chi'(G - M_i) \leq 3$ for $i = 1, 2$ and $M_1 \cup M_2$ is the edge-set of a cycle. They used this to verify the Berge–Fulkerson conjecture for certain families of snarks.

The relation between the Berge–Fulkerson conjecture and shortest cycle cover problems has been investigated by Fan and Raspaud [53]. They also drew attention to the following implication of the Berge–Fulkerson conjecture which has become known as the Fan–Raspaud conjecture:

Every bridgeless cubic graph contains three perfect matchings with empty intersection.

If a cubic graph has a family of six perfect matchings covering each edge twice, then the intersection of any three of them must be empty. On the other hand, even the existence of an integer $k \geq 3$ for which every bridgeless cubic graph contains k perfect matchings with empty intersection is not clear.

The list-chromatic index

The following list-colouring conjecture, suggested independently by various researchers (including V. G. Vizing, M. O. Albertson, K. Collins, A. Tucker, R. P. Gupta, B. Bollobás and A. J. Harris) is the most popular problem in edge-colouring theory. It first appeared in print in a paper by Bollobás and Harris [22].

Let G be a graph with $\chi'(G) = k$, and let each edge of G be assigned a list of k colours. Then it is possible to colour the edges of G so that each edge receives a colour from its list.

The study of list-colouring problems for graphs was initiated in the 1970s by Vizing [168], and independently by Erdős, Rubin and Taylor [49]. The *list-chromatic index* $\chi'^{\ell}(G)$ of a graph G is the least number k for which, whenever we assign a list of k colours to each edge of G, there is a proper colouring of the edges of G for which each edge receives a colour from its list. By definition, every graph G satisfies $\chi'(G) \leq \chi'^{\ell}(G)$ and the list-colouring conjecture suggests that equality holds. In contrast, the gap between the chromatic number and the list-chromatic number can be arbitrarily large, even for bipartite graphs, as observed by Vizing [168] and Erdős, Rubin and Taylor [49].

One of the most spectacular results about edge-colourings from the 1990s is the verification by Galvin [62] of the list-colouring conjecture for the class of bipartite graphs – that is, the proof that $\chi'(G) = \chi'^{\ell}(G)$ for every bipartite graph G. This particular case of the list-colouring conjecture is due to J. Dinitz (see Chetwynd and Häggkvist [37]).

Borodin, Kostochka and Woodall [28] obtained an attractive generalization of Galvin's landmark theorem, allowing lists of different lengths, where an edge e with ends u and v has a list of length $\max\{d_G(u), d_G(v)\}$. Based on this they proved that

$$\chi'^{\ell}(G) \leq \left\lfloor \tfrac{3}{2}\Delta(G) \right\rfloor$$

for all graphs G. The list-colouring conjecture has been verified for the class of simple graphs, at least asymptotically, by Kahn [86], who proved that every simple graph G satisfies $\chi'^{\ell}(G) = \Delta(G) + o(\Delta(G))$. The following conjecture, posed by Bojan Mohar, deals with list-colourings of χ'-critical class 2 graphs – that is, simple graphs G for which $\chi'(G) = \Delta(G)+1$ and $\chi'(H) < \chi'(G)$, whenever H is a proper subgraph of G.

Suppose that G is a χ'-critical graph of class 2, and for each edge of G there is a list of $\Delta(G)$ colours. Then the edges of G can be properly coloured so that each edge receives a colour from its list unless all the lists are equal to each other.

6. Flow problems

A *k-flow* in a graph (also called a 'nowhere-zero k-flow' for emphasis) is an assignment to each edge of a direction and of a flow value in $\{1, 2, \ldots, k-1\}$, so that the net flow entering and leaving each vertex is 0. The concept of a k-flow was introduced by Tutte [164], who showed that a plane graph allows a k-flow if and only if its dual graph is vertex-k-colourable. A comprehensive survey of k-flows written by Lovász [115] is available on-line (see also Chapter 9).

Tutte's 5-flow conjecture

The following conjecture was posed by Tutte [164] in 1954:

Every graph without bridges allows a 5-flow.

Kochol [108] proved that if the 5-flow conjecture is false, then a smallest counter-example is a cyclically 6-edge-connected cubic graph – that is, a 3-regular graph for which, whenever fewer than six edges are deleted, the remaining graph has at most one component that contains a cycle.

A bound on the possible order of a counter-example was obtained by Steffen [151], who proved Tutte's 5-flow conjecture for all graphs with fewer than 44 vertices. As a consequence, the conjecture is true for all graphs embeddable on a surface with Euler characteristic at least -3. A result by Kochol [109] shows that the girth of a smallest counter-example is at least 9. Since the order of a $(3,9)$-cage is known (see [32]), this improves Steffen's lower bound to 58 vertices, and the bound on the Euler characteristic to -10.

An interesting equivalent reformulation of the 5-flow conjecture, stated in terms of a parity condition, is due to Matamala and Zamora [118]. Let F be a subgraph of a graph G. Then F is called a $(1,2)$-*factor* of G if each vertex of G is a vertex of F of degree 1 or 2. An edge of G is called F-*balanced* if it is either an edge of F or its ends have the same degree in F. If every cycle of G contains an even number of F-balanced edges, then F is called an *even subgraph*. Matamala and Zamora [118] proved that a cubic graph without bridges allows a 5-flow if and only if it contains an even $(1,2)$-factor.

As a common weakening of Tutte's 5-flow and 4-flow conjectures (see below), Kochol [106] observed that if a cubic graph has no bridge and does not contain the Petersen graph as a minor, then it allows a 5-flow.

Hochstättler [80] related Tutte's k-flow problems to Hadwiger's conjecture by extending both problems to the more general setting of regular matroids (see, for example, Oxley [128]).

Tutte's 4-flow conjecture

The following conjecture, formally a strengthening of the four-colour theorem, was made by Tutte [165] in 1966.

Every graph without a bridge and without a subdivision of the Petersen graph as a subgraph allows a 4*-flow.*

When restricted to cubic graphs, the existence of a 4-flow becomes equivalent to the 3-colourability of edges. The truth of the conjecture in this special case was proved by Robertson, Seymour and Thomas [139], with a proof extending the proof by Robertson, Sanders, Seymour and Thomas [137] of the four-colour theorem.

A 4-flow in a graph is equivalent to a covering of its edges by two subgraphs, each of which is an edge-disjoint union of cycles. Thus, many statements about 4-flows have equivalent reformulations in terms of cycle covers. The excellent monograph by Zhang [173] presents several additional connections between Tutte's k-flow conjectures and other theorems and conjectures on flows and cycle covers in graphs. Such connections were also explored by Kaiser and Škrekovski [88].

Extending these ideas from graphs to regular matroids, Lai, Li and Poon [114] showed (using the four-colour theorem) that if a regular matroid without a co-loop (a singleton circuit in the dual matroid) does not allow a 4-flow, then it contains a minor that is either K_5 or the matroid dual of K_5. In particular, this proves the conjecture of Jensen and Toft [85] that every graph without a bridge and without a K_5-minor has a 4-flow.

Tutte's 3-flow conjecture

According to a survey by Steinberg [153] on the three-colour problem, the following 3-flow conjecture was made by Tutte in 1972. Its restriction to the planar case is dually equivalent to Grötzsch's theorem [69].

Every graph without a bridge and without an edge-cut of size 3 *allows a* 3*-flow.*

A weaker conjecture by Jaeger [83], called 'the weak 3-flow conjecture', states that there exists a number λ for which every graph of edge-connectivity at least λ allows a 3-flow. This conjecture has been proved by Thomassen [161] for $\lambda = 8$, and improved to $\lambda = 6$ by Lovász, Thomassen, Wu and Zhang [116]. Tutte's conjecture corresponds to $\lambda = 4$ in the weak 3-flow conjecture. However, Kochol [107] proved that Tutte's conjecture already follows from the case $\lambda = 5$ in the weak 3-flow conjecture.

Similar to a k-flow in a graph, except with non-integral flow values, a *circular r-flow* for a real number $r \geq 2$ is defined as an assignment to each edge of a direction and of a flow value in the interval $[1, r - 1]$, for which the net flow entering and leaving each vertex is 0. According to Zhang [174], L. A. Goddyn and P. D. Seymour have conjectured that, for each $\varepsilon > 0$, there exists a number λ for which every graph of edge-connectivity at least λ allows a circular $(2 + \varepsilon)$-flow. Klostermeyer and Zhang [104] proved Goddyn and Seymour's conjecture for planar graphs. The general theorem on circular flows by Thomassen [161] implies, for each positive natural number k, that if a graph has large enough edge-connectivity, then it allows a circular flow in which all flow values are either 1 or $1 + 1/k$, thus proving the conjecture also for the general case (personal communication from C. Thomassen).

7. Concluding remarks

As indicated in this chapter, even though many important graph-colouring problems have been solved, there is a large variety of interesting unsolved problems still around. We made a selection, mostly guided by our own taste, and mostly from our book [85] in updated form. As for historical remarks about the problems, the original book is usually a more comprehensive source.

The state of the art for edge-colourings is described in Stiebitz, Scheide, Toft and Favrholdt [156]. This book also contains a chapter with twenty attractive edge-colouring conjectures. Here we have used formulations from [156], as well as from [85].

Probabilistic methods play a fast-growing role in the development of combinatorial mathematics. These we have to some degree neglected. For those wishing insight into this side of colouring theory we strongly recommend the excellent book by Molloy and Reed [123], as well as the well-written book by Beck [16], which is a rich source of how the probabilistic method, in a game-theoretic setting, may be used to obtain exact results by 'derandomization'.

Graph colouring theory flourishes as never before, and we hope and expect to see solutions and partial solutions to several of the above problems in the near future. The area continues to show surprising developments, and it is hard to predict which problems will find solutions. Tutte's iceberg is still there, with many beautiful hidden secrets waiting for discovery just beneath the surface.

References

1. H. L. Abbott and B. Zhou, On small faces in 4-critical planar graphs, *Ars Combin.* **32** (1991), 203–207.
2. M. Aigner and S. Brandt, Embedding arbitrary graphs of maximum degree two, *J. London Math. Soc.* **48** (1993), 39–51.
3. V. A. Aksionov and L. S. Melnikov, Essay on the theme: the three-color problem, *Combinatorics* (eds. A. Hajnal and V. T. Sós), Colloq. Math. Soc. János Bolyai **18**, North-Holland (1978), 23–34.
4. B. Albar and D. Gonçalves, On triangles in K_r-minor free graphs, *arXiv:1304.5468v1*, 2013.
5. M. O. Albertson, Open problem 2, *Theory and Applications of Graphs* (eds. G. Chartrand *et al.*), Wiley (1981), 609.
6. M. O. Albertson, D. L. Boutin and E. Gethner, The thickness and chromatic number of r-inflated graphs, *Discrete Math.* **310** (2010), 2725–2734.
7. M. O. Albertson, D. L. Boutin and E. Gethner, More results on r-inflated graphs: Arboricity, thickness, chromatic number and fractional chromatic number, *Ars Math. Contemp.* **4** (2011), 5–24.
8. R. E. L. Aldred, S. Bau, D. A. Holton and B. D. McKay, Nonhamiltonian 3-connected cubic planar graphs, *SIAM J. Discrete Math.* **13** (2000), 25–32.
9. L. D. Andersen, On edge-colourings of graphs, *Math. Scand.* **40** (1977), 161–175.
10. K. Appel and W. Haken, Every planar map is four colorable, Part I: Discharging, *Illinois J. Math.* **21** (1977), 429–490.

11. K. Appel, W. Haken and J. Koch, Every planar map is four colorable, Part II: Reducibility, *Illinois J. Math.* **21** (1977), 491–567.
12. J. Balogh, A. V. Kostochka, N. Prince and M. Stiebitz, The Erdős–Lovász Tihany conjecture for quasi-line graphs, *Discrete Math.* **309** (2012), 3985–3991.
13. J. Barát, G. Joret and D. R. Wood, Disproof of the list Hadwiger conjecture, *Electron. J. Combin.* **18** (2011), 232.
14. D. Barnette, Conjecture 5, *Recent Progress in Combinatorics* (ed. W. T. Tutte), Proc. Third Waterloo Conference on Combinatorics, May 1968, Academic Press (1969), 343.
15. T. Bartnicki, J. Grytczuk, H. A. Kierstead and X. Zhu, The map-coloring game, *Amer. Math. Monthly* **114** (2007), 793–803.
16. J. Beck, *Combinatorial Games*, Encyclopedia of Mathematics and its Applications **114**, Cambridge University Press, 2008; paperback edn., 2011.
17. M. Behzad, *Graphs and Their Chromatic Numbers*, Ph.D. thesis, Michigan State University, 1965.
18. A. Bernhart, Six rings in minimal five color maps, *Amer. J. Math.* **69** (1947), 391–412.
19. H.-G. Bigalke, *Heinrich Heesch, Kristallgeometrie, Parkettierungen, Vierfarbenforschung*, Birkhäuser Verlag (1988), 155.
20. B. Bollobás and S. E. Eldridge, Problem, *Proc. Fifth British Combinatorial Conference* (Aberdeen, 1975) (eds. C. St.J. A. Nash-Williams and J. Sheehan), Utilitas Math. (1976), 690.
21. B. Bollobás and S. E. Eldridge, Packings of graphs and applications to computational complexity, *J. Combin. Theory (B)* **25** (1978), 105–124.
22. B. Bollobás and A. J. Harris, List-colourings of graphs, *Graphs Combin.* **1** (1985), 115–127.
23. O. V. Borodin, Colorings of plane graphs: a survey, *Discrete Math.* **313** (2013), 517–539.
24. O. V. Borodin, Z. Dvořák, A. V. Kostochka, B. Lidický and M. P. Yancey, Planar 4-critical graphs with four triangles, *Europ. J. Combin.* **41** (2014), 138–151.
25. O. V. Borodin, A. N. Glebov, T. R. Jensen and A. Raspaud, Planar graphs without triangles adjacent to cycles of length from 3 to 9 are 3-colorable, *Sib. Elektron. Mat. Izv.* **3** (2006), 428–440.
26. O. V. Borodin, A. N. Glebov, M. Montassier and A. Raspaud, Planar graphs without 5- and 7-cycles and without adjacent triangles are 3-colorable, *J. Combin. Theory (B)* **99** (2009), 668–673.
27. O. V. Borodin, A. N. Glebov, A. Raspaud and M. R. Salavatipour, Planar graphs without cycles of length from 4 to 7 are 3-colorable, *J. Combin. Theory (B)* **93** (2005), 303–311.
28. O. V. Borodin, A. V. Kostochka and D. R. Woodall, List edge and list total colourings of multigraphs, *J. Combin. Theory (B)* **71** (1997), 184–204.
29. M. Borowiecki, Research problem 172, *Discrete Math.* **121** (1993), 235–236.
30. D. L. Boutin, E. Gethner and T. Sulanke, Thickness-two graphs, part one: new nine-critical graphs, permuted layer graphs, and Catlin's graphs, *J. Graph Theory* **57** (2008), 198–214.
31. G. Brinkmann, B. D. McKay and U. von Nathusius, Backtrack search and look-ahead for the construction of planar cubic graphs with restricted face sizes, *MATCH Commun. Math. Comput. Chem.* **48** (2003), 163–177.
32. G. Brinkmann, B. D. McKay and C. Saager, The smallest cubic graphs of girth nine, *Combin. Probab. Comput.* **5** (1995), 1–13.
33. H. Broersma, P. A. Golovach, D. Paulusma and J. Song, Updating the complexity status of coloring graphs without a fixed induced linear forest, *Theoret. Comput. Sci.* **414** (2012), 9–19.
34. R. L. Brooks, On colouring the nodes of a network, *Proc. Cambridge Philos. Soc.* **37** (1941), 194–197.

35. P. A. Catlin, *Embedding Subgraphs and Coloring Graphs under Extremal Degree Conditions*, Ph.D. thesis, Ohio State University, 1976.
36. P. A. Catlin, Embedding subgraphs under extremal degree conditions. *Proc. 8th South-Eastern Conference on Combinatorics, Graph Theory and Computing* (Baton Rouge, 1977), *Congr. Numer.* **19** (1977), 139–145.
37. A. G. Chetwynd and R. Häggkvist, A note on list colourings, *J. Graph Theory* **13** (1989), 87–95.
38. M. Chudnovsky, A. D. King, M. Plumettaz and P. D. Seymour, A local strengthening of Reed's omega, Delta, chi conjecture for quasi-line graphs, *SIAM J. Discrete Math.* **27** (2013), 95–108.
39. M. Chudnovsky, N. Robertson, P. D. Seymour and R. Thomas, The strong perfect graph theorem, *Ann. of Math. (2)* **164** (2006), 51–229.
40. M. Chudnovsky and P. D. Seymour, Extending the Gyárfás–Sumner conjecture, *J. Combin. Theory (B)* **105** (2014), 11–16.
41. B. Csaba, A. Shokoufandeh and E. Szemerédi, Proof of a conjecture of Bollobás and Eldridge for graphs of maximum degree three, *Combinatorica* **23** (2003), 35–72.
42. M. DeVos, K. Kawarabayashi and B. Mohar, Locally planar graphs are 5-choosable, *J. Combin. Theory (B)* **98** (2008), 1215–1232.
43. G. A. Dirac, A property of 4-chromatic graphs and some remarks on critical graphs, *J. London Math. Soc.* **27** (1952), 85–92.
44. A. A. Dobrynin, L. S. Melnikov and A. V. Pyatkin, Regular 4-critical graphs of even degree, *J. Graph Theory* **46** (2004), 103–130.
45. A. A. Dobrynin, L. S. Melnikov and A. V. Pyatkin, Erdős regular graphs of even degree, *Discuss. Math. Graph Theory* **27** (2007), 269–279.
46. P. Duchet and H. Meyniel, On Hadwiger's number and the stability number, *Graph Theory* (ed. B. Bollobás), Annals of Discrete Mathematics, North-Holland (1982), 71–73.
47. K. Edwards and A. D. King, A superlocal version of Reed's conjecture, *arXiv:1208.5188*, 2012.
48. P. Erdős, Graph theory and probability, *Canad. J. Math.* **11** (1959), 34–38.
49. P. Erdős, A. I. Rubin and H. Taylor, Choosability in graphs, *Proc. West-Coast Conf. on Combinatorics, Graph Theory and Computing, Congr. Numer.* **XXVI** (1979), 125–157.
50. P. Erdős, On some aspects of my work with Gabriel Dirac, *Graph Theory in Memory of G. A. Dirac* (eds. L. D. Andersen *et al.*), Annals of Discrete Mathematics **41**, North-Holland (1989), 111–116.
51. L. Esperet, M. Montassier and X. Zhu, Adapted list coloring of planar graphs, *J. Graph Theory* **62** (2009), 127–138.
52. S. Fadnavis, A generalization of the birthday problem and the chromatic polynomial, *arXiv:1105.0698v2*, 2011.
53. G. Fan and A. Raspaud, Fulkerson's conjecture and circuit covers, *J. Combin. Theory (B)* **61** (1994), 133–138.
54. T. Feder and C. Subi, On Barnette's conjecture, *Electron. Colloq. Comput. Complexity* **15** (2006).
55. U. Feige, E. Ofek and U. Wieder, *Approximating Maximum Edge Coloring in Multigraphs*, Lecture Notes in Comput. Sci. **2462** (2002), 108–121.
56. J. L. Fouquet and J. M. Vanherpe, On the perfect matching index of bridgeless cubic graphs, *arXiv:0904.1296*, 2009.
57. O. Frink, A proof of Petersen's theorem, *Ann. Math.* **27** (1926), 491–493.
58. R. Fritsch and G. Fritsch, *The Four Color Theorem: History, Topological Foundations and Idea of Proof*, Springer, 1998.
59. D. R. Fulkerson, Blocking and anti-blocking pairs of polyhedra, *Math. Program.* **1** (1971), 168–194.

60. F. G., Tinting maps, *Athenæum* **1389** (June 1854), 726.
61. T. Gallai, Kritische Graphen I, *Publ. Math. Inst. Hungar. Acad. Sci.* **8** (1963), 165–192.
62. F. Galvin, The list chromatic index of a bipartite multigraph, *J. Combin. Theory (B)* **63** (1995), 153–158.
63. M. Gardner, *Martin Gardner Papers*, Stanford University Department of Special Collections, SC 647, Box 52, Folder 6, 1980.
64. M. Gardner, Mathematical games, *Scientific American* **242** (February 1980), 14–21.
65. M. K. Goldberg, On multigraphs of almost maximal chromatic class (in Russian), *Diskret. Analiz* **23** (1973), 3–7.
66. M. K. Goldberg, Edge-coloring of multigraphs; recoloring technique, *J. Graph Theory* **8** (1984), 123–137.
67. G. Gonthier, Formal proof – the four-color theorem, *Notices Amer. Math. Soc.* **55** (11) (2008), 1382–1393.
68. P. R. Goodey, Hamiltonian circuits in polytopes with even sides, *Israel J. Math.* **22** (1975), 52–56.
69. H. Grötzsch, Zur Theorie der diskreten Gebilde, VII: Ein Dreifarbensatz für dreikreisfreie Netze auf der Kugel, *Wiss. Z. Martin-Luther-U., Halle-Wittenberg, Math.-Nat. Reihe* **8** (1959), 109–120.
70. R. P. Gupta, On the chromatic index and the cover index of a multigraph, *Theory and Applications of Graphs* (Michigan, 1976) (eds. A. Donald and B. Eckmann), Lecture Notes in Math. **642**, Springer (1978), 91–110.
71. A. Gyárfás, On Ramsey covering-numbers, *Infinite and Finite Sets,* (Keszthely 1973) (eds. A. Hajnal, R. Rado and V. T. Sós), Coll. Math. Soc. János Bolyai **10**, North-Holland (1975), 801–816.
72. A. Gyárfás, Problems from the world surrounding perfect graphs, *Zastos. Mat.* **XIX** (1987), 413–441.
73. A. Gyárfás and P. Erdős, Split and balanced colorings of complete graphs, *Discrete Math.* **200** (1999), 79–86.
74. H. Hadwiger, Über eine Klassifikation der Streckenkomplexe, *Vierteljschr. Naturforsch. Ges. Zürich* **88** (1943), 133–143.
75. A. Hajnal and E. Szemerédi, Proof of a conjecture of Erdős, *Infinite and Finite Sets* (eds. P. Erdős, A. Rényi and V. T. Sós), Coll. Math. Soc. János Bolyai **10**, North Holland (1970), 601–623.
76. R. Hao, J. Niu, X. Wang, C. Q. Zhang and T. Zhang, A note on Berge–Fulkerson coloring, *Discrete Math.* **309** (2009), 4235–4240.
77. P. Hell, Z. Pan, T.-L. Wong and X. Zhu, Adaptable chromatic number of graph products, *Discrete Math.* **309** (2009), 6153–6159.
78. P. Hell and X. Zhu, On the adaptable chromatic number of graphs, *Europ. J. Combin.* **29** (2008), 912–921.
79. D. S. Hochbaum, T. Nishizeki and D. B. Shmoys, A better than 'best possible' algorithm to edge color multigraphs, *J. Algorithms* **7** (1986), 79–104.
80. W. Hochstättler, *A Hadwiger Conjecture for Hyperplane Arrangements*, Technical report feu-dmo018.09, FernUniversität Hagen, 2009.
81. D. A. Holton, B. Manvel and B. D. McKay, Hamiltonian cycles in cubic 3-connected bipartite planar graphs, *J. Combin. Theory (B)* **38** (1985), 279–297.
82. B. Jackson and G. Ringel, Variations on Ringel's earth–moon problem, *Discrete Math.* **211** (2000), 233–242.
83. F. Jaeger, Nowhere-zero flow problems, *Selected Topics in Graph Theory* **3** (eds. L. W. Beineke and R. J. Wilson), Academic Press (1988), 71–95.
84. T. R. Jensen, Dense critical and vertex-critical graphs, *Discrete Math.* **258** (2002), 63–84.

85. T. R. Jensen and B. Toft, *Graph Coloring Problems*, Wiley, 1995.
86. J. Kahn, Asymptotically good list-colorings, *J. Combin. Theory (A)* **73** (1996), 1–59.
87. T. Kaiser, D. Král' and S. Norine, Unions of perfect matchings in cubic graphs, *Topics in Discrete Mathematics* (eds. M. Klazar *et al.*), Springer (2006), 225–230.
88. T. Kaiser and R. Škrekovski, Cycles intersecting edge-cuts of prescribed sizes, *SIAM J. Discrete Math.* **22** (2008), 861–874.
89. H. Kaul, A. V. Kostochka and G. Yu, On a graph packing conjecture by Bollobás, Eldridge and Catlin, *Combinatorica* **28** (2008), 469–485.
90. K. Kawarabayashi, Minors in 7-chromatic graphs, preprint.
91. K. Kawarabayashi, A. S. Pedersen and B. Toft, Double-critical graphs and complete minors, *Electron. J. Combin.* **17** (2010), R87.
92. K. Kawarabayashi, M. D. Plummer and B. Toft, Improvements of the theorem of Duchet and Meyniel on Hadwiger's conjecture, *J. Combin. Theory (B)* **95** (2005), 152–167.
93. K. Kawarabayashi and B. A. Reed, Hadwiger's conjecture is decidable, *Proc. 41st Annual ACM Symposium on Theory of Computing, STOC '09* (2009), 445–454.
94. K. Kawarabayashi and Z.-X. Song, Independence number and clique minors, *J. Graph Theory* **56** (2007), 219–226.
95. K. Kawarabayashi and C. Thomassen, From the plane to higher surfaces, *J. Combin. Theory (B)* **102** (2012), 852–868.
96. K. Kawarabayashi and B. Toft, Any 7-chromatic graph has K_7 or $K_{4,4}$ as a minor, *Combinatorica* **25** (2005), 327–353.
97. A. K. Kelmans, Constructions of cubic bipartite 3-connected graphs without Hamiltonian cycles, *Amer. Math. Soc. Transl.* **158** (1994), 127–140.
98. H. A. Kierstead and A. V. Kostochka, A short proof of the Hajnal–Szemerédi theorem on equitable colouring, *Combin. Probab. Comput.* **17** (2008), 265–270.
99. H. A. Kierstead and S. G. Penrice, Recent results on a conjecture of Gyárfás, *Proc. 21st S.-E. Conference on Combinatorics, Graph Theory and Computing* (Baton Rouge, 1990) (eds. F. Hoffman *et al.*), *Congr. Numer.* **79** (1990), 182–186.
100. H. A. Kierstead and S. G. Penrice, Radius two trees specify χ-bounded classes, *J. Graph Theory* **18** (1994), 119–129.
101. H. A. Kierstead and W. T. Trotter, Planar graph coloring with an uncooperative partner, *Planar Graphs* (ed. W. T. Trotter), DIMACS Series in Discrete Mathematics and Theoretical Computer Science **9**, American Mathematical Society (1993), 85–93.
102. A. D. King, *Claw-free Graphs and Two Conjectures on* ω, Δ, *and* χ, Ph.D. thesis, McGill University, 2009.
103. H. Klein and M. Margraf, A remark on the conjecture of Erdős, Faber and Lovász, *J. Geom.* **88** (2008), 116–119.
104. W. Klostermeyer and C. Q. Zhang, $(2+\varepsilon)$-coloring of planar graphs with large odd-girth, *J. Graph Theory* **33** (2000), 109–119.
105. W. Knight, Computer generates verifiable mathematics proof, *New Scientist Breaking News*, 19 April 2005.
106. M. Kochol, Cubic graphs without a Petersen minor have nowhere-zero 5-flows, *Acta Math. Univ. Comenian. (N.S.)* **68** (1999), 249–252.
107. M. Kochol, An equivalent version of the 3-flow conjecture, *J. Combin. Theory (B)* **83** (2001), 258–261.
108. M. Kochol, Reduction of the 5-flow conjecture to cyclically 6-edge-connected snarks, *J. Combin. Theory (B)* **90** (2004), 139–145.
109. M. Kochol, Restrictions on smallest counterexamples to the 5-flow conjecture, *Combinatorica* **26** (2006), 83–89.

110. D. König, *Theorie der Endlichen und Unendlichen Graphen*, Akademische Verlagsgesellschaft M. B. H. Leipzig, 1936; reprinted by Chelsea, 1950, and B. G. Teubner, 1986; English transl., Birkhäuser, 1990.
111. A. V. Kostochka and M. P. Yancey, Ore's conjecture on color-critical graphs is almost true, *arXiv:1209.1050*, 2012.
112. A. V. Kostochka and X. Zhu, Adapted list coloring of graphs and hypergraphs, *SIAM J. Discrete Math.* **22** (2008), 398–408.
113. F. Kramer and H. Kramer, A survey on the distance-colouring of graphs, *Discrete Math.* **308** (2008), 422–426.
114. H.-J. Lai, X. Li and H. Poon, Nowhere zero 4-flow in regular matroids, *J. Graph Theory* **49** (2005), 196–204.
115. L. Lovász, Tutte's flow conjectures, *tlovering.files.wordpress.com/2012/06/ laszloessay.pdf*, 2012.
116. L. Lovász, C. Thomassen, Y. Wu and C. Q. Zhang, Nowhere-zero 3-flows and modulo k-orientations, *J. Combin. Theory (B)* **103** (2013), 587–598.
117. A. Mansfield, Determining the thickness of graphs is NP-hard, *Math. Proc. Cambridge Philos. Soc.* **93** (1983), 9–23.
118. M. Matamala and J. Zamora, Nowhere-zero 5-flows and even (1,2)-factors, *Graphs Combin.* **29** (2013), 609–616.
119. G. Mazzuoccolo, The equivalence of two conjectures of Berge and Fulkerson, *J. Graph Theory* **68** (2011), 125–128.
120. B. D. McKay, A note on the history of the four-colour conjecture, *J. Graph Theory* **72** (2013), 361–363.
121. M. Mirzakhani, A small non-4-choosable planar graph, *Bull. Inst. Combin. Appl.* **17** (1996), 15–18.
122. M. Molloy and B. A. Reed, A bound on the total chromatic number, *Combinatorica* **18** (1998), 241–280.
123. M. Molloy and B. A. Reed, *Graph Colouring and the Probabilistic Method*, Algorithms and Combinatorics **23**, Springer-Verlag, 2002.
124. M. Molloy and G. Thron, The adaptable choosability number grows with the choosability number, *Discrete Math.* **311** (2011), 2268–2271.
125. M. Molloy and G. Thron, An asymptotically tight bound on the adaptable chromatic number, *J. Graph Theory* **71** (2012), 331–351.
126. M. Montassier, A. Raspaud and X. Zhu, An upper bound on adaptive choosability of graphs, *Europ. J. Combin.* **30** (2009), 351–355.
127. J. Nešetřil and P. Ossona de Mendez, Grad and classes with bounded expansion III. Restricted graph homomorphism dualities, *Europ. J. Combin.* **29** (2008), 1012–1024.
128. J. G. Oxley, *Matroid Theory* (2nd edn.), Oxford University Press, 2006.
129. V. Patel, *Unions of perfect matchings in cubic graphs and implications of the Berge–Fulkerson conjecture*, CDAM Research Report LSE-CDAM-2006-06, 2006.
130. A. S. Pedersen and B. Toft, A basic elementary extension of the Duchet-Meyniel theorem, *Discrete Math.* **310** (2010), 480–488.
131. J. Petersen, Die Theorie der regulären Graphs, *Acta Math.* **15** (1891), 193–220.
132. M. D. Plummer, M. Stiebitz and B. Toft, On a special case of Hadwiger's conjecture, *Discuss. Math. Graph Theory* **23** (2003), 333–363.
133. B. A. Reed, ω, Δ, and χ, *J. Graph Theory* **27** (1998), 177–212.
134. B. A. Reed, A strengthening of Brooks' theorem, *J. Combin. Theory (B)* **76** (1999), 136–149.
135. G. Ringel, *Färbungsprobleme auf Flächen und Graphen*, Math. Monographien **2**, VEB Deutscher Verlag der Wissenschaften, 1959.

136. R. Rizzi, Approximating the maximum 3-edge-colorable subgraph problem, *Discrete Math.* **309** (2009), 4166–4170.
137. N. Robertson, D. P. Sanders, P. D. Seymour and R. Thomas, The four-colour theorem, *J. Combin. Theory (B)* **70** (1997), 2–44.
138. N. Robertson, P. D. Seymour and R. Thomas, Hadwiger's conjecture for K_6-free graphs, *Combinatorica* **13** (1993), 279–361.
139. N. Robertson, P. D. Seymour and R. Thomas, Tutte's edge-colouring conjecture, *J. Combin. Theory (B)* **70** (1997), 166–183.
140. V. Rödl, A. Ruciński and A. Taraz, Hypergraph packing and graph embedding, *Combin. Probab. Comput.* **8** (1999), 363–376.
141. D. Romero and A. Sánchez-Arroyo, Advances on the Erdős–Faber–Lovász conjecture, *Combinatorics, Complexity and Chance: A Tribute to Dominic Welsh* (eds. G. Grimmett and C. McDiarmid), Oxford Univ. Press (2007), 285–298.
142. M. R. Salavatipour, *The three color problem for planar graphs*, Technical report CSRG-458, Department of Computer Science, University of Toronto, July 2002.
143. D. P. Sanders and Y. Zhao, A note on the three color problem, *Graphs Combin.* **11** (1995), 91–94.
144. N. W. Sauer, Vertex partition problems, *Combinatorics, Paul Erdős is Eighty*, **1** (eds. D. Miklós, V. T. Sós and T. Szőnyi), Bolyai Society Mathematical Studies, János Bolyai Mathematical Society (1993), 361–377.
145. N. W. Sauer and J. H. Spencer, Edge-disjoint placements of graphs, *J. Combin. Theory (B)* **15** (1978), 295–302.
146. A. Schrijver, *Combinatorial Optimization: Polyhedra and Efficiency*, Algorithms and Combinatorics **24**, Springer, 2003.
147. P. D. Seymour, Some unsolved problems on 1-factorizations of graphs, *Graph Theory and Related Topics* (Waterloo, Ontario, 1977) (eds. J. A. Bondy and U. S. R. Murty), Academic Press (1979), 367–368.
148. P. D. Seymour, On multi-colorings of cubic graphs, and conjectures of Fulkerson and Tutte, *Proc. London Math. Soc.* **38** (1979), 423–460.
149. M. Simonovits, On colour-critical graphs, *Studia Sci. Math. Hungar.* **7** (1972), 67–81.
150. A. Soifer, *The Mathematical Coloring Book*, Springer, 2009.
151. E. Steffen, Tutte's 5-flow conjecture for graphs of nonorientable genus 5, *J. Graph Theory* **22** (1996), 309–319.
152. M. Stehlik, Minimal connected τ-critical hypergraphs, *Graphs Combin.* **22** (2006), 421–426.
153. R. Steinberg, The state of the three color problem, *Quo Vadis, Graph Theory?* (eds. J. Gimbel, J. W. Kennedy and L. V. Quintas), Annals of Discrete Math. **55**, North-Holland (1993), 211–248.
154. J. P. Steinberger, An unavoidable set of D-reducible configurations, *Trans. Amer. Math. Soc.* **362** (2010), 6633–6661.
155. M. Stiebitz, A relaxed version of the Erdős–Lovász Tihany conjecture, manuscript, TU Ilmenau, 2011.
156. M. Stiebitz, D. Scheide, B. Toft and L. M. Favrholdt, *Graph Edge Coloring*, Wiley, 2012.
157. W. R. Stromquist, *Some Aspects of the Four Color Problem*, Ph.D. thesis, Harvard University, 1975.
158. D. P. Sumner, Subtrees of a graph and chromatic number, *Theory and Applications of Graphs*, Proc. 4th. International Graph Theory Conference (Kalamazoo, 1980) (eds. G. Chartrand *et al.*), Wiley (1981), 557–576.
159. C. Thomassen, Every planar graph is 5-choosable, *J. Combin. Theory (B)* **62** (1994), 180–181.

160. C. Thomassen, A short list color proof of Grötzsch's theorem, *J. Combin. Theory (B)* **88** (2003), 189–192.
161. C. Thomassen, The weak 3-flow conjecture and the weak circular flow conjecture, *J. Combin. Theory (B)* **102** (2012), 521–529.
162. B. Toft, On the maximal number of edges of critical k-chromatic graphs, *Studia Sci. Math.* **5** (1970), 461–470.
163. B. Toft, Two theorems on critical 4-chromatic graphs, *Studia Sci. Math.* **7** (1972), 83–89.
164. W. T. Tutte, A contribution to the theory of chromatic polynomials, *Canad. J. Math.* **6** (1954), 80–91.
165. W. T. Tutte, On the algebraic theory of graph colourings, *J. Combin. Theory* **1** (1966), 15–50.
166. W. T. Tutte, Colouring problems, *Math. Intelligencer* **1** (1978), 72–75.
167. V. G. Vizing, Some unsolved problems in graph theory (in Russian), *Uspekhi Mat. Nauk.* **23** (1968), 117–134; English transl. in *Russian Math. Surveys* **23** (1968), 125–141.
168. V. G. Vizing, Colouring the vertices of a graph in prescribed colours (in Russian), *Diskret. Analiz* **29** (1976), 3–10.
169. M. Voigt, List colourings of planar graphs, *Discrete Math.* **120** (1993), 215–219.
170. K. Wagner, Über eine Eigenschaft der ebenen Komplexe, *Math. Ann.* **114** (1937), 570–590.
171. G. Wegner, *Graphs with Given Diameter and a Coloring Problem*, Technical report, University of Dortmund, 1977.
172. R. Wilson, *Four Colors Suffice: How the Map Problem was Solved* (revised color edn.), Princeton Science Library, 2014.
173. C. Q. Zhang, *Integer Flows and Cycle Covers of Graphs*, Marcel Dekker, 1997.
174. C. Q. Zhang, Circular flows of nearly Eulerian graphs and vertex-splitting, *J. Graph Theory* **40** (2002), 147–161.
175. Y. Zhao, 3-coloring graphs embedded in surfaces, *J. Graph Theory* **32** (2000), 140–143.

Notes on contributors

Lowell Beineke is Schrey Professor of Mathematics at Indiana University–Purdue University Fort Wayne (IPFW), where he has worked since receiving his Ph.D. degree from the University of Michigan under the guidance of Frank Harary. His graph theory interests include topological graph theory, line graphs, tournaments, decompositions and vulnerability. He has published over a hundred papers in graph theory and has served as Editor of the *College Mathematics Journal*. With Robin Wilson he has co-edited five books in addition to the three earlier volumes in this series. Recent honours include an award instituted in his name by the College of Arts and Sciences at IPFW and a Certificate of Meritorious Service from the Mathematical Association of America.

Csilla Bujtás received her Ph.D. degree in 2009. She is currently an associate professor at the University of Pannonia in Veszprém in Hungary, and is a member of the research laboratory 'Discrete Mathematical Structures and Algorithms'. Her main field of interest is graph and hypergraph theory from both a structural and algorithmic viewpoint: this includes graph colouring, general models of hypergraph colouring, independence and domination.

Dominique de Werra is Honorary Professor of Operations Research at the Ecole Polytechnique Fédérale de Lausanne in Switzerland, where he taught graph and network theory and operations research. His special interests are in the applications of graphs to scheduling and timetabling, with an emphasis on sports-league schedules. He serves as an associate editor of several journals, including *Discrete Applied Mathematics*, and is the author of more than two hundred papers on graphs and their applications.

Alain Hertz holds a diploma in mathematical engineering and obtained his Ph.D. degree in operations research from the École Polytechnique Fédérale de Lausanne in Switzerland. Since 2001, he has been a professor at the department of mathematics and industrial engineering at the École Polytechnique in Montréal. He is also member of the multi-disciplinary GERAD research group that includes nearly sixty researchers and experts in operations research and discrete mathematics. He is the author of almost two hundred scientific publications, and his main research areas are combinatorial optimization, graph theory, algorithmics, and the development of decision aid systems for scheduling and distribution problems.

Thore Husfeldt received his Ph.D. degree in computer science at Aarhus University, Denmark, for a thesis on the computational complexity of data structures. More recently, his research has focused on algorithms for computationally hard problems – in particular, fundamental graph problems. He is a professor of computer science at Lund University in Sweden and an associate professor at IT University of Copenhagen in Denmark.

Bill Jackson received his Ph.D. degree from the University of Waterloo, Canada, in 1978 and then returned to England as a postdoctoral research fellow at the University of Reading. He has lectured at the University of London since 1980 and is currently Professor of Mathematical Sciences at Queen Mary. He has served on the British Combinatorial Committee and is a member of the Egerváry Research Group at Eötvös University in Budapest. His research interests lie in graph theory, matroid theory and discrete geometry.

Tommy R. Jensen is an associate professor of mathematics at Kyungpook National University in Daegu, Republic of Korea. He received his Ph.D. degree from the University of Southern Denmark and his habilitation degree from the University of Hamburg in Germany. His main research interests include cycles and colourings of graphs.

Ross J. Kang received his D.Phil. degree in 2008 from the University of Oxford. Following brief stints at McGill University, Durham University, CWI Amsterdam and Utrecht University, he is currently assistant professor at the Institute of Mathematics, Astrophysics and Particle Physics (IMAPP) at Radboud University Nijmegen, Netherlands. Aside from graph colouring, his research interests include probabilistic and extremal combinatorics, discrete geometry and random graph theory.

Ken-ichi Kawarabayashi is a professor at the National Institute of Informatics in Japan. His research interests are anything related to graphs, especially algorithmic graph theory, structural graph theory, sports scheduling, graph database, AI and machine learning. More recently, he has tackled several practical problems for large graphs, including social networks and web graphs. Since 2012 he has been leading a research centre with more than 65 people, whose purpose is to apply theoretical research to hard practical problems.

Hong-Jian Lai is a professor of mathematics at West Virginia University. He received his Ph.D. degree in mathematics at Wayne State University under Paul Catlin, and was a post-doctoral fellow at the University of Waterloo under Adrian Bondy. His research interests are mainly in graph theory, with an emphasis on supereulerian graphs, Hamiltonian line graphs and claw-free graphs, nowhere-zero flows, group connectivity and group colorings.

Rong Luo is an associate professor of mathematics at West Virginia University, from where he received his Ph. D. degree in 2002. Before returning there as a faulty member, he worked for ten years at Middle Tennessee State University. His primary research interests are in graph colorings, flows, and combinatorial matrix theory.

Colin McDiarmid is Professor of Combinatorics at the University of Oxford and a tutorial fellow of Corpus Christi College. He served for several years on the British Combinatorial Committee, was recently Senior Tutor at Corpus and Head of the Department of Statistics in Oxford, and has been on the Editorial Board of many journals, including *Combinatorics Probability and Computing*, and *Random Structures and Algorithms*. He has had a good crop of graduate students, and has published around 150 papers in mathematical journals. Recurring themes in his research over many years have been graph colouring, algorithms, and random graphs of various kinds.

Jessica McDonald is an assistant professor in the Department of Mathematics and Statistics at Auburn University. She received her Ph.D. degree from the University of Waterloo in 2009 and held an NSERC postdoctoral fellowship at Simon Fraser University, Canada, prior to joining the faculty at Auburn. Her research interests are centred on graph theory, with two topics of particular interest being edge colouring and immersion.

Bojan Mohar has a Canada Research Chair and is a professor at Simon Fraser University, Canada, and at the University of Ljubljana, Slovenia. He has published over two hundred research papers, in topological graph theory, graph minors, graph colouring, graph algorithms and algebraic graph theory, and is co-author with Carsten Thomassen of a monograph *Graphs on Surfaces*. In 1990 he was awarded the Slovenian national prize for exceptional achievements in science, and in 2009 received the Ambassador of Science Award of the Republic of Slovenia. He is an elected member of the Engineering Academy of Slovenia.

Alexander Soifer is a professor of mathematics, film studies and art history at the University of Colorado at Colorado Springs. In 1978 he left Russia in search of freedom, and in 1988 was charmed by Paul Erdős into switching from group theory to Euclidean Ramsey theory and Erdősian discrete geometry. His three hundred papers and ten books include *The Mathematical Coloring Book*, and he co-authored and edited *Ramsey Theory Yesterday, Today, and Tomorrow*. He is President of the World Federation of National Mathematics Competitions, which presented him with the Paul Erdős Award at its 2006 Congress in Cambridge University. He is publisher and editor of the quarterly journal *Geombinatorics*.

Michael Stiebitz is a graduate of the Humboldt University in Berlin, and since 1992 he has been Professor of Combinatorics at the Institute for Mathematics at the Technical University Ilmenau. His main research interests lie in graph colouring theory, and particularly in the structure of critical graphs. He has co-authored *Graph Edge Coloring: Vizing's Theorem and Goldberg's Conjecture.*

BjarneToft obtained his Ph.D. degree from the University of London in 1970 and has taught mathematics at the University of Southern Denmark in Odense for more than forty years. His main mathematical interests are in graph theory and the history and popularization of mathematics. He is the author of around seventy publications, including the book *Graph Coloring Problems*, written with Tommy Jensen, containing two hundred unsolved problems. His popular lectures included a 'Golden Days in Copenhagen' lecture at the Royal Library and a lecture for long-term inmates in Nyborg State Prison. Since 2012 he has been chairman of the Danish Mathematical Society.

Nicolas Trotignon graduated in economics and statistics in 1995 and in mathematics (French aggregation) in 1997. From 1998 to 2005 he was a teacher of mathematics. His Ph.D. degree, defended in 2004, was about perfect graphs and was written under the supervision of Frédéric Maffray. In 2005 he became an assistant professor in computer science at the Université Panthéon-Sorbonne, and obtained a position as a CNRS researcher in 2008, in LIAFA, University Paris Diderot. Since 2011 he has been a CNRS researcher at LIP, Ecole Normale Supérieure de Lyon. His main interests are in algorithmic graph theory. He has published about twenty papers, about half of which deal directly with perfect graphs.

Zsolt Tuza is a research professor at the Alfréd Rényi Institute of Mathematics, Hungarian Academy of Sciences (formerly at the Computer and Automation Institute), and also professor and head of the Research Laboratory 'Discrete Mathematical Structures and Algorithms' at the University of Pannonia, Veszprém, Hungary. He won a gold medal at the International Mathematical Olympiad in 1972, and has published over 350 research papers on a wide range of topics: graph and hypergraph colouring, extremal set theory, packing and covering, structural graph domination, scheduling and bin packing, structural process control, developmental systems, and more.

Margit Voigt is a professor of mathematics at the University of Applied Sciences in Dresden, Germany. After studying at the University of Jena she worked there for several years in the optimization of optical systems. She moved to the Technical University Ilmenau in 1991, where she acquired her Ph.D. degree. After a post-doctoral position (habilitation) she was a lecturer in Ilmenau until 2004. Her research interests and publications are in graph theory, and more specifically in graph colourings.

Vitaly I. Voloshin is a professor of Troy University, Alabama. For many years he worked at Moldova State University and at the Institute of Mathematics and Informatics of the Moldovan Academy of Sciences in the former Soviet Union, before moving to the United States in 2002. His research areas include graph and hypergraph colouring, theory, algorithms and applications. In 1993 he originated the direction of mixed hypergraph colouring, and there are now more than two hundred publications on the subject, including the author's Fields Institute Monograph *Coloring Mixed Hypergraphs: Theory, Algorithms and Applications*.

Robin Wilson is an emeritus professor of pure mathematics at the Open University, UK, and emeritus professor of geometry at Gresham College, London. After graduating from Oxford, he received his Ph.D. degree in number theory from the University of Pennsylvania. He has written and co-edited many books on graph theory and the history of mathematics, including *Introduction to Graph Theory*, *Four Colors Suffice* and *Combinatorics: Ancient & Modern*. His combinatorial research interests formerly included graph colourings and now focus on the history of combinatorics. An enthusiastic populariser of mathematics, he has won two awards for his expository writing from the Mathematical Association of America.

Cun-Quan Zhang is Eberly Professor of Mathematics at West Virginia University, where he has taught since receiving his Ph.D. degree from Simon Fraser University under the guidance of Brian Alspach. His graph theory interests are mainly in chromatic graph theory and structural graph theory including graph colouring, integer flow, cycle covering and decomposition, matching covering, spanning trees, Hamiltonian cycles and paths, and related optimisation problems. He has published over eighty papers in graph theory, and has written two monographs, *Integer Flows and Cycle Covers of Graphs* and *Circuit Double Cover of Graphs*.

Xuding Zhu is a Chinese National Chair Professor at Zhejiang Normal University. He received his Ph.D. degree from the University of Calgary in 1991. He did postdoctoral research at Simon Fraser University and the University of Bielefeld and worked for many years at National Sun Yat-sen University in Taiwan before moving to Zhejiang Normal University in mainland China in 2010. His research interests are mainly in chromatic graph theory and he has published about two hundred research papers on a wide range of topics, including circular colouring and circular flow, fractional colouring, game colouring, online list-colouring, graph homomorphisms, graph decomposition, chromatic Ramsey theory, etc.

Index